Thermal Stresses and Temperature Control of Mass Concrete

Thermal Stresses and Temperature Control of Mass Concrete

Zhu Bofang

China Institute of Water Resources and Hydropower Research and Chinese Academy of Engineering

TSINGHUA UNIVERSITY PRESS

ELSEVIER

AMSTERDAM • BOSTON • HEIDELBERG • LONDON
NEW YORK • OXFORD • PARIS • SAN DIEGO
SAN FRANCISCO • SINGAPORE • SYDNEY • TOKYO
Butterworth-Heinemann is an imprint of Elsevier

Butterworth-Heinemann is an imprint of Elsevier
The Boulevard, Langford Lane, Kidlington, Oxford OX5 1GB, UK
225 Wyman Street, Waltham, MA 02451, USA

First edition 2014

Library of Congress Cataloging-in-Publication Data
A catalogue record for this book is available from the Library of Congress.

British Library Cataloguing-in-Publication Data
A catalogue record for this book is available from the British Library.

ISBN: 978-0-12-407723-2

Contents

Preface

The cracking of massive concrete structures due to thermal stresses is a problem which had puzzled engineers for a long time. "No dam without crack" is the actual state of concrete dams in the world. The theory of thermal stress and temperature control of mass concrete is established by the writer in this book under the direction of which the problem of cracking of massive concrete structures had been solved, and several concrete dams without crack have been successfully constructed in China in recent years which indicates that the history of "No dam without crack" has ended.

Mass concrete is important for the economical construction of a country. For example, more than 10 million cubic meters of mass concrete are placed in the hydraulic engineering projects in China every year. In addition, a large amount of mass concrete is placed every year in the engineering of harbors, foundation of high buildings heavy machines, nuclear reactors, etc.

The thickness of a massive concrete structure is immense, e.g., the thickness of a concrete dam may be 100–200 m, the depth of the region under tension may be 10–30 m; if all the tensile stresses are undertaken by steel reinforcement, the amount of steel will be considerable, and the cost will be very high. In the process of construction, if there are many vertical steel reinforcements on the top of a concrete block, the spreading and placing of the new concrete lift will be very difficult. Thus in the design of massive concrete structures, such as concrete dams, generally it is required that the tensile stresses do not exceed the allowable tensile stress of concrete so that no steel reinforcement is used. If there are only concrete weight and water pressure acting on the dam, the above-mentioned requirement is easy to achieve, but the period of construction of a high concrete dam may be several years. Due to the heat of hydration of cement and the variation of the ambient temperature, large tensile stresses may appear in the massive concrete structure. As a result, cracks developed in almost all the concrete dams.

The concrete dams are divided into blocks and each block is constructed in horizontal lifts with thickness 1–3 m. The intermissions between two lifts are 5–10 days. As the mechanical and thermal properties of concrete vary with age and have different values in different layers, so the computing of thermal stresses in concrete dams is rather complicated. In the past, there were no methods to compute the thermal stresses in the period of construction of concrete dams, although some temperature control measures had been adopted, but the thermal stresses in the dam are unknown. Actually the tensile stresses are so large that many cracks developed in almost all the dams.

Now a perfect system of the theory of thermal stress and temperature control is established by the writer in this book which includes the following parts:

1. A series of methods for computing the temperature field and the thermal stress field, especially the simulation method for computing the temperature field and stress field of the structure taking account of the influences of all the factors including (a) the process of construction, (b) the mechanical and thermal properties varying with the age of concrete, (c) the variation of ambient air and water temperature, (d) the various measures of temperature control.
2. The law of variation and peculiarity of thermal stresses of different types of massive concrete structures, such as gravity dams, arch dams, buttress dams, concrete blocks, locks, sluices, concrete beams on elastic foundations, concrete pipes, and concrete linings of tunnels. Understanding these issues by engineers is favorable for the construction of massive concrete structures without crack.
3. Various technical measures to prevent cracking of mass concrete, such as choice of raw materials, precooling, pipe cooling, and superficial thermal insulation.
4. The experiences of many practical concrete dams, particularly the success of the construction of several concrete dams without crack in China in recent years.
5. Many new ideas and new methods for prevention of cracking and temperature control of mass concrete.
6. Comprehensive analysis of different schemes of construction of concrete dams with different combinations of the measures of temperature control.

In the design stage of a massive concrete structure, several schemes of temperature control may be given and computed the temperature field and stress field in detail by the methods given in this book, after comprehensive analysis, a rational scheme may be obtained. Otherwise, a new scheme with improved combination of temperature control may be given and analyzed, until a good scheme of temperature control is obtained which will lead to the possibility that there will be no crack in the dam in the construction and operation period. By this method, several concrete dams without crack have been constructed in China in recent years. This is an important and valuable experience in the construction of massive concrete structures.

Cracks in massive concrete structures, such as concrete dams, will reduce the safety, integrity, and durability of the structure. The repair of cracks in concrete dams is very difficult, e.g., a big crack developed in Norfork dam, the engineers had attempted to repair the dam by grouting, but due to the worry that the crack may develop further under the pressure of grouting, the crack was not repaired and the dam has been working with a big crack in the dam body; as a result, the safety and durability of the dam are reduced remarkably.

The successful construction of several concrete dams without crack in China is an important achievement in technical science in the world.

Due to the needs of flood control, irrigation, and hydropower, many concrete dams have been constructed in China in the past 60 years. At present the amount of concrete dams higher than 15 m in China is over 40% of those elsewhere and the three highest concrete dams in the world (Jingping 305 m, Xiaowan 295 m, Xiluodu 284 m) are in China. In the process of large-scale construction of concrete dams in China, besides learning abroad experiences, systematic research works had been carried out and new theory and new experiences were created; hence, the problem of cracking in mass concrete has been solved and several concrete dams without crack have been constructed in recent 10 years.

After graduating from university in 1951, the writer participated in the design and construction of the first three concrete dams in China (Fuzhiling dam, Meishang dam, and Xiang hongdian dam) in 1951–1957. Although some measures to prevent cracking had been adopted, cracks still appeared in these dams, which indicates that cracking of mass concrete is a complex problem. The writer began to research the problem in 1955 and published two papers in 1956 and 1957 which triggered research of thermal stress and temperature control of mass concrete in China.

In 1958, the writer was transferred to the China Institute of Water Resources and Hydropower Research where he was engaged in the research work of high concrete dams, particularly the thermal stresses and temperature control of concrete dams.

A vast amount of research works had been carried out under the direction of the writer for a series of important concrete dams in China, such as Three Gorges, Xiaowan, Longtan, Xiluodu, Sanmenxia, Liujiaxia, Xin'anjiang, and Gutian.

More than 120 papers had been published putting forward a series of new ideas, new calculating methods, and new technical measures, including (1) a new idea of "long time superficial thermal insulation together with comprehensive temperature control" which may prevent crack in mass concrete effectively, (2) methods for calculating the temperature field and thermal stresses in dams, docks, sluices, tunnels, concrete blocks, and beams on elastic foundations; (3) simulation thermal stress computation taking into account the influences of all the factors and simulating the process of construction; (4) method of back analysis for determining the practical thermal and mechanical properties of concrete from the observed results; (5) the new idea of numerical monitoring of mass concrete; (6) the new idea of semi-mature age of concrete; and (7) formulas for determining the water temperature in reservoirs and temperature loading of arch dams.

Hence, a perfect system of the theory of thermal stress and temperature control of mass concrete is established whereby several concrete dams without crack have been successfully constructed in China in the past 10 years, including the Sangianghe concrete arch dam and the third stage of the famous Three Gorges concrete gravity dam and hence "no dam without crack" is no longer a problem.

The solution of the problem of cracking is an important achievement in the technology of mass concrete.

More than 10 results of the author's scientific research were adopted in the specifications for design and construction of gravity dams, arch dams, docks, and massive concrete structures in China.

In order to summarize the experiences, the author published the book *Thermal Stresses and Temperature Control of Mass Concrete* (in Chinese) in 1999.

The Information Center of the China Academy of Science published two statistics in 2011: (1) According to the number of quotations, the first 10 books of each profession of China, *Thermal Stresses and Temperature Control of Mass Concrete* is one of the 10 most widely quoted books of civil engineering in China. (2) According to the number of quotations, the first 20 authors of scientific papers of each profession in

China, the writer is the first one of the 20 most widely quoted authors of hydraulic engineering.

The author was awarded the China National Prize of Natural Science in 1982 for research work in thermal stresses in mass concrete, the China National Prize of Scientific Progress in 1988 for research work in the optimum design of arch dams, the China National Prize of Scientific Progress in 2000 for research work in simulating computation and thermal stresses, and the International Congress on Large Dams Honorary Member at Saint Petersburg in 2007.

Outside China there are two books on temperature control of mass concrete: (1) US Bureau of Reclamation, *Cooling of Concrete Dams,* 1949, (2) Stuky A, Derron MH, *Problemes Thermiques Poses Par La Construction des Barrages-Reservoirs,* Lausanne, Sciences & Technique, 1957. Theoretical solutions and many graphs for determining the temperatures of concrete dams are given in these two books which are useful to engineers, but there is no method for computing the thermal stresses, no method for preventing crack except pipe cooling, no criterion for temperature control, no experiences for preventing cracks, particularly the successful experiences in China, thus, they are insufficient for engineers to design and construct mass concrete structures without crack.

A vast amount of mass concrete is placed in the world every year. How to prevent crack is still an important problem, thus Thermal Stresses and *Temperature Control of Mass Concrete* in English will be useful for engineers and professors of civil engineering.

In this book, consideration is given to both the theory and the practice. On one side, the methods for computing the temperature fields, thermal stresses, and the variation of temperatures and thermal stresses in various types of mass concrete structures are introduced in detail; on the other side, the technical measures to control temperature and to prevent cracking, the criterion of temperature control and the experiences of practical engineering projects, particularly, the successful experiences in China in the construction of several concrete dams without crack, are described. A series of new ideas and new techniques, e.g., the idea of "long time superficial thermal insulation together with comprehensive temperature control," MgO self-expansive concrete, etc., many useful methods, formulas, graphs, charts, and figures are given.

Apart from causing cracks, the change of temperature is an important and complex loading which has great influence on the stress state of concrete structures, particularly the arch dam. In the design and construction of mass concrete structures, particular attention should be paid to thermal stress and temperature control. I hope the publication of this book will give useful help to the engineers engaged in the design and construction of mass concrete structures and the professors and students of the department of civil engineering of universities.

I am grateful to Mr. Wu Longshen, Miss Hao Wengqian, and Mrs. Li Yue for their help given to me in the preparation of this book.

Zhu Bofang
July 2013

About the Author

Zhu Bofang, the academician of the Chinese Academy of Engineering and a famous scientist of hydraulic structures and solid mechanics in China, was born in October 17, 1928 in Yujiang country, Jiangxi Province. In 1951, he graduated in civil engineering from Shanghai Jiaotong University, and then participated in the design of the first three concrete dams in China (Foziling dam, Meishan dam, and Xianghongdian dam). In 1957, he was transferred to the China Institute of Water Resources and Hydropower Research where he was engaged in the research work of high concrete dams. He was elected the academician of the Chinese Academy of Engineering in 1995. He is now the consultant of the technical committee of the Ministry of Water Resources of China, the consultant of the technical committee of water transfer from south part to north part of China, and a member of the consultant group of the Xiaowang dam, the Longtan dam, and the Baihetan dam. He was a member of the Eighth and the Ninth Chinese People's Consultative Conference, the board chairman of the Computer Application Institute of China Civil Engineering Society, and a member of the standing committee of the China Civil Engineering Society and the standing committee of the China Hydropower Engineering Society.

He is the founder of the theory of thermal stresses of mass concrete, the shape optimization of arch dams, the simulating computation of concrete dams, and the theory of creep of concrete in China.

He has established a perfect system of the theory of thermal stress and temperature control of mass concrete, including two basic theorems of creep of nonhomogeneous concrete structures, the law of variation, and the methods of computation of the thermal stresses of arch dams, gravity dams, docks, sluices, tunnels, and various massive concrete structures, the method of computation of temperature in reservoirs and pipe cooling, thermal stress in beams on foundation, cold wave, heightening of gravity dams, and the methods and criteria for control of temperatures. He proposed the idea of "long time thermal insulation as well as comprehensive temperature control" which ended the history of "no concrete dam without crack" and some concrete dams without crack had been constructed in China in recent years, including the Sanjianghe concrete arch dam and the third stage of the famous Three Gorges concrete gravity dam.

He proposed the mathematical model and methods of solution for shape optimization of arch dams, which was realized for the first time in the world and to date has been applied to more than 100 practical dams, resulting in a 10–30% saving of dam concrete and raising enormously the efficiency of design.

He developed the simulating computation of concrete dams and proposed a series of methods, including the compound element, different time increments in different regions, the equivalent equation of heat conduction for pipe cooling, and the implicit method for computing elastocreeping stresses by FEM.

He proposed the equivalent stress for FEM and its allowable values which had been adopted in the design specifications of arch dams in China, thus the condition for substituting the trial load method by FEM is provided.

The reason why houses and bridges were destroyed but no concrete dam was destroyed by strong earthquakes is explained. It is due to the fact that concrete dams must resist large horizontal water loads with large coefficients of safety in the ordinary loading case.

The instrumental monitoring can give the displacement field but cannot give the stress field and the coefficient of safety of concrete dams. In order to overcome this defect, a new idea for numerical monitoring has been proposed which can give the stress field and the coefficient of safety and raise the level of safety control of concrete dams.

The new idea for the semimature age of concrete has been proposed. The crack resistance of concrete may be promoted by changing its semimature age.

A vast amount of scientific research work has been conducted under his direction for a series of important concrete dams in China, such as Three Gorges, Xiaowan, Longtan, Xiluodu, Sanmenxia, Liujiaxia, and Xing'anjiang. More than 10 results of his scientific research were adopted in the design specifications of gravity dams, arch dams, docks, and hydraulic concrete structures.

He has published eight books: *Theory and Applications of the Finite Element Method* (1st ed. in 1979, 2nd ed. in 1998, 3rd ed. in 2009), *Thermal Stresses and Temperature Control of Mass Concrete* (1999), *Thermal Stresses and Temperature Control of Hydraulic Concrete Structures* (1976), *Theory and Applications of Structural Optimization* (1984), *Design and Research of Arch Dams* (2002), *Collected Works on Hydraulic Structures and Solid Mechanics* (1988), *Selected Papers of Academician Zhu Bofang* (1997), and *New Developments in Theory and Technology of Concrete Dams* (2009). He has published more than 200 scientific papers.

Academician Zhu was awarded the China National Prize of Natural Science in 1982 for his research work in thermal stresses in mass concrete, the China National Prize of Scientific Progress in 1988 for his research work in the optimum design of arch dams, and the China National Prize of Scientific Progress in 2001 for his research works in simulating computation and thermal stresses. He was awarded the ICOLD Honorary Member at Saint Petersburg in 2007.

1 Introduction

1.1 The Significance of Thermal Stress in Mass Concrete

Mass concrete plays an important role in modern construction, especially in hydraulic and hydroelectric construction. In China, more than 10 million m^3 mass concrete are poured every year in hydraulic and hydroelectric engineering. Besides, the structure of harbor engineering and foundations of heavy machines are often built with mass concrete.

The following are the peculiarities of a massive concrete structure:

1. Concrete is a kind of brittle material, the tensile strength of which is only about 8% of its compressive strength and the tensile deformability is poor. For short-time loading, the ultimate tensile strain is about $(0.6 \sim 1.0) \times 10^{-4}$, which is equal to the strain caused by 6–10°C temperature drop. For long-time load, the ultimate tensile strain is about $(1.2 \sim 2.0) \times 10^{-4}$.
2. As the section size of a massive concrete structure is quite large, after the pouring of concrete, the internal temperature increases dramatically due to the heat of hydration. As the modulus of elasticity of concrete is relatively small and the creep is relatively large at this time, the compressive stress caused by the temperature rise is not large; however, when the temperature gradually decreases with time later on, the modulus of elasticity is large and the creep is small, it will cause considerable tensile stress.
3. Mass concrete is often exposed to the air or water, the changes of air and water temperature will cause considerable tensile stress in a massive concrete structure.
4. In a reinforced concrete structure, tensile stresses are undertaken by steel reinforcement and concrete only bears the compressive stresses. Due to the immense thickness, if the tensile stresses in a massive concrete structure are undertaken by steel reinforcement, the volume and cost of steel reinforcement will be very big, thus generally there is no steel reinforcement in mass concrete and the tensile stresses must be undertaken by concrete itself.

Based on the features above, in the design of a massive concrete structure, it is required to have no or little tensile stress. For the external load like deadweight and water pressure, this requirement is not difficult to achieve. But in the process of construction and operation, the changes of temperature will cause large tensile stress in mass concrete, and it is not easy to control the tensile stress in an allowable value, so cracks often appeared in mass concrete.

As shown in Figure 1.1, the cracks in mass concrete can be classified into three kinds, namely through cracks, deep cracks and surface cracks. Through cracks cut the structure section and may probably destroy the stability and integrity of the structure. Leakage may occur if the cracks reach to the upstream surface. They are

Thermal Stresses and Temperature Control of Mass Concrete. DOI: http://dx.doi.org/10.1016/B978-0-12-407723-2.00001-4

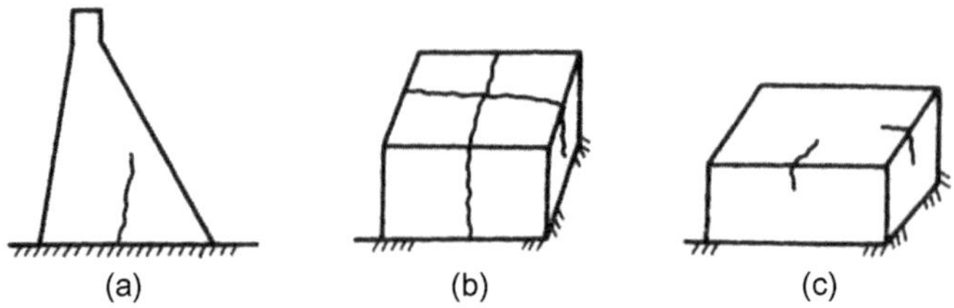

Figure 1.1 Sketch of different kinds of cracks in a massive concrete structure: (a) through crack, (b) deep crack or surface crack, and (c) surface crack.

very dangerous. Deep cracks partly cut the structure, and they are also dangerous. For surface cracks, if they do not extend, the impact is not serious. But upon reservoir impoundment, pressurized water enters the cracks, and the surface crack at upstream face of the dam may extend to a deep crack or even a through crack. Surface cracks in the region above foundation or old concrete may also develop to deep cracks or even through cracks during the cooling process of the internal concrete.

Cracks in concrete can also come from dry shrinkage, but the changes of humidity are small in mass concrete, and these changes are limited to a very shallow range near the surface, so it is not difficult to solve the problem by curing.

Experiences show that it is possible but not easy to prevent hazardous cracks of mass concrete. For the project of Qingtongxia Hydropower Station, which was built during the early stage of new China, because the engineers lacked experience and did not fully realize the importance of thermal stress, the riverbed power plants constructed in cold areas are designed with thin-wall structure and lack of effective temperature control measure. As a result, severe cracks occurred after construction started. The construction was subsequently stopped and delayed for several years. In the 1950s, several slotted gravity dams were built by the Soviet Union in the cold Siberi region. There were severe cracks in all these dams. Consequently, the hydropower stations are all built with solid gravity dams, and the Toktogul method was developed for preventing cracks.

In a massive concrete structure, the changes of temperature can not only lead to cracks but also have an important impact on the stress state of the structure. Sometimes, the thermal stress can exceed the sum of the stresses caused by other external loads. For instance, as is shown in the study of the stress state around the orifice of the Sanmenxia gravity dam, the alignment of stress values caused by different loads from high to low is caused by the temperature, the internal water pressure, self-weight, and external water pressure, and the thermal stress is larger than the sum of stresses caused by all other loads. The changes of temperature also have a remarkable impact on the stress state of arch dams.

The thermal stress is closely related to the type of structure, the weather conditions, the construction process, the properties of material, and the operating conditions. The variation of thermal stress is very complicated. It is more complex to analyze the thermal stress than the stresses caused by water, self-weight, and other external loads.

In conclusion, the analysis of the thermal stress, the temperature control, and the measures to prevent cracking are the crucial topics in the design and construction of massive concrete structures [104–110].

1.2 The Features of Thermal Stresses in Concrete Structures

Here we use an example to explain the features of the thermal stresses in concrete structures. As is shown in Figure 1.2(a), we assume that there is a steel bar AB whose ends are fixed. The temperature change is $T(\tau)$ which is a function of time: when $\tau = 0$, $T(0) = 0$, at the beginning, $T(\tau)$ increases as the time proceeds; after it reaches the highest temperature T_0, the steel bar gradually cools down, and finally $T(\infty)$ equals 0. The elastic modulus of steel is a constant E_s. Since the steel bar is fixed at both ends, the thermal stress of steel bar AB is

$$\sigma_s(\tau) = -E_s\alpha_s T(\tau) \tag{1.1}$$

The thermal stress $\sigma_s(\tau)$ of the steel bar is proportional to $T(\tau)$, and the proportional factor is $-E_s\alpha_s$, where α_s is the linear expansion coefficient of steel. When the temperature reaches its highest from the original 0°C, the stress also reaches its highest from the zero stress. When the temperature gradually cools down to 0°C, the stress also decreases to 0, and finally they return to the initial state.

As for the concrete bar AB, since the elastic modulus of concrete is varying with age τ, the thermal stress cannot be calculated using Eq. (1.1). Instead, we should use an incremental method to calculate. Dividing the time τ into a series of time intervals $\Delta\tau_i$ $(i = 1 - n)$ in the ith time interval $\Delta\tau_i$, the increment of temperature is ΔT_i, the average elastic modulus is $E(\tau_i)$, so the increment of elastic stress should be

$$\Delta\sigma_i = -\alpha E(\tau_i)\Delta T_i$$

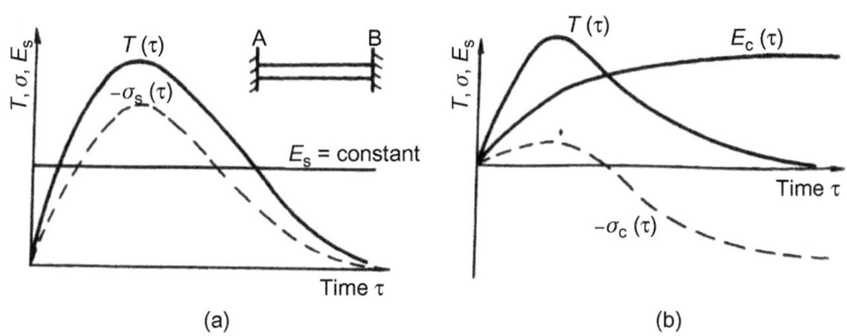

Figure 1.2 Comparison of the thermal stress between steel structure and concrete structure: (a) thermal stress in a steel bar with fixed ends and (b) thermal stress in a concrete bar with fixed ends.

After accumulation, the elastic stress is

$$\sigma_c(t) = -\alpha \sum E(\tau_i)\Delta T_i \tag{1.2}$$

Considering the influence of creep of the concrete, we should use the following equation to calculate:

$$\sigma_c(\tau) = -\alpha \sum E(\tau_i)K(t,\tau_i)\Delta T_i \tag{1.3}$$

where $K(t,\tau_i)$ is the stress relaxation coefficient, its definition is referring to Eq. (8.72). Assuming that the concrete is subjected to stress $\sigma(\tau)$ at age τ, if the strain remains at a constant, because of the creep effect, at time t, the stress will decrease to $\sigma(t) = \sigma(\tau)K(t,\tau)$, and the relaxation coefficient is the ratio of $\sigma(t)$ to $\sigma(\tau)$, namely

$$K(t,\tau) = \sigma(t)/\sigma(\tau) \tag{1.4}$$

Figure 1.2(b) shows the changes of temperature $T(t)$ and stress $\sigma(t)$ with time τ. In the early stage of temperature rising, compressive stress is developed in the bar. But since in early stage the elastic modulus of concrete and the relaxation coefficient is small, the compressive stress is not large. In the later cooling stage, the elastic modulus of concrete is relatively large, as are the relaxation coefficient and the increment of stress produced by unit temperature difference. As the temperature of the bar gradually decreases, not only the early compressive stress is canceled, but large tensile stress will be created in the bar. Finally, when the time $\rightarrow \infty$, the temperature $T(\infty)\rightarrow 0$. If the stress is not 0, there will be a large surplus tensile stress. In practice, when the temperature drop reaches 12–20°C, as for the fully restrained concrete bar, the later tensile stress is big enough to pull the concrete to failure.

We can conclude from the above examples that the changing pattern of thermal stress between the concrete structure and steel structure is totally different, the reasons accounting for this being (1) the elastic modulus of concrete is changing with age τ and (2) the impact of the creep effect of the concrete.

1.3 The Variation of Temperature and Thermal Stress of Mass Concrete with Time

1.3.1 The Variation of Temperature of Mass Concrete with Time

Because of the large size, the variation of temperature in a mass concrete structure is shown in Figure 1.3; the placing temperature T_p is the concrete temperature just after pouring. If the concrete cannot be completely cooled, it would be in an adiabatic state, and the temperature will increase according to the adiabatic rise of

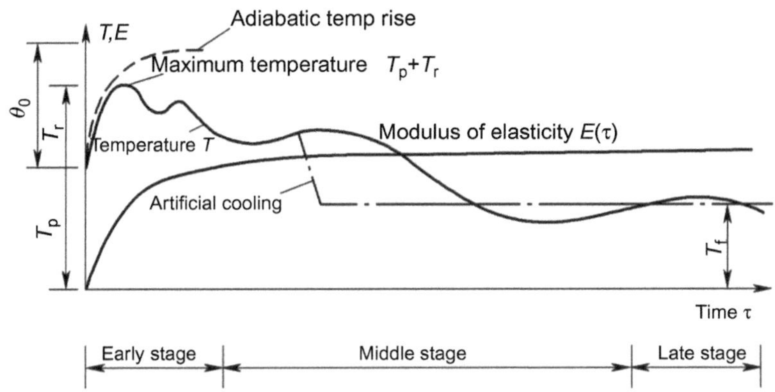

Figure 1.3 The variation of the temperature and elastic modulus of mass concrete with time.

temperature curve, as shown by the dotted line in the figure. In practice, since some heat may be lost from the top and the sides of the pouring layer, the concrete temperature will change along the solid line in the figure. The temperature rises to its highest $T_p + T_r$ and then decreases. T_r is the temperature rise due to the heat of hydration of cement. After being covered with newly poured concrete, the old concrete will be influenced by the heat of hydration produced by the newly poured concrete, and temperature recovers slightly. After the second peak temperature, the temperature will continue to decrease. If the point is more than 7 m far from the lateral surface, the temperature of this point will not be affected by the external temperature changes and is influenced only by the placing temperature, the hydration heat, and the temperature of the top of the placing layer. As is shown by the solid line in the figure, finally the temperature will vary with a small difference about the steady temperature T_f and is called the quasi-steady temperature.

In the concrete dam, the interior temperature cools down from the highest temperature to the steady temperature very slowly. It normally takes several decades or hundreds of years. In order to accelerate the cooling process, cooling pipes are adopted.

1.3.2 The Variation of the Thermal Stress in Mass Concrete

Since the elastic modulus of concrete varies with age, in a massive concrete structure, the development of thermal stress can be divided into three stages:

1. *Early stage*: It is about 1 month from the start of concrete pouring to the finish of the heat release of cement. There are two features in this stage: Firstly the temperature field will change dramatically because of the intense heat of cement hydration. And secondly, the elastic modulus of the concrete will change rapidly with time.
2. *Mid stage*: This stage starts from the end of heat release of cement and ends when the concrete is cooled down to a final steady temperature. The thermal stress in this stage is

caused by the cooling of the concrete and the changes of external temperature. In the mid stage, the elastic modulus will change slightly with time.

3. *Late stage*: The operation stage after the concrete is completely cooled down. Thermal stress is mainly caused by the changes of external air temperature and water temperature. The stresses of the three stages accumulated to form the final stress state of concrete.

1.4 Kinds of Thermal Stress

There are two kinds of thermal stress in mass concrete:

1. Self-stress

   For structures without any external constraint or statically determinate structure, if the internal temperature is linearly distributed, no stress will appear; if the internal temperature is nonlinearly distributed, the stress caused by restraint of the structure itself is called self-stress. For instance, when a concrete wall is cooled in the air, the surface temperature is low and the inner temperature is high. The shrink of the surface is restrained by the inner concrete. The tensile stress appears at the surface, and the compressive stress appears in the interior. At any section, the area of tensile stress must be equal to the area of compressive stress, as shown in Figure 1.4(a).

2. Restraint stress

   When the whole or part of the boundaries of the structure is restrained, the structure cannot deform freely with the change of temperature. The stress produced by this reason is called restraint stress, for instance, the stress in a concrete block caused by the restraint of the rock foundation when the concrete is cooling as shown in Figure 1.4(b).

   In the statically determinate structure, only self-stress will appear, but in the statically indeterminate structure, both self-stress and restraint stress will appear.

1.5 Analysis of Thermal Stress of a Massive Concrete Structure

1. Analysis of temperature field of mass concrete

   The temperature field of mass concrete depends on the weather conditions and the construction process. The problem can be treated by solving a heat conduction equation with given boundary condition and initial condition. For the simple cases, theoretical

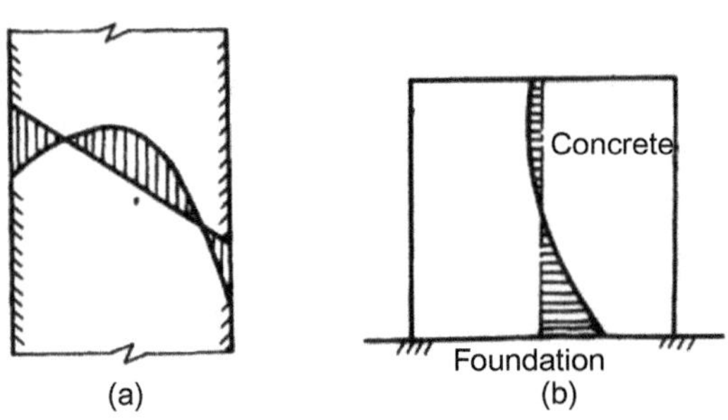

Figure 1.4 Sketch of two types of thermal stress: (a) self-stress and (b) restraint stress.

solution can be found; as for the practical complex cases, the finite difference method or finite element method can be used.

2. Analysis of thermal stress field of mass concrete

   It is more difficult to analyze the thermal stress in a given temperature field. A theoretical solution can be found only in simple cases. The numerical method is mostly used. The finite element method is commonly used at present.

   The creep of concrete will influence thermal stress. When calculating the concrete thermal stress, impact of concrete creep must be considered.

   Shrinkage stress is similar to thermal stress. The method used to analyze thermal stress can also be used to analyze shrinkage stress.

1.6 Thermal Stress—The Cause of Crack

The cracks will appear when the tensile stress of concrete exceeds its tensile strength. The tensile stress depends not only on temperature difference but also on the constraint condition. As shown in Figure 1.5, there are concrete plates on rock foundation and soil foundation. Since rock foundation has a large deformation modulus, the restraint to the deformation of the concrete plate is large; however, soil foundation has small deformation modulus, and the restraint to the deformation of the concrete plate is small. Even though the thickness and temperature drops of the two concrete plates are the same, the concrete plate on the rock foundation may crack, but the concrete plate on the soil foundation may not crack.

The thermal stress of concrete can be approximately represented as

$$\sigma = RK_{\mathrm{p}}E\alpha\,\Delta T \tag{1.5}$$

where

σ—thermal stress
R—restraint coefficient
K_{p}—stress relaxation coefficient caused by the creep of concrete
E—elastic modulus of concrete
α—coefficient of linear expansion of concrete
ΔT—temperature difference of concrete.

To prevent cracks, we must control the thermal stress so that it does not exceed the allowable tensile stress, as

$$\sigma = RK_{\mathrm{p}}E\alpha\,\Delta T \le \frac{R_{\mathrm{t}}}{K} \tag{1.6}$$

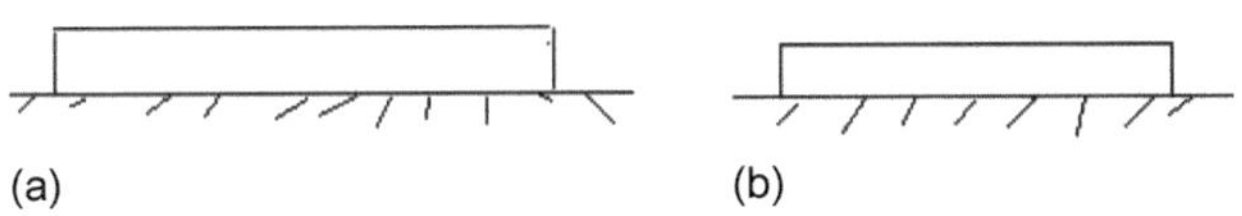

Figure 1.5 Concrete plate on (a) rock and (b) soil foundation.

where

R_t—tensile strength of concrete
K—safety factor.

From the above equation, it is clear that to prevent concrete crack, the following three aspects should be considered:

1. Control temperature difference ΔT
2. Minimize the restraint coefficient R
3. Enhance the tensile strength R_t.

The restraint factor R includes the external restraint and the internal restraint.

1.7 Technical Measures for Control of Thermal Stress and Prevention of Cracking

Once cracks appear in a massive concrete structure, it is difficult to restore the integrity of the structure by repairing. Experiences show that it is possible but not easy to prevent cracking in mass concrete. It requires careful design, careful study, and careful construction.

The following aspects should be considered when dealing with cracks in a massive concrete structure:

1. Rational choice of the type of structure and joint spacing.

   As experience shows, the type of structure has a great impact on the thermal stress and cracks. In the 1950s and 1960s, the Soviet Union constructed several slotted gravity dams in the cold Siberia area, such as the Mamakansky dam, Bratsky dam and the Boohtarminsky dam. Cracks emerged in all of these dams. The engineers learnt from this experience. They constructed solid gravity dams instead in later projects.

   The size of pouring block may influence the thermal stress. The bigger the pouring block, the larger the thermal stress. So rational joint spacing is important to prevent cracks. Practical experience and theoretical analysis have shown that when the size of the pouring block is controlled for about 15 m × 15 m, the thermal stress is low, and the constraint height of the foundation is only about 3–4 m. In temperate areas, cracks are less likely to happen. But in cold areas, because of the extensive temperature difference, a pouring block of this size is still difficult to prevent cracks, so some rigorous heat preservation actions are needed.

   Elevation difference of foundation should be avoided in the same pouring block. Stress concentration should be avoided or reduced in structure design.
2. Choosing the raw material of concrete and optimizing the mix of concrete.

   The purpose of choosing the raw material of concrete and optimizing the mix ratio of concrete is to improve the crack resistance of the concrete. Specifically, it requires concrete to have low adiabatic temperature rise, large extensibility, and low linear expansion coefficient. It is fine if the autogenous volume deformation is micro-expansion or at least low shrinkage:
   a. Choice of cement. Crack resistance, low heat, and high strength are important factors for choice of cement for internal concrete. As for external concrete, despite the crack

resistance, it requires resistance to freezing, thawing and erosion, high strength, and low shrinkage.

b. Mixed with admixture to lower the adiabatic temperature rise and to improve crack resistance of concrete. At present, fly ash is widely used.

c. Mixed with agent, including water reducing agent, air entraining agent, retarder, early strength agent, etc. Water reducing agent is the most commonly used. It can help to reduce water and to increase plasticity. With the same level of slumps and strength, it can help to reduce the water consumption, save cement, and reduce the adiabatic temperature rise. Air entraining agent helps to create large quantities of small bubbles in concrete in order to improve the freezing–thawing resisting durability of concrete. Retarder is used in summer construction and early strength agent is used in winter construction.

d. Optimize the concrete mix. To guarantee of the strength and fluidity of concrete, efforts should be made to save cement to reduce the adiabatic temperature rise of concrete.

3. Rigorous control of the temperature of concrete.

Rigorous control of the concrete temperature is the most important measure to prevent crack.

a. Reduce the placing temperature of concrete. Cooling the mixing water, adding ice to mixing water, pre-cooling aggregate, and other methods are used to reduce the concrete temperature at the exit of the concrete mixer. Increasing the strength of concrete pouring, cooling of pouring surface, and other methods are used to reduce the temperature rise during the pouring process.

b. Pipe cooling. Cooling pipes are embedded in the concrete to reduce the concrete temperature.

c. Superficial heat insulation. Insulation material is used to cover the surface of the concrete to reduce the inside and outside temperature difference and reduce the surface temperature gradient of concrete.

4. Emphasis on the preparation work before construction.

In the early stage of the construction, preparation work of temperature control of concrete must be emphasized, such as the installation and testing of the cooling plant and ice machine, cooling water pipe, and preparation for heat insulation material.

5. Strengthen the management of construction.

a. Improve the quality of concrete construction. To prevent cracks, despite the rigorous control of concrete temperature, reinforcement of construction management and improvement for construction quality are also needed. Obviously, the strength distribution in a concrete block is nonuniform. Cracks emerge firstly at the most vulnerable place. A survey was conducted at the Dan Jiang Kou dam site, and hundreds of concrete layers were investigated. The results showed that the emergence of the cracks had significant connection with the strength distribution of the concrete. When the quality of concrete is poor and the deviation coefficient of concrete strength C_v is large, there will be more cracks. Projects with good concrete construction quality may have fewer cracks; otherwise, there will be many cracks. So it is important to strengthen the construction management to improve the concrete construction quality.

b. Even rising with thin layer and short interval. For the schedule of concrete pouring, it is better to pour concrete with thin layer, short interval (5–10 days) and even rising. Avoid pouring concrete in a rush and resting for a long time; avoid large height difference between adjacent dam blocks; especially avoid "thin block, long interval."

c. Better to pour concrete above foundation in cold weather.

d. Strengthen curing and prevent shrinkage.

1.8 The Experience of the Temperature Control and Crack Prevention of Mass Concrete in the Last 30 Years

1. Enhancement of pipe cooling in the local area makes the control of the maximum tensile stress in concrete dams no longer a challenge.

   In the past, since the steel water pipe has too many connections, it takes time to set up and can only be set on the surface of the old concrete layer. The vertical spacing of water pipe is equal to the thickness of the pouring layer. In recent years, steel water pipe is substituted by the plastic water pipe. The plastic water pipe is flexible and can be paved during the pouring process. The vertical spacing of the water pipe can be reduced to the thickness of pouring layer, which is about 0.3−0.7 m; the horizontal space can be reduced to about 1.0 m. Cooling of water pipe with small spacing can greatly reduce the temperature rise caused by the heat of hydration. The combination of pipe cooling and pre-cooling of the concrete makes the control of the maximum tensile stress in the concrete dam no longer a challenge. Moreover, the height of cooling area with closely arranged cooling pipes is only 0.1−0.2 the length of the pouring block. The range of cooling area with closely arranged cooling pipes is not large, and the cost is low.
2. Application of the plastic foam board can effectively help to control the surface tensile stress.

   In the past, straw bags were mainly used to insulate the surface of the concrete dam. But the straw bags become damp and rot. Moreover, the straw bags are inflammable and their heat insulation effect is poor. The poor heat insulation at the surface is the significant cause of "No dam without cracks" in the past. After 1980, plastic foam boards were applied in the superficial thermal insulation of mass concrete, and the effect is excellent. Construction with plastic foam board is easy, and the cost is not high. Plastic foam board can be used for long-term heat insulation.
3. It is a trend to built concrete dams without cracks.

   In the past, there were cracks in almost all concrete dams. It is an objective fact that there is "No dam without cracks." Nowadays, with the remarkable development of the technique of temperature control, several concrete dams have been constructed without cracks, such as the third stage of Three Gorges concrete gravity dam and the San Jiang He arch dam.

   Today, if the dam is well designed, well studied, and well constructed, a concrete dam can be constructed without cracking and the cost is not high. Thus the trend in the future is to construct mass concrete without cracking.

2 Conduction of Heat in Mass Concrete, Boundary Conditions, and Methods of Solution

2.1 Differential Equation of Heat Conduction, Initial and Boundary Conditions

2.1.1 *Differential Equation of Heat Conduction [1–4, 7–11]*

As shown in Figure 2.1, an elementary parallelepiped $dx\,dy\,dz$ is taken from the interior of a mass concrete structure. The sum of the heat fluxes across the six surfaces of the elementary parallelepiped is

$$Q_1 = \lambda\left(\frac{\partial^2 T}{\partial x^2} + \frac{\partial^2 T}{\partial y^2} + \frac{\partial^2 T}{\partial z^2}\right) \tag{2.1}$$

The heat emitted by the hydration of cement is

$$Q_2 = c\rho\frac{\partial\theta}{\partial\tau}\mathrm{d}t \tag{2.2}$$

The heat absorbed by concrete due to the rise of temperature is

$$Q_3 = c\rho\frac{\partial T}{\partial\tau}\mathrm{d}\tau \tag{2.3}$$

From the balance of heat, $Q_3 = Q_1 + Q_2$, namely

$$c\rho\frac{\partial T}{\partial\tau}\mathrm{d}\tau = \lambda\left(\frac{\partial^2 T}{\partial x^2} + \frac{\partial^2 T}{\partial y^2} + \frac{\partial^2 T}{\partial z^2}\right) + c\rho\frac{\partial\theta}{\partial\tau}\mathrm{d}\tau \tag{2.4}$$

Dividing Eq. (2.4) by $c\rho\,\mathrm{d}\tau$, the differential equation of heat conduction is derived in the following:

$$\frac{\partial T}{\partial\tau} = a\left(\frac{\partial^2 T}{\partial x^2} + \frac{\partial^2 T}{\partial y^2} + \frac{\partial^2 T}{\partial z^2}\right) + \frac{\partial\theta}{\partial\tau} \tag{2.5}$$

Thermal Stresses and Temperature Control of Mass Concrete. DOI: http://dx.doi.org/10.1016/B978-0-12-407723-2.00002-6

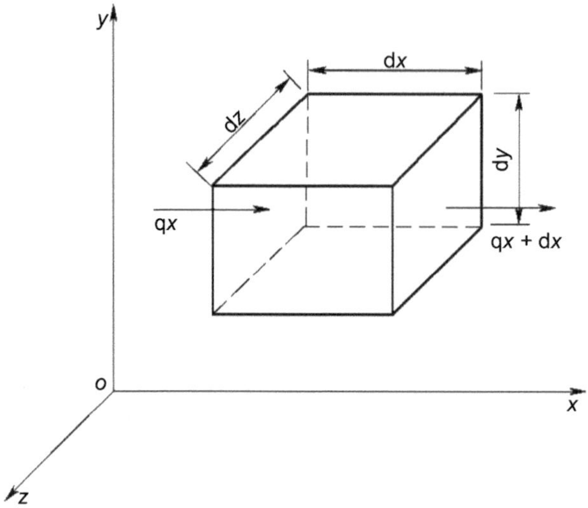

Figure 2.1 An elementary parallelepiped.

where

$a = \lambda/c\rho$—the diffusivity of concrete
λ—the conductivity of concrete
c—the specific heat of concrete
ρ—the density of concrete
T—the temperature
θ—the adiabatic temperature rise due to hydration of cement
τ—time
x, y, z—the coordinates.

2.1.2 Initial Condition

The initial temperature of the structure is given as follows:

when $\tau = 0$

$$T(x, y, z, 0) = T_0(x, y, z) \tag{2.6a}$$

or

$$T(x, y, z, 0) = T_0 \tag{2.6b}$$

where $T_0(x,y,z)$ is a continuous function of x,y,z and T_0 is a constant.

On the contact surface between the concrete and the rock or between the new and the old concrete, the initial temperature may be discontinuous; in this case, two numbers relating to different initial temperatures must be given to one point on the contact surface.

2.1.3 Boundary Conditions

There are four kinds of boundary conditions.

1. First kind of boundary condition: prescribed surface temperature.
   The surface temperature is prescribed as follows:
   on the surface,

$$T_s(\tau) = f_1(\tau) \tag{2.7}$$

   where $f_1(\tau)$ is a function of τ.

2. Second kind of boundary condition: prescribed heat flux across the surface.
   The flux of heat across the surface is a known function of time, namely
   on the surface,

$$-\lambda \frac{\partial T}{\partial n} = f_2(\tau) \tag{2.8}$$

   where
   n—the outward normal of the surface
   λ—the conductivity of concrete
   $f_2(\tau)$—a known function of time τ.
   When there is no flux across the surface, Eq. (2.8) will become:
   on the surface

$$\frac{\partial T}{\partial n} = 0 \tag{2.9}$$

3. Third kind of boundary condition: linear heat transfer on the surface.
   The flux across the surface is proportional to the temperature difference between the surface and the surrounding medium, namely
   on the surface

$$-\lambda \frac{\partial T}{\partial n} = \beta(T_s - T_a) \tag{2.10}$$

   where
   β—the surface conductance, kJ/(m^2 h °C)
   T_s—surface temperature
   T_a—the air temperature.
   As shown in Figure 2.2, if the radiation heat from the sun on unit surface in unit time is S, the portion absorbed by the concrete is R and the left part $S-R$ is reflected, then

$$R = \alpha_s S \tag{2.11}$$

   in which
   α_s—the coefficient of absorption, generally $\alpha_s \cong 0.65$.
   The boundary condition considering the sun radiation is

$$-\lambda \frac{\partial T}{\partial n} = \beta(T_s - T_a) - R \tag{2.12}$$

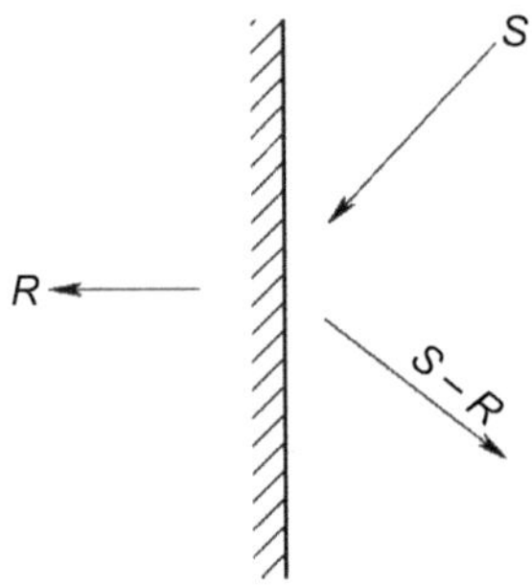

Figure 2.2 Boundary of mass concrete.

or

$$-\lambda\frac{\partial T}{\partial n} = \beta\left[T_s - \left(T_a + \frac{R}{\beta}\right)\right] \tag{2.13}$$

After comparing Eq. (2.13) with Eq. (2.10), it is clear that the influence of sun radiation is approximately equal to the increment of the surrounding air temperature:

$$\Delta T_a = R/\beta \tag{2.14}$$

4. Fourth kind of boundary condition: contact surface between two different solids.

If the contact is good, the temperature will be continuous on the surface, the boundary condition is

on the contact surface:

$$\left.\begin{aligned} T_1 &= T_2 \\ \lambda_1\frac{\partial T_1}{\partial n} &= \lambda_2\frac{\partial T_2}{\partial n} \end{aligned}\right\} \tag{2.15}$$

where λ_1 and λ_2—the conductivities of the two solids.

If the contact is imperfect, the temperature will be discontinuous and the boundary conditions will be

$$\left.\begin{aligned} \lambda_1\frac{\partial T_1}{\partial n} &= \frac{1}{R_c}(T_2 - T_1) \\ \lambda_1\frac{\partial T_1}{\partial n} &= \lambda_2\frac{\partial T_2}{\partial n} \end{aligned}\right\} \tag{2.16}$$

where

R_c—the thermal resistance due to the imperfectness of contact.

2.1.4 The Approximate Treatment of the Third Kind of Boundary Condition

Equation (2.10) may be transformed into the following form:

$$-\frac{\partial T}{\partial n} = \frac{T_s - T_a}{\lambda/\beta} \tag{2.17}$$

When the surface temperature T_s is changed from T_1 to T_2, the negative temperature gradient will be

$$-\frac{\partial T_1}{\partial n} = \tan\varphi_1 = \frac{T_1 - T_a}{\lambda/\beta}$$

and

$$-\frac{\partial T_2}{\partial n} = \tan\varphi_2 = \frac{T_2 - T_a}{\lambda/\beta}$$

As shown in Figure 2.3, the tangents to the temperature curves at the surface will always pass through point B and the distance between point B and the surface of concrete is

$$d = \lambda/\beta \tag{2.18}$$

For the third kind of boundary condition, if a virtual thickness $d = \lambda/\beta$ is added to the plate, a virtual boundary is obtained. The temperature on the virtual boundary is equal to the air temperature T_a. It means that, the virtual boundary is a first kind of boundary with prescribed temperature T_a. If thickness $d = \lambda/\beta$ is added to the plate on both sides, we get a new plate with thickness

$$L' = L + 2d \tag{2.19}$$

the temperature field which may be computed with the first kind of boundary condition.

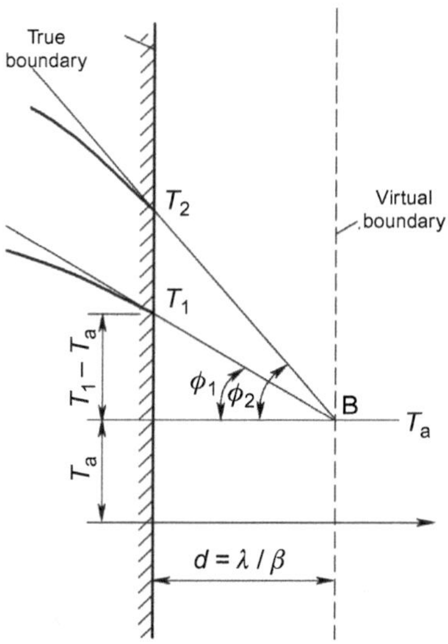

Figure 2.3 The third boundary condition.

The conductivity of concrete is $\lambda = 8-12$ kJ/(m h °C), if the concrete is in contact with water, the surface conductance is $\beta = 8000-16{,}000$ kJ/(m^2 h °C), the virtual thickness $\lambda/\beta = 0.5-1.0$ mm, the influence of which may be neglected and the surface temperature of concrete is equal to the water temperature.

If the concrete is in contact with air, the surface conductance is $\beta = 40-80$ kJ/(m h °C), $\lambda/\beta = 0.1-0.2$ m. When the air temperature changes rapidly (as a cold wave or the variation in 1 day), the influence of $\lambda/\beta = 0.1-0.2$ is remarkable. When the air temperature T_a changes slowly (as the annual variation), the influence of $\lambda/\beta = 0.1-0.2$ m is so small that the surface may be computed as it is the first kind of boundary condition.

Equation (2.19) is generally applied only in the manual computation and Eq. (2.10) is used in the numerical analysis on a computer.

2.2 Surface Conductance and Computation of Superficial Thermal Insulation

2.2.1 Surface Conductance β

The surface conductance β of solid in air depends on the wind speed as given in Table 2.1. β may also be computed by the writer's following formulas:

$$\left.\begin{aligned} \text{For rough surface:}\quad & \beta = 21.06 + 17.58 v_a^{0.910} \\ \text{or}\quad & \beta = 21.06 + 14.60 F^{1.38} \\ \text{For smooth surface:}\quad & \beta = 18.46 + 17.36 v_a^{0.885} \\ \text{or}\quad & \beta = 18.46 + 13.60 F^{1.36} \end{aligned}\right\} \tag{2.20}$$

where

v_a—wind speed, m/s
F—wind class
β—surface conductance, kJ/(m^2 h °C) (Table 2.2).

Table 2.1 Surface Conductance β of Solid in Air

Wind Speed v_a (m/s)	β (kJ/(m^2 h °C))		Wind Speed v_a (m/s)	β (kJ/(m^2 h °C))	
	Smooth Surface	Rough Surface		Smooth Surface	Rough Surface
0.0	18.46	21.06	5.0	90.14	96.71
0.5	28.68	31.36	6.0	103.25	110.99
1.0	35.75	38.64	7.0	116.06	124.89
2.0	49.40	53.00	8.0	128.57	138.46
3.0	63.09	67.57	9.0	140.76	151.73
4.0	76.70	82.23	10.0	152.69	165.13

Table 2.2 Wind Class F and Wind Speed v_a

Wind Class *F*	**0**	**1**	**2**	**3**	**4**	**5**
Wind speed v_a (m/s)	0–0.2	0.3–1.5	1.6–3.3	3.4–5.4	5.5–7.9	8.0–10.7
Wind class F	6	7	8	9	10	
Wind speed v_a (m/s)	10.8–13.8	13.9–17.1	17.2–20.7	20.8–24.4	24.5–28.4	

Table 2.3 Influence Coefficient of Wind Velocity, b1

Wind Velocity v_a	**v_a < 4 m/s**	**v_a > 4 m/s**
Material pervious to wind (straw, sawdust)		
Without protection layer	2.6	3.0
With protection layer	1.6	1.9
Non-pervious material	1.3	1.5

2.2.2 Computation of the Effect of Superficial Thermal Insulation

The surface of mass concrete with thermal insulation layer may be computed as the third kind of boundary condition with equivalent surface conductance β_s. As shown in Figure 2.4, when the concrete surface is covered by several thermal insulation layers, the thermal resistance of each insulation layer is

$$R_i = \frac{h_i}{\lambda_i b_1 b_2} \tag{2.21}$$

where

h_i—the thickness of the ith layer
λ_i—the conductivity of the ith layer
R_i—the thermal resistance of the ith layer
b_1—the influence coefficient of wind velocity, see Table 2.3
b_2—the influence coefficient of humidity of the material, $b_2 = 3-5$ for wet material, $b_2 = 1$ for dry material.

The thermal resistance between the air and the insulation layer in contact with air is $1/\beta$, so the total thermal resistance of all the insulation layers is

$$R_s = \frac{1}{\beta} + \sum \frac{h_i}{\lambda_i b_1 b_2} \tag{2.22}$$

The thermal capacity of the insulation layers is small and may be neglected and the equivalent surface conduction of concrete surface is given by

$$\beta_s = \frac{1}{R_s} = \frac{1}{(1/\beta) + \sum(h_i/\lambda_i b_1 b_2)} \tag{2.23}$$

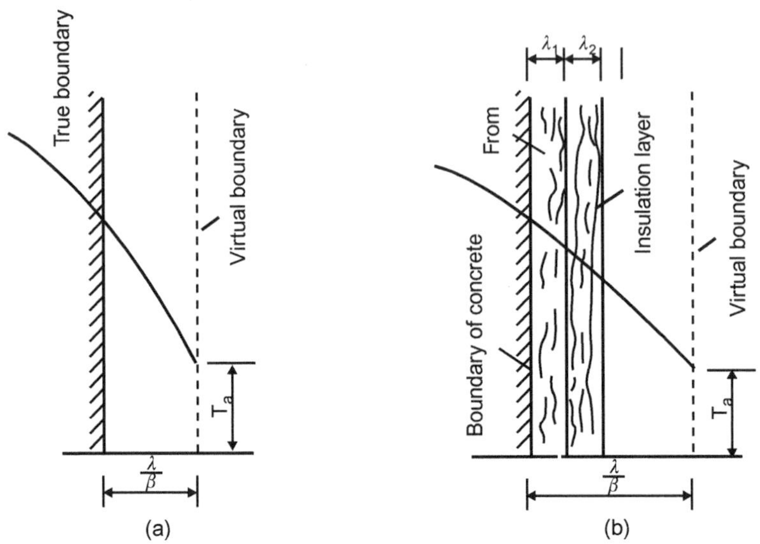

Figure 2.4 Superficial thermal insulation: (a) uncovered surface and (b) surface with thermal insulation layer.

Table 2.4 Conductivities of Thermal Insulation Materials, λ (kJ/(m h °C))

Insulation Material	λ	Insulation Material	λ
Foamed plastics	0.1256	Dry sand	1.172
Wood plate	0.837	Wet sand	4.06
Wood crambs	0.628	Mineral wool	0.209
Straw mat	0.502	Paper board	0.628
Foamed concrete	0.377	Gunny felt	0.188
Asphalt	0.938	Asphalt felt	0.167

If the results of thermal stress analysis show that β_s is the required surface conductance for preventing cracking of concrete, then the necessary thermal resistance provided by the insulation layers is

$$\sum \frac{h_i}{\lambda_i b_1 b_2} = \frac{1}{\beta_s} - \frac{1}{\beta} \tag{2.24}$$

If there is only one insulation layer with conductivity λ_1, its thickness h_1 may be computed by

$$h_1 = b_1 b_2 \lambda_1 \left(\frac{1}{\beta_s} - \frac{1}{\beta} \right) \tag{2.25}$$

The conductivities of various insulation materials are given in Table 2.4.

2.3 Air Temperature

2.3.1 Annual Variation of Air Temperature

The variation of air temperature in 1 year or 1 day generally can be expressed by the following cosine series:

$$T_a(\tau) = T_{am} + \sum_{i=1}^{n} A_i \cos\left[\frac{2i\pi}{P}(\tau - \tau_0)\right] \tag{2.26}$$

$$A_i = \frac{2}{P}\int_0^P T_a(\tau)\cos\left[\frac{2i\pi}{P}(\tau - \tau_0)\right]d\tau \tag{2.27}$$

where

$T_a(\tau)$—the air temperature
T_{am}—the mean air temperature
P—the period of variation, P is equal to 1 year or 1 day
τ—time
τ_0—the time for the maximum air temperature
n—the number of terms, generally $n = 1$ or 2.

Fox example, the air temperature at Three Gorges dam is given by ($n = 1$):

$$T_a = 17.30 + 11.35\cos\left[\frac{\pi}{6}(\tau - 6.50)\right]\ (^\circ\text{C}) \tag{2.28}$$

and the air temperature at Baise dam is expressed by ($n = 2$):

$$T_a = 22.1 + 7.57\cos\left[\frac{\pi}{6}(\tau - 6.50)\right] - 1.22\cos\left[\frac{\pi}{3}(\tau - 6.50)\right]\ (^\circ\text{C}) \tag{2.29}$$

2.3.2 Cold Wave

The cold wave, a rapid drop of air temperature in 2–6 days, is an important cause for cracking of mass concrete. The variation of the air temperature during a cold wave may be expressed as follows (Figure 2.5):

$$T_a = T_0 - A_c \sin\left[\frac{\pi(\tau - \tau_1)}{2Q}\right] \tag{2.30}$$

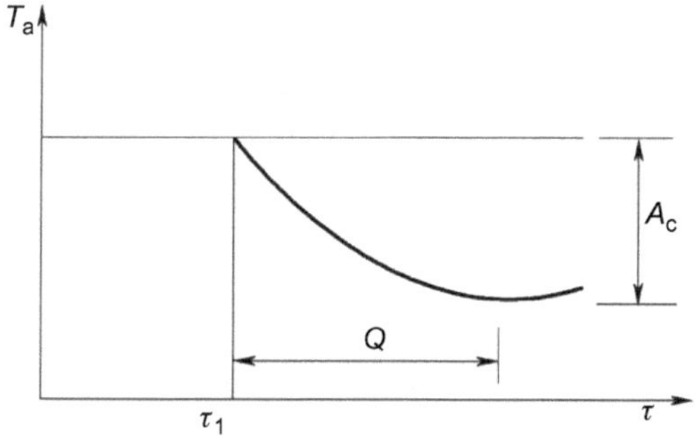

Figure 2.5 Variation of air temperature in a cold wave.

where

T_0—the initial air temperature
A_c—the maxima drop of air temperature
Q—the duration of drop of air temperature in the cold wave.

2.4 Temperature Increments due to Sunshine

It is clear from Eq. (2.13) that the increment of air temperature due to sunshine is $\Delta T_a = R/\beta$, where R is the sun radiation and β is the surface conductance of concrete.

2.4.1 Sun Radiation on Horizontal Surface

The data about sun radiation should be obtained from the meteorological station at the damsite. Some data for reference are given in Table 2.5, where S_0 is the heat of radiation in a sunny day. The radiation heat S in a cloudy day may be computed by

$$S = S_0(1 - k_1 n) \tag{2.31}$$

where

S—the radiation heat of sun in a cloudy day
S_0—the radiation heat of sun in a sunny day, see Table 2.5
n—cloud coefficient
k_1—latitude coefficient given in Table 2.6.

Example Latitude 30°, cloud coefficient $n = 0.20$, surface conductance $\beta = 80.0$ kJ/(m^2 h °C), from Table 2.5, the annual mean value $S_0 = 1066.8$ kJ/(m^2 h), From Table 2.6, $k_1 = 0.68$.

By Eq. (2.31):

$$S = 1066.8(1 - 0.68 \times 0.20) = 921.7 \text{ kJ/(m}^2\text{ h)}$$

Let the coefficient of absorption $\alpha_s = 0.65$, then

$$\frac{R}{\beta} = \frac{\alpha_s S}{\beta} = \frac{0.65 \times 921.7}{80.0} = 7.49°\text{C}$$

Thus, the annual mean value of air temperature is increased to 7.49°C due to the solar radiation.

The minimum solar radiation in December: $S_0 = 641.5$ kJ/(m^2 h),

$$R/\beta = 4.50°\text{C}$$

Table 2.5 Monthly Mean Value of Radiation Heat of Sun S_0 (kJ/(m^2 h)) in a Sunny Day

Latitude	Month												Annual Mean Value
	1	2	3	4	5	6	7	8	9	10	11	12	
80	0	0	140.0	558.2	1007.3	1180.4	1063.6	607.8	209.3	22.5	0	0	401.5
75	5.6	37.4	225.1	651.3	1052.3	1215.3	1108.6	692.2	308.2	95.7	11.6	0	453.1
70	11.3	87.2	326.4	738.5	1091.7	1244.4	1142.4	771.0	407.1	168.8	11.6	0	504.7
65	45.0	155.8	427.7	819.9	1131.1	1273.5	1181.8	849.7	511.7	253.2	87.2	22.5	567.8
60	95.7	243.0	540.2	895.5	1170.5	1296.7	1215.5	922.9	610.6	343.3	151.2	67.5	630.9
55	168.8	348.9	647.2	965.3	1209.9	1320.0	1243.7	996.0	715.2	433.3	238.4	129.4	705.4
50	264.5	467.3	759.7	1035.1	1243.7	1337.5	1266.2	1058.0	825.7	540.2	337.3	213.8	780.0
45	371.4	585.6	866.6	1104.9	1271.8	1354.9	1288.7	1131.1	930.4	652.8	447.8	320.8	860.3
40	489.6	716.5	956.7	1163.0	1288.7	1366.5	1305.6	1187.4	1023.4	754.1	564.1	433.3	940.6
35	607.7	847.3	1041.1	1221.2	1294.3	1366.5	1311.2	1226.8	1093.2	849.7	686.2	540.2	1009.4
30	714.6	947.0	1097.3	1256.0	1294.3	1366.5	1311.2	1249.3	1151.4	928.5	790.8	641.5	1066.8
25	804.7	1028.0	142.4	1267.7	1288.7	1360.7	1299.9	1254.9	1192.1	990.4	872.3	737.2	1106.9
20	872.2	1090.3	1170.5	1267.7	1271.8	1331.6	1277.4	1136.7	1221.2	1041.1	947.8	816.0	1129.8
15	934.1	1140.1	1181.8	1256.0	1238.0	1290.9	1243.7	1226.8	1227.0	1080.5	1006.0	883.5	1141.3
10	979.2	1183.8	1181.8	1238.6	1193.0	1232.8	1193.0	1193.0	1227.0	1103.0	1046.7	934.1	1141.3
5	1012.9	1214.9	1170.5	1209.5	1148.0	1151.4	1131.1	1153.6	1209.5	1119.9	1081.6	973.5	1129.9
0	1041.1	1233.6	1147.9	1174.6	1080.5	1046.7	1052.3	1103.0	1186.3	1125.5	1104.9	1012.9	1106.9

Table 2.6 Latitude Coefficient k_1

Latitude	**75**	**70**	**65**	**60**	**55**	**50**	**45**	**40**	**35**	**30**	**25**	**20**	**15**	**10**	**5**	**0**
k_1	0.45	0.50	0.55	0.60	0.62	0.64	0.66	0.67	0.68	0.68	0.68	0.67	0.67	0.66	0.66	0.65

The maximum solar radiation in June: $S_0 = 1366.5\ \text{kJ/(m}^2\ \text{h)}$,

$$R/\beta = 9.59°\text{C}$$

The amplitude of annual variation

$$A = (9.59 - 4.50)/2 = 2.55°\text{C}$$

Thus, due to the solar radiation, the annual mean temperature is increased to 7.49°C and the amplitude of annual variation is increased to 2.55°C.

If the cloud coefficient varies with month, the increments of air temperature due to sunshine may be computed month by month.

2.4.2 Temperature Increment of the Dam Surface due to Sunshine

Due to solar radiation, the temperature of the dam surface above the water level is higher than the air temperature.

On the basis of observed temperatures on some dams and combined with some theoretical analysis, Figure 2.6 is proposed by the U.S. Bureau of Reclamation [3], which may be used to compute the increments of annual mean temperature. When the valley is narrow, the sunshine may be partly sheltered by the hills on both sides, a topography coefficient k is introduced to consider its influence.

Example 1 Refer to Figure 2.7, the angle between the north and the normal to dam surface is $\phi = 156°$, the slope of the downstream face of dam is 0.26, the latitude of damsite is 33°. From Figure 2.6(a), the increment of the annual mean temperature for 100% exposed surface is 5.1°C, the topography coefficient $k = 132/180$, so the temperature increment of the dam surface due to sunshine is

$$\Delta T_s = 5.1 \times 132/180 = 3.7°\text{C}$$

2.4.3 Influence of Sunshine on the Temperature of Horizontal Lift Surface

If there is observed value of solar radiation R, the increment of temperature of the lift surface will be $\Delta T = R/\beta$. When there is no observed R, it may be estimated as follows.

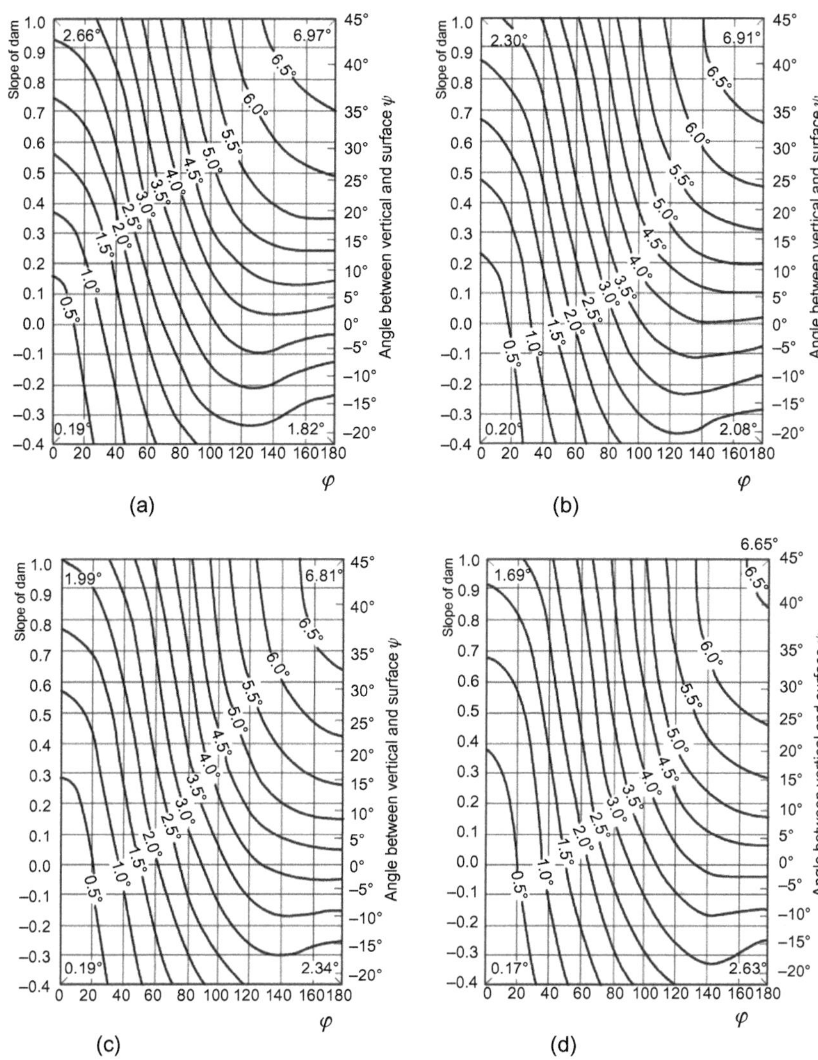

Figure 2.6 Increments of annual mean temperature due to sunshine (ϕ is the angle between normal to dam surface and direction of north, ψ is the angle between the vertical and dam surface): (a) latitude 30–35°, (b) latitude 35–40°, (c) latitude 40–45°, (d) latitude 45–50°.

As shown in Figure 2.8, take the origin of time at noon, suppose that the sun radiation S varies with time τ in the following manner:

$$\left.\begin{aligned} &\text{When} \quad P_S/2 \le \tau \le P_S/2, \quad S = A_S \cos\left(\frac{\pi\tau}{P_S}\right) \\ &\text{When} \quad |\tau| > P_S/2, \quad S = 0 \end{aligned}\right\} \tag{2.32}$$

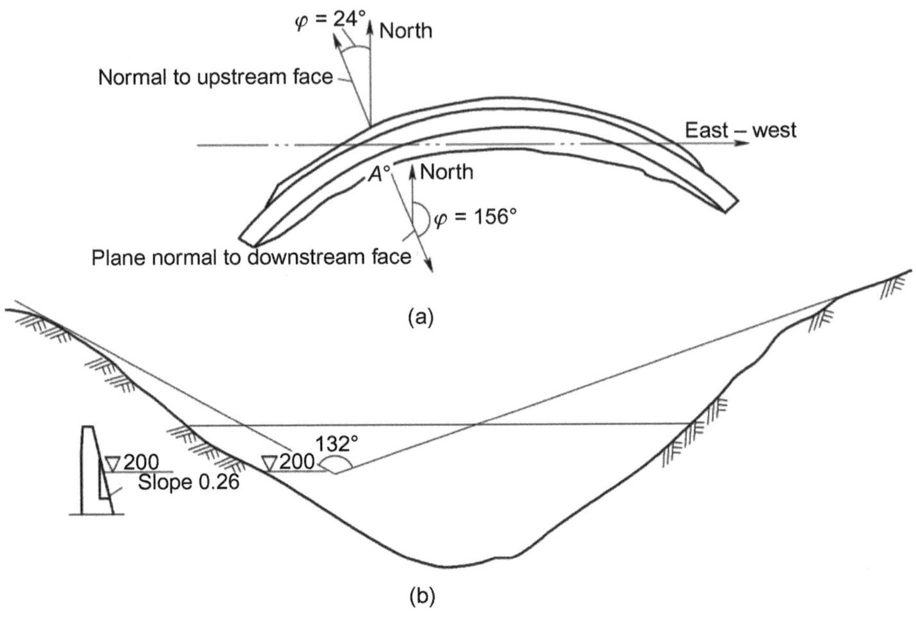

Figure 2.7 Example of influence of sunshine: (a) plane and (b) cross-section along east-west.

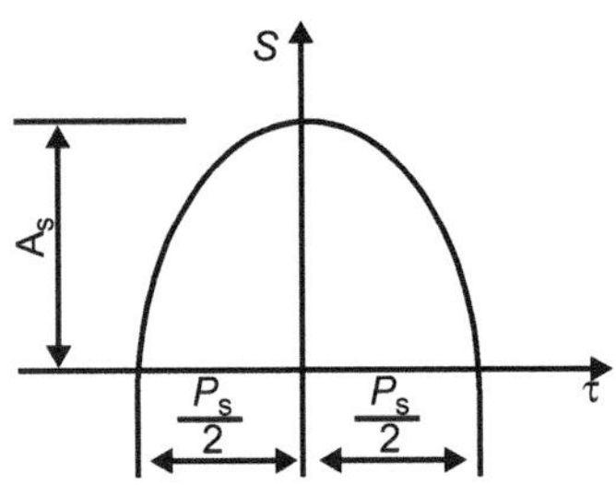

Figure 2.8 Variation of sun radiation with time.

After integration, the mean sun radiation in a day is

$$S_0 = \frac{P_S A_S}{12\pi} \tag{2.33}$$

where P_S is the time of sunshine of a day, given in Table 2.7.

Example Latitude 30°, cloudy coefficient $n = 0.20$, from Table 2.4, $S_0 = 1366.5$ kJ/(m^2 h), $P_S = 14$ h, from Eq. (2.33).

$$A_S = \frac{12\pi}{P_S} S_0 = \frac{12\pi}{14} \times 1366.5 = 3679 \text{ kJ}/(\text{m}^2\text{ h})$$

Table 2.7 Time of Sunshine P_S(h)

Latitude	Spring Equinox, Autumn Equinox	Winter Solstice	Summer Solstice
30°	12	10	14
40°	12	9	15
50°	12	8	16

From Table 2.6, the latitude coefficient $k_1 = 0.68$, by Eq. (2.31).

$$A' = 3679(1 - 0.68 \times 0.20) = 3179\ \mathrm{kJ/(m^2\ h)}$$

Supposing the surface conductance $\beta = 85.0$ kJ/(m^2 h °C) and coefficient of absorption $\alpha_s = 0.65$, the maximum temperature increment due to sunshine is

$$R/\beta = 0.65 \times 3179/85.0 = 24.3°\mathrm{C}$$

From Eq. (2.32), the increment of air temperature due to sunshine in 1 day is

$$\Delta T_a = \frac{R}{\beta} = \begin{cases} 24.3 \cos\left(\dfrac{\pi}{14}\tau\right), & \text{when } -7 \le \tau \le 7 \\ 0, & \text{when } |\tau| > 7 \end{cases}$$

2.5 Estimation of Water Temperature in Reservoir

In a natural river, the velocity of water is high and the flow is turbulent, so the water temperature is nearly uniform in the cross-section of the river. In a reservoir, the velocity of water is very small so the flow is laminar. The density of water is highest at 4°C, thus in a large reservoir, the water temperature is lowest at the bottom and higher in a higher elevation. The water at the same elevation will have the same temperature [93–95].

The variations of water temperature with time and depth of water for two reservoirs are shown in Figures 2.9 and 2.10 [21].

The water temperature in a reservoir may be computed by the following formulas proposed by the writer in 1985 [43, 51]:

the water temperature at depth y:

$$T(y, \tau) = T_m(y) + A(y) \cos \omega(\tau - \tau_0 - \varepsilon) \tag{2.34}$$

the annual mean temperature at depth y:

$$T_m(y) = c + (T_s - c)e^{-\alpha y} \tag{2.35}$$

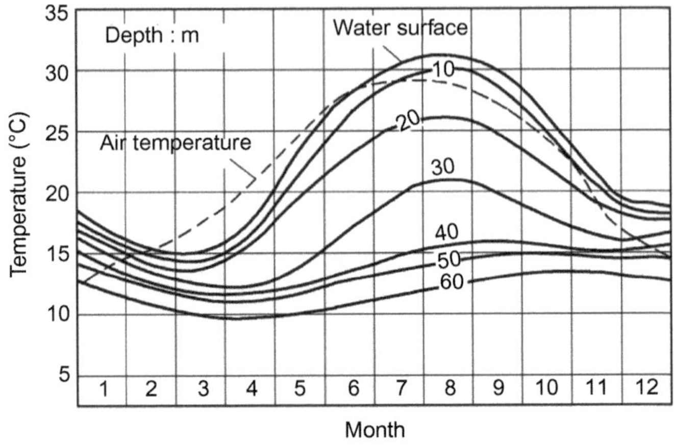

Figure 2.9 Observed water temperature in Xingfengjiang reservoir in south China.

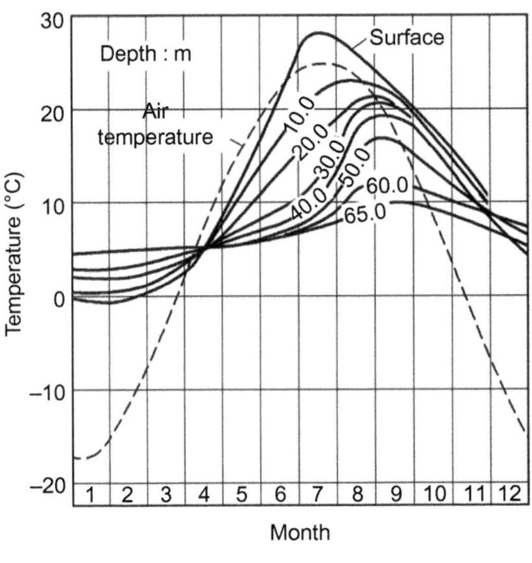

Figure 2.10 Observed water temperature in Fengman reservoir in northeast China.

the amplitude of annual variation of water temperature:

$$A(y) = A_0\, e^{-\beta y} \tag{2.36}$$

the phase difference of water temperature:

$$\varepsilon = 2.15 - 1.30\, e^{-0.085y} (\text{month}) \tag{2.37}$$

where

y—depth of water, m
τ—time, month

$\omega = 2\pi/P$—circular frequency of variation of water temperature
P—period of variation, $P = 12$ month
ε—phase difference, month
$T(y,\tau)$—water temperature at depth y and time τ, °C
$T_m(y)$—annual mean temperature of water at depth y, °C
$A(y)$—the amplitude of annual variation, °C
τ_0—the time for maximum air temperature, month.

The constants α, β, and A_0 are determined by the observed water temperatures in practical reservoirs.

Fox example, the water temperature in Xinfengjiang reservoir is expressed by

$$T_m(y) = 11.37 + 10.33\, e^{-0.04y}\ ^\circ C \tag{2.38}$$

$$A(y) = 7.4\, e^{-0.018y}\ ^\circ C \tag{2.39}$$

and those for Fengman reservoir are

$$T_m(y) = 5.67 + 5.43\, e^{-0.04y}\ ^\circ C \tag{2.40}$$

$$A(y) = 13.4\, e^{-0.018y}\ ^\circ C \tag{2.41}$$

A comparison between the computed and the observed annual mean temperatures of water in two reservoirs is shown in Figure 2.11. The computed temperatures agree well with the observed values.

The temperatures on the upstream face of concrete dams are influenced by the elevation of water surface which is discussed in Section 2.9.

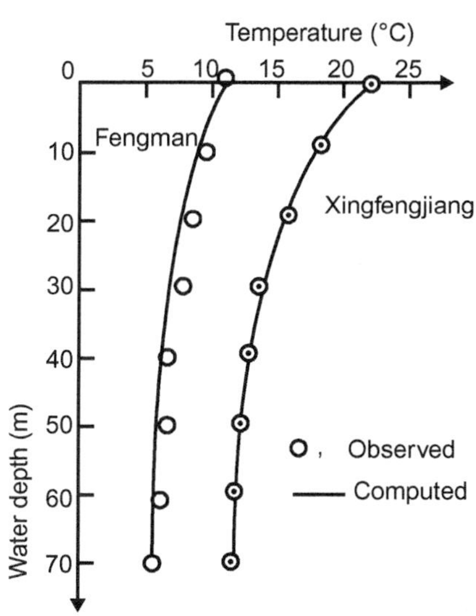

Figure 2.11 Comparison between the computed and the observed annual mean temperatures of water in two reservoirs.

2.6 Numerical Computation of Water Temperature in Reservoir

As shown in Figure 2.12, the reservoir is divided into horizontal layers. Assuming that each layer has the same temperature, i.e., the water temperature is a function of y, from balance of heat, the fundamental equation for computing water temperature is derived as follows [93–95]:

$$\frac{\partial T}{\partial \tau} + v\frac{\partial T}{\partial y} = \frac{1}{A}\frac{\partial}{\partial y}\left(Aa\frac{\partial T}{\partial y}\right) + \frac{q_i(T_i - T)}{A} + \frac{1}{c\rho A}\frac{\partial(AR)}{\partial y} \tag{2.42}$$

where

a—diffusivity of water, $a = \lambda/c\rho$
T—temperature of water
τ—time
A—horizontal area
q_i—horizontal inlet discharge in unit height
q_o—horizontal outlet discharge in unit height
v—vertical mean velocity of water, can be calculated by $q_i - q_o$
R—sun radiation
c—specific heat of water
ρ—density of water.

The bottom of the reservoir is an insulated surface, so

when

$$y = 0, \quad \frac{\partial T}{\partial y} = 0 \tag{2.43}$$

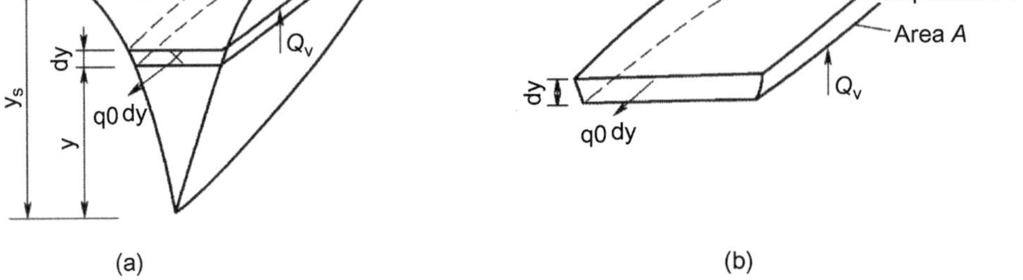

Figure 2.12 Sketch for computing the reservoir temperature: (a) the reservoir and (b) an elementary layer.

At the surface of the reservoir:

when

$$y = y_s, \quad -\lambda\frac{\partial T}{\partial y} = \varphi_s + \varphi_a - \varphi_L \tag{2.44}$$

in which

ϕ_s—sun radiation absorbed on the surface
ϕ_a—atmosphere radiation on the surface
ϕ_L—loss of heat on the surface due to vaporization, etc.

Taking the early spring as the origin of time, when the water temperature in the reservoir is nearly uniform, thus the initial condition is

when

$$\tau = 0, \quad T(y, 0) = T_0 \tag{2.45}$$

Solving Eq. (2.42) by numerical method with the above initial and boundary conditions, the water temperature in the reservoir will be obtained.

2.7 Thermal Properties of Concrete

The thermal properties of concrete include the diffusivity a (m^2/h), the conductivity λ (kJ/(m h °C)), the specific heat c (kJ/(kg °C)) and the density ρ (kg/m^3). From the definition of diffusivity, a is given by

$$a = \frac{\lambda}{c\rho} \tag{2.46}$$

There are four thermal properties of concrete, namely λ, c, ρ, a, three of them must be determined by tests and the fourth one may be calculated by Eq. (2.46).

The thermal properties of some conventional concrete dams are given in Table 2.8 and those of some RCC dams are given in Table 2.9.

In the preliminary design, if there are no test results, the thermal properties of concrete can be estimated with the aid of Table 2.10 [4]. Experiences show that the specific heat estimated with Table 2.10 is somewhat lower than the practical value and it is suggested to multiply it by a coefficient $k_c = 1.05$.

Example The mix of concrete is given in Table 2.11. The thermal properties at 32°C are estimated by the percentages of weights with a coefficient of modification $k_c = 1.05$ for the specific heat:

$$\begin{aligned}\lambda &= \sum p_i\lambda_i = (8.35\times4.593+18.51\times11.099+68.96\times10.467+4.18\times2.16)/100\\ &= 9.75\,\text{kJ}/(\text{mh}^\circ\text{C})\end{aligned}$$

Table 2.8 Thermal Properties of Conventional Concrete Dams

Name of Dam	Conductivity λ (kJ/(m h °C))	Specific Heat c (kJ/(kg °C))	Density ρ (kg/m^3)	Diffusivity a (m^2/h)
Xiaowan	8.26	1.036	2500	0.00319
Laxiwa	8.44	0.996	2395	0.00354
Xilodu	7.28	0.997	2635	0.00280
Jingping	7.74	0.850	2475	0.00360
Ertan	5.72	0.936	2478	0.00247
Three Gorges	9.04	0.959	2450	0.00347
Danjiangkou	12.94	1.00	2469	0.00520
Seminoe	12.18	0.925	2483	0.00550
Norris	12.98	1.034	2570	0.00490
Wheeler	11.10	0.984	2330	0.00480
Hoover	10.43	0.925	2500	0.00460
Hiwassee	9.21	0.976	2490	0.00379
Grand Coulee	6.55	0.950	2530	0.00270
Average	9.37	0.967	2485	0.00390

Table 2.9 Thermal Properties of Roller-Compacted Concrete Dams

Name of Dam	Conductivity λ (kJ/(m h °C))	Specific Heat c (kJ/(kg °C))	Density ρ (kg/m^3)	Diffusivity a (m^2/h)
Longtan	8.78	0.967	2450	0.00370
Mianhuatan	8.59	0.980	2400	0.00365
Jinghong	9.27	0.967	2400	0.00394
Guangzhao	8.14	0.935	2445	0.00380
Linghekou	8.31	1.004	2480	0.00341
Zhaolaihe	8.70	1.04	2400	0.00349
Bailiangyan	8.50	0.921	2400	0.00385
Shapai	8.28	0.934	2507	0.00363
Yantan	8.31	0.990	2400	0.00350
Average	8.54	0.970	2431	0.00366

$$c=\sum kp_{ici}=1.05(8.35\times0.536+18.51\times0.745+68.96\times0.708+4.18\times4.187)/100$$
$$=0.888\text{kJ}/(\text{kg}°\text{C})$$

$$\rho=2539\text{kg}/\text{m}^3$$

$$a=\frac{\lambda}{c\rho}=\frac{9.75}{0.888\times2539}=0.00432\text{m}^2/\text{h}$$

Table 2.10 Coefficients for Estimating the Thermal Properties of Mass Concrete

Material	Density ρ (kg/m^3)	Conductivity λ (kJ/(m h °C))				Specific Heat c (kJ/(kg °C))			
		21°C	32°C	43°C	54°C	21°C	32°C	43°C	54°C
Water	1000	2.160	2.160	2.160	2.160	4.187	4.187	4.187	4.187
Cement	3100	4.446	4.593	4.735	4.865	0.456	0.536	0.662	0.825
Quartz sand	2660	11.129	11.099	11.053	11.036	0.699	0.745	0.795	0.867
Basalt gravel	2660	6.891	6.871	6.858	6.837	0.766	0.758	0.783	0.837
Dolomite gravel	2660	15.533	15.261	15.014	14.336	0.804	0.821	0.854	0.888
Granite gravel	2680	10.505	10.467	10.442	10.379	0.716	0.708	0.733	0.775
Lime gravel	2670	14.528	14.193	13.917	13.657	0.749	0.758	0.783	0.821
Quartz gravel	2660	16.910	16.777	16.638	16.475	0.691	0.724	0.758	0.791
Rhyolite gravel	2660	6.770	6.812	6.862	6.887	0.766	0.775	0.800	0.808

Table 2.11 Example of Concrete Mix

	Cement	Sand	Gravel (Granite)	Water	Total
Weight (kg)	212	470	1751	106	2539
Percentage	8.35	18.51	68.96	4.18	100

2.8 Heat of Hydration of Cement and the Adiabatic Temperature Rise of Concrete

2.8.1 Heat of Hydration of Cement

The relation between the heat of hydration of cement and the age may be expressed by one of the following formulas:

Exponential formula [3]

$$Q(\tau) = Q_0(1 - e^{-m\tau}) \tag{2.47}$$

Hyperbolic formula [4, 8]

$$Q(\tau) = Q_0\tau/(n + \tau) \tag{2.48}$$

Complex exponential formula [4, 8]

$$Q(\tau) = Q_0(1 - e^{-a\tau^b}) \tag{2.49}$$

Table 2.12 Constants for Heat of Hydration of Cement in Eq. (2.49)

Kind of Cement	Q_0 (kJ/kg)	a	b
Silicate cement 425#, 525#	330	0.69	0.56
	350	0.36	0.74
Silicate cement for dam 525#	270	0.79	0.70
Slag silicate cement for dam 425#	285	0.29	0.76

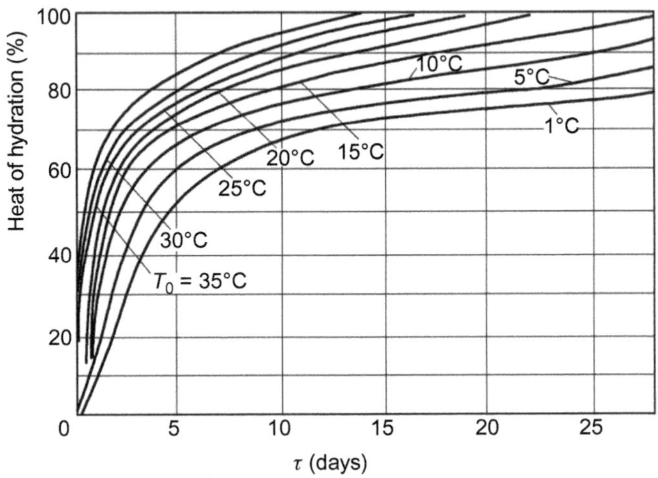

Figure 2.13 Heat of hydration at different initial temperatures.

where

$Q(\tau)$—the accumulated heat of hydration of cement at age τ, kJ/kg
Q_0—the final heat of hydration of cement as $\tau \to \infty$, kJ/kg
τ—age, day
m, n, a, b—constants.

Formulas (2.48) and (2.49) agree well with the test results. Some constants for formula (2.49) are given in Table 2.12.

The rate of heat of hydration is influenced by the curing temperature, as shown in Figure 2.13.

2.8.2 Adiabatic Temperature Rise of Concrete

The adiabatic temperature rise of concrete is influenced by the kind and amount of cement and admixture and the placing temperature. Generally it is determined by tests. In case of no test results, it can be estimated as follows:

$$\theta(\tau) = \frac{Q(\tau)(W + kF)}{c\rho} \tag{2.50}$$

where

$\theta(\tau)$—the adiabatic temperature rise of concrete
W—amount of cement
F—amount of admixture
$Q(\tau)$—heat of hydration of cement
c—specific heat of concrete
ρ—density of concrete
k—coefficient of reduction, $k = 0.25$ for fly ash.

The curves of the adiabatic temperature rises of concretes with the same amount of cements of different kinds are shown in Figure 2.14.

Curve	Kind of Cement	C_3S (%)	C_3A (%)	Fineness (cm^2/g)
1	Early strength cement, Type III	56	12	2030
2	Common cement, Type I	43	11	1790
3	Moderate heat cement, Type II	40	8	1890
4	Type II cement 75% + fly ash 25%			
5	Low heat cement, Type IV	20	6	1910

The relation between $\theta(\tau)$ and age τ can be expressed by the following formulas:

Exponential formula

$$\theta(\tau) = \theta_0(1 - e^{-m\tau}) \tag{2.51}$$

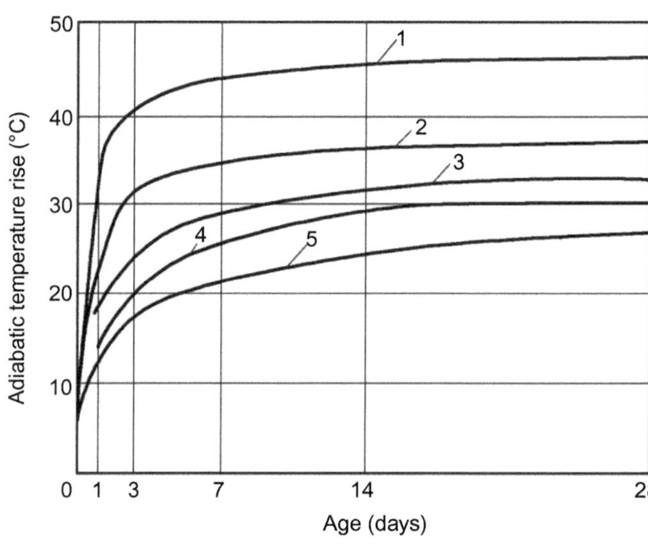

Figure 2.14 The curves of adiabatic temperature rise of concretes with the same amount of cements of different kinds.

Hyperbolic formula

$$\theta(\tau) = \theta_0\tau/(n+\tau) \tag{2.52}$$

Complex exponential formula

$$\theta(\tau) = \theta_0(1 - e^{-a\tau^b}) \tag{2.53}$$

where θ_0 is the final temperature rise of concrete as $\tau \to \infty$.

Equations (2.52) and (2.53) agree well with test results. Some examples are given in the following:

Xiaowang dam, conventional concrete,

$$\left.\begin{array}{ll} C_{180}45, & \theta(\tau) = 27.0\tau/(1.35+\tau) \quad ^\circ\text{C} \\ C_{180}40, & \theta(\tau) = 26.0\tau/(1.25+\tau) \quad ^\circ\text{C} \end{array}\right\} \tag{2.54}$$

Longtan dam, RCC,

$$\text{C20} \quad \theta(\tau) = 20.7\tau/(3.26+\tau) \quad ^\circ\text{C} \tag{2.55}$$

Yangtan dam, RCC,

$$\theta(\tau) = 20.33(1 - e^{-0.1232\tau^{0.7558}}) \tag{2.56}$$

Mangwan dam, RCC,

$$\theta(\tau) = 21.59(1 - e^{-0.319\tau^{0.5972}}) \tag{2.57}$$

The exponential formula (2.51) is suitable for mathematical manipulation, but it does not agree with test results. A new formula, which is suitable for mathematical manipulation and agrees well with test results, is given by the writer in the following:

$$\theta(\tau) = \theta_0[s(1 - e^{-m_1\tau}) + (1-s)(1 - e^{-m_2\tau})] \tag{2.58}$$

where s, m_1 and m_2 are constants, and the initial values of them may be

$$s = 0.60, \quad m_1 = 1.55/n, \quad m_2 = 0.55/n \tag{2.59}$$

where n is the constant in the hyperbolic formula (2.52).

2.9 Temperature on the Surface of Dam

As shown in Figure 2.15, above the maximum water level, the dam is in contact with air, the surface temperature is given in the following:

$$T_s(\tau) = T_a(\tau) + T_R(\tau) \tag{2.60}$$

where

$T_s(\tau)$—the surface temperature of the dam
$T_a(\tau)$—the air temperature
$T_R(\tau)$—the temperature increment due to sun radiation.

Below the maximum water level, for a fixed point on the dam surface, sometimes it is in contact with air and sometimes it is in contact with water; thus, the surface temperature of dam can be computed in the following:

$$\left.\begin{array}{lll} T_s(\tau) = T_w(y, \tau) & \text{when} & y = z - z_0 \geq 0 \\ T_s(\tau) = T_a(\tau) + T_R(\tau) & \text{when} & z \leq z_0 \end{array}\right\} \tag{2.61}$$

where

$T_w(y,\tau)$—water temperature
y—depth of water
z_0—elevation of the point on dam surface.

Assuming that the surface temperature of the dam varies periodically with time, it may be expressed by Fourier series as follows:

$$T_u(\tau) = T_{um} + \sum_{n=1}^{\infty} \left\{ B_n \cos\left[\frac{2n\pi}{P}(\tau - \tau_0)\right] + C_n \sin\left[\frac{2n\pi}{P}(\tau - \tau_0)\right] \right\} \tag{2.62}$$

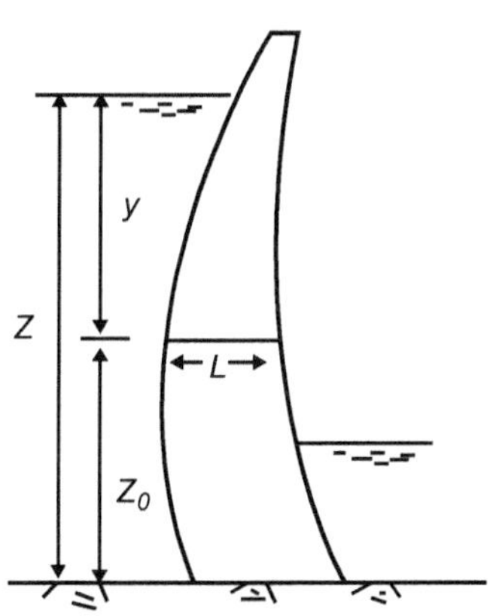

Figure 2.15 Cross-section of a dam.

The coefficients B_n and C_n are determined by

$$\left.\begin{aligned} B_n &= \frac{2}{P}\int_0^P T_u(\tau)\cos\left[\frac{2n\pi}{P}(\tau-\tau_0)\right]d\tau \\ C_n &= \frac{2}{P}\int_0^P T_u(\tau)\sin\left[\frac{2n\pi}{P}(\tau-\tau_0)\right]d\tau \end{aligned}\right\} \tag{2.63}$$

2.10 The Autogenous Deformation of Concrete

The autogenous deformation of concrete is the volume change caused by the hydration of the cementing materials under the condition of constant temperature and constant humidity. The autogenous deformations of concrete made by different kinds of cement are shown in Figure 2.16 and the influences of the amount of admixture of fly ash on the autogenous deformation are shown in Figure 2.17.

2.11 Semi-Mature Age of Concrete

The adiabatic temperature rise, strength, and modulus of elasticity of concrete increase gradually with age and finally reach their ultimate values [7]. At present, there is no appropriate index for expressing the rate of growth of these properties of concrete.

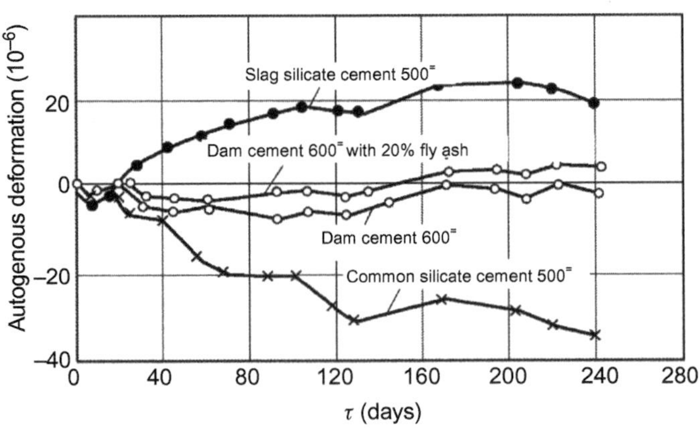

Figure 2.16 The autogenous deformation of concretes made by different kinds of cements.

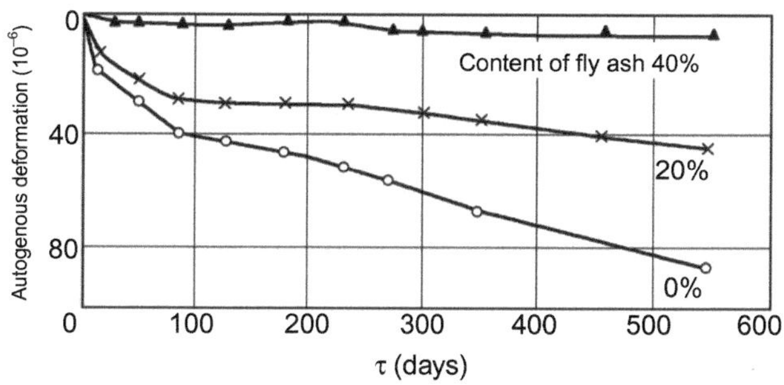

Figure 2.17 The influence of the amount of admixture of fly ash on the autogenous deformation of concrete.

For expressing the rate of growth of these concrete properties, it is suggested by the writer [79] that the ages when the adiabatic temperature rise, strength, and the modulus of elasticity reach the halves of their final values are defined as the semi-mature ages, which can be denoted by $\tau_{1/2}$. The value of $\tau_{1/2}$ reflects the rate of maturity of concrete. The smaller the value of $\tau_{1/2}$, the greater the rate of maturity of concrete. Conversely, the bigger the value of $\tau_{1/2}$, the smaller the rate of maturity of concrete. For mass concrete structures like concrete dams, the age of concrete for undertaking full loads is rather high. There are no problems for the strength of concrete in early age. But there are requirements for the prevention of cracks in the structures. Generally, the temperatures in concrete dams are controlled by natural dissipation of heat from the surface and pipe cooling in the interior. If the semi-mature age of concrete is too small, the adiabatic temperature rise will increase rapidly and the temperatures in the interior of concrete will also increase quickly. When the measures of natural surface cooling and internal pipe cooling have not performed their function yet, the temperature in the concrete has already reached its maximum value. On the contrary, if the semi-mature age of concrete is higher, the adiabatic temperature rise will increase slowly. There is enough time for the natural cooling and pipe cooling to perform their function. The maximum temperature in the concrete will be lower and the thermal stresses will be smaller. Therefore, semi-mature age is an important index for mass concrete.

2.11.1 Method for Determining the Semi-Mature Age of Concrete

The semi-mature age $\tau_{1/2}$ can be determined directly from the test results of concrete. For example, the test results of adiabatic temperature rise of mass concrete are shown in Figure 2.18. By a smooth curve $\theta(\tau)$, draw a straight line $y = \theta_0/2$ which intersects the curve of $\theta(\tau)$ at $\tau = \tau_{1/2}$, then $\tau_{1/2}$ is the semi-mature age for the adiabatic temperature rise of concrete.

The semi-mature age of concrete for the modulus of elasticity, the tensile strength, and the extensibility may be determined in the same way.

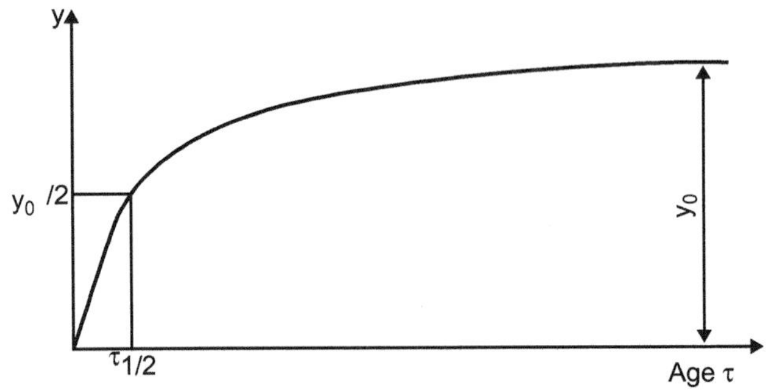

Figure 2.18 Determination of $\tau_{1/2}$.

2.11.2 Formulas for Computing the Semi-Mature Age of Concrete

Let one property of concrete be expressed by the following equation:

$$y(\tau) = y_0 f(\tau) \tag{2.64}$$

where y_0 is the final value of $y(\tau)$; then, the semi-mature age $\tau_{1/2}$ is determined by

$$f(\tau_{1/2}) = 1/2 \tag{2.65}$$

For some function expressing property of concrete, $\tau_{1/2}$ are given in the following:

1. Exponential function $f(\tau) = 1 - e^{-m\tau}$
   From Eq. (2.65), $\exp(-m\tau_{1/2}) = 0.5$, by taking logarithm on both sides, we get

$$\tau_{1/2} = \frac{0.693}{m} \quad \text{or} \quad m = \frac{0.693}{\tau_{1/2}} \tag{2.66}$$

2. Hyperbolic function $f(\tau) = \tau/(n + \tau)$

$$\tau_{1/2} = n \tag{2.67}$$

3. Compound exponential function $f(\tau) = 1 - \exp(-g\tau^d)$
   From Eq. (2.65), $\exp(g\tau_{1/2}^d) = 0.5$, taking logarithm on both sides, $g\tau_{1/2}^d = -\ln 0.500 = 0.693$, taking logarithm again,

$$\ln \tau_{1/2} = \frac{1}{d}\ln\left(\frac{0.693}{g}\right)$$

thus

$$\tau_{1/2} = \exp\left[\frac{1}{d}\ln\left(\frac{0.693}{g}\right)\right] \tag{2.68}$$

The semi-mature ages of concrete for adiabatic temperature rise for modulus of elasticity, for tensile and compressive strength, and for extensibility of some concrete dams are given in Table 2.13.

Table 2.13 Semi-Mature Ages of Concrete of Some Concrete Dams

Dam	Concrete Mix	Concrete Mark	Semi-Mature Age $\tau_{1/2}$ (days)				
			For Adiabatic Temperature Rise	For Modulus of Elasticity	For Tensile Strength	For Compressive Strength	For Extensibility
Xiaowan	1	$c_{180}45$	1.35	3.30	6.80	10.50	1.20
	2	$c_{180}40$	1.25	4.79	7.43	14.90	1.70
	3	$c_{180}35$	1.30	6.86	6.87	14.95	2.40
	4	$c_{180}30$	1.30	6.70	–	15.30	3.00
Xiluodu	1	$c_{180}35$	2.58	7.76	1.95	11.2	3.71
	2	$c_{180}30$	2.46	6.92	5.72	11.8	2.65
	3	$c_{180}25$	2.58	5.71	6.67	11.1	3.51
Jingping1	1	$c_{180}40$	3.63	5.41	–	10.8	7.76
	2	$c_{180}35$	3.59	5.72	–	10.7	7.64
	3	$c_{180}30$	3.47	5.26	–	12.4	7.71

Remark: (1) The concrete mark $c_{180}40$ means the compressive strength of concrete at 180-day age is 40 MPa. (2) The extensibility of concrete is the maximum tensile strain before its failure under tension.

2.11.3 Meaning of Semi-Mature Age in Engineering

If the semi-mature age of concrete is too short, the adiabatic temperature rise will increase rapidly and there is not enough time for the natural surface cooling and pipe cooling in the interior of the dam. So the maximum temperature and thermal stress in the interior of the dam due to the heat of hydration of cement will be high. On the contrary, if the semi-mature age of concrete is longer, the adiabatic temperature rise will increase slowly, and there is enough time for the natural cooling and pipe cooling to reduce the maximum temperature of concrete. The maximum temperature rise in the dam and the thermal stresses will be lower. Thus, the semi-mature age of concrete is an important index for mass concrete.

2.11.4 Example of the Influence of Semi-Mature Age

The influence of semi-mature age on the temperature field and stress field of massive concrete structures will be shown by an example.

Considering two kinds of concrete with different semi-mature ages, the mechanical and thermal properties of them are given in Table 2.14, the coefficient of heat diffusivity $a = 0.10\ \mathrm{m^2/day}$, and the coefficient of surface conductance $\beta = 70$ kJ/(m h °C).

A massive concrete block on rock foundation (Figure 2.19) cooled naturally by dissipation of heat from the top surface to the air, length × width × thickness = 20 m × 20 m × 1.5 m. Two kinds of concrete are considered as given in Table 2.14. The initial temperature of the concrete and rock is 20°C. The top and lateral surfaces of the concrete block are free from external restraint, but the bottom surface is restrained by the rock foundation. The temperature field and stress field are

Table 2.14 Properties of Mass Concrete

Properties	Concrete A	Concrete B	Rock Foundation
Adiabatic temperature rise (°C)	$\theta = \dfrac{25.0\tau}{1.2+\tau}$	$\theta = \dfrac{25.0\tau}{3.6+\tau}$	$\theta = 0$
Modulus of elasticity (MPa)	$E = \dfrac{35,000\tau}{1.5+\tau}$	$E = \dfrac{35,000\tau}{4.5+\tau}$	$E = 35,000$

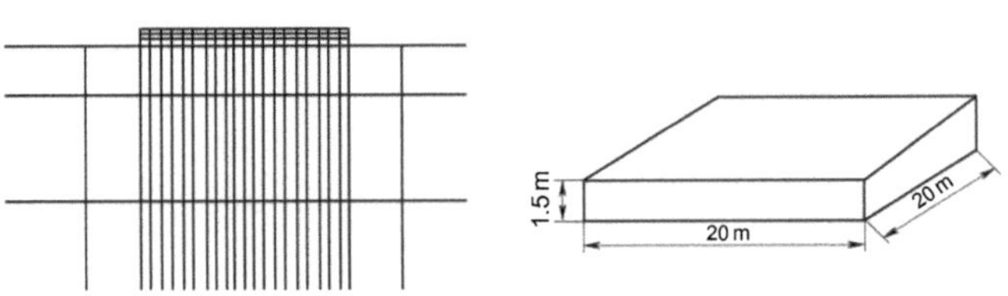

Figure 2.19 Massive concrete block on rock foundation and the network of FEM computation.

computed by the finite element method (FEM) taking into account the effect of creep of concrete. The variation of the temperature at the center of the concrete block is shown in Figure 2.20. The envelope of the temperatures along the centerline of the concrete block is shown in Figure 2.21. The maximum temperature of concrete B with $\tau_{1/2} = 3.6$ days is 3.8°C lower than that of concrete A with $\tau_{1/2} = 1.2$ days. This is approximately equal to the effect of mixing the cement with 50% of fly ash. The variations of thermal stresses at the center of block. The envelopes of the stresses along the centerline of the concrete block. The maximum tensile stress in concrete B is 30.4% (0.38 MPa) lower than that in concrete A.

2.11.5 Measures for Adjusting the Semi-Mature Ages of Concrete

As mentioned above, if the semi-mature age of concrete is long, it will be favorable for prevention of cracking in mass concrete. The following methods may be used to change the semi-mature age of concrete.

1. Change the mineral composition and the fineness of cement

   In order to prevent cracking, the reasonable properties of mass concrete are high strength in late age and low total quantity of discharge of heat of hydration. It is necessary to reduce the content of C_3A and C_3S and increase the content of C_2S. The finer the cement, the quicker is the rate of hydration, so it is reasonable to reduce the specific surface of cement.

   In the 1930s, the common Portland cement contained only about 30% C_3S and its specific surface was about 220 m^2/kg. The rate of hydration of cement and the rate of growth of strength are rather low which is favorable for crack prevention. As the designed age of the conventional reinforced concrete structure is generally 28 days, under the drive for catering to the need of market, the cement factories pursue high strength at 28-days age. The content of C_3S and the specific surface of cement become higher and higher.

   For the Portland cement manufactured in China at present, the content of C_3S exceeds 50% and the specific surface is 340−370 m^2/kg; the rates of discharge of heat of hydration and growth of strength are high, which is unfavorable for crack prevention.

   It is necessary to reduce the content of C_3S and the specific surface of cement.

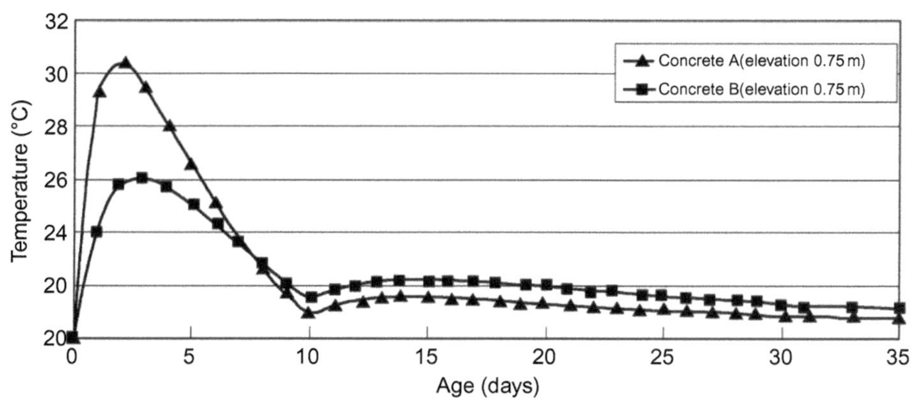

Figure 2.20 The variation of temperature at the center of concrete block cooled naturally.

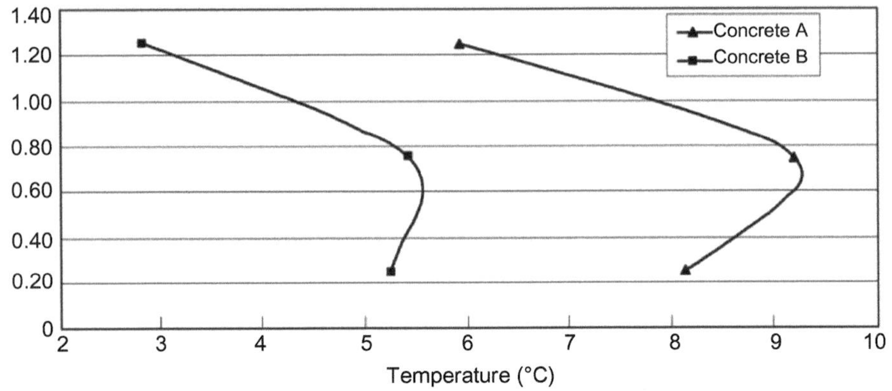

Figure 2.21 The envelopes of temperatures along the vertical centerline of concrete block on rock foundation cooled naturally.

2. Fly ash and admixture agent

   By mixing fly ash with cement, the rate of discharge of heat of hydration will be slower and the semi-mature age of concrete will be longer. Formerly the admixture agent was used to change the initial setting time; now it may be used to change the semi-mature age of concrete.

2.11.6 Conclusions

1. The age when the adiabatic temperature rises, the strength or the modulus of elasticity reaches one-half of its final value is defined as the semi-mature age of concrete.
2. If the semi-mature age of adiabatic temperature rise is higher, the temperature in the interior of a mass concrete structure will increase more slowly, the effect of natural superficial cooling and artificial pipe cooling will be greater, and the thermal stresses may be reduced remarkably. This is favorable for the prevention of cracks.
3. The semi-mature age introduced in this chapter is a new and important index for mass concrete.
4. The semi-mature age of concrete may be adjusted by reducing the content of C_3S and C_3A and the specific surface of cement and mixing the cement with fly ash and admixture agent.

2.12 Deformation of Concrete Caused by Change of Humidity

The loss of water in concrete will cause shrinkage and the absorption of water will cause expansion. The differential equation of diffusivity of humidity in concrete is

$$\frac{\partial U}{\partial \tau} = k_1 \left(\frac{\partial^2 U}{\partial x^2} + \frac{\partial^2 U}{\partial y^2} + \frac{\partial^2 U}{\partial z^2} \right) \tag{2.69}$$

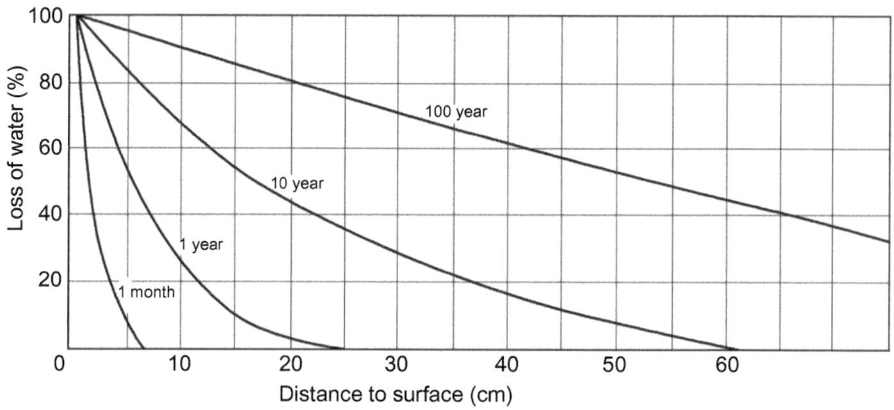

Figure 2.22 Loss of water (%) in mass concrete.

For the dry concrete in contact with air, the boundary condition is

$$k_1 \frac{\partial U}{\partial n} = \beta_1 (U_F - U_B) \tag{2.70}$$

where

U—humidity (kg/m^3), the weight of water contained in concrete of unit volume
U_F—surface humidity of concrete
U_B—balance humidity
k_1—coefficient of diffusivity of humidity (m^2/h), the weight of water (kg) passed through unit area (m^2) in unit time (h) when the gradient of U is 1 unit (kg/m^3 m)
β_1—coefficient of water exchange on the surface (m/h), it is the loss of water (kg) in unit time and unit area when the difference of water content between air and concrete surface is 1 unit (kg/m^3)
n—normal to surface of concrete.

The law of change of humidity of concrete is similar to the change of temperature. The differential equation of the diffusion of humidity of concrete is the same as the equation of conduction of heat, but the coefficient of diffusion of humidity of concrete is only 1/1200–1/1600 of the coefficient of conductivity of heat of concrete. The speed of drying of concrete is very small. The computed results of the drying process of mass concrete with one surface exposed in the air of 50% relative humidity is shown in Figure 2.22.

2.13 Coefficients of Thermal Expansion of Concrete

The coefficients of thermal expansion of different kinds of aggregates are given in Table 2.15 and the coefficients of 1:6 concretes made by various aggregates are

Table 2.15 Coefficients of Thermal Expansion of Various Aggregates

Kind of Aggregate	Quartz	Sandstone	Basalt	Granite	Limestone
α (10^{-5}/°C)	1.02–1.34	0.61–1.17	0.61–0.75	0.55–0.85	0.36–0.60

Table 2.16 Coefficients of Thermal Expansion of 1:6 Concrete Made by Various Aggregates

Kind of Aggregate	Quartz	Sandstone	Granite	Basalt	Limestone
α (10^{-5}/°C)	1.22	1.01	0.86	0.85	0.61

Table 2.17 Coefficient of Thermal Expansion of Concrete in Some Dams

Type of Dam	Name of Dam	Kind of Aggregate	Test Condition	Coefficient of Expansion α (10^{-5}/°C)
RCC	Guanyinge	Gravel	Room test	0.906
	Yantan	Limestone	Room test	0.580
	Mangwan	Rhyolite	Room test	0.524
	Longtan		Room test	0.70
	Mianhuatan		Room test	0.66
	Jinghong		Room test	1.00
Conventional concrete	Xilodu	Limestone, basalt	Room test	0.63
	Yoojiangdu	Limestone	Observed in dam	0.52
	Zaxi		Observed in dam	0.96–1.00
	Danjiangkou		Observed in dam	0.90
	Salamond (Portugal)		Observed in dam	0.80–0.83
	Dworshak (USA)			0.94–0.99

given in Table 2.16. The coefficients of expansion measured in some concrete dams are given in Table 2.17.

2.14 Solution of Temperature Field by Finite Difference Method

The two-dimensional and the three-dimensional problems of heat conduction are generally computed by FEM, which will be described in Chapter 8, and the one-dimensional problem is generally computed by the finite difference method described in the following.

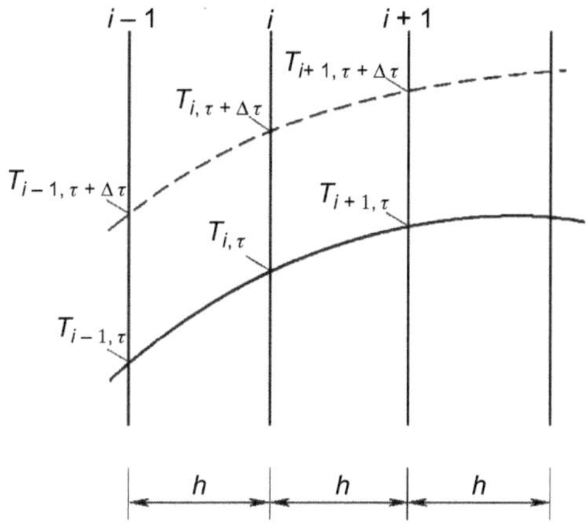

Figure 2.23 The finite difference method.

The one-dimensional equation of conduction of heat is

$$\frac{\partial T}{\partial \tau} = a\frac{\partial^2 T}{\partial x^2} + \frac{\partial \theta}{\partial \tau} \tag{2.71}$$

As shown in Figure 2.23, an infinite plate with thickness L is divided into $n-1$ layers with equal thickness $\Delta x = L/(n-1)$. The partial derivative of temperature T in the x-direction is

$$\left(\frac{\partial T}{\partial x}\right)_{i+0.5\Delta x,\tau} = \frac{1}{\Delta x}(T_{i+1,\tau} - T_{i,\tau})$$

$$\left(\frac{\partial T}{\partial x}\right)_{i-0.5\Delta x,\tau} = \frac{1}{\Delta x}(T_{i,\tau} - T_{i-1,\tau})$$

Thus

$$\left(\frac{\partial^2 T}{\partial x^2}\right)_{i,\tau} = \frac{1}{\Delta x}\left[\left(\frac{\partial T}{\partial x}\right)_{i+0.5\Delta x,\tau} - \left(\frac{\partial T}{\partial x}\right)_{i-0.5\Delta x,\tau}\right] = \frac{1}{\Delta x^2}(T_{i-1,\tau} + T_{i+1,\tau} - 2T_{i,\tau}) \tag{2.72}$$

The finite differential of T and θ with respect to time τ are

$$\left(\frac{\partial T}{\partial \tau}\right)_{i,\tau} = \frac{1}{\Delta \tau}(T_{i,\tau+\Delta\tau} - T_{i,\tau}) \tag{2.73}$$

$$\left(\frac{\partial \theta}{\partial \tau}\right)_{i,\tau} = \frac{\theta(\tau+\Delta\tau) - \theta(\tau)}{\Delta \tau} = \frac{\Delta \theta}{\Delta \tau} \tag{2.74}$$

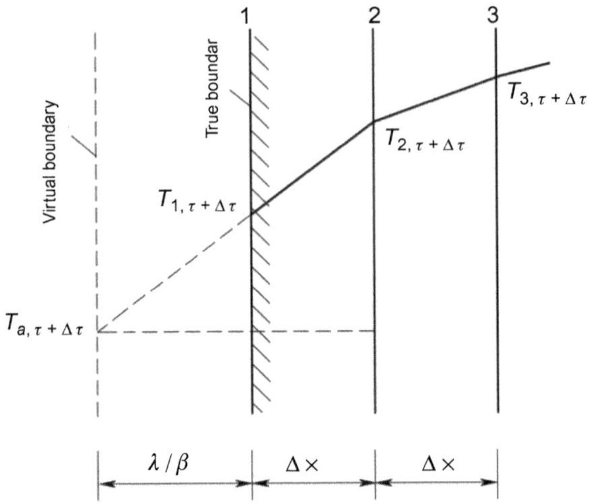

Figure 2.24 The boundary condition on the left side.

Substituting Eqs. (2.72)–(2.74) into Eq. (2.5), we get the finite difference equation of conduction of heat in the following:

$$T_{i,\tau+\Delta\tau} = (1-2r)T_{i,\tau} + r(T_{i-1,t} + T_{i+1,\tau}) + \Delta\theta \tag{2.75}$$

where

$$r = a\Delta\tau/\Delta x^2 \tag{2.76}$$

$T_{i,\tau}$—the temperature of point i at time τ.

The initial temperature at point i is

$$T_{i,0} = T_0 \tag{2.77}$$

As shown in Figure 2.24, the temperature of the point 1 on the left boundary is

$$T_{1,\tau+\Delta\tau} = T_{a,\tau+\Delta\tau} + \frac{\lambda}{\beta}\frac{T_{2,\tau+\Delta\tau} - T_{a,\tau+\Delta\tau}}{\Delta x + \lambda/\beta} = \frac{\Delta x\, T_{a,\tau+\Delta\tau} + (\lambda/\beta)T_{2,\tau+\Delta\tau}}{\Delta x + \lambda/\beta} \tag{2.78}$$

Similarly, the temperature of point n on the right boundary is

$$T_{n,\tau+\Delta\tau} = \frac{\Delta x\, T_{a,\tau+\Delta\tau} + (\lambda/\beta)T_{n-1,\tau+\Delta\tau}}{\Delta x + \lambda/\beta} \tag{2.79}$$

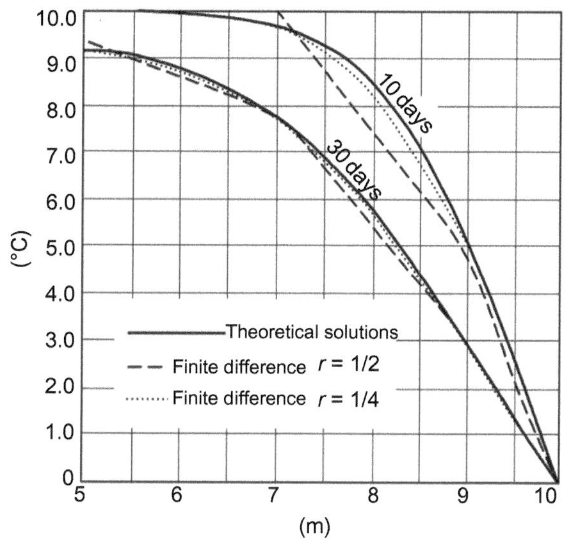

Figure 2.25 Example of finite difference method.

With the initial temperature T_0 and the boundary conditions (2.78) and (2.79), the temperatures of all the points $i = 1{-}n$ at time $\tau = 0, \Delta\tau, 2\Delta\tau, 3\Delta\tau, \ldots$ may be computed by Eq. (2.75) step by step. The magnitude of $\Delta\tau$ is limited by the stability condition:

$$r = \frac{a\,\Delta\tau}{\Delta x^2} \leq \frac{1}{2} \tag{2.80}$$

Example Infinite plate, thickness $L = 10$ m, initial temperature $T_0 = 10°\text{C}$, the boundary temperatures on both sides are 0°C, the diffusivity $a = 0.10$ m^2/day, on account of symmetry, only one-half is taken in the computation, take (1) $\Delta x = 1.0$ m, $\Delta\tau = 5$ days, $r = a\,\Delta\tau/\Delta x^2 = 0.10 \times 5.0/1.0^2 = 0.50$; (2) $\Delta x = 1.0$ m, $\Delta\tau = 2.5$ days, $r = 0.10 \times 2.5/1.0^2 = 0.25$. The results of computing are shown in Figure 2.25.

3 Temperature Field in the Operation Period of a Massive Concrete Structure

3.1 Depth of Influence of the Variation of Exterior Temperature in the Operation Period

The surface of a dam is in contact with water and air in the operation period, so the periodical variations of the temperatures of water and air will exert influence on the temperature field of the dam. The depth of influence depends on the period of variation.

3.1.1 *Depth of Influence of Variation of Water Temperature*

It is assumed that the surface temperature of the dam is equal to water temperature and the depth of influence is very small in comparison with the thickness of the dam, so the temperature field of a semi-infinite solid is analyzed in the following. The temperature $T(x,\tau)$ must satisfy the following conditions:

$$\left.\begin{aligned} &\frac{\partial T}{\partial \tau} = a\frac{\partial^2 T}{\partial x^2} && (0 < x < \infty) \\ &T = A\sin\frac{2\pi\tau}{P}, && \text{when } x = 0 \\ &T = 0, && \text{when } x \to \infty \\ \text{and}\quad &T = 0, && \text{when } \tau = 0 \end{aligned}\right\} \tag{3.1}$$

where

P—period of variation of temperature
A—amplitude of variation of the surface temperature.

The solution of Eq. (3.1) is

Thermal Stresses and Temperature Control of Mass Concrete. DOI: http://dx.doi.org/10.1016/B978-0-12-407723-2.00003-8

$$T(x,\tau)=A\,e^{-x\sqrt{\pi/aP}}\sin\left(\frac{2\pi\tau}{P}-x\sqrt{\frac{\pi}{aP}}\right)+\frac{2aA}{\pi}\int_0^\infty\frac{(2\pi/P)e^{-a\xi^2\tau}}{(2\pi/P)^2+a^2\xi^4}\sin\xi x\,d\xi \tag{3.2}$$

where ξ is a variable of integration. The second term on the right part of the above equation represents the influence of the initial temperature and will vanish as $\tau\rightarrow\infty$. Finally, only the first term is left, namely

$$T(x,\tau)=A\,e^{-x\sqrt{\pi/aP}}\sin\left(\frac{2\pi\tau}{P}-x\sqrt{\frac{\pi}{aP}}\right) \tag{3.3}$$

which varies periodically with time and is called the quasi-steady temperature field. The amplitude of variation in the interior is

$$\Delta T(x)=A\,e^{-x\sqrt{\pi/aP}} \tag{3.4}$$

so

$$\Delta T(x)/A=e^{-x\sqrt{\pi/aP}} \tag{3.5}$$

Taking the natural logarithm on both sides of Eq. (3.5), we get

$$x=-\sqrt{aP/\pi}\,\ln[\Delta T(x)/A] \tag{3.6}$$

Let the diffusivity of concrete $a=0.10$ m^2/day, the depths of influence of variation of water temperature computed by Eq. (3.6) are given in Table 3.1 and Figure 3.1.

Taking $\Delta T(x)=0.05A$, from Table 3.1, the depth of influence of water temperature is 0.53 and 10.21 m for the daily and annual variation, respectively.

3.1.2 Depth of Influence of Variation of Air Temperature

When the mass concrete is in contact with air, the differential equation for the temperature field is the same as Eq. (3.1), but the boundary condition becomes

$$-\lambda\frac{\partial T}{\partial x}=\beta\left(T-A\sin\frac{2\pi\tau}{P}\right),\quad \text{when } x=0 \tag{3.7}$$

where A is the amplitude of variation of air temperature and β is the surface conductance of concrete. The quasi-steady temperature solution satisfying Eq. (3.7) is

$$T(x,\tau)=k\,Ae^{-qx}\sin\left(\frac{2\pi\tau}{P}-qx-M\right) \tag{3.8}$$

Table 3.1 Depth of Influence of Variation of Water Temperature (m)

Period of Variation (d)		**1**	**15**	**365**
Amplitude of variation $\Delta T(x)$	$0.10A$	0.41	1.59	7.85
	$0.05A$	0.53	2.07	10.21
	$0.01A$	0.82	3.19	15.75

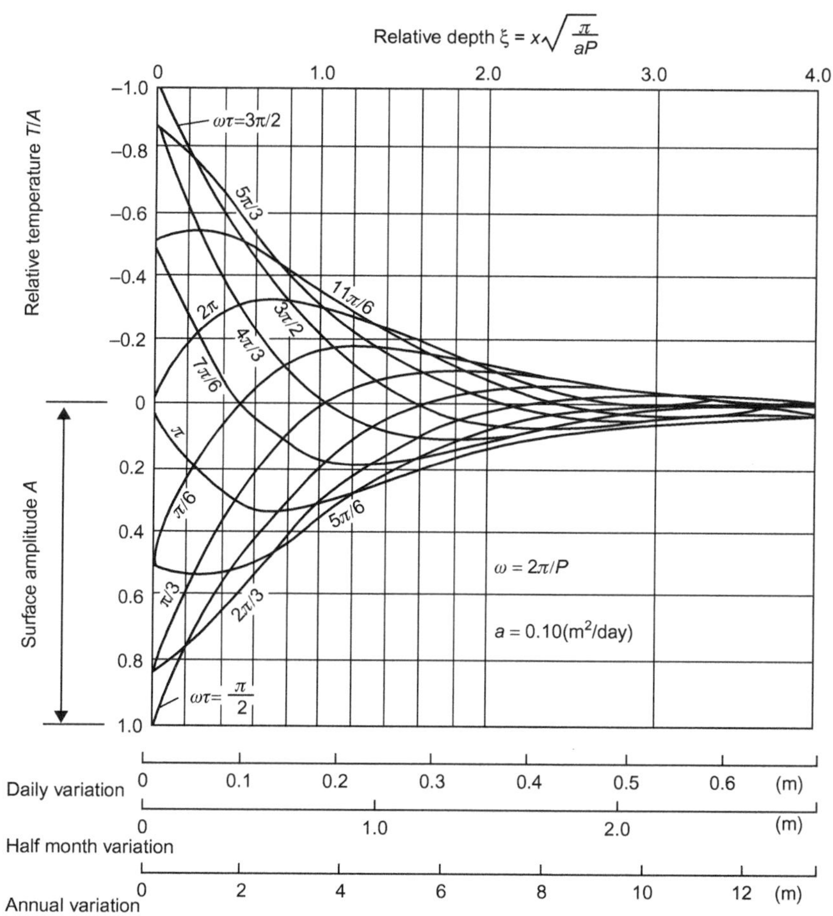

Figure 3.1 Influence of variation of water temperature on the interior temperature of concrete.

$$\left.\begin{aligned} k &= [1+2q\lambda/\beta+2(q/\lambda\beta)^2]^{-1/2} \\ q &= \sqrt{\pi/aP} \\ M &= \tan^{-1}\left(\frac{1}{1+\beta/\lambda q}\right) \end{aligned}\right\} \tag{3.9}$$

The surface temperature ($x = 0$) of concrete is

$$T_s = kA\sin\left(\frac{2\pi\tau}{P} - M\right) \tag{3.10}$$

The amplitude of variation of surface temperature of mass concrete with $a = 0.10$ m²/day is kA which is given in Table 3.2.

The depth of influence of variation of air temperature is

$$x = -\sqrt{\frac{aP}{\pi}}\ln[\Delta T(x)/kA] \tag{3.11}$$

Let $a = 0.10$ m²/day, the depth of influence x computed by Eq. (3.11) is given in Table 3.3.

Let $\lambda/\beta = 0.10$ m and $P = 365$ days, from Table 3.3, the depth of influence of variation of air temperature is 7.75 m for $\Delta T(x) = 0.10A$ and 10.11 m for $\Delta T(x) = 0.05A$.

Table 3.2 Amplitude of Variation of Surface Temperature of Concrete Due to Change of Air Temperature

λ/β	Period of Variation of Temperature		
	1 day	**15 days**	**365 days**
10 m	0.61A	0.87A	0.97A
0.20 m	0.42A	0.77A	0.94A

Table 3.3 Depth of Influence of Variation of Air Temperature

λ/β	Amplitude of Variation	Period of Variation		
		1 day	**15 days**	**365 days**
0.10 m	0.10A	0.32 m	1.49 m	7.75 m
	0.05A	0.44 m	1.97 m	10.11 m
	0.01A	0.73 m	3.08 m	15.60 m
0.20 m	0.10A	0.25 m	1.40 m	7.65 m
	0.05A	0.38 m	1.88 m	10.01 m
	0.01A	0.67 m	2.99 m	15.50 m

In conclusion, in the operation period of massive concrete structure, the depth of influence of variation of exterior temperature is about 8–10 m.

3.2 Variation of Concrete Temperature from the Beginning of Construction to the Period of Operation

The horizontal thickness of mass concrete is generally about 10–150 m. The massive concrete structures are constructed layer by layer. The vertical thickness of a layer is 1.50–3.00 m, thus the maximum temperature in the construction period is dependent on the vertical thickness of the layer and is practically independent on the horizontal thickness of the block. Therefore in the construction period, the maximum temperature of a block with horizontal thickness >30 m is equal to that of a block with horizontal thickness <30 m, but those in the operation period are different, as shown in Figure 3.2.

As shown in Figure 3.2, in the construction period, the temperature of concrete will increase from the placing temperature T_P to the maximum temperature $T_P + T_r$, where T_r is the temperature rise due to heat of hydration of cement; thereafter, the temperature will decrease owing to natural or artificial cooling. After the disappearance of the influence of the initial conditions, the temperature field in the structure depends only on the boundary temperature. Since the depth of influence of the variation of the surface temperature is 8–10 m, the temperature in the interior of the structure does not vary with time, if the thickness of the structure is greater than 30 m and it is called the steady temperature. If the thickness of the structure is less than 30 m and the surface temperature varies periodically with time, the temperature in the interior also varies with time with the same period but with smaller amplitude and it is called the quasi-steady temperature.

Since the influence of the initial temperature disappeared with the elapse of time, the temperature in the mass concrete structure depends only on the boundary temperature in the operation period. As shown in Figure 3.3, the surface temperature is $T_m(s) + A(s)\sin \omega\tau$, where $T_m(s)$ is the annual mean temperature relating to

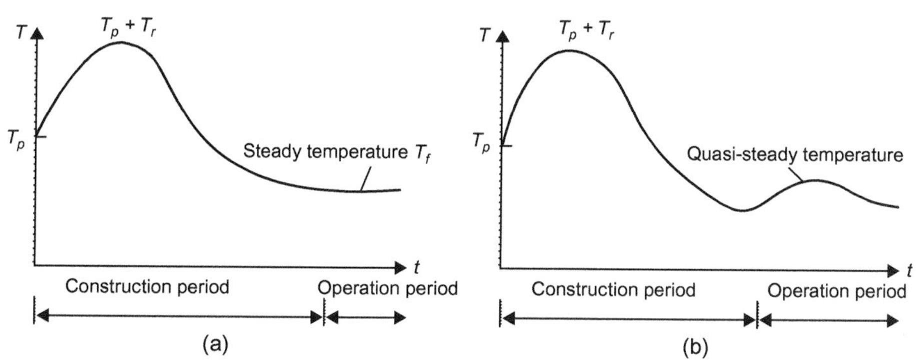

Figure 3.2 Variation of temperatures in the interior of mass concrete structure: (a) structure with horizontal thickness ≥ 30 m and (b) structure with horizontal thickness <30 m.

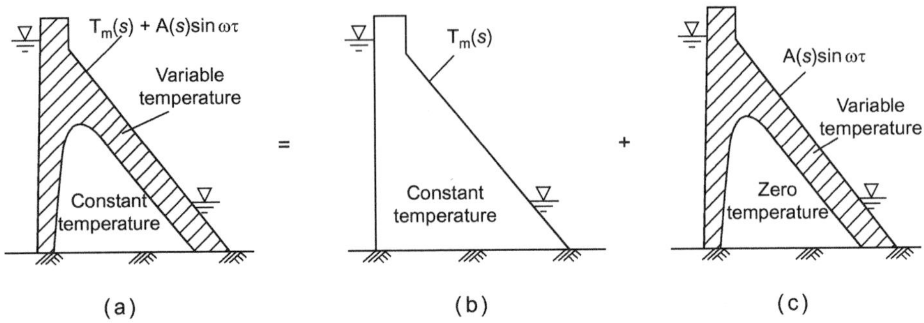

Figure 3.3 Temperature field in operation period and its resolution: (a) practical temperature, (b) steady temperature, and (c) quasi-steady temperature.

the position s of the surface, $A(s)$ is the amplitude of variation, $\omega = 2\pi/P$, and P is the period of variation. The temperature field may be resolved into two parts as follows (Figure 3.3):

$$T_a = T_f + T_q \tag{3.12}$$

where

T_a—the practical temperature
T_f—the steady temperature relating to the surface temperature $T_m(s)$
T_q—the quasi-steady temperature relating to the variation of surface temperature $A(s)\sin\omega\tau$.

T_f and T_q satisfy the following equations:

1. Steady temperature T_f

$$\left.\begin{aligned} &\frac{\partial^2 T}{\partial x^2} + \frac{\partial^2 T}{\partial y^2} + \frac{\partial^2 T}{\partial z^2} = 0 \\ &T = T_m(s) \text{ on the surface} \end{aligned}\right\} \tag{3.13}$$

2. Quasi-steady temperature T_q

$$\left.\begin{aligned} &\frac{\partial T}{\partial \tau} = a\left(\frac{\partial^2 T}{\partial x^2} + \frac{\partial^2 T}{\partial y^2} + \frac{\partial^2 T}{\partial z^2}\right) \\ &T = A(s)\sin\omega\tau \qquad \text{on the surface} \end{aligned}\right\} \tag{3.14}$$

3.3 Steady Temperature Field of Concrete Dams

The steady temperature T_f determined by Eq. (3.13) represents the final temperature of the concrete dam after the disappearance of the initial conditions. The two-

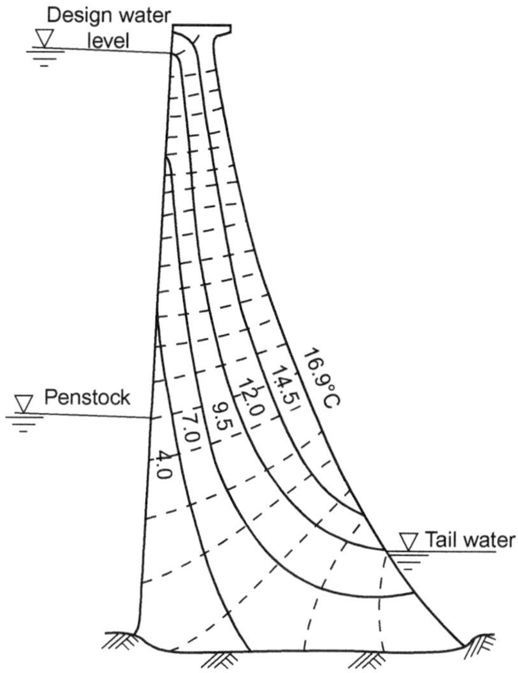

Figure 3.4 Steady temperature field of a gravity-arch dam.

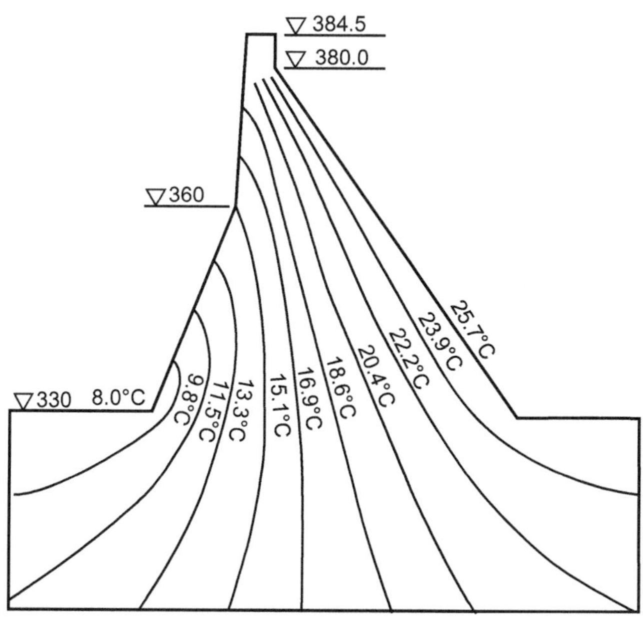

Figure 3.5 Steady temperature field of a gravity dam.

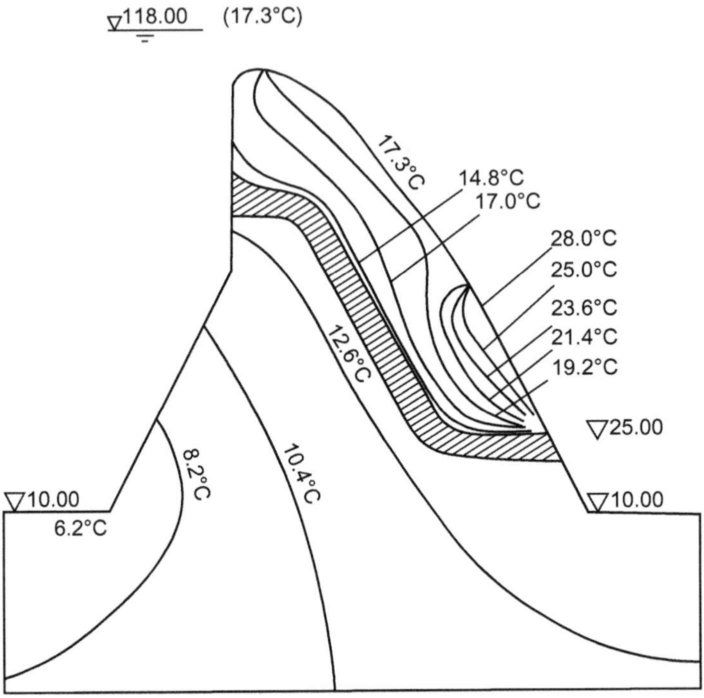

Figure 3.6 Steady temperature field of a gravity dam with a penstock in the dam and an overflowing plant behind the dam.

dimensional T_f may be given by the method of flow net. The two- and the three-dimensional T_f can be computed by the finite element method (FEM).

Figure 3.4 is the steady temperature field of a gravity-arch dam given by the writer in 1955 with the flow net method. Figure 3.5 is the steady temperature field of a solid gravity dam, and Figure 3.6 is the steady temperature field of a gravity dam with a penstock in the dam body and a top-overflow power plant after the dam.

4 Placing Temperature and Temperature Rise of Concrete Lift due to Hydration Heat of Cement

The following three characteristic temperatures are important in dam engineering:

1. The placing temperature T_p which is the starting point for the variation of temperature in a concrete structure.
2. The maximum temperature $T_p + T_r$, where T_r is the temperature rise due to hydration heat of cement.
3. The steady temperature T_f which is the final state of temperature in a concrete dam.

4.1 Mixing Temperature of Concrete—T_0

The mixing temperature is the temperature of concrete at the exit of mixer after mixing which is computed by

$$T_0 = \frac{(c_s + c_w q_s)W_s T_s + (c_g + c_w q_g)W_g T_g + c_c W_c T_c + c_w(W_w - q_s W_s - q_g W_g)T_w + H_c}{c_s W_s + c_g W_g + c_c W_c + c_w W_w} \tag{4.1}$$

where

T_0—the mixing temperature of concrete
c_s, c_g, c_c, c_w—the specific heat of sand, gravel, cement, and water
q_s, q_g—the water content (%) of sand and gravel
W_s, W_g, W_c, W_w—the weight of sand, gravel, cement, and water in 1 m^3 of concrete
T_s, T_g, T_c, T_w—the temperature of sand, gravel, cement, and water
H_c—the mechanical heat (kJ/m^3) produced in the mixing process.

The mechanical heat H_c is given by the following formula:

$$H_c = 42Pt/V \tag{4.2}$$

where

P—the power of motor of the concrete mixer, kW

Thermal Stresses and Temperature Control of Mass Concrete. DOI: http://dx.doi.org/10.1016/B978-0-12-407723-2.00004-X

t—the time of mixing, min
V—the effective volume of the concrete mixer, m^3.

Taking $c_s = c_g = c_c = 0.837$ kJ/(kg °C), $c_w = 4.19$ kJ/(kg °C), replacing part of water by ice and considering the latent heat of ice -335 kJ/kg, the mixing temperature of concrete at the exit of the mixer is

$$T_0 = \frac{(0.837 + 4.19q_s)W_sT_s + (0.837 + 4.19q_g)W_gT_g + 0.837W_cT_c}{0.837(W_s + W_g + W_c) + 4.19W_w} + \frac{4.19(1-p)(W_w - q_sW_s - q_gW_g)T_w - 335\eta p(W_w - q_sW_s - q_gW_g) + H_c}{0.837(W_s + W_g + W_c) + 4.19W_w} \tag{4.3}$$

where

p—percentage (%) of water replaced by ice
η—effective coefficient of ice, generally $\eta = 0.75–0.85$.

For example, let $W_w = 110$ kg/m^3, $W_c = 130$ kg/m^3, $W_s = 480$ kg/m^3, $W_g = 1750$ kg/m^3, $q_s = 3\%$, $q_g = 1\%$, from Eq. (4.3), we get

$$T_0 = 0.1896T_s + 0.631T_g + 0.0446T_c + 0.1342(1-p)T_w - 10.74\eta p \tag{4.4}$$

If there is no artificial cooling, the original temperatures of sand and gravel will be higher than the air temperature; as a result, the mixing temperature of concrete is higher than that of air, namely

$$T_0 = T_a + \Delta T \tag{4.5}$$

where

T_a—the mean daily air temperature
ΔT—increment of concrete temperature due to sunshine, e.g., $\Delta T \cong 5°C$ in middle China in the summer.

4.2 The Forming Temperature of Concrete T_1

The forming temperature of concrete T_1 is the temperature of concrete when it is discharged into the forms after being transported from the mixer. Considering the loss or gain of heat in the process of transport from the mixer to the forms, the forming temperature of concrete T_1 may be estimated by the following formula:

$$T_1 = T_0 + (T_a + R/\beta - T_0)(\phi_1 + \phi_2 + \phi_3 + \cdots + \phi_n) \tag{4.6}$$

where

T_1—the forming temperature of concrete
T_a—the air temperature
R—the solar radiation
β—the surface conductance
T_0—the mixing temperature of concrete.

ϕ_1, ϕ_2, ϕ_3, ϕ_n are the coefficients of experience, and they are determined by observed temperatures in the process of construction of important projects. Some reference values are given in the following:

1. For the loading, unloading, and change of transporter, $\phi = 0.032$ for each time.
2. In the process of transport, $\varphi = A\tau$, where τ is the time of transport, min; and the coefficients A are given in Table 4.1.

In the summer, the air temperature is higher than the mixing temperature of concrete, $T_a > T_0$, heat is absorbed in the process of transport, so the forming temperature of concrete will be higher than the mixing temperature, i.e., $T_1 > T_0$. On the contrary, $T_1 < T_0$ in the winter because $T_a < T_0$.

Example Constructing in summer, the air temperature $T_a = 25°\text{C}$, influence of solar radiation $R/\beta = 15°\text{C}$, the mixing temperature of concrete $T_0 = 7°\text{C}$, transporting by 2 m^3 automatic truck in 10 min. Try to compute the forming temperature of concrete. The coefficients φ_i in Eq. (4.6) are given in the following:

Loading	$\phi_1 = 0.032$
10 min transport by 2 m^3 truck	$\phi_2 = 0.0030 \times 10 = 0.030$
Change of transporter	$\phi_3 = 0.032$
10 min in rectangular bucket in crane	$\phi_4 = 0.0013 \times 10 = 0.013$
Discharge	$\phi_5 = 0.032$
Total	$\sum_{i=1}^{5} \varphi_i = 0.139$

Table 4.1 Coefficients A for Loss of Heat in the Process of Transport

Means of Transport	**Volume of Concrete (m^3)**	***A***	**Means of Transport**	**Volume of Concrete (m^3)**	***A***
Automatic truck	1.0	0.0040	Cylindrical bucket	1.6	0.0009
	1.4	0.0037		3.0	0.0007
	2.0	0.0030		6.0	0.0005
	4.0	0.0022	Hand cart (insulated, covered)	0.15	0.0070
Rectangular bucket	1.6	0.0013	Trolley (insulated)	0.75	0.0100

$$T_1 = 7.0 + (25 + 15 - 7) \times 0.139 = 7.0 + 4.6 = 11.6°C$$

If it is changed to place the concrete in the night, the air temperature $T_a = 18.0°C$, the solar radiation $R = 0$, and the forming temperature of concrete will be

$$T_1 = 7.0 + (18.0 - 7.0) \times 0.139 = 7.0 + 1.53 = 8.53°C$$

4.3 Placing Temperature of Concrete T_p

The concrete dam is constructed lift by lift with a 5–10-day time interval between two lifts, and the thickness of each lift is generally 1.5–3.0 m.

In fact, each lift of concrete is divided into several thin layers with thickness $L \cong 0.30$ m. The concrete mixture discharged in the form is spread and vibrated to form a new concrete layer, then new mix will be discharged on it. The temperature in the concrete layer just before it is covered by new concrete is called the placing temperature and denoted by T_p.

A sketch is shown in Figure 4.1 to show the construction process, and five 0.3 m layers are placed one by one to form a lift of 1.50 m. After a time interval of 5–10 days, the next lift of 1.50 is constructed by 5 m × 0.3 m layers. The process of placing of each layer consists of two steps, i.e., step 1 for spreading the concrete mixture and step 2 for vibrating the concrete. The placing temperature T_p is computed as follows:

$$T_p = T_1 + (T_a + R/\beta - T_1)(\varphi_1 + \varphi_2) \tag{4.7}$$

where

T_p—placing temperature
T_1—forming temperature
T_a—air temperature

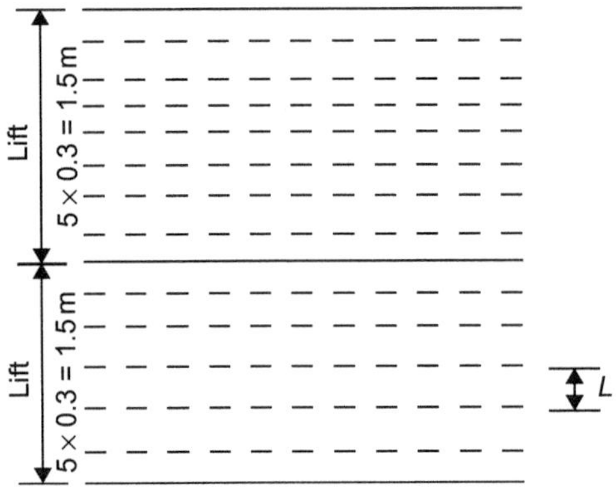

Figure 4.1 Sketch of the rise of a concrete dam.

R—solar radiation
β—surface conductance
ϕ_1—coefficient of step 1
ϕ_2—coefficient of step 2.

The coefficient φ_1 is estimated by the following formula:

$$\varphi_1 = k\tau_1 \tag{4.8}$$

where τ_1 is the time in step 1 for spreading the mixture and k is the experience coefficient, which should be estimated from observed temperatures in the practical construction process. We may take $k = 0.0030$ (1/min) when there is no practical observed data.

In order to compute the coefficient φ_2 in step 2, consider a placing layer with thickness L as shown in Figure 4.1, the bottom surface is insulated and the top surface is in contact with air. A theoretical solution is given in Section 5.3 and $\psi = T_m/T_0$ in Figure 5.9 is related to the coefficient φ_2 as follows:

$$\varphi_2 = 1 - \psi \tag{4.9}$$

φ_2 for $\lambda/\beta = 0.10$ or 0.20 m are shown in Figure 4.2.

Example 1 Placing concrete in a sunny day in the summer, air temperature $T_a = 25°C$, influence of solar radiation $R/\beta = 15°C$, forming temperature of concrete $T_1 = 11.60°C$, thickness of placing layer $L = 0.30$ m, $\lambda/\beta = 0.10$ m, $a = 0.0040$ m^2/h, the time for spreading the concrete mixture $\tau_1 = 10$ min, let $k = 0.0030$. From Eq. (4.8), we get

$$\varphi_1 = 0.0030 \times 10 = 0.030$$

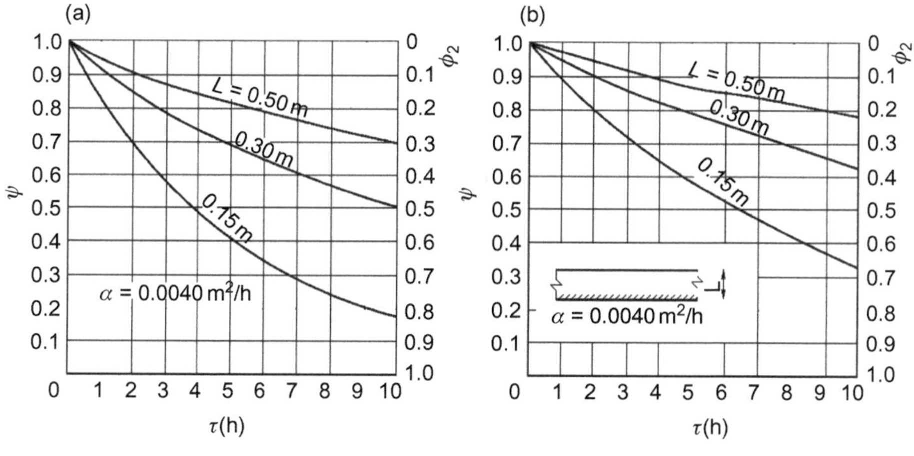

Figure 4.2 Temperature coefficient ϕ_2 in a placing layer with thickness L and no superficial insulation: (a) $\lambda/\beta = 0.10$ m and (b) $\lambda/\beta = 0.20$ m.

The top of the layer is covered at $\tau = 2$ h, let $L = 0.30$ m, $\lambda/\beta = 0.10$ m, $\tau = 2$ h. From Figure 4.2, we get $\varphi_2 = 0.160$. From Eq. (4.7), the placing temperature of concrete is

$$T_p = 11.60 + (25.0 + 15.0 - 11.60)(0.030 + 0.160) = 17.00°\text{C}$$

If it is changed to place the concrete at night, $T_a = 18°$C, $R/\beta = 0$, then the placing temperature will be $T_p = 12.8°$C, which is 5.2°C lower than 17.0°C.

Example 2 A concrete layer with thickness $L = 0.30$ m, diffusivity $a = 0.00355$ m^2/h, conductivity $\lambda = 9.40$ kJ/(m h °C), the bottom surface is insulated, the time of placing $\tau_1 = 2$ h, the top surface of the layer is covered by foamed plastics with thickness h and conductivity $\lambda_s = 0.1256$ kJ/(m h °C) after placing ($\tau \geq 2h$), the equivalent surface conductance of the top is $\beta_e = 1/(1/\beta + h/\lambda_s)$. The temperature coefficient φ_2 of the concrete layer is computed by the finite difference method and is shown in Figure 4.3.

4.4 Theoretical Solution of Temperature Rise of Concrete Lift due to Hydration Heat of Cement

4.4.1 Temperature Rise due to Hydration Heat in Concrete Lift with First Kind of Boundary Condition

As shown in Figure 4.4, a lift of concrete with thickness L, the bottom surface is insulated, the temperature of the top surface is zero, the adiabatic temperature rise due to hydration heat of cement is $\theta(\tau) = \theta_0(1 - \mathrm{e}^{-m\tau})$, and $\partial\theta/\partial\tau = \theta_0 m\,\mathrm{e}^{-m\tau}$.

The problem to be solved is as follows [8, 61]:

$$\left.\begin{array}{llll} \text{Differential equation} & \dfrac{\partial T}{\partial \tau} = a\dfrac{\partial^2 T}{\partial x^2} + \theta_0 m\,\mathrm{e}^{-m\tau} & & \\ \text{Initial condition} & \tau = 0, & T = 0 & \\ \text{Boundary condition} & \tau > 0, & x = 0, & T = 0 \\ & \tau > 0, & x = L, & \dfrac{\partial T}{\partial x} = 0 \end{array}\right\} \tag{4.10}$$

The solution given by the writer in 1956 is in the following:

$$T(x,\tau) = \theta\left(\frac{\cos(L-x)\sqrt{m/a}}{\cos L\sqrt{m/a}} - 1\right)\mathrm{e}^{-m\tau} - \frac{4m\theta_0}{\pi}\sum_{n=1,3,5,\ldots}\frac{\sin(n\pi x/2L)}{n(an^2\pi^2/4L^2 - m)}\exp\left(-\frac{an^2\pi^2\tau}{4L^2}\right) \tag{4.11}$$

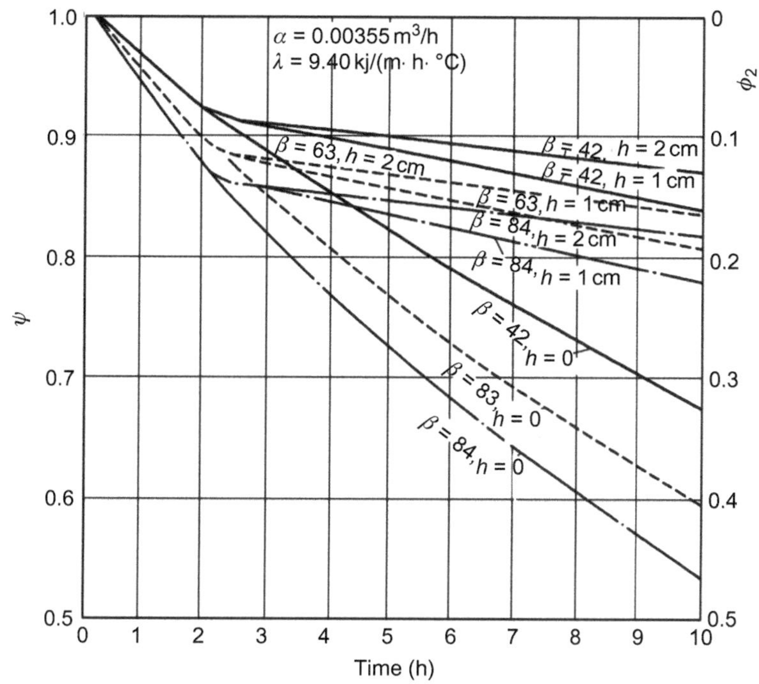

Figure 4.3 The temperature coefficient ϕ_2 for concrete layer covered by foamed plastics with thickness $h = 0$, 1, and 2 cm.

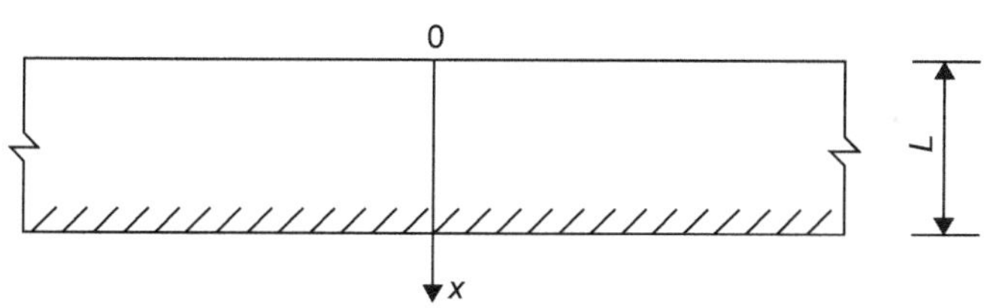

Figure 4.4 A lift of concrete.

The mean temperature is

$$T_m = \theta_0 \left(\frac{\sin L\sqrt{m/a}}{L\sqrt{m/a}\ \cos L\sqrt{m/a}} - 1 \right) e^{-m\tau} - \frac{8m\theta_0}{\pi^2} \sum_{n=1,3,5,\ldots} \frac{1}{n^2 \left(\frac{an^2\pi^2}{4L^2} - m \right)} \exp\left(-\frac{an^2\pi^2\tau}{4L^2} \right) \tag{4.12}$$

The temperature rise due to hydration heat of cement in concrete lifts of different thickness is shown in Figure 4.5. It is assumed that the top surface is insulated

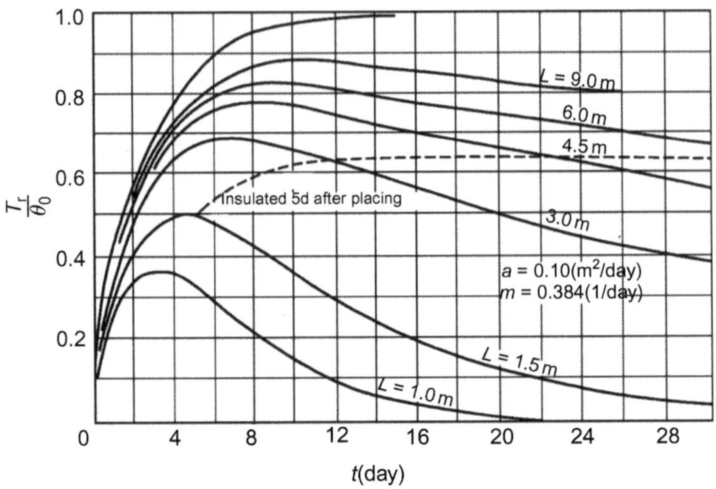

Figure 4.5 Temperature rise due to hydration heat of cement in lifts of concrete.

when it is covered by new concrete after the time interval and then the hydration heat cannot be dissipated and the final temperature rise of the lift due to hydration heat is shown in Figure 4.6.

4.4.2 Temperature Rise due to Hydration Heat in Concrete Lift with Third Kind of Boundary Condition

The differential equation to be solved is the same as Eq. (4.10), but the boundary condition on the top surface is changed to the following:

when

$$\tau > 0, \quad x = L, \quad \lambda\frac{\partial T}{\partial x} = \beta T \tag{4.13}$$

The problem is solved by means of the solution of the temperature in an infinite plate with initial temperature T_0 and without internal source of heat in Section 5.3; in this case the mean temperature of the plate is

$$\frac{T_{\mathrm{m}}(\tau)}{T_0} = \sum_{n-1}^{\infty} B_n \mathrm{e}^{-s_n\tau} = \eta(\tau) \tag{4.14}$$

where

$$B_n = \frac{2B_i^2}{\mu_n^2(B_i^2 + B_i + \mu_n^2)}, \quad B_i = \frac{\beta L}{\lambda} \tag{4.15}$$

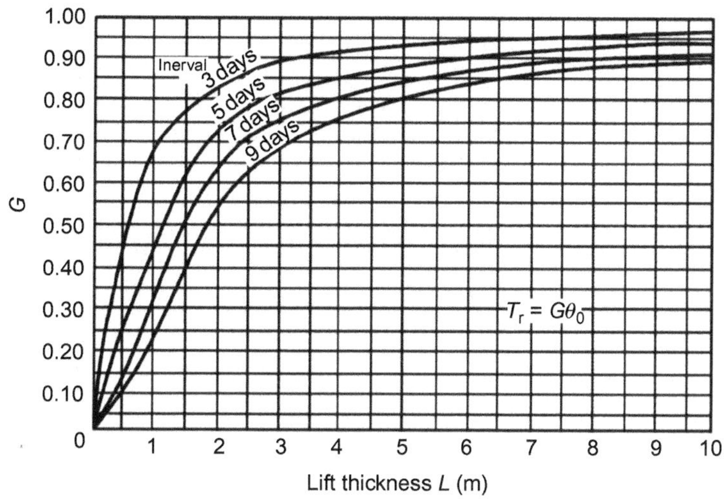

Figure 4.6 Final temperature rise in lifts of concrete due to hydration heat of cement.

$$s_n = \frac{\mu_n^2 a}{L^2} \tag{4.16}$$

μ_n is the root of the characteristic Eq. (5.31).

Now, returning to our original problem (4.13), if there is a temperature increment $\Delta\theta(\tau)$ at age τ, the mean temperature rise at time t induced by $\Delta\theta(\tau)$ will be $\Delta\theta(\tau)\eta(t-\tau)$. Integrating from 0 to t, we get the mean temperature of the concrete lift:

$$\begin{aligned} T_m(t) &= \int_o^t \eta(t-\tau)\frac{\partial\theta}{\partial\tau}\mathrm{d}\tau \\ &= \theta_0 m \sum_{n=1}^{\infty} \frac{B_n}{s_n - m}(\mathrm{e}^{-mt} - \mathrm{e}^{-s_n t}) \end{aligned} \tag{4.17}$$

where B_n and s_n are given in Eqs. (4.15) and (4.40).

Example 1 $L = 1.50\,\text{m}$, $\lambda/\beta = 0.10\,\text{m}$, $B_i = \beta L/\lambda = 1.50/0.10 = 15.0$, $a = 0.10\,\text{m}^2/\text{day}$, $m = 0.384$ (l/day), from Eq. (5.31), Eqs. (4.15) and (4.16), the first three terms of B_n, μ_n, and S_n are as follows:

$$\begin{aligned} &\mu_1 = 1.4729, \quad \mu_2 = 4.4255, \quad \mu_3 = 7.3959 \\ &B_1 = 0.8565, \quad B_2 = 0.0885, \quad B_3 = 0.0279 \\ &s_1 = 0.09642, \quad s_2 = 0.8704, \quad s_3 = 2.4311 \end{aligned}$$

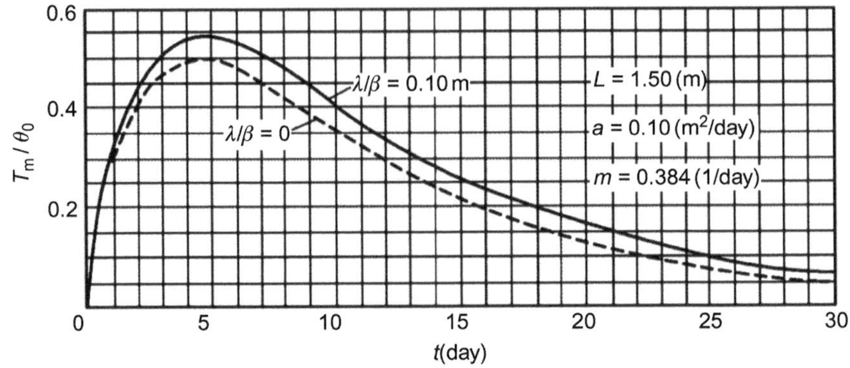

Figure 4.7 Example 1: temperature rise due to hydration heat of a concrete lift with third kind of boundary condition.

Substituting them in Eq. (4.17), the mean temperature of the lift due to hydration heat is

$$\frac{T_m(t)}{\theta_0} = 1.1437(e^{-0.09642t} - e^{-0.384t}) + 0.0696(e^{-0.384t} - e^{-0.8704t}) + 0.0052(e^{-0.384t} - e^{-2.4311t})$$

The results of computation are shown in Figure 4.7.

4.4.3 Temperature Rise due to Hydration Heat with Adiabatic Temperature Rise Expressed by Compound Exponentials

The temperature rise of concrete expressed by exponential formula is convenient for integration and differentiation but does not agree well with test results. In order to avoid this drawback, the following compound exponential formula proposed by the writer may be used:

$$\theta(\tau) = \theta_1(1 - e^{-m_1\tau}) + \theta_2(1 - e^{-m_2\tau}) \tag{4.18}$$

$$\frac{\partial\theta}{\partial\tau} = \theta_1 m_1 e^{-m_1\tau} + \theta_2 m_2 e^{-m_2\tau} \tag{4.19}$$

Substituting in Eq. (4.17), the following formula for mean temperature rise of concrete lift is derived:

$$T_m(t) = \sum_{n=1}^{\infty} B_n \left[\frac{\theta_1 m_1}{s_n - m_1}(e^{-m_1 t} - e^{-s_n t}) + \frac{\theta_2 m_2}{s_n - m_2}(e^{-m_2 t} - e^{-s_n t})\right] \tag{4.20}$$

Example 2 A concrete lift with thickness $L = 1.50\text{m}$, $\lambda/\beta = 0.10\text{m}$, $a = 0.10\text{m}^2/\text{day}$, $\theta(\tau) = 16.4(1 - \mathrm{e}^{-0.50\tau}) + 10.9(1 - \mathrm{e}^{-0.55\tau})$, B_n and S_n are the same as in example 1. Taking the first three terms, Eq. (4.20) gives the mean temperature rise of the concrete lift as shown in Figure 4.8.

4.5 Theoretical Solution of Temperature Field of Concrete Lift due to Simultaneous Action of Natural Cooling and Pipe Cooling

The adiabatic temperature rise of concrete due to hydration heat of cement is

$$\theta(\tau) = \sum \theta_i(1 - e^{-m_i\tau}) \tag{4.21}$$

where θ_i and m_i are constants. The mean temperature of a concrete cylinder with insulated surface and cooling pipe in the interior is

$$U(t) = T_w + (T_0 - T_w)\varphi(t) + \int_0^t \varphi(t - \tau)\frac{\partial\theta}{\partial\tau}\mathrm{d}\tau \tag{4.22}$$

$$\varphi(t) = \mathrm{e}^{-pt} \tag{4.23}$$

where

T_0—initial temperature of concrete
T_{w}—temperature of water at the inlet
$\varphi(t) = \mathrm{e}^{-pt}$—function of pipe cooling given by Eq. (17.65).

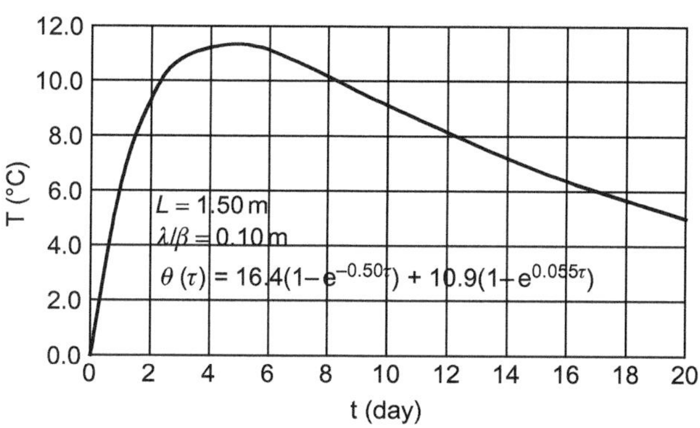

Figure 4.8 Example 2: temperature rise in a concrete lift with the adiabatic temperature rise $\theta(\tau)$ expressed by compound exponential formula.

Substituting Eqs. (4.21) and (4.23) into Eq. (4.22), we get

$$U(t) = T_w + (T_0 - T_w)\varphi(t) + \sum \theta_i \psi_i(t) \tag{4.24}$$

where

$$\psi_i(t) = \frac{m_i}{m_i - p}(e^{-pt} - e^{-m_i t}) \tag{4.25}$$

Differentiating Eq. (4.24), we get

$$\frac{\partial U}{\partial t} = \sum A_i e^{-m_i t} - B e^{-pt} \tag{4.26}$$

$$A_i = \frac{\theta_i m_i^2}{m_i - p}, \quad B = \sum \frac{\theta_i m_i p}{m_i - p} + (T_0 - T_w)p \tag{4.27}$$

Substituting $\partial U/\partial t$ for $\partial\theta/\partial t$ in the equation of heat conduction, the equivalent equation of heat conduction considering the effect of pipe cooling is derived as follows:

$$\frac{\partial U}{\partial \tau} = a\frac{\partial^2 T}{\partial x^2} + \sum A_i e^{-m_i t} - B e^{-pt} \tag{4.28}$$

From Eqs. (4.15) and (4.17), the mean temperature of the concrete lift under the simultaneous action of pipe cooling in the interior and natural cooling of the top surface is

$$\begin{aligned} T_m(t) &= \int_o^t \eta(t-\tau)\frac{\partial U}{\partial \tau}\mathrm{d}\tau \\ &= \sum_{n=1,3,5,\ldots} \int_0^t B_n e^{-s_n(t-\tau)}\frac{\partial U}{\partial \tau}\mathrm{d}\tau \end{aligned} \tag{4.29}$$

Substituting Eq. (4.26) into the above equation, we get

$$T_m(t) = \sum_{n=1,3,5,\ldots} B_n\left[\sum \frac{A_n}{s_n - m_i}(e^{-m_i t} - e^{-s_n t}) - \frac{B}{s_n - p}(e^{-pt} - e^{-s_n t})\right] \tag{4.30}$$

Example A concrete lift with thickness $L = 1.5$ m, and first kind of boundary condition on the top, $a = 0.10\ \mathrm{m^2/day}$, $\theta(\tau) = 25.0[1 - \exp(-0.40\,\tau)]$, the length of cooling pipe $L_1 = 300$ m, $\xi = \lambda L_1/(C_w \rho_w q_w) = 0.50$, $T_p = 0°C$, $T_w = 0°C$. The mean temperatures given by Eq. (4.30) are shown in Figure 4.9, from which the influence of the spacing of cooling pipes is remarkable.

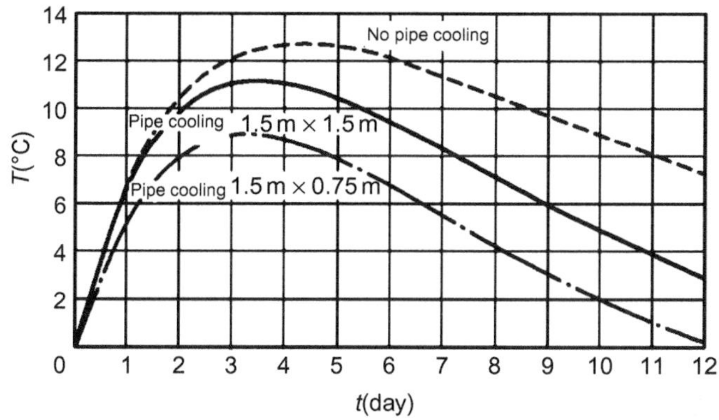

Figure 4.9 The mean temperature of a concrete lift with thickness $L = 1.5$ m under the simultaneous action of pipe cooling and natural cooling of top surface.

4.6 Temperature Field in Concrete Lift Computed by Finite Difference Method

4.6.1 Temperature Field in Concrete Lift due to Hydration Heat Computed by Finite Difference Method

The one-dimensional finite difference method for temperature field is given in Section 2.14. Two examples for temperature field in concrete lifts on rock foundation due to hydration heat of cement are given in the following.

Example 1 Place mass concrete on rock foundation one lift every 4 days. The diffusivity $a = 0.10$ m^2/day, the adiabatic temperature rise $\theta(\tau) = 27.3(1 - e^{-0.384\tau})$°C, $\lambda/\beta = 0.10$. Taking $\Delta x = 0.50$ m, $\Delta\tau = 1$ day, $r = a\Delta\tau/\Delta x^2 = 0.10 \times 1.0/0.50^2 = 0.40$, from Eq. (2.75)

$$T_{i,\Delta+\Delta\tau} = 0.20T_{i,\tau} + 0.40(T_{i-1,\tau} + T_{i+1,\tau}) + \Delta\theta \tag{4.31}$$

On the contact surface of the old and new concrete, the increments of adiabatic temperature rise are given by

$$\Delta\theta = (\Delta\theta_{\text{new}} + \Delta\theta_{\text{old}})/2$$

Two cases are computed, in one case the thickness of concrete lift is $L = 1.50$ m; in the other case, $L = 3.0$ m. The computed results are shown in Figures 4.10 and 4.11.

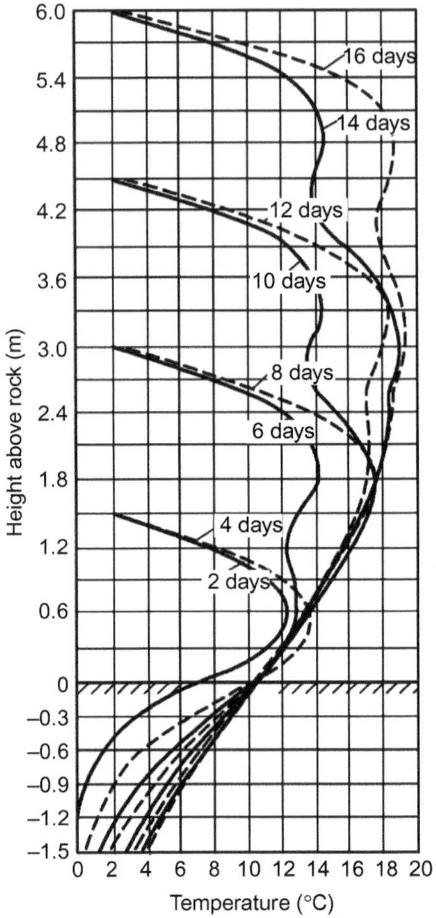

Figure 4.10 Temperature rise in concrete lifts on rock foundation (thickness of lift 1.5 m, time interval 4 days).

4.6.2 Temperature Field due to Hydration Heat in Concrete Lift with Cooling Pipe Computed by Finite Difference Method

The adiabatic temperature rise due to hydration heat is expressed by Eq. (4.21). The mean temperature of a concrete cylinder with insulated surface and cooling pipe in the interior is expressed by Eq. (4.24); the increment of $U(t)$ is

$$\Delta U(t) = (T_0 - T_w)\Delta\varphi(t) + \sum \theta_i \Delta\psi(t) \tag{4.32}$$

Substituting $\Delta U(t)$ for $\Delta\theta(t)$ in Eq. (2.75), the finite difference equation considering the effect of pipe cooling is derived in the following:

$$T_{i,\tau+\Delta\tau} = (1-2r)T_{i,\tau} + r(T_{i-1,\tau} + T_{i+1,\tau}) + \Delta U \tag{4.33}$$

which may be used to compute the temperature in concrete lift with pipe cooling.

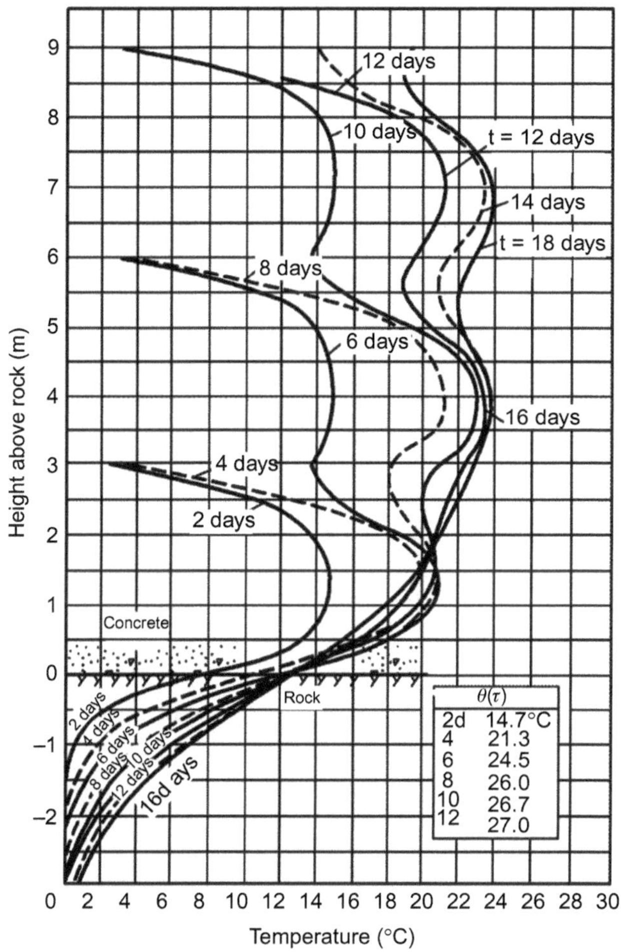

Figure 4.11 Temperature rise in concrete lifts on rock foundation (thickness of lift 3.0 m, time interval 4 days).

Example 2 Mass concrete is placed on rock foundation, one lift every 4 days, $L = 1.5$ m, $a = 0.10$ m^2/day, $\lambda/\beta = 0.10$ m, $T_p = 0°C$, $T_a = 0°C$, $p = 0.0437$ (1/day); the adiabatic temperature rise is

$$\theta(\tau) = 16.4(1 - e^{-0.50\tau}) + 10.9(1 - e^{-0.055\tau})$$

Taking $\Delta x = 0.30$ m, $\Delta\tau = 0.333$ day, $r = a\,\Delta\tau/\Delta x^2 = 0.10 \times 0.333/0.30^2 = 0.370$, and substituting into Eq. (4.33), we get

$$T_{i,\tau+\Delta\tau} = 0.260T_{i,\tau} + 0.370(T_{i-1,\tau} + T_{i+1,\tau}) + \Delta U$$

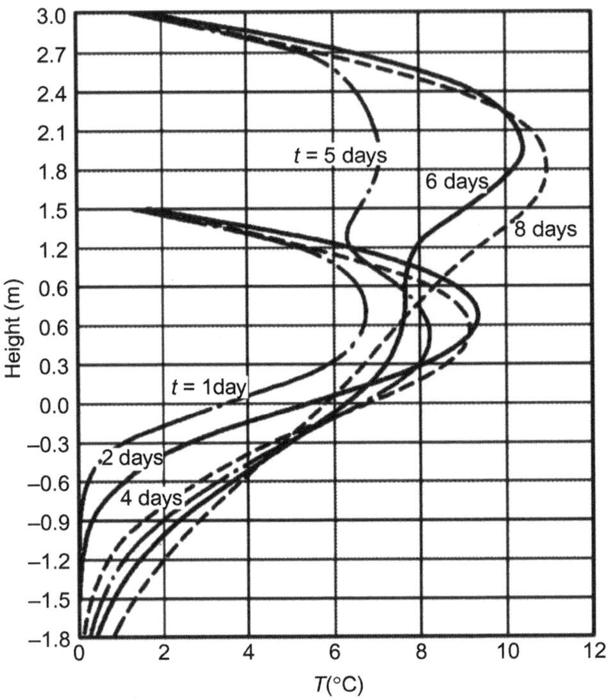

Figure 4.12 Temperature rise in concrete lifts on rock foundation due to hydration heat, pipe cooling, and natural cooling of lift surface.

The temperature on the surface point 1 is

$$T_{1,\tau+\Delta\tau} = \frac{\Delta x\, T_a + (\lambda/\beta)T_{2,\tau+\Delta\tau}}{\Delta x + \lambda/\beta}$$

and

$$U(t) = 17.97(e^{-.0437t} - e^{-.50t}) + 53.1(e^{-.0437t} - e^{-0.055t})$$

$$\Delta U(t_i) = \frac{\partial U(t_i + 0.5\Delta t_i)}{\partial t}\Delta t = 2.995\, e^{-0.50(t_i+0.167)}$$

$$+ 0.972\, e^{-.055(t+0.167)} - 1.0346\, e^{-.0437(t+0.167)}$$

The computed results are shown in Figure 4.12.

4.7 Practical Method for Computing Temperature Field in Construction Period of Concrete Dams

The adiabatic temperature rise of concrete due to hydration heat is

$$\theta(\tau) = \theta_0 f(\tau) \tag{4.34}$$

$$f(\tau) = s(1 - e^{-m_1\tau}) + (1 - s)(1 - e^{-m_2\tau}) \tag{4.35}$$

where

θ_0—the final temperature rise
s, m_1, and m_2—constants.

The mean temperature of a concrete cylinder with diameter D, length L, insulated surface, and cooling pipe is

$$U(t) = T_W + (T_P - T_W)\varphi(t) + \theta_0\psi(t) \tag{4.36}$$

$$\varphi(t) = e^{-pt} \tag{4.37}$$

$$\psi(t) = \frac{sm_1}{m_1 - p}(e^{-pt} - e^{-m_1 t}) + \frac{(1 - s)m_2}{m_2 - p}(e^{-pt} - e^{-m_2 t}) \tag{4.38}$$

$$p = \frac{kga}{D^2} = \frac{kga}{1.362 S_1 S_2} \tag{4.39}$$

$$k = 2.09 - 1.35\xi + 0.320\xi^2 \tag{4.40}$$

$$g = 1.67 \exp\left\{-0.0628\left[\frac{b}{c}\left(\frac{c}{r_0}\right)^{\eta} - 20\right]^{0.48}\right\} \tag{4.41}$$

or approximately

$$g = \frac{\ln 100}{\ln(b/c) + (\lambda/\lambda_1)\ln(c/r_0)} \tag{4.42}$$

$$\xi = \lambda L/(c_w \rho_w q_w); \quad \eta = \lambda/\lambda_1 \tag{4.43}$$

where

S_1, S_2—spacing of cooling pipes
$b = D/2$—exterior radius of the concrete cylinder
c—the exterior radius of the pipe
r_0—the inner radius of pipe
λ—conductivity of concrete
λ_1—conductivity of pipe
c_w, ρ_w, q_w—the specific heat, density, and discharge of the cooling water.

When there is no cooling pipe, the equation of heat conduction is

$$\frac{\partial T}{\partial \tau} = a\frac{\partial^2 T}{\partial x^2} + \frac{\partial \theta}{\partial \tau} \tag{4.44}$$

Substituting $\partial U/\partial t$ for $\partial \theta/\partial t$ in the above equation, the equivalent equation of heat conduction with pipe cooling is derived as follows:

$$\frac{\partial T}{\partial \tau} = a\frac{\partial^2 T}{\partial x^2} + (T_p - T_W)\frac{\partial \varphi}{\partial \tau} + \theta_0\frac{\partial \psi}{\partial \tau} \tag{4.45}$$

4.7.1 Practical Method for Computing Temperature Field in Concrete Lift without Pipe Cooling

As shown in Figure 4.13(a), when the new concrete is placed, the temperature in the old concrete is T_D, the temperature in the new concrete is 0°C, and the air temperature is 0°C. The concrete dam is analyzed as it is a semi-infinite solid, the boundary condition is $T(0,t)=0$ when $x=0$; the initial condition is when $t=0$, $T(0 \le x \le l, 0)=0$, $T(x>l,0)=T_D$.

The solution to this problem is

$$T(x,t) = T_D F_1(x,t) \tag{4.46}$$

$$F_1(x,t) = \frac{1}{2}P\left(\frac{l+x}{2\sqrt{at}}\right) - \frac{1}{2}P\left(\frac{l-x}{2\sqrt{at}}\right) \tag{4.47}$$

$$P(u) = \frac{2}{\sqrt{\pi}}\int_0^u e^{-u^2}\,du \tag{4.48}$$

where

$P(u)$—the probability function
l—the thickness of concrete lift.

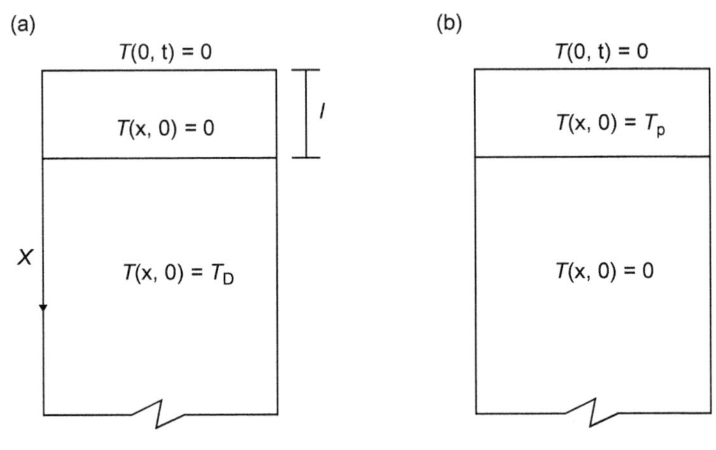

Figure 4.13 Temperature in the concrete block: (a) T_D—temperature in old concrete and (b) T_p—temperature in new concrete.

Let $x = l/2$, we get the temperature of the center point of the lift:

$$T_C(t) = T_D C_1(t) \tag{4.49}$$

$$C_1(t) = \frac{1}{2}P(0.75v) - \frac{1}{2}P(0.25v) \tag{4.50}$$

$$v = l/\sqrt{at} \tag{4.51}$$

Integrating in $x = 0 - l$ in Eq. (4.46) and dividing by l, we get the mean temperature of the lift:

$$T_m(t) = T_D E_1(t) \tag{4.52}$$

$$E_1(t) = P(v) - P(0.5v) + \frac{1}{v\sqrt{\pi}}\left[\exp(-v)^2 + 1 - 2\exp\left(-\frac{v^2}{4}\right)\right] \tag{4.53}$$

$C_1(t)$ and $E_1(t)$ are shown in Figure 4.14.

4.7.2 Influence of the Placing Temperature T_p of the New Concrete

As shown in Figure 4.13(b), the placing temperature of the new concrete is T_p, the initial temperature of the old concrete is 0°C, and the surface temperature is 0°C. In other words, the boundary condition is $T(0, t) = 0$ when $x = 0$; the initial

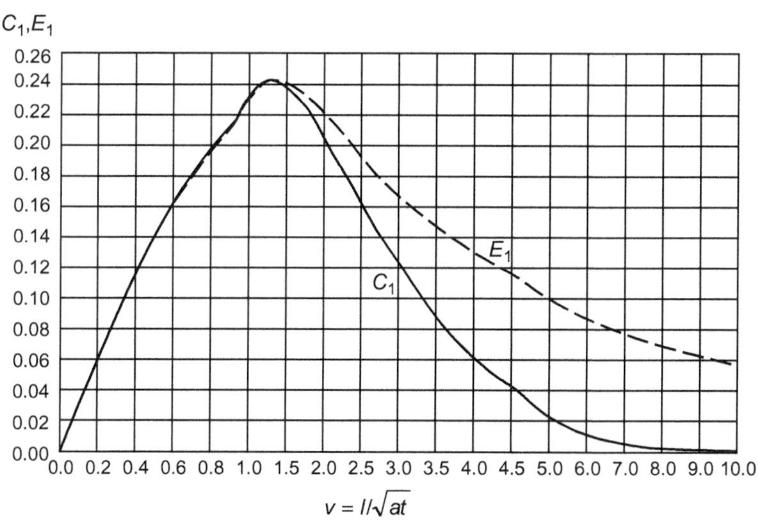

Figure 4.14 C_1 and E_1.

conditions are $T(0 \le x \le l, 0) = T_p$ and $T(x > l, 0) = 0$ when $t = 0$. The solution for this problem is

$$T(x,t) = T_p F_2(x,t) \tag{4.54}$$

$$F_2(x,t) = P\left(\frac{x}{2\sqrt{at}}\right) - \frac{1}{2}P\left(\frac{l+x}{2\sqrt{at}}\right) + \frac{1}{2}P\left(\frac{l-x}{2\sqrt{at}}\right) \tag{4.55}$$

The temperature of the center point is

$$T_c(t) = T_p C_2(t) \tag{4.56}$$

$$C_2(t) = \frac{3}{2}P(0.25v) - \frac{1}{2}P(0.75v) \tag{4.57}$$

The mean temperature of the lift is

$$T_m(t) = T_p E_2(t) \tag{4.58}$$

$$E_2(t) = 2P\left(\frac{v}{2}\right) - P(v) + \frac{1}{v\sqrt{\pi}}\left[4\exp\left(-\frac{v^2}{4}\right) - 3 - \exp(-v^2)\right] \tag{4.59}$$

$C_2(t)$ and $E_2(t)$ are shown in Figure 4.15.

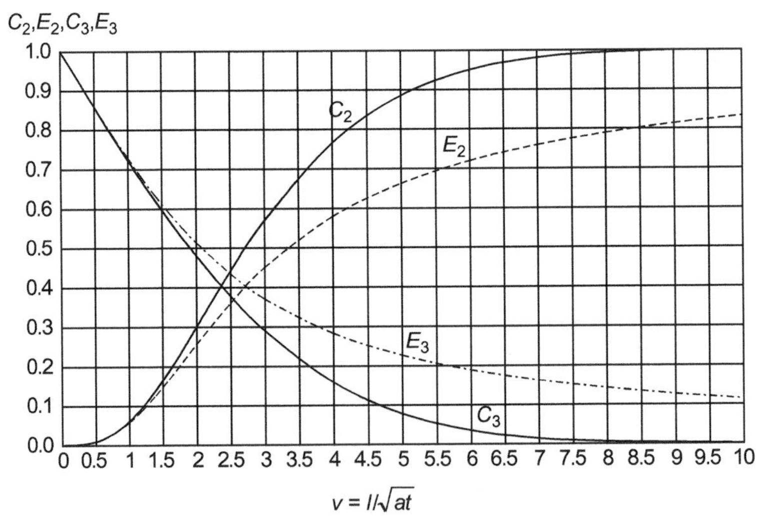

Figure 4.15 C_2, E_2, C_3, E_3.

4.7.3 Practical Method for Computing Temperature in Concrete Lift without Pipe Cooling

The initial temperature of new concrete is the placing temperature T_{p}, the initial temperature of the old concrete is T_{D}, and the adiabatic temperature rise of the new concrete is $\theta(\tau)$. The problem to be solved is as follows:

Equation of heat conduction

$$\frac{\partial T}{\partial \tau} = a\frac{\partial^2 T}{\partial x^2} + \frac{\partial \theta}{\partial \tau} \tag{4.60}$$

$$\left.\begin{array}{lll} \text{Initial condition:} & \text{when} & t=0, T(0\leq x\leq l,0)=T_{\mathrm{p}} \\ & & T(x>l,0)=T_{\mathrm{D}} \\ \text{Boundary condition:} & \text{when} & x=0, T(0,\tau)=T_{\mathrm{s}} \end{array}\right\} \tag{4.61}$$

The solution is

$$T(x,t) = T_{\mathrm{s}} + (T_{\mathrm{D}} - T_s)F_1(x,t) + (T_{\mathrm{p}} - T_{\mathrm{s}})F_2(x,t) + \int_0^t F_2(x,t-\tau)\frac{\partial \theta}{\partial \tau}\mathrm{d}\tau \tag{4.62}$$

or

$$T(x,t) = T_{\mathrm{s}} + (T_{\mathrm{D}} - T_{\mathrm{s}})F_1(x,t) + (T_{\mathrm{p}} - T_{\mathrm{s}})F_2(x,t) + \sum_{i=1}^{n} \Delta\theta_i F_2(x,t-t_i-0.5\Delta t_i) \tag{4.63}$$

The mean temperature of the lift is

$$T_{\mathrm{m}}(t) = T_{\mathrm{s}} + (T_{\mathrm{D}} - T_{\mathrm{S}})E_1(t) + (T_{\mathrm{p}} - T_{\mathrm{s}})E_2(t) + \sum \Delta\theta_i E_2(t - t_i - 0.5\Delta t_i) \tag{4.64}$$

The temperature of the center point of lift is

$$T_{\mathrm{c}}(t) = T_{\mathrm{s}} + (T_{\mathrm{D}} - T_{\mathrm{s}})C_1(t) + (T_{\mathrm{p}} - T_{\mathrm{s}})C_2(t) + \sum \Delta\theta_i C_2(t - t_i - 0.5\Delta t_i) \tag{4.65}$$

4.7.4 Practical Method for Computing Temperature Field in Concrete Lift with Pipe Cooling

The initial temperature of concrete is shown in Figure 4.16(a). The mean temperature of concrete lift without pipe cooling is given by Eq. (4.64). When there is pipe

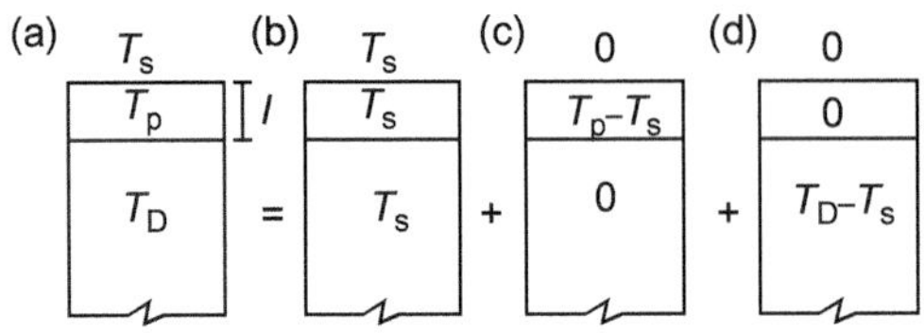

Figure 4.16 Decomposition of the initial and boundary conditions.

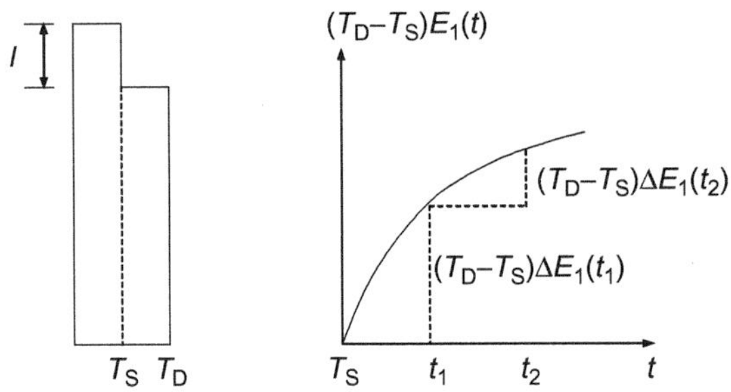

Figure 4.17 Temperature difference $(T_D - T_s)E_1(t)$.

cooling, the difference between the mean temperature of concrete lift and the temperature of cooling water is

$$\Delta T = (T_D - T_s)E_1(t) + (T_p - T_s)E_2(t) + T_s - T_w + \sum \Delta\theta_i E_2(t - t_i - 0.5\Delta t_i) \tag{4.66}$$

which may be resolved into four parts, namely $\Delta T_1 = (T_D - T_s)E_1(t)$, $\Delta T_2 = (T_p - T_s)E_2(t)$, $\Delta T_3 = T_s - T_w$, and $\Delta T_4 = \sum \Delta\theta_i E_2(t - t_i - 0.5\Delta t_i)$. The contributions of each part to the mean temperature of concrete under the simultaneous action of pipe cooling and natural cooling are derived in the following.

1. Contribution of $\Delta T_1 = (T_D - T_s)E_1(t)$

   Considering the effect of pipe cooling, the contribution of ΔT_1 to the mean temperature of concrete lift is (refer to Figure 4.17)

$$T_{m1}(t) = (T_D - T_s)\sum \Delta E_1(t_i)\phi(t - t_i - 0.5\Delta t_i) \tag{4.67}$$

2. Contribution of $\Delta T_2 = (T_p - T_s)E_2(t)$

   When $t = 0$, $E_2(0) = 1.0$, and $\Delta T_2 = T_p - T_s$ and when $t = t_1$, the temperature difference is $\Delta T_2 = (T_p - T_s)E_2(t) = (T_p - T_s)[1 + \Delta E_2(t_1)]$ which can be resolved into two parts; the first part is the constant temperature difference $T_p - T_s$ which will become $(T_p - T_s)\varphi(t_1)$ due to pipe cooling and the second part is the variable temperature difference $(T_p - T_s)\Delta E_2(t_1)$ which will become $(T_p - T_s)\Delta E_2(t_1)\varphi(t - 0.5\Delta t_1)$ owing to pipe cooling. Thus, the contribution of ΔT_2 is (Figure 4.18)

$$T_{m2}(t) = (T_p - T_s)[\varphi(t) + \sum \Delta E_2(t_i)\varphi(t - t_i - 0.5\Delta t_i)] \tag{4.68}$$

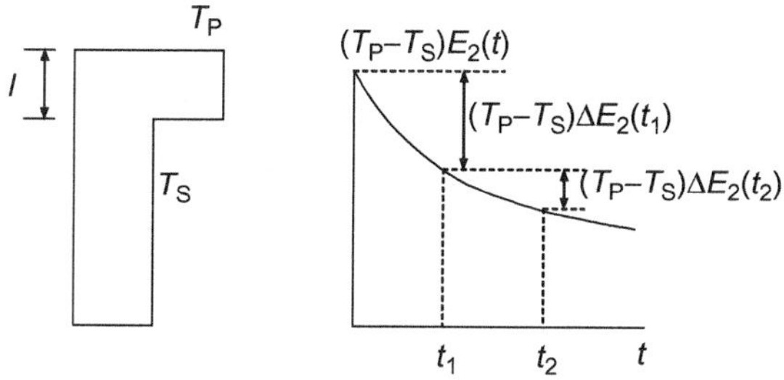

Figure 4.18 Temperature difference $(T_p - T_s)E_2(t)$.

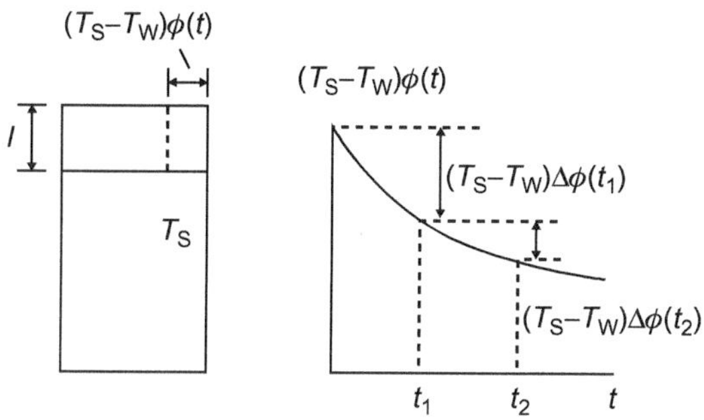

Figure 4.19 Contribution of temperature difference $T_s - T_w$.

3. Contribution of $\Delta T_3 = T_s - T_w$

   As shown in Figure 4.19, when $t = 0$, $\varphi(0) = 1.0$, there is no temperature difference between the upper and lower part of concrete; when $t > 0$, the temperature difference between the upper and lower part of concrete will be $(T_s - T_w)\varphi(t)$. Thus the contribution of ΔT_3 is

$$T_{m3} = (T_s - T_w)[1 + \sum \Delta\varphi(t_i)E_2(t - t_i - 0.5\Delta t_i)] \tag{4.69}$$

4. Contribution of $\Delta T_4 = \sum \Delta\theta_i E_2(t - t_i - 0.5\Delta t_i)$

   Taking account of pipe cooling, the contribution of ΔT_4 is

$$T_{m4} = T_{rm}(t) = \sum \Delta\theta(\tau_i)\varphi(t - t_i - 0.5\Delta t_i)E_2(t - t_i - 0.5\Delta t_i) \tag{4.70}$$

To sum up, the mean temperature of a concrete lift is as follows:

$$\begin{aligned} T_m(t) = {} & T_s + (T_D - T_s)\sum \Delta E_1(t_i)\varphi(t - t_i - 0.5\Delta t_i) + (t_p - T_s) \\ & [\varphi(t) + \sum \Delta E_2(t_i)\varphi(t - t_l - 0.5\Delta t_i] + (T_s - T_w)\sum \Delta\varphi(t_i)E_2(t - t_i - 0.5\Delta t_i)] \\ & + T_{rm}(t) \end{aligned} \tag{4.71}$$

If only one time increment $\Delta t_1 = t$ is taken in Eq. (4.71), then $\Delta E_1(t) = E_1(t)$, $\Delta E_2(t) = E_2(t) - 1$, and $\Delta\varphi(t) = \varphi(t) - 1$, and the simplified formula for the mean temperature of concrete lift is derived in the following:

$$T_m(t) = T_s + (T_D - T_s)E_1(t)\varphi(0.5t) + (T_p - T_s)\{\varphi(t) + [E_2(t) - 1]\varphi(0.5t)\} \\ + (T_s - T_w)[\varphi(t) - 1]E_2(0.5t) + T'_{rm}(t) \tag{4.72}$$

in which

$$T'_{rm}(t) = \theta_0 \sum \Delta f(t_i)E_2(t - t_i - 0.5\Delta t_i)\varphi(t - t_i - 0.5t_i) \tag{4.73}$$

4.7.5 Practical Treatment of Boundary Condition on the Top Surface

In the above-mentioned computing method, it is assumed that the temperature of the top surface of concrete lift is equal to air temperature. The following approximate method may be adopted to consider the influence of the third kind of boundary condition:

1. Additional thickness λ/β

   An additional thickness λ/β is put on the exposed surface of the concrete lift, the new thickness of which will be

   $$l = l' + \lambda/\beta \tag{4.74}$$

   where

   l—new thickness of the concrete lift

   l'—the original thickness of the concrete lift

   λ—the conductivity of concrete

   β—the surface conductance of concrete.

   The temperature of the new surface is equal to the air temperature.
2. Estimated surface temperature

   For the semi-infinite solid $x \geq 0$ with initial temperature T_0, on basis of the third kind of boundary condition, the temperature of the concrete surface is given by Eq. (5.6).

Example Concrete lift with thickness $l = 1.50$ m and initial temperature $T_p = 12°C$, the initial temperature of the lower old concrete $T_D = 20°C$, the spacing of cooling pipes 1.50 m × 1.50 m, the water temperature at the inlet $T_w = 10°C$, the temperature of concrete surface $T_s = 15°C$, the coefficient of pipe cooling p = 0.0437 (1/day), and the adiabatic temperature rise of concrete is

$$\theta(t) = 25.0[0.60(1 - e^{-.78t}) + 0.40(1 - e^{-.078t})] \tag{4.75}$$

Table 4.2 Example of Temperature Rise and Mean Temperature of Concrete Lift (°C)

Time (days)		**0**	**1**	**2**	**3**	**4**	**5**
Temperature rise due to hydration heat, Eq. (4.70)		0	6.45	7.83	7.33	6.50	5.40
Mean temperature of concrete lift	Eq. (4.71)	12.00	20.04	21.87	21.89	21.29	20.33
	Eq. (4.72)	12.00	19.66	21.64	21.92	22.11	21.31

The function of pipe cooling is

$$\varphi(t) = e^{-.0437t} \tag{4.76}$$

The temperature rise due to hydration heat considering the effect of pipe cooling is given by Eq. (4.70), the mean temperature of the concrete lift is given by Eq. (4.71) or the simplified formula (4.72). The results of computation are given in Table 4.2. It is clear that the precision of the simplified formula (4.72) is good so it is extensively used in engineering.

5 Natural Cooling of Mass Concrete

5.1 Cooling of Semi-Infinite Solid, Third Kind of Boundary Condition

Let the solid be bounded by the plane $x = 0$ and extend to infinity in the direction of x positive, as shown in Figure 5.1, the problem to be solved is

Equation of heat conduction

$$\frac{\partial T}{\partial \tau} = a\frac{\partial^2 T}{\partial x^2} \tag{5.1}$$

$$\left.\begin{array}{lll} \text{Initial condition} & T(x,0) = T_0, & \text{when } \tau = 0 \\ \text{Boundary condition} & \lambda\dfrac{\partial T}{\partial x} + \beta(T_a - T) = 0, & \text{when } x = 0 \\ & \dfrac{\partial T}{\partial x} = 0, & \text{when } x = \infty \end{array}\right\} \tag{5.2}$$

where

T_a—air temperature
T_0—initial temperature
β—surface conductance
λ—conductivity.

The solution is

$$T(x,\tau) = T_0 + (T_a - T_0)f(x,\tau) \tag{5.3}$$

$$f(x,\tau) = \text{erfc}\left(\frac{x}{2\sqrt{a\tau}}\right) - \exp\left(\frac{\beta x}{\lambda} + w^2\right)\cdot \text{erfc}\left(\frac{x}{2\sqrt{a\tau}} + w\right) \tag{5.4}$$

where

$$w = \beta\sqrt{a\tau}/\lambda \tag{5.5}$$

Thermal Stresses and Temperature Control of Mass Concrete. DOI: http://dx.doi.org/10.1016/B978-0-12-407723-2.00005-1

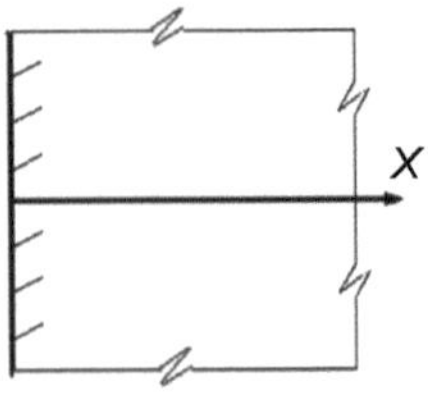

Figure 5.1 Semi-infinite solid.

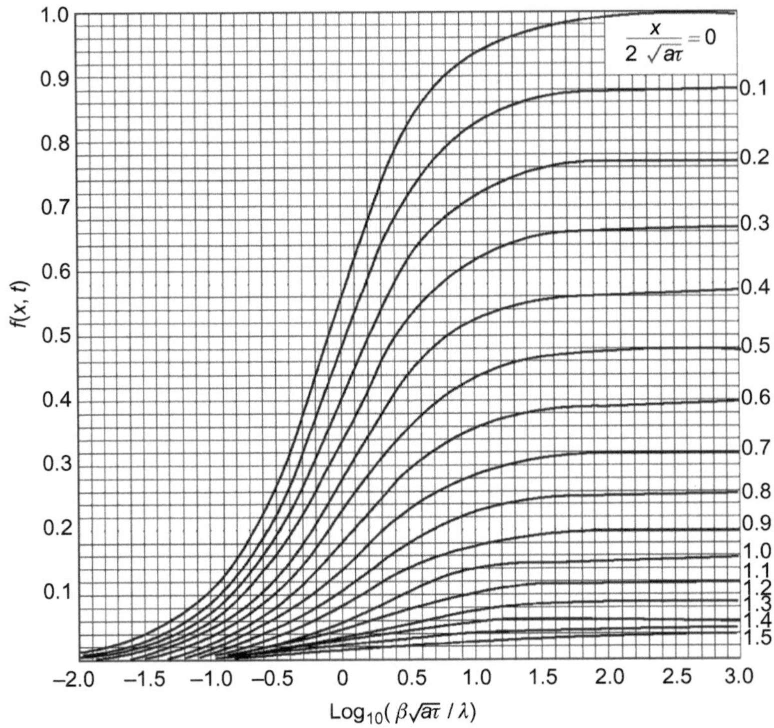

Figure 5.2 Cooling function $f(x,\tau)$ of semi-infinite solid with third kind of boundary condition.

erfc $x = 1 - \text{erf } x = 1 - P(x)$, $P(x)$ is the probability function, $f(x,\tau)$ is shown in Figure 5.2.

If the air temperature varies with time, the concrete temperature is given by

$$T(x,\tau) = T_0 + (T_a - T_0)f(x,\tau) + \sum \Delta T_a(\tau_i)f(x,\tau - \tau_i) \tag{5.6}$$

Let $x = 0$ in Eq. (5.4), we get the surface temperature

$$T_s = T_0 + (T_a - T_0)G(t) \tag{5.7}$$

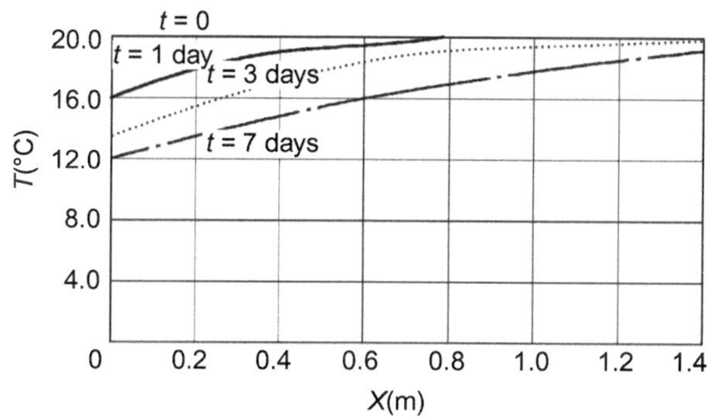

Figure 5.3 Example of semi-infinite solid with third boundary condition (surface insulated by 2 cm foamed polystyrene plate, $T_0 = 20°C$).

$$G(t) = 1 - \exp(w^2) \cdot \operatorname{erfc} w \tag{5.8}$$

where

T_s—the surface temperature of concrete lift
T_0—the initial temperature of concrete
T_a—the air temperature
T—time.

It is more convenient to compute $G(t)$ by the following approximate formula given by the writer:

$$G(t) = 1 - \exp(-0.85 w^{0.76}) \tag{5.9}$$

Example Semi-infinite concrete solid, conductivity $\lambda = 10.0$ kJ/(m h °C), diffusivity $a = 0.10$ m^2/day, the concrete surface is insulated by 2 cm foamed polystyrene plate with conductivity $\lambda_1 = 0.14$ kJ/(m h °C) and surface conductance $\beta_0 = 70$ kJ/(m^2 h °C), the initial temperature of the solid is $T_0 = 20°C$, and the air temperature $T_a = 0°C$. The results of computation are shown in Figure 5.3.

5.2 Cooling of a Slab with First Kind of Boundary Condition

Consider a slab of thickness L whose surfaces are kept at zero temperatures, as shown in Figure 5.4. The problem to be solved is as follows:

Equation of heat conduction

$$\frac{\partial T}{\partial \tau} = a \frac{\partial^2 T}{\partial x^2} \tag{5.10}$$

Initial condition

$$\tau = 0, 0 \le x \le L, T(x, 0) = T_0 \tag{5.11}$$

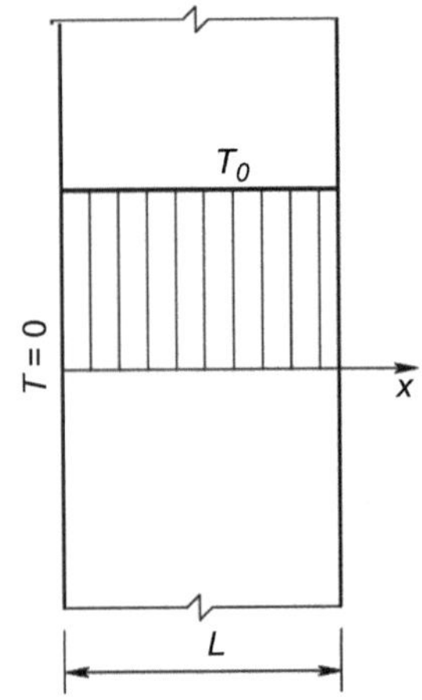

Figure 5.4 A slab with uniform initial temperature.

Boundary condition

$$\left.\begin{array}{l} t > 0, x = 0, T(0, \tau) = 0 \\ \tau > 0, x = L, T(L, \tau) = 0 \end{array}\right\} \tag{5.12}$$

The solution is

$$T(x, \tau) = \frac{4T_0}{\pi} \sum_{n-1}^{\infty} \frac{1}{2n-1} \sin \frac{(2n-1)\pi x}{L} \mathrm{e}^{-(2n-1)^2 \pi^2 a\tau/L^2} \tag{5.13}$$

The mean temperature is

$$T_{\mathrm{m}}(\tau) = \frac{8T_0}{\pi^2} \sum_{n-1}^{\infty} \frac{1}{(2n-1)^2} \mathrm{e}^{-(2n-1)^2 \pi^2 a\tau/L^2} \tag{5.14}$$

Let

$$F_0 = a\tau/L^2 \tag{5.15}$$

which is a dimensionless number and is called the Fourier number.

From Eqs. (5.13) and (5.14), an important conclusion is derived: for two slabs with different diffusivities a_1, a_2, different thicknesses L_1, L_2, and the same initial temperature T_0, if they have the same Fourier number F_0, namely

$$\frac{a_1 \tau_1}{L_1^2} = F_0 = \frac{a_2 \tau_2}{L_2^2} \tag{5.16}$$

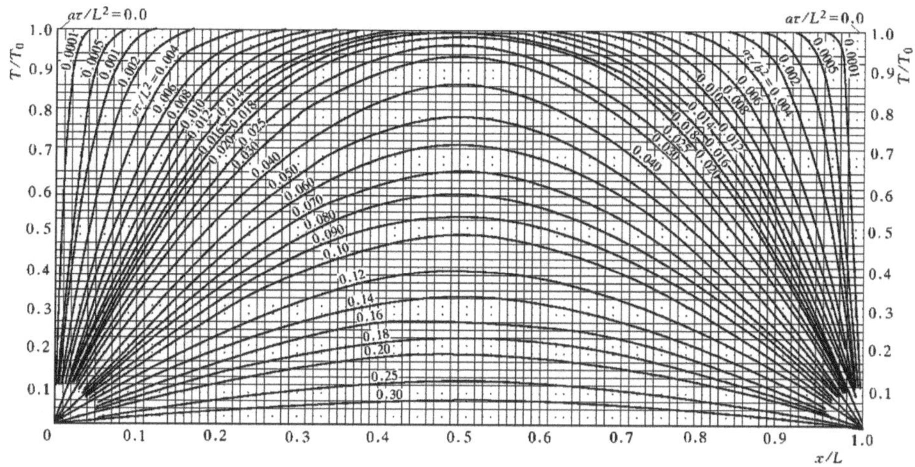

Figure 5.5 Distribution of temperature in a slab with uniform initial temperature T_0 and zero surface temperature.

$$T_{m1}(\tau_1) = T_{m2}(\tau_2) \tag{5.17}$$

and

$$T_1(x_1, \tau_1) = T_2(x_2, \tau_2)$$

when

$$x_1/L_1 = x_2/L_2 \tag{5.18}$$

The distribution of temperature in the slab is shown in Figure 5.5. In order to give an obvious idea to the reader, the distribution of temperature in a slab of thickness 10 m is shown in Figure 5.6. The mean temperature of the slab is shown in Figure 5.7.

If the external temperature is T_c, the initial temperature is T_B, then

$$T_0 = T_B - T_c \tag{5.19}$$

and the mean temperature T_m of the slab is given by

$$T_m = T_c + HT_0 = T_c + H(T_B - T_c) \tag{5.20}$$

The coefficient H is shown in Figure 5.7.

Example Concrete slab, $L = 40$ m, $a = 0.10$ m^2/day, the initial temperature $T_B = 40°C$, and external temperature $T_c = 25°C$, compute the mean temperature T_m after 40-day cooling.

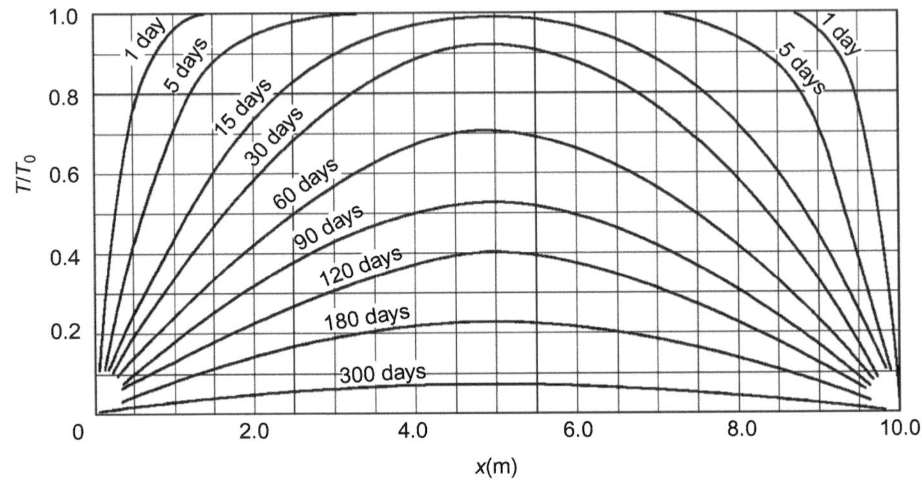

Figure 5.6 Distribution of temperature in a 10 m slab with uniform initial temperature T_0 and zero surface temperature ($a = 0.10\ m^2/day$).

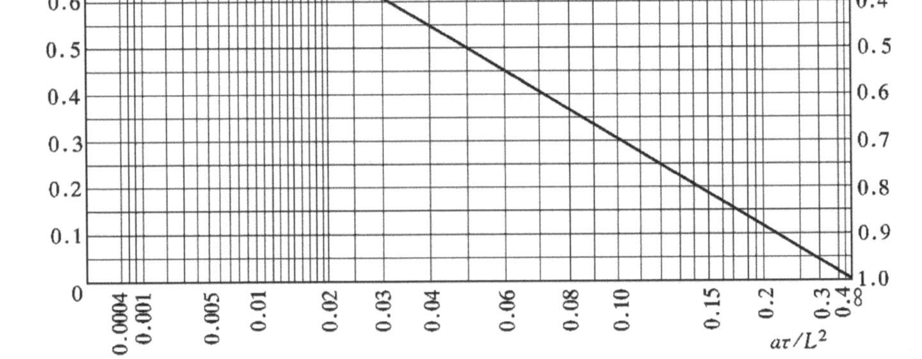

Figure 5.7 The mean temperature T_m of a slab with uniform initial temperature T_0 and zero surface temperature.

$$F_0 = \frac{a\tau}{L^2} = \frac{0.10 \times 40}{40^2} = 0.0025$$

From Figure 5.7, $H = 0.885$, by Eq. (5.20), the mean temperature is

$$T_m = 25 + 0.885(40 - 25) = 25 + 13.3 = 38.3°\text{C}$$

5.3 Cooling of a Slab with Third Kind of Boundary Condition

As shown in Figure 5.8, consider an infinite slab with thickness $L = 2l$, uniform initial temperature T_0, conductivity λ, diffusivity a, surface conductance β, and air temperature $T_a = 0$.

Because of symmetry, take one half to analysis:

Equation of heat conduction

$$\frac{\partial T}{\partial \tau} = a\frac{\partial^2 T}{\partial x^2} \tag{5.21}$$

Initial condition

$$\tau = 0, \quad T(x, 0) = T_0 \tag{5.22}$$

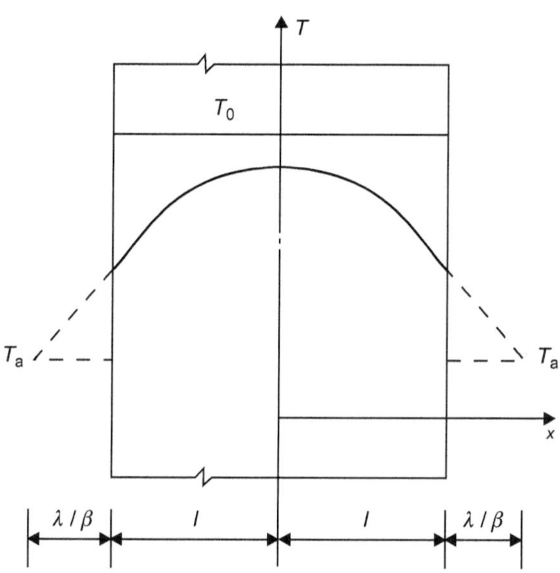

Figure 5.8 Cooling of a slab, third kind of boundary condition.

Boundary condition

$$t > 0,\ \ x = 0,\ \ \frac{\partial T}{\partial x} = 0 \tag{5.23}$$

$$t > 0,\ \ x = l,\ \ -\lambda\frac{\partial T}{\partial x} = \beta T(l, x) \tag{5.24}$$

The solution is

$$T(x, \tau) = T_0 \sum_{n=1}^{\infty} A_n \cos\frac{\mu_n x}{l} \mathrm{e}^{-\mu^2 n a \tau / l^2} \tag{5.25}$$

where

$$A_n = 2 \sin \mu_n / (\mu_n + \sin \mu_n \cos \mu_n) \tag{5.26}$$

The mean temperature is

$$T_{\mathrm{m}}(\tau) = T_0 \sum_{n=1}^{\infty} B_n\, \mathrm{e}^{-\mu_n^2 a \tau / l^2} = T_0 \sum_{n=1}^{\infty} B_n\, \mathrm{e}^{-s_n \tau} = T_0 \eta(\tau) \tag{5.27}$$

where

$$B_n = \frac{A_n \sin \mu_n}{\mu_n} = \frac{2B_i^2}{\mu_n^2(B_i^2 + B_i + \mu_n^2)} \tag{5.28}$$

$$\mathrm{s}_n = \mu_n^2 a / L^2 \tag{5.29}$$

$$B_i = \beta l / \lambda \tag{5.30}$$

where

B_i—the Biot number
μ_n—the root of the characteristic equation:

$$\cot \mu_n - \mu_n / B_i = 0 \tag{5.31}$$

Putting $x = l$ in Eq. (5.25), the surface temperature is

$$T(l, \tau) = T_0 \sum_{n=1}^{\infty} A_n \cos \mu_n\, \mathrm{e}^{-\mu^2 n a \tau / l^2} \tag{5.32}$$

The ratio T_{m}/T_0 in Eq. (5.27) is shown in Figure 5.9.

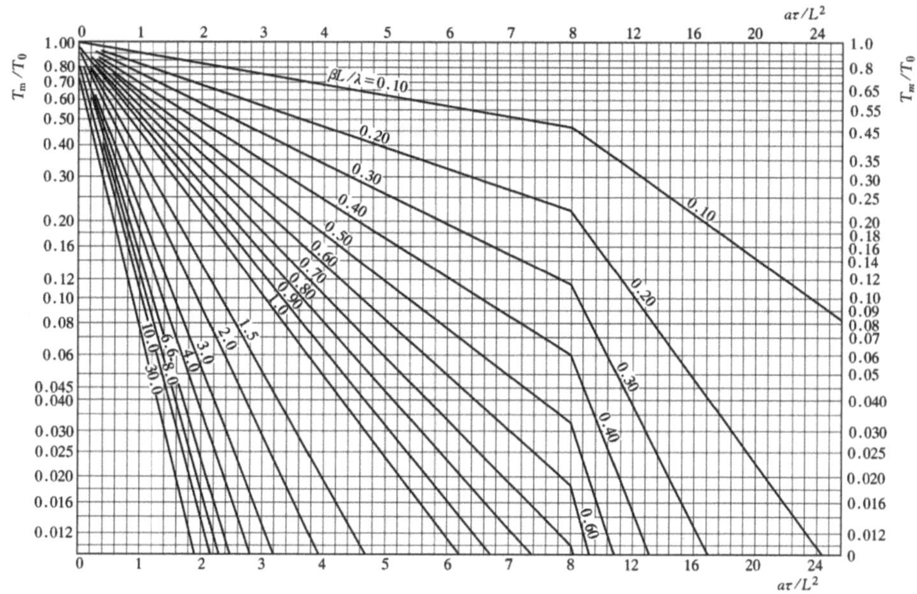

Figure 5.9 The mean temperature of a slab, initial temperature T_0, zero air temperature, third kind of boundary condition.

5.4 Temperature in a Concrete Slab with Harmonic Surface Temperature

5.4.1 Concrete Slab with Zero Initial Temperature and Harmonic Surface Temperature

As the surface temperature of mass concrete structure generally can be expressed by cosine function of time, the temperature field of a slab with zero initial temperature and surface temperature varying in cosine law of time as shown in Figure 5.10 is analyzed in the following. The problem is

$$\left.\begin{array}{ll} \text{Equation of heat conduction} & \dfrac{\partial T}{\partial \tau} = a\dfrac{\partial^2 T}{\partial x^2} \\ \text{Initial condition} & \tau = 0,\ T(x,0) = 0 \\ \text{Boundary condition} & x = 0 \text{ and } x = L,\ T = -A_0 \cos\dfrac{2\pi}{P}(\tau_0 + \tau) \end{array}\right\} \tag{5.33}$$

where

A_0—the amplitude of variation of surface temperature

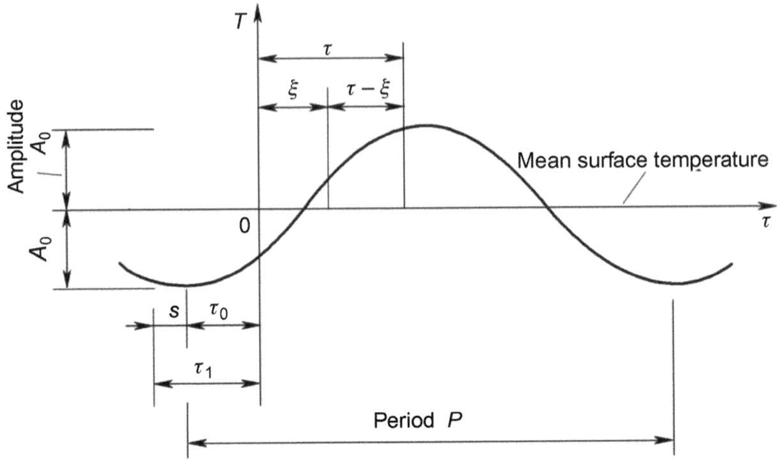

Figure 5.10 The variation of surface temperature with time.

P—the period of variation of surface temperature
τ_0—the interval between the time of minimum surface temperature and the placing of concrete as shown in Figure 5.10.

The solution of Eq. (5.33) is

$$T(x,\tau) = -A_0\cos\frac{2\pi}{P}(\tau_0-\tau) + \frac{4A_0}{\pi}\cos\frac{2\pi\tau_0}{P}\sum_{n=1}^{\infty}\frac{e^{-(2n-1)^2\pi^2 a\tau/L^2}}{(2n-1)}\sin\frac{(2n-1)\pi x}{L}$$
$$-\frac{8A_0}{P}\int_0^{\tau}\sin\frac{2\pi}{P}(\tau_0-\xi)\sum_{n=1}^{\infty}\sin\frac{e^{-(2n-1)^2\pi^2 a(\tau-\xi)/L^2}}{(2n-1)}\sin\frac{(2n-1)\pi x}{L}d\xi \tag{5.34}$$

The mean temperature is

$$T_m = -A'A_0\cos\frac{2\pi}{P}(\tau_0+\tau) - B'A_0\sin\frac{2\pi}{P}(\tau_0+\tau)$$
$$+C'A_0\cos\frac{2\pi}{P}\tau_0 + D'A_0\sin\frac{2\pi}{P}\tau_0 \tag{5.35}$$

or

$$T_m = -\rho A_0\cos\frac{2\pi}{P}(\tau_0+\tau-\gamma) + C'A_0\cos\frac{2\pi}{P}\tau_0 + D'A_0\sin\frac{2\pi}{P}\tau_0 \tag{5.35a}$$

where

$$A' = 1 - \frac{32}{\pi^4} L_c^4 \sum_{n=1}^{\infty} \frac{1}{(2n-1)^2[(2n-1)^4 + 4L_c^4/\pi^2]}$$

$$B' = \frac{16}{\pi^3} L_c^4 \sum_{n=1}^{\infty} \frac{1}{(2n-1)^4 + 4L_c^4/\pi^2}$$

$$C' = \frac{8}{\pi^2} L_c^4 \sum_{n=1}^{\infty} \frac{(2n-1)^2}{(2n-1)^4 + 4L_c^4/\pi^2} e^{-(2n-1)^2/\pi^2 F_0}$$

$$D' = \frac{16}{\pi^3} L_c^4 \sum_{n=1}^{\infty} \frac{1}{(2n-1)^4 + 4L_c^4/\pi^2} e^{-(2n-1)^2/\pi^2 F_0}$$

$$\gamma = \frac{P}{2\pi} \tan^{-1}\left(\frac{B'}{A'}\right), \rho = \sqrt{A'^2 + B'^2}$$

$$L_c = \frac{L}{\sqrt{aP}}, \quad F_0 = \frac{a\tau}{L^2} \tag{5.36}$$

C' and D' represent the influence of initial temperature. When $\tau \to \infty$, $C' = D' = 0$, then the mean temperature becomes

$$T'_m = -\rho A_0 \cos\frac{2\pi}{P}(\tau_0 + \tau - \gamma) \tag{5.37}$$

which is a quasi-steady temperature varying periodically with time, the amplitude and phase of variation is, respectively, ρA_0 and γ. The coefficients A', B', C', D', ρ, and γ are shown in Figures 5.11–5.14.

It is clear from Figure 5.14 that when $L/\sqrt{aP} \geq 1.6$, the lag to the mean temperature will be

$$\gamma = \frac{1}{8}P \tag{5.38}$$

Taking $a = 3.0$ m^2/month, for annual variation, $\sqrt{aP} = \sqrt{3 \times 12} = 6$, so the lag of the mean temperature is

$$\gamma = \frac{1}{8} \times 12 = 1.5 \text{ month}$$

when $L \geq 10$ m, because $L/\sqrt{aP} = 10/\sqrt{3.0 \times 12} = 1.66$.

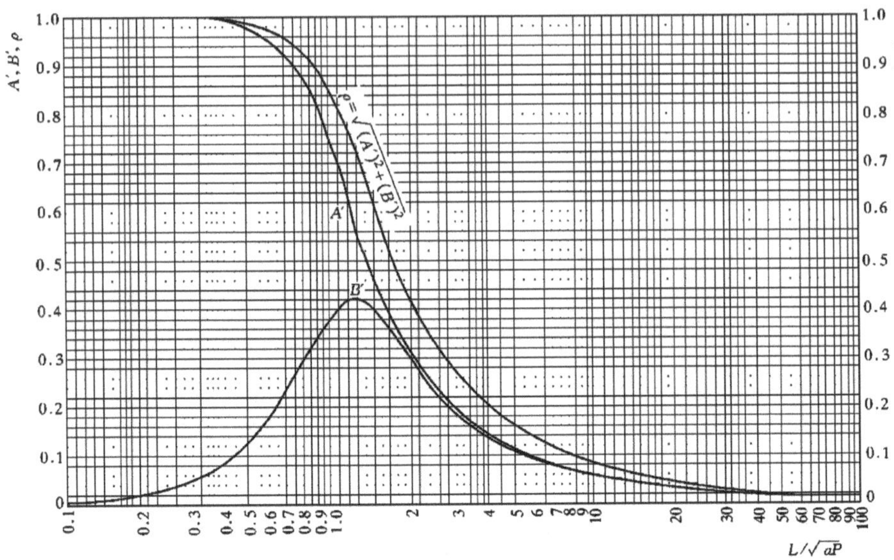

Figure 5.11 Coefficients A', B', ρ.

5.4.2 Concrete Slab, Initial Temperature T_0, Harmonic Surface Temperature

The concrete slab with uniform initial temperature T_0 and harmonic surface temperature is analyzed. Let τ_1 be the calendar time, $\tau_1 = s$ being the time of minimum external temperature, generally $s = 0.5$ month, from Figure 5.15,

$$\tau_0 = \tau_1 - s \tag{5.39}$$

The problem to be solved is

$$\left.\begin{array}{ll}\text{Equation of heat conduction} & \dfrac{\partial T}{\partial \tau} = a\dfrac{\partial^2 T}{\partial x^2} \\ \text{Initial condition} & \tau = 0,\ T(x,0) = T_0 \\ \text{Boundary condition} & x = 0 \text{ and } L,\ T = T_{am} - A_0 \cos\dfrac{2\pi}{P}(\tau_0 + \tau)\end{array}\right\} \tag{5.40}$$

Superposition of Eqs. (5.13) and (5.34) will give the solution of Eq. (5.40). The mean temperature is

$$\begin{aligned}T_m = {} & T_{am} + (T_0 - T_{am})H \\ & + A_0[-A' \cos \omega(\tau_0 + \tau) - B' \sin \omega(\tau_0 + \tau) \\ & + C' \cos \omega\tau_0 + D' \sin \omega\tau_0]\end{aligned} \tag{5.41}$$

where $\omega = 2\pi/P$.

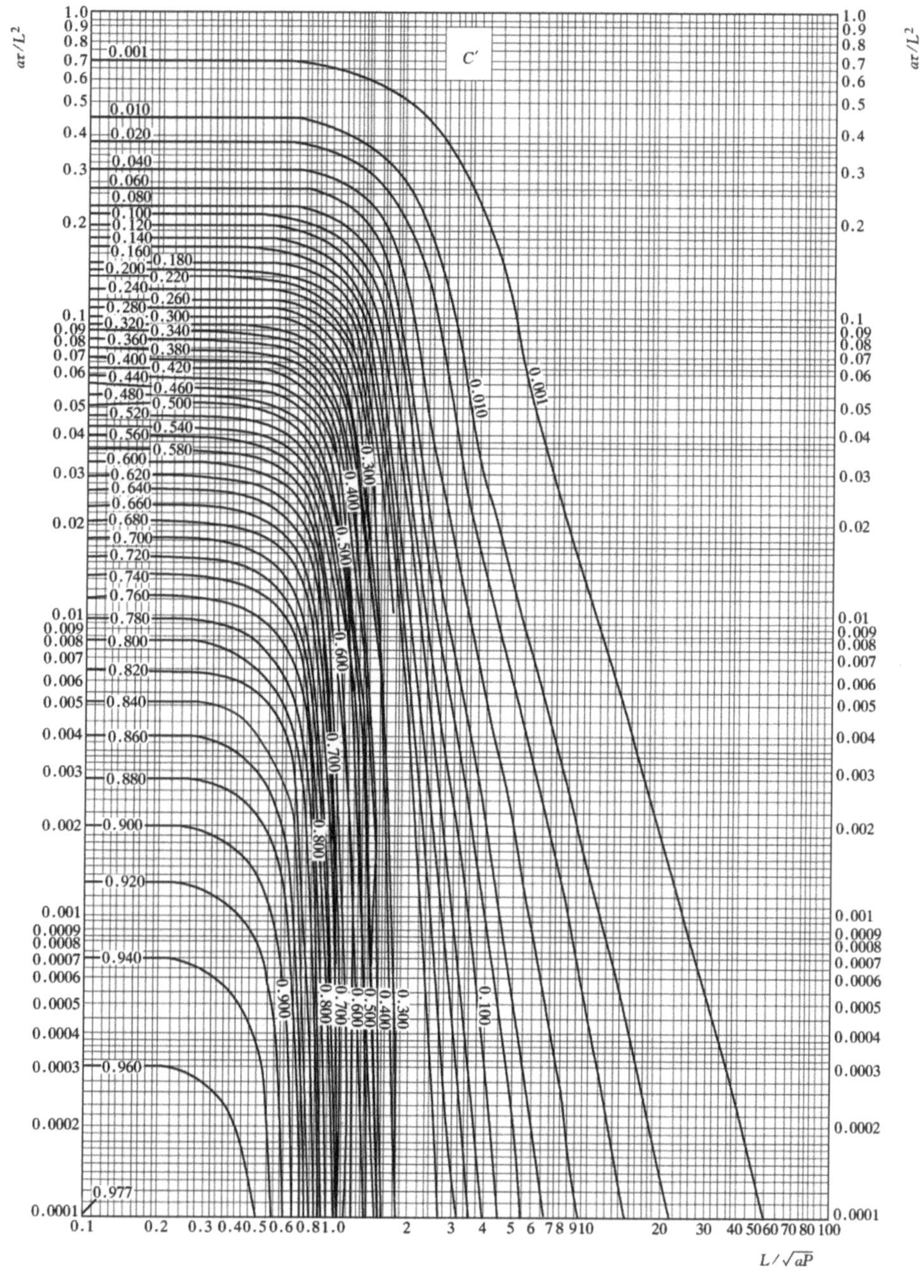

Figure 5.12 Coefficient C'.

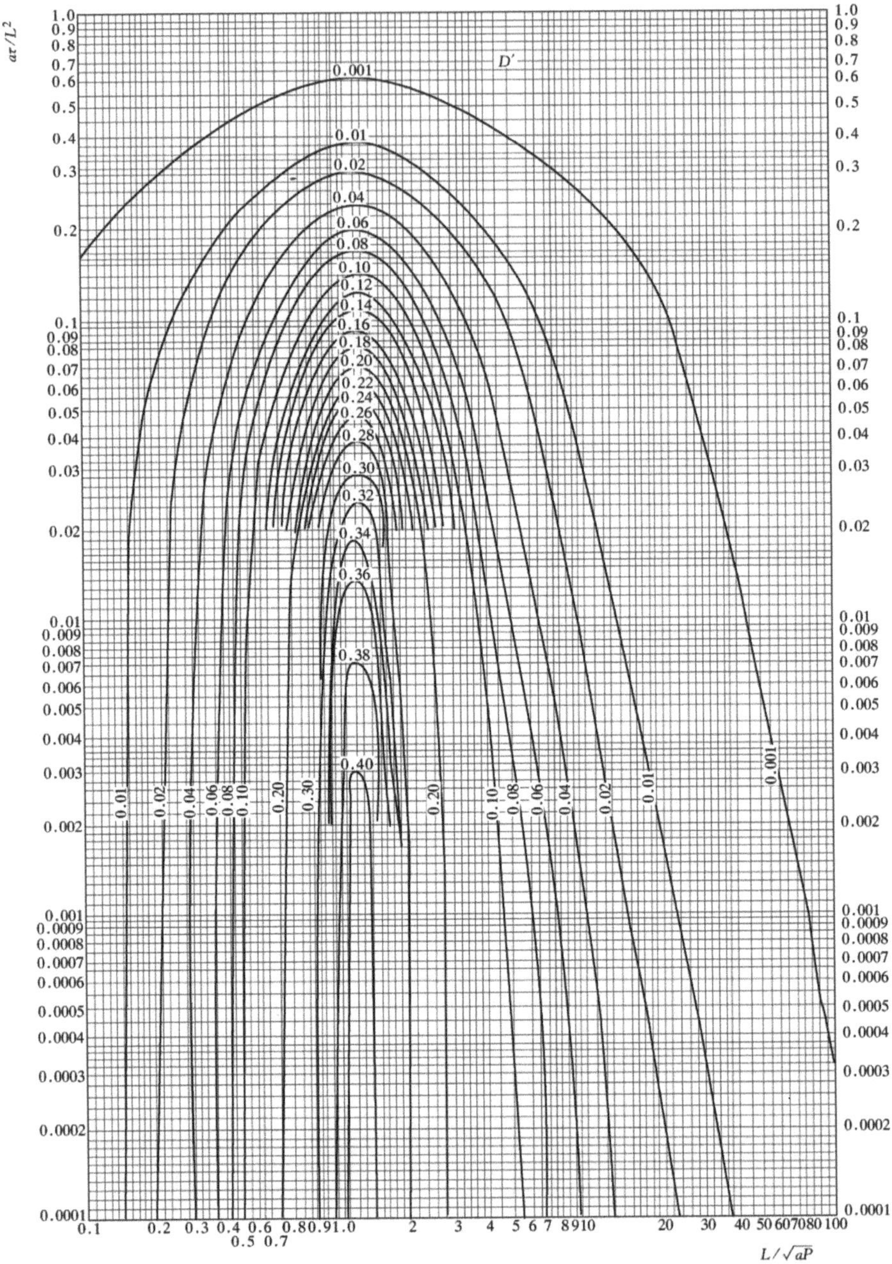

Figure 5.13 Coefficient D'.

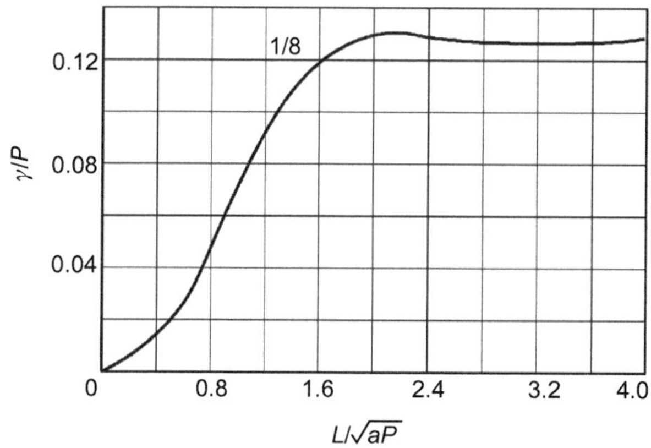

Figure 5.14 The lag of mean temperature.

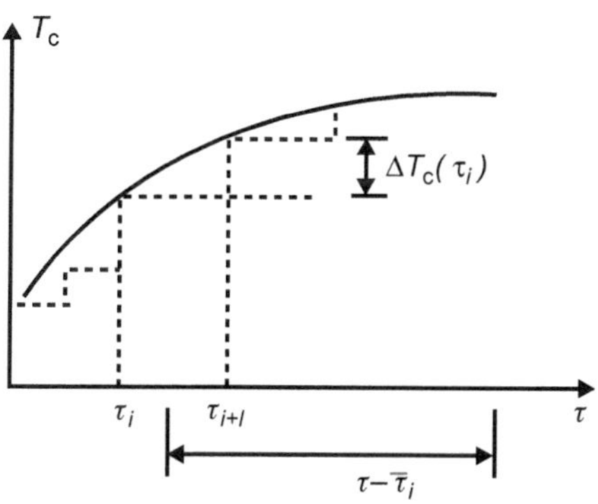

Figure 5.15 Variation of external temperature T_c.

Example Concrete slab, thickness $L = 30$ m, diffusivity temperature $a = 0.0035$ m^2/day $= 2.56$ m^2/month, the concrete is placed in the middle of May ($\tau_1 = 4.5$ month), placing temperature $T_p = 20°C$, the temperature rise due to hydration heat $T_r = 18°C$, the annual mean air temperature $T_{am} = 12°C$, the amplitude of variation of air temperature $A_0 = 13°C$. The time of minimum air temperature $s = 0.5$ month. It is required to grout the dam joints in the middle of March of the second year ($\tau_1 = 12 + 2.5 = 14.5$ month) with mean temperature of dam 12°C. Try to check whether the requirements can be satisfied by means of natural cooling.

From Eq. (5.39),

$$\tau_0 = \tau_1 - s = 4.5 - 0.5 = 4.0 \text{ month}$$

The time of natural cooling is

$$\tau = 14.5 - 4.5 = 10.0 \text{ month}$$

The initial temperature of concrete is

$$\tau_0 = 20.0 + 18.0 = 38.0°\text{C}$$

The period of variation $P = 12$ month

$$L/\sqrt{aP} = 30/\sqrt{2.56 \times 12} = 5.41$$
$$a\tau/L^2 = 2.56 \times 10/30^2 = 0.02844$$

From Figures 5.7 and 5.11–5.13, the following constants are obtained

$$A' = 0.100,\ \ B' = 0.105,\ \ C' = 0.005,\ \ D' = 0.035,\ \ H = 0.615$$

By formula (5.41), the mean temperature of concrete in March of the next year is

$$\begin{aligned} T_m &= 12.0 + 0.615(38 - 12) - 0.100 \times 13 \cos\frac{2\pi}{12}(4.0 + 10.0) \\ &\quad -0.105 \times 13 \times \sin\frac{2\pi}{12}(4.0 + 10.0) + 0.005 \times 13 \times \cos\frac{2\pi}{12} \times 4.0 \\ &\quad +0.035 \times 13 \times \sin\frac{2\pi}{12} \times 4.0 \\ &= 12.0 + 16.0 - 0.65 - 1.18 - 0.03 + 0.39 = 26.5 > 12 \end{aligned}$$

The results of computation show that the requirement cannot be satisfied by natural cooling.

5.5 Temperature in a Slab with Arbitrary External Temperature

The temperature in a slab with arbitrary external temperature can be derived from the solution of a slab with constant external temperature in Section 5.2 (first kind of boundary condition) or 5.3 (third kind of boundary condition) by the principle of superposition.

From Eq. (5.13), the temperature in a slab with initial temperature T_0, external temperature T_c = constant, and the first kind of boundary condition is

$$T(x,\tau) = T_c + (T_0 - T_c)h(x,\tau) \tag{5.42}$$

where

$$h(x,\tau) = \frac{4}{\pi}\sum_{n=1}^{l}\frac{1}{2n-1}\sin\frac{(2n-1)\pi x}{L}e^{-(2n-1)^2\pi^2 a\tau/L^2} \tag{5.43}$$

Let

$$f(x,\tau) = 1 - h(x,\tau) \tag{5.44}$$

and substitute it into Eq. (5.42), we get

$$T(x,\tau) = T_0 + (T_c - T_0)f(x,\tau) \tag{5.45}$$

The mean temperature is

$$T_m(\tau) = T_0 + (T_c - T_0)F(\tau) \tag{5.46}$$

in which $F(\tau) = 1 - H(\tau)$ and $H(\tau)$ is given by Eq. (5.14). $F(\tau)$ is shown in Figure 5.7.

As shown in Figure 5.15, the external temperature T_c may be transformed into a series of increments $\Delta T_c(\tau_i)$ and the mean temperature of the slab with arbitrary external temperature is given by

$$T_m(\tau) = T_0 + (T_c - T_0)F(\tau) + \sum \Delta T_c(\tau_i)F(\tau - \overline{\tau}_i) \tag{5.47}$$

where

$$\Delta T_c(\tau_i) = T_c(\tau_{i+1}) - T_c(\tau_i), \overline{\tau}_i = \frac{\tau_i + \tau_{i+1}}{2} \tag{5.48}$$

The process of computation is given in Table 5.1.

If the surface temperature of the two sides of the slab is different, e.g., on one side is air temperature $T_a(\tau)$ and on the other side is water temperature $T_w(\tau)$, Eq. (5.47) can still be used to compute the mean temperature of the slab, but it is necessary to take the external temperature $T_c(\tau)$ as the following:

$$T_c(\tau) = \frac{1}{2}[T_a(\tau) + T_w(\tau)] \tag{5.49}$$

Example Concrete slab, thickness $L = 15$ m, diffusivity $a = 3.0$ m^2/month, placed in June, placing temperature $T_p = 25°C$, temperature rise due to hydration heat $T_r = 10°C$, initial temperature $T_0 = T_p + T_r = 25 + 10 = 35°C$, try to compute the mean temperature of slab under natural cooling by Eq. (5.47). When $\tau = 0$, the external temperature $T_{c0} = 26.0°C$, $T_{c0} - T_0 = 26.0 - 35.0 = -9.0°C$, the first increment of external temperature$\Delta T_{c1} = T_{c1} - T_{c0} = 27.3 - 26.0 = 1.3°C, ...$ The process and results of computation are given in Table 5.2.

Table 5.1 Calculation of Mean Temperature of a Slab with Arbitrary External Temperature

i	T_c	ΔT_c	$\frac{a\tau}{L^2}$	$F(\tau)$	$\frac{a(\tau-\overline{\tau}_i)}{L^2}$	$F(\tau-\overline{\tau}_i)$	$\Delta T_c(\tau_i)F(\tau-\overline{\tau}_i)$				
							τ_0	τ_1	τ_2	τ_3	τ_4
0	T_{c0}	$T_{c0}-T_0$	0	F_0		0	$(T_c-T_0)F_0$	$(T_c-T_0)F_1$	$(T_c-T_0)F_2$	$(T_c-T_0)F_3$	$(T_c-T_0)F_4$
1	T_{c0}	ΔT_{c1}	$a\Delta\tau/L^2$	F_1	$0.5a\ \Delta\tau/L^2$	$F_{0.5}$		$\Delta T_{c1}F_{0.5}$	$\Delta T_{c1}F_{1.5}$	$\Delta T_{c1}F_{2.5}$	$\Delta T_{c1}F_{3.5}$
2	T_{c2}	ΔT_{c2}	$2a\Delta\tau/L^2$	F_2	$1.5a\ \Delta\tau/L^2$	$F_{1.5}$			$\Delta T_{c2}F_{0.5}$	$\Delta T_{c2}F_{1.5}$	$\Delta T_{c2}F_{2.5}$
3	T_{c3}	ΔT_{c3}	$3a\Delta\tau/L^2$	F_3	$2.5a\ \Delta\tau/L^2$	$F_{2.5}$				$\Delta T_{c3}F_{0.5}$	$\Delta T_{c3}F_{1.5}$
4	T_{c4}	ΔT_{c4}	$4a\Delta\tau/L^2$	F_4	$3.5a\ \Delta\tau/L^2$	$F_{3.5}$				—	$\Delta T_{c4}F_{0.5}$
…	…	…	…	…	…	…				—	
Σ											
$T_m(\tau)=T_0+\Sigma$											

Table 5.2 Example of Mean Temperature in Slab with Arbitrary External Temperature T_c

Month	T_c (°C)	ΔT_c (°C)	τ (month)	$\frac{a\tau}{L^2}$	F (τ)	$\frac{a(\tau-\overline{\tau}_i)}{L^2}$	$F(\tau-\overline{\tau}_i)$	$\Delta T_c(\tau_i)F(\tau-\tau_i)$ For Different Months: 6	7	8	9
6	26.0	−9.0	0	0	0			0	−2.34	−3.33	−4.05
7	27.3	+1.3	1	0.0133	0.26	0.0067	0.20		0.26	0.42	0.55
8	26.3	−1.0	2	0.0267	0.37	0.0200	0.32			−0.20	−0.32
9	22.6	−3.7	3	0.0400	0.45	0.0333	0.42				−0.68
Σ								0	−2.08	−3.11	−4.50
$T_m = 35.0 + \Sigma$								35.0	32.92	31.89	30.5

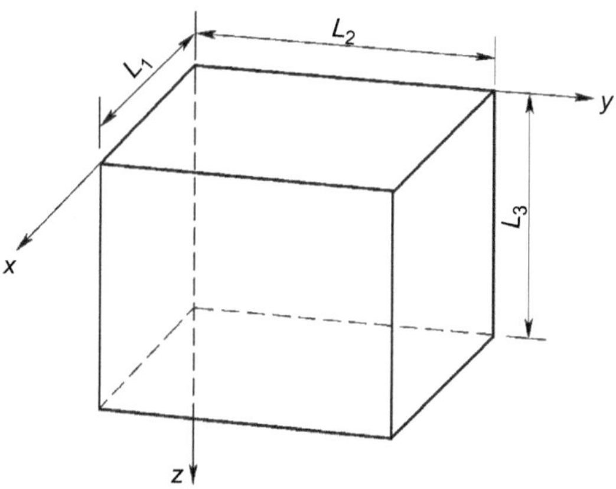

Figure 5.16 A rectangular parallelepiped.

5.6 Cooling of Mass Concrete in Two and Three Directions, Theorem of Product

The problem of cooling of a rectangular parallelepiped as shown in Figure 5.16 is as follows:

$$\left.\begin{array}{ll}\text{Equation of heat conduction} & \dfrac{\partial T}{\partial \tau} = a\left(\dfrac{\partial^2 T}{\partial x^2} + \dfrac{\partial^2 T}{\partial y^2} + \dfrac{\partial^2 T}{\partial z^2}\right) \\ \text{Initial condition} & T(x,y,z,0) = T_0 = \text{constant} \\ \text{Boundary condition} & T(0,y,z,\tau) = T(L_1,y,z,\tau) = T_c \\ & T(x,0,z,\tau) = T(x,L_2,z,\tau) = T_c \\ & T(x,y,0,\tau) = T(x,y,L_3,\tau) = T_c \end{array}\right\} \quad (5.50)$$

It has been proved that the solution $T(x,y,z,\tau)$ of Eq. (5.50) is the product of the solutions of three one-variable problems and may be expressed in the following:

$$\frac{T(x,y,z,\tau)-T_c}{T_0-T_c}=\frac{T(x,\tau)-T_c}{T_0-T_c}\cdot\frac{T(y,\tau)-T_c}{T_0-T_c}\cdot\frac{T(z,\tau)-T_c}{T_0-T_c} \tag{5.51}$$

where $T(x,\tau)$, $T(y,\tau)$, and $T(z,\tau)$ are respectively the one-dimensional solutions in x, y, z directions. For example, $T(x,\tau)$ is the solution of the following problem:

$$\left.\begin{array}{ll} \text{Equation of heat conduction} & \dfrac{\partial T(x,\tau)}{\partial\tau}=a\dfrac{\partial^2 T(x,\tau)}{\partial x^2} \\ \text{Initial condition} & T(x,0)=T_0 \\ \text{Boundary condition} & T(0,\tau)=T(L_1,\tau)=T_c \end{array}\right\} \tag{5.52}$$

Equation (5.51) is called the theorem of product. The mean temperature T_m of the rectangular parallelepiped may be computed as follows:

$$\frac{T_m-T_c}{T_0-T_c}=\frac{T_{mx}-T_c}{T_0-T_c}\times\frac{T_{my}-T_c}{T_0-T_c}\times\frac{T_{mz}-T_c}{T_0-T_c}=H_xH_yH_z \tag{5.53}$$

or

$$T_m=T_c+H(T_0-T_c) \tag{5.54}$$

in which

$$H=H_xH_yH_z \tag{5.55}$$

where

T_m—mean temperature of parallelepiped
T_{mx}, T_{my}, T_{mz}—mean temperature of one-variable problem in x, y, z directions, respectively
H—residual ratio of three-dimensional problem
H_x, H_y, H_z—one-dimensional residual ratio in x, y, z directions, respectively.

From Eq. (5.55), it is clear that the residual ratio H of a three-dimensional problem is equal to the product of three residual ratios H_x, H_y, H_z of one-dimensional problems in x, y, z directions.

For a two-dimensional problem, the temperature $T(x,y,\tau)$ may be expressed as follows:

$$\frac{T(x,y,\tau)-T_c}{T_0-T_c}=\frac{T(x,\tau)-T_c}{T_0-T_c}\times\frac{T(y,\tau)-T_c}{T_0-T_c} \tag{5.56}$$

or

$$T(x, y, \tau) = T_c + h(x, \tau)h(y, \tau)(T_0 - T_c) \tag{5.57}$$

where

$$h(x, \tau) = \frac{T(x, \tau) - T_c}{T_0 - T_c}, \quad h(y, \tau) = \frac{T(y, \tau) - T_c}{T_0 - T_c} \tag{5.58}$$

The mean temperature T_m is given by

$$T_m = T_c + H_x H_y (T_0 - T_c) \tag{5.59}$$

6 Stress–Strain Relation and Analysis of Viscoelastic Stress of Mass Concrete

6.1 Stress–Strain Relation of Concrete

Under the action of stress in one direction, the total deformation $\varepsilon(t)$ of concrete at time t may be expressed in the following [5, 17-19, 37-40, 79, 96-98]:

$$\varepsilon(t) = \varepsilon^{e}(t) + \varepsilon^{c}(t) + \varepsilon^{s}(t) + \varepsilon^{T}(t) + \varepsilon^{g}(t) \tag{6.1}$$

where

$\varepsilon^{e}(t)$—the instantaneous elastic strain due to stress
$\varepsilon^{c}(t)$—the creep strain due to stress
$\varepsilon^{s}(t)$—the shrinkage strain due to loss of water in concrete
$\varepsilon^{T}(t)$—the thermal strain due to change of temperature
$\varepsilon^{g}(t)$—the autogenous strain.

In the right part of Eq. (6.1), the first two terms, $\varepsilon^{e}(t)$ and $\varepsilon^{c}(t)$ are caused by the stress and the last three terms, $\varepsilon^{s}(t)$, $\varepsilon^{T}(t)$, and $\varepsilon^{g}(t)$, are independent of stress. The computing method of the elastic strain $\varepsilon^{e}(t)$ and the creep strain $\varepsilon^{c}(t)$ due to stress are described in this chapter.

6.1.1 *Strain of Concrete due to Constant Stress*

Assuming that the one-dimensional stress $\sigma(\tau)$ is applied to concrete at age τ, the instantaneous elastic strain $\varepsilon^{e}(\tau)$ produced at the instant of application of stress is

$$\varepsilon^{e}(\tau) = \frac{\sigma(\tau)}{E(\tau)} \tag{6.2}$$

where

$E(\tau)$—the instantaneous modulus of elasticity of concrete at age τ.

Thermal Stresses and Temperature Control of Mass Concrete. DOI: http://dx.doi.org/10.1016/B978-0-12-407723-2.00006-3

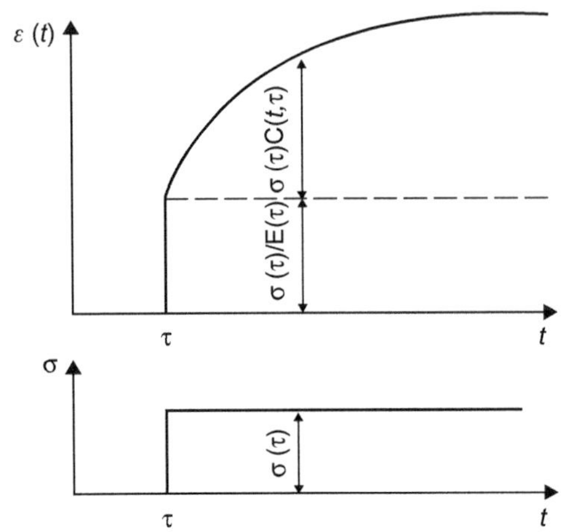

Figure 6.1 The strain of concrete under the action of constant stress.

If the stress does not vary with time t, then the strain will increase slowly with time, as shown in Figure 6.1; this part of strain is called creep of concrete. Experimental results show that when the tensile stress does not exceed 70% of tensile strength and the compressive stress does not exceed 50% of compressive strength, the principle of superposition is valid, and the creep strain $\varepsilon^c(t)$ may be computed by [5]

$$\varepsilon^c(t) = \sigma(\tau)C(t,\tau) \tag{6.3}$$

where $C(t,\tau)$—the creep produced under the action of unit stress and is called the unit creep of concrete whose dimension is MPa^{-1}.

Thus, if the stress is applied at age τ, the total strain at time t is the sum of the instantaneous elastic strain $\varepsilon^e(t)$ and the creep strain $\varepsilon^c(t)$, namely

$$\varepsilon(t) = \varepsilon^e(\tau) + \varepsilon^c(t) = \frac{\sigma(\tau)}{E(\tau)} + \sigma(\tau)C(t,\tau) = \sigma(\tau)J(t,\tau) \tag{6.4}$$

where

$$J(t,\tau) = \frac{1}{E(\tau)} + C(t,\tau) \tag{6.5}$$

where $J(t,\tau)$—the creep compliance whose dimension is MPa^{-1}.

If unit stress $\sigma = 1$ MPa is applied at $t = \tau_0, \tau_1, \tau_2, \ldots$, a group of strain curves will be obtained, as shown in Figure 6.2.

The creep compliance of concrete $J(t,\tau)$ can also be expressed in the following:

$$J(t,\tau) = \frac{1 + \varphi(t,\tau)}{E(\tau)} \tag{6.6}$$

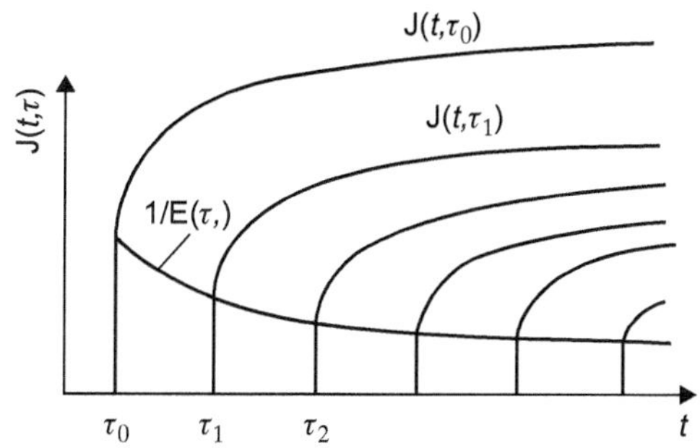

Figure 6.2 The creep compliance of concrete for stress applied at different ages.

where

$$\varphi(t,\tau) = E(\tau)C(t,\tau) = C(t,\tau):[1/E(\tau)] \tag{6.7}$$

where $\phi(t,\tau)$—the coefficient of creep and is equal to the ratio of creep to elastic strain.

The reciprocal of creep compliance is called the sustained modulus of elasticity or the effective modulus of elasticity and is denoted by $E^*(t,\tau)$ and

$$E^*(t,\tau) = \frac{1}{J(t,\tau)} = \frac{E(\tau)}{1+\varphi(t,\tau)} = \frac{E(\tau)}{1+E(\tau)C(t,\tau)} \tag{6.8}$$

6.1.2 Strain of Concrete due to Variable Stress

If the stress $\sigma(\tau_0)$ is applied at $t = \tau_0$ and then the stress $\sigma(t)$ varies with time, the strain at time t is given by

$$\varepsilon(t) - \varepsilon^0(t) = \sigma(\tau_0)J(t,\tau_0) + \int_{\tau_0}^{t} J(t,\tau)\frac{d\sigma}{d\tau}d\tau \tag{6.9}$$

where $\varepsilon^0(t)$—the strain independent of stress.

6.1.3 Modulus of Elasticity and Creep of Concrete

The modulus of elasticity $E(\tau)$ of concrete is dependent on the age τ and can be expressed by one of the following three formulas:

Exponential formula

$$E = E_0(1 - e^{-m\tau}) \tag{6.10}$$

Complex exponential formula

$$E = E_0(1 - e^{-a\tau^b}) \tag{6.11}$$

Hyperbolic formula

$$E = E_0\tau/(n+\tau) \tag{6.12}$$

The exponential formula (6.10) is suitable for mathematical manipulation but does not agree well with test results. The complex exponential formula (6.11) and the hyperbolic formula (6.12), all proposed by the writer, agree well with test results and are now applied extensively in practice.

The creep of mass concrete may be expressed by the following formula:

$$C(t,\tau) = (f_1 + g_1\tau^{-p_1})[1-e^{-r_1(t-\tau)}] + (f_2 + g_2\tau^{-p_2})[1-e^{-r_2(t-\tau)}] + D(e^{-s\tau} - e^{-st}) \tag{6.13}$$

where $f_1, g_1, r_1, \ldots, D$, and s—constants.

For example, the modulus of elasticity and unit creep of mass concrete of Gongzui gravity dam are expressed by the following formulas:

$$E(\tau) = 38,500[1 - \exp(-0.402\tau^{0.335}] \text{ (MPa)} \tag{6.14}$$

$$\begin{aligned} C(t,\tau) &= (7.0 + 64.4\tau^{-0.45})[1 - e^{-0.30(t-\tau)}] \\ &\quad + (16.0 + 27.2\tau^{-0.45})[1 - e^{-0.005(t-\tau)}] \ (10^{-6}/\text{MPa}) \end{aligned} \tag{6.15}$$

Owing to the property of exponential function, the exponential formula (6.13) is very suitable for the numerical stress analysis of mass concrete structure when the influence of creep is considered.

The creep compliance of concrete can also be expressed by logarithmic function of time as follows:

$$J(t,\tau) = \frac{1}{C\ln(\tau^b + 1)} + (A_0 + A_1\tau^{-B_1})\ln(t - \tau + 1) \tag{6.16}$$

For example, the creep compliance of concrete of Felly Cangon dam is expressed by

$$J(t,\tau) = \frac{1000}{24.0\ln(\tau^{0.30} + 1)} + (0.20 + 11.0\tau^{-0.29})\ln(t - \tau + 1) \ (10^{-6}/\text{MPa}) \tag{6.17}$$

The logarithmic formula (6.16) is simple and agrees well with the test results (Figure 6.3), but it is not suitable for stress analysis of concrete when the influence of creep is considered by the finite element method.

Some formulas for $E(\tau)$ and $C(t,\tau)$ adopted in practical concrete dams are given below for reference. All of them are based on the results of creep tests in the laboratory.

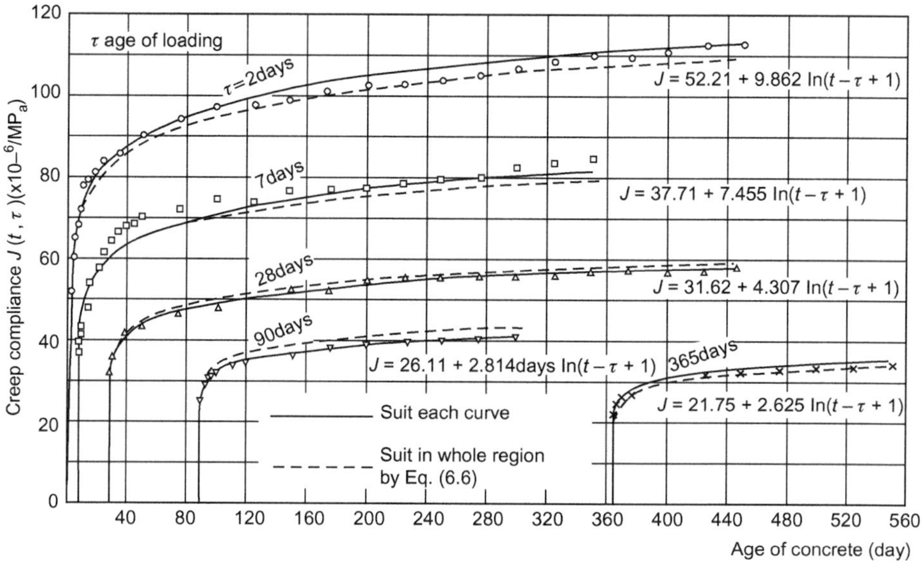

Figure 6.3 Creep compliance of concrete of Felly Cangon dam.

1. Tongfeng arch dam

$$E(\tau) = 41.4[1 - \exp(-0.52\tau^{0.35})]\ \text{(GPa)} \tag{6.18}$$

$$C(t,\tau) = (3.0 + 70\tau^{-0.50})[1 - e^{-0.30(t-\tau)}] + (1.0 + 4.0\tau^{-0.50})[1 - e^{-0.0050(t-\tau)}]\ (10^{-6}/\text{MPa}) \tag{6.19}$$

2. Xiaowang arch dam, $C_{180}40$ ($C_{180}40$ means that the compressive strength at age 180 days is 40 MPa)

$$E(\tau) = 35{,}000[1 - \exp(-0.44\tau^{0.29})]\ \text{(MPa)} \tag{6.20}$$

$$\begin{aligned} C(t,\tau) = {} & 56.5\tau^{-0.405}[1 - e^{-0.604(t-\tau)}] + (3.75 + 24.5\tau^{-6.53})[1 - e^{-0.034(t-\tau)}] \\ & + 26.1e^{-0.0213\tau}[1 - e^{-0.0213(t-\tau)}]\ (10^{-6}/\text{MPa}) \end{aligned} \tag{6.21}$$

3. Xilaodu arch dam, $C_{180}35$

$$E(\tau) = 40{,}800[1 - \exp(-0.22\tau^{0.56})]\ \text{(MPa)} \tag{6.22}$$

$$\begin{aligned} C(t,\tau) = {} & (2.02 + 74.1\tau^{-0.697})[1 - e^{-0.306(t-\tau)}] + (5.30 + 24.6\tau^{-3.0})[1 - e^{0.109(t-\tau)}] \\ & + 23.3e^{-0.0373\tau}[1 - e^{-0.00373(t-\tau)}]\ (10^{-6}/\text{MPa}) \end{aligned} \tag{6.23}$$

4. Jingping-I arch dam

$$E(\tau) = 36{,}800[1 - \exp(-0.354\tau^{0.398})]\ \text{(MPa)} \tag{6.24}$$

$$C(t,\tau) = (3.99 + 74.7\tau^{-0.727})[1 - e^{-0.925(t-\tau)}] + (5.74 + 24.8\tau^{-1.25})[1 - e^{-0.0139(t-\tau)}]$$
$$+22.4e^{-0.0334\tau}[1 - e^{-0.0334(t-\tau)}]\,(10^{-6}/\text{MPa}) \tag{6.25}$$

5. Xiangjiaba gravity dam, $C_{90}20$

$$E(\tau) = 44{,}500\tau/(3.37 + \tau)\ (\text{MPa}) \tag{6.26}$$

$$C(t,\tau) = (5.97 + 68.1\tau^{-0.902})[1 - e^{-0.862(t-\tau)}] + (0.140 + 32.7\tau^{-0.288})$$
$$[1 - e^{-0.0298(t-\tau)}]\,(10^{-6}/\text{MPa}) \tag{6.27}$$

6.1.4 Lateral Strain and Poisson's Ratio of Concrete

Under action of constant stress $\sigma(\tau)$ applied at time τ in the x direction, the elastic strain is $\sigma_x(\tau)/E(\tau)$ and the creep strain is $\sigma_x(\tau)/C(t,\tau)$. Experimental results show that in the y and z directions, there will be lateral elastic strain

$$\varepsilon_y^e(t) = \varepsilon_z^e(t) = -\mu_1(\tau)\frac{\sigma_x(\tau)}{E(\tau)} \tag{6.28}$$

and the lateral creep strain

$$\varepsilon_y^e(t) = \varepsilon_z^c(t) = -\mu_2(t,\tau)\sigma_x(\tau)C(t,\tau) \tag{6.29}$$

The total lateral strain will be

$$\varepsilon_y(t) = \varepsilon_z(t) = -\sigma_x(\tau)\left[\frac{\mu_1(\tau)}{E(\tau)} + \mu_2(t,\tau)C(t,\tau)\right] \tag{6.30}$$

where $\mu_1(\tau)$ is the Poisson's ratio for elastic strain and $\mu_2(t,\tau)$ is the Poisson's ratio for creep strain. Theoretically, $\mu_1(\tau)$ is a function of τ and $\mu_2(t,\tau)$ is a function of t and τ.

Experimental results show that $\mu_1(\tau)$ and $\mu_2(t,\tau)$ are equal to 0.10–0.25 for concrete with different aggregates. In engineering analysis, generally Poisson's ratio is taken

$$\mu_1(\tau) = \mu_2(t,\tau) = \mu \tag{6.31}$$

Thus the total lateral strain under long time action of constant stress $\sigma_x(\tau)$ is

$$\varepsilon_y(t) = \varepsilon_z(t) = -\mu\sigma_x(\tau)\left[\frac{1}{E(\tau)} + C(t,\tau)\right] = -\mu\sigma_x J(t,\tau) \tag{6.32}$$

When there are no test results, we may take $\mu = 1/6$ for mass concrete.

6.2 Stress Relaxation of Concrete

6.2.1 Stress Relaxation of Concrete Subjected to Constant Strain

When the strain $\varepsilon(\tau)$ is imposed to concrete at age τ, the stress at the instant of application of strain is

$$\sigma(\tau) = E(\tau)\varepsilon(\tau) \tag{6.33}$$

If the strain $\varepsilon(\tau)$ does not change with time, namely

$$\varepsilon(t) = \varepsilon(\tau) = \text{constant} \tag{6.34}$$

as shown in Figure 6.4, under the action of constant strain, the stress will decrease slowly with time and the stress at time t will be

$$\sigma(t) = R(t,\tau)\varepsilon(\tau) \tag{6.35}$$

where $R(t,\tau)$—the relaxation modulus of concrete.

Refer to Figure 6.4, under the action of constant strain $\varepsilon(t) = \text{constant}$, the ratio of the stress $\sigma(t)$ at time t to the stress $\sigma(\tau)$ at time τ when the strain is imposed is called the relaxation coefficient and denoted by $K(t,\tau)$, thus

$$K(t,\tau) = \frac{\sigma(t)}{\sigma(\tau)} \tag{6.36}$$

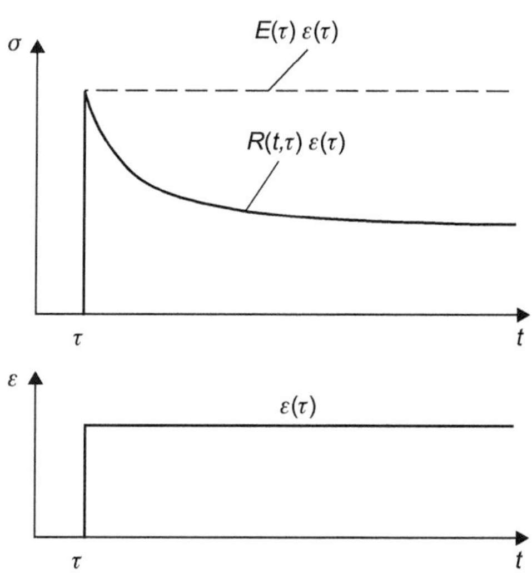

Figure 6.4 The stress relaxation of concrete.

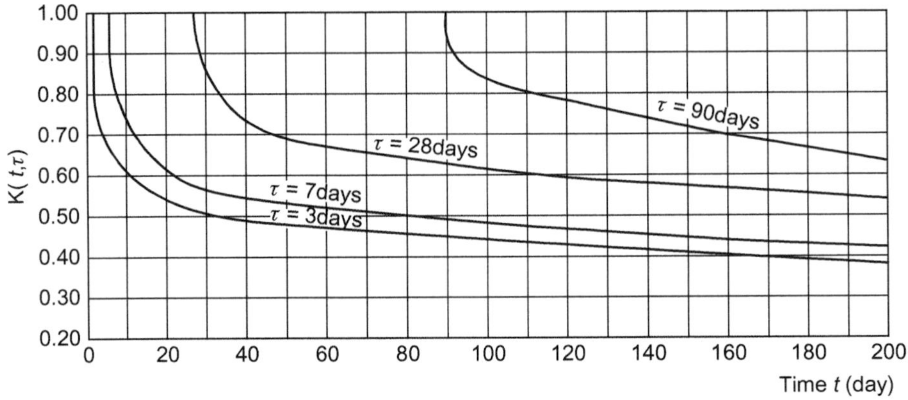

Figure 6.5 The relaxation coefficients of the conventional concrete of Liujiaxia gravity dam.

Substituting Eqs. (6.33) and (6.35) into Eq. (6.36), the following equation is derived:

$$K(t,\tau) = \frac{R(t,\tau)}{E(\tau)} \tag{6.37a}$$

or

$$R(t,\tau) = K(t,\tau)E(\tau) \tag{6.37b}$$

The relaxation coefficients of conventional concrete of Liujiaxia gravity dam are shown in Figure 6.5 and those of roller-compacted concrete (RCC) of Longtang gravity dam are shown in Figure 6.6.

6.2.2 Method for Computing the Relaxation Coefficient from Creep of Concrete

As the creep test is simpler than the relaxation test, generally only creep test is conducted and the relaxation coefficients are calculated from the results of creep test of concrete. The method of calculation is given in the following.

Let the stress-free strain $\varepsilon^0(t) = 0$ in Eq. (6.9), the strain caused by stress is given by

$$\varepsilon(t) = \frac{\sigma(\tau_0)}{E(\tau_0)} + \sigma(\tau_0)C(t,\tau_0) + \int_{\tau_0}^{t} J(t,\tau)\frac{d\sigma}{d\tau}d\tau \tag{6.38}$$

If the strain is kept constant with time:

$$\varepsilon(t) = \frac{\sigma(\tau_0)}{E(\tau_0)} = \text{constant} \tag{6.39}$$

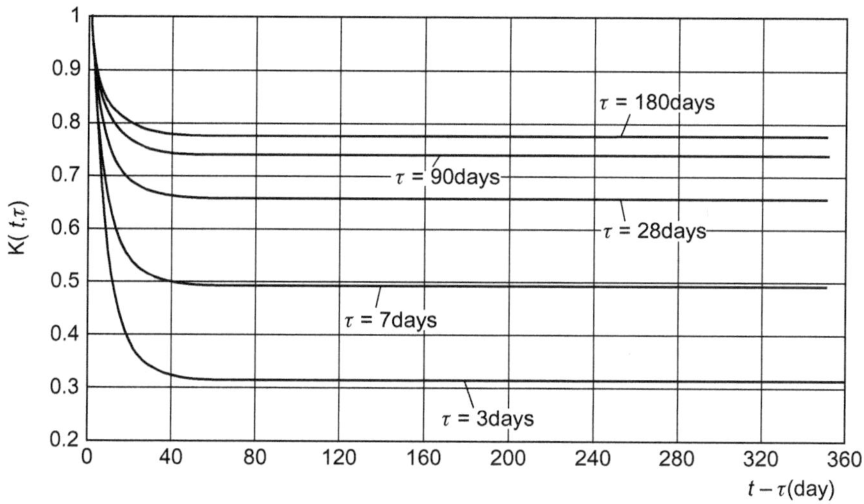

Figure 6.6 The relaxation coefficients of RCC of Longtang RCC gravity dam.

Subtracting Eq. (6.39) from Eq. (6.38), we get the fundamental equation for computing relaxation coefficient as follows:

$$\sigma(\tau_0)C(t,\tau_0) + \int_{\tau_0}^{t} J(t,\tau)\frac{d\sigma}{d\tau}d\tau = 0 \tag{6.40}$$

The ratio of $\sigma(t)$ given by the above equation to $\sigma(\tau_0)$ is the relaxation coefficient.

Taking the initial stress $\sigma(\tau_0) = 1$, we get

$$C(t,\tau_0) + \int_{\tau_0}^{t} J(t,\tau)d\sigma = 0 \tag{6.41}$$

The stress $\sigma(t)$ given by the above equation is the relaxation coefficient. Dividing $t - \tau_0$ into a series of time increments $\Delta\tau_i$, Eq. (6.41) is transformed into the following equation:

$$C(\tau_n,\tau_0) + J(\tau_n,\overline{\tau}_1)\Delta\sigma_1 + J(\tau_n,\overline{\tau}_2)\Delta\sigma_2 + \cdots + J(\tau_n,\overline{\tau}_{n-1})\Delta\sigma_{n-1} + J(\tau_n,\overline{\tau}_n)\Delta\sigma_n = 0 \tag{6.42}$$

Hence the nth stress increment is

$$\Delta\sigma_n = -\frac{C(\tau_n,\tau_0) + \sum_{i=1}^{n-1} J(\tau_n,\overline{\tau}_i)\Delta\sigma_i}{J(\tau_n,\overline{\tau}_n)} \tag{6.43}$$

where

$$J(\tau_n, \overline{\tau}_i) = \frac{1}{E(\overline{\tau}_i)} + C(\tau_n, \overline{\tau}_i)$$

$$\overline{\tau}_i = (\tau_{i-1} + \tau_i)/2 \tag{6.44}$$

The relaxation coefficient is

$$K(t, \tau) = 1 + \sum \Delta\sigma_i \tag{6.45}$$

6.2.3 Formulas for Relaxation Coefficient

The relaxation coefficient $K(t,\tau)$ is dependent on two variables τ and t and may be expressed by the following formulas proposed by the writer:

1. Exponential formula

$$K(t, \tau) = 1 - \sum_{i=1}^{n} (a_i + b_i \tau^{-d_i})[1 - \mathrm{e}^{-h_i(t-\tau)}] \tag{6.46}$$

where a_i, b_i, d_i, and h_i are constants. For the conventional mass concrete, the relaxation coefficient may be given by

$$K(t, \tau) = 1 - (0.25 + 0.25\tau^{-0.40})[1 - \mathrm{e}^{-0.20(t-\tau)}] - (0.15 + 0.30\tau^{-0.40})[1 - \mathrm{e}^{-0.0060(t-\tau)}] \tag{6.47}$$

2. Compound exponential formula

$$\left.\begin{aligned} K(t, \tau) &= 1 - \psi(\tau)\{1 - \exp[-m(\tau)(t-\tau)^{n(\tau)}]\} \\ \psi(\tau) &= \psi_0 + (1 - \psi_0)\exp(-f\tau^{g}) \\ m(\tau) &= m_0 + m_1\tau^{-\lambda} \\ n(\tau) &= n_0 + n_1\tau^{-\beta} \end{aligned}\right\} \tag{6.48}$$

where $\psi_0, f, g, m_0, m_1, \lambda, n_1$, and β—material constants.

For example, the relaxation coefficient of mass concrete of Gongzui gravity dam is

$$\begin{aligned} K(t, \tau) = 1 &- [0.47 + 0.53\exp(-0.623\tau^{0.170})] \\ &\times \{1 - \exp[-(0.20 + 0.271\tau^{-0.225})(t-\tau)^{(0.326+0.125\tau^{-0.583})}]\} \end{aligned} \tag{6.49}$$

6.3 Modulus of Elasticity, Unit Creep, and Relaxation Coefficient of Concrete for Preliminary Analysis [39, 40]

The modulus of elasticity $E(\tau)$, unit creep $C(t,\tau)$, and relaxation coefficient $K(t,\tau)$ are required in the viscoelastic stress analysis of mass concrete. They may be estimated by the following formulas when there are no test results:

1. $E(\tau)$, $C(t,\tau)$, $K(t,\tau)$ for conventional mass concrete

$$E(\tau) = E_0[1 - \exp(-0.40\tau^{0.34})] \tag{6.50}$$

$$C(t,\tau) = C_1(1 + 9.20\tau^{-0.45})[1 - e^{-0.30(t-\tau)}] + C_2(1 + 1.70\tau^{-0.45})[1 - e^{-0.0050(t-\tau)}] \tag{6.51}$$

$$K(t,\tau) = 1 - (0.25 + 0.25\tau^{-0.40})[1 - e^{-0.20(t-\tau)}] - (0.15 + 0.30\tau^{0.40})[1 - e^{-0.0060(t-\tau)}] \tag{6.52}$$

where $C_1 = 0.23/E_0$, $C_2 = 0.52/E_0$, $E_0 = 1.05E(360)$ (or $E_0 = 1.20E(90)$, or $E_0 = 1.45E(28)$). The unit of time is day.

2. $E(\tau)$, $C(t,\tau)$, $K(t,\tau)$ for RCC.

$$E(\tau) = E_0[1 - \exp(-0.31\tau^{0.35})] \tag{6.53}$$

$$C(t,\tau) = \frac{0.10}{E_0}(1 + 30.0\tau^{-0.50})\left[1 - e^{-0.30(t-\tau)}\right] + \frac{0.15}{E_0}(1 + 1.0\tau^{-0.40})\left[1 - e^{-0.0020(t-\tau)}\right] \tag{6.54}$$

$$K(t,\tau) = 1 - (0.23 + 0.24\tau^{-0.40})[1 - e^{0.20(t-\tau)}] - (0.13 + 0.29\tau^{-0.40})[1 - e^{-0.0060(t-\tau)}] \tag{6.55}$$

6.4 Two Theorems About the Influence of Creep on the Stresses and Deformations of Concrete Structures [17, 18, 33, 35]

The influence of creep on the stresses and deformations of concrete structure depends on the boundary conditions shown in Table 6.1 and whether the structure is homogeneous. Alfrey had studied the homogeneous structure under simple boundary conditions, namely cases (1) and (2) in Table 6.1 [17]. The writer had studied the nonhomogeneous structure under mixed boundary condition, i.e., case 6 in Table 6.1 (cases 1–5 are special cases of case 6) and proposed two theorems [33].

Consider the structure consisting of two kinds of materials. The material constants in region I are $E_1(\tau)$, $C_1(t, \tau)$, $\mu_1(t, \tau)$, and those in region II are $E_2(\tau)$, $C_2(t, \tau)$, $\mu_2(t, \tau)$.

Table 6.1 Boundary Conditions of Structures

	First Kind Problem	Second Kind Problem	Mixed Problem
Homogeneous	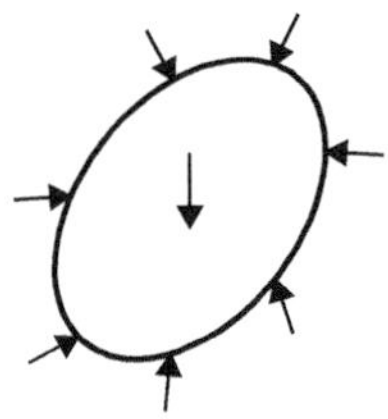 (1)	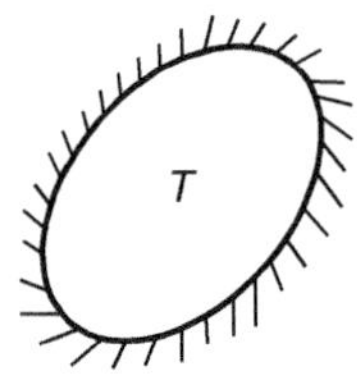 (2)	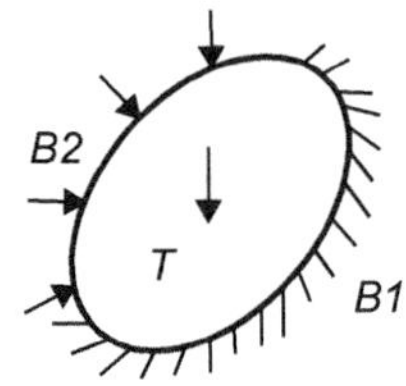 (3)
Nonhomogeneous	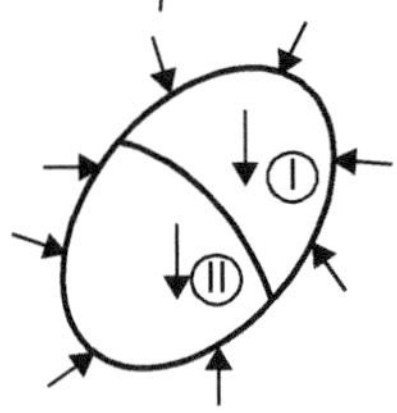 (4)	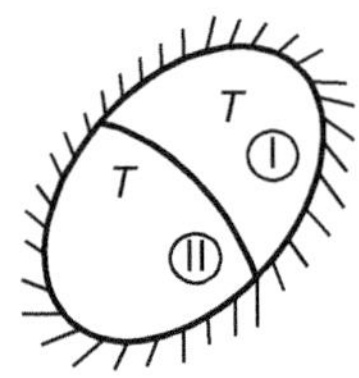 (5)	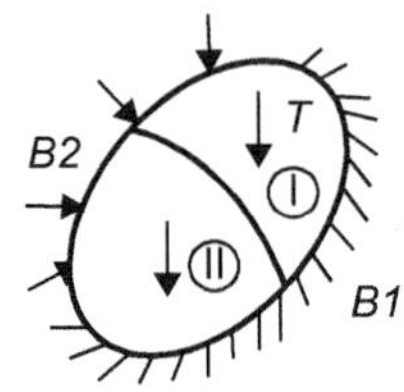 (6)

If the material constants tally with the proportional relations,

$$\left.\begin{aligned} E_1(\tau)C_1(t,\tau) &= E_2(\tau)C_2(t,\tau) \\ \text{or} \quad C_1(t,\tau):C_2(t,\tau) &= \frac{1}{E_1(\tau)}:\frac{1}{E_2(\tau)} \\ \mu_1(t_1\tau) = \mu_2(t_2\tau) &= \mu \end{aligned}\right\} \tag{6.56}$$

it is called the structure of proportional material constants.

Theorem I For the homogeneous structure or the nonhomogeneous viscoelastic structure whose material constants satisfy the proportional relations (6.56), if change of temperature in the structure is zero, part of its boundary is subjected to external force, and part of its boundary is fixed (no displacement), then the stresses in the structure due to the action of body force and external force are not influenced by the creep of concrete and the strain $\varepsilon_x(x,y,z,t)$ and displacement $u(x,y,z,t)$ of any point (x,y,z) at time t may be determined from the elastic strain $\varepsilon_x^{\mathrm{e}}(x,y,z,\tau_0)$ and elastic displacement $u^{\mathrm{e}}(x,y,z,\tau_0)$ at time τ_0 by the following equations:

$$\left.\begin{aligned} \varepsilon_x(x,y,z,t) &= \varepsilon_x^{\mathrm{e}}(x,y,z,\tau_0)[1+E(\tau_0)C(t,\tau_0)] \\ u(x,y,z,t) &= u^{\mathrm{e}}(x,y,z,\tau_0)[1+E(\tau_0)C(t,\tau_0)] \end{aligned}\right\} \tag{6.57}$$

Theorem II For the homogeneous structure or the nonhomogeneous viscoelastic structure whose material constants satisfy the proportional relation (6.56), if the body force is zero, the change of temperature in the structure is not zero, part of the boundary is free, and part of the boundary is subjected to known displacements, then under the action of the temperature and the imposed boundary displacements, the displacements of the structure is the same as those of the elastic structure and the stress of the structure may be computed with relaxation coefficients by the following equation:

$$\sigma(x,y,z,t) = \sum \Delta\sigma^{e}(x,y,z,\tau_i)K(t,\tau_i) \tag{6.58}$$

where

$K(t,\tau_0)$—the relaxation coefficient
$\Delta\sigma^{e}(x,y,z,\tau_i)$—the increment of elastic stress.

6.5 Classification of Massive Concrete Structures and Method of Analysis

According to the two theorems in Section 6.4, the massive concrete structures may be classified into two kinds as follows:

1. Structure of proportional deformation
   This first kind of structure includes
   a. homogeneous structure supported on rigid foundation;
   b. nonhomogeneous structure with material constants satisfying proportional relation (6.56) and supported on rigid foundation;
   c. nonhomogeneous structure supported on viscoelastic foundation and all the material constants of the structure and foundation satisfy the proportional relation (6.56).
   For the structure of proportional deformation, the viscoelastic stresses and deformations may be computed from the elastic stress and deformations by Eqs. (6.57) and (6.58).
2. Structure of nonproportional deformation
   This second kind of structure includes all the structures other than the structure of proportional deformation. The stresses and deformation of this kind of structure generally must be computed by the finite element method.

6.6 Method of Equivalent Modulus for Analyzing Stresses in Matured Concrete due to Harmonic Variation of Temperature

The modulus of elasticity and unit creep for matured concrete can be expressed by

$$\left.\begin{aligned} E(\tau) &= E = \text{constant} \\ C(t,\tau) &= C_1[1 - e^{-r_1(t-\tau)}] \end{aligned}\right\} \tag{6.59}$$

where E, C_1, and r_1 are constants. Considering the quasi-steady temperature field and taking $\tau_0 = -\infty$, the strain at time t is given by

$$\varepsilon(t) = \frac{\sigma(t)}{E} + \int_{-\infty}^{t} C_1 r_1 \, \mathrm{e}^{-r_1(t-\tau)} \sigma(t) \mathrm{d}\tau \tag{6.60}$$

Because the air temperature and the water temperature generally vary with cosine function of time, after disappearance of the initial influence, the temperature and stresses of any point in the interior of the structure will vary with cosine function of time, so the stress may be expressed by

$$\sigma(t) = \sigma_0 \cos \omega(t + \eta) \tag{6.61}$$

where

$\omega = 2\pi/P$—circular frequency
P—period of variation
σ_0—amplitude of variation
η—phase difference.

Substituting Eq. (6.61) into Eq. (6.60), we get

$$\varepsilon(t) = \frac{\sigma_0}{\rho E} \cos \omega(1 + \eta - \xi) \tag{6.62}$$

where

$$\left.\begin{aligned} \rho &= 1/\sqrt{a^2 + b^2} \\ \xi &= \frac{1}{\omega} \tan^{-1}\left(\frac{b}{a}\right) \\ a &= 1 + \frac{Ec_1 r_1^2}{r_1^2 + \omega^2} \\ b &= \frac{Ec_1 r_1 \omega}{r_1^2 + \omega^2} \end{aligned}\right\} \tag{6.63}$$

where ξ—the phase difference between the maximum strain and the maximum stress.

Let

$$\left.\begin{aligned} E^* &= \rho E \\ t^* &= t - \xi \end{aligned}\right\} \tag{6.64}$$

and substitute them into Eq. (6.62), we get

$$\varepsilon(t) = \frac{\sigma_0}{E^*} \cos \omega(t + \eta^*) \tag{6.65}$$

The above equation is consistent with the stress–strain relation of elastic body in shape, thus the following conclusion is derived: The viscoelastic body under the action of harmonic stress may be replaced by an equivalent elastic body with equivalent modulus of elasticity $E^* = \rho E$ and analyzed by elastic method which gives rise to stress σ_0, strain ε_0, and phase difference η^*, then, σ_0, ε_0, and η^* are equal to the maximum stress, maximum strain, and phase difference of the original viscoelastic body. This method proposed by the writer is called the equivalent modulus method [37].

7 Thermal Stresses in Fixed Slab or Free Slab

The fixed and free slabs are two simple structural types for studying the thermal stresses in massive concrete structures.

The fixed slab is a thin concrete slab placed on a rigid foundation with infinite plane dimensions. Due to the restraint of the rigid foundation, the horizontal displacements and turn of the slab are zero. The free slab is free of external restraint and can deform freely in all directions. The thermal stresses in a free slab are induced by the nonuniform distribution of temperature in the slab. Through the fixed and free slabs, readers can understand some basic rules and important characteristics of the thermal stresses in massive concrete structures.

7.1 Thermal Stresses in Fixed Slab [7, 8]

7.1.1 Computation of the Temperature Field

As shown in Figure 7.1, it is assumed that the temperature varies only in the z-direction, i.e., $T = T(z,t)$. The top surface of the slab is exposed in the air and the bottom surface is in contact with rock foundation. There is hydration heat of cement in the slab, and there is no hydration heat in the rock. Generally, the temperature field is computed by the finite difference method.

7.1.2 The Elastic Thermal Stress

Due to the restraint of rigid foundation, the slab is perfectly restrained in the x- and y-directions, thus

$$\varepsilon_x = \varepsilon_y = 0 \tag{7.1}$$

The slab is free in the z-direction, so

$$\sigma_z = 0 \tag{7.2}$$

Thermal Stresses and Temperature Control of Mass Concrete. DOI: http://dx.doi.org/10.1016/B978-0-12-407723-2.00007-5

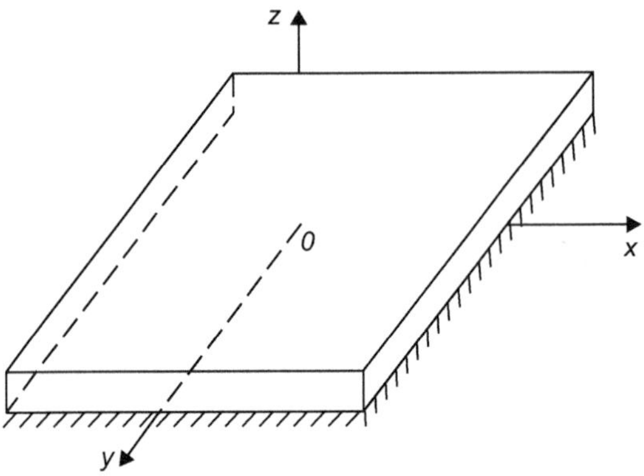

Figure 7.1 A fixed slab.

According to Hooke's law, we have

$$\varepsilon_x = \frac{\sigma_x - \mu\sigma_y}{E} + \alpha T = 0 \tag{7.3}$$

$$\varepsilon_y = \frac{\sigma_y - \mu\sigma_x}{E} + \alpha T = 0 \tag{7.4}$$

Multiplying Eq. (7.4) by μ and adding to Eq. (7.3), we get

$$\frac{(1-\mu^2)\sigma_x}{E} + (1+\mu)\alpha T = 0 \tag{7.5}$$

Dividing Eq. (7.5) by $(1+\mu)$, the fundamental equation for computing elastic thermal stresses in a fixed slab with constant modulus of elasticity is derived

$$\sigma = \sigma_x = \sigma_y = -\frac{E\alpha T}{1-\mu} \tag{7.6}$$

Practically, the modulus of elasticity of concrete varies with age. In order to consider this important factor, the incremental method is adopted. The time t is divided into a series of time increments Δt_i, $i = 1,2, \ldots ,n$. As shown in Figure 7.2, the temperature increment in Δt_i is

$$\mathrm{D}T_i = T(t_i) - T(t_{i-1}) \tag{7.7}$$

The elastic stress increment induced by Δt_i is

$$\Delta\sigma_i^e == -\frac{E_i\alpha\Delta T_i}{1-\mu} \tag{7.8}$$

where

$$E_i = [E(t_{i-1}) + E(t_i)]/2 \tag{7.9}$$

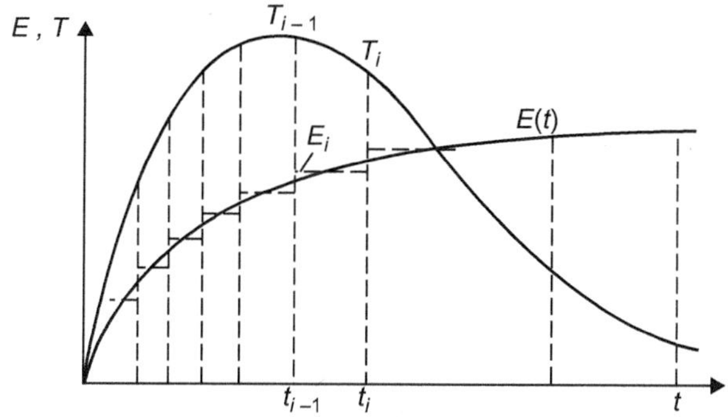

Figure 7.2 The variation of temperature and modulus of elasticity.

7.1.3 The Viscoelastic Thermal Stresses

According to Theorem II in Section 6.4, the viscoelastic thermal stresses in a fixed slab may be computed by the method of relaxation coefficient as follows:

$$\sigma(t) = \sum_{i=1}^{n} \Delta\sigma_i^e K(t, \overline{\tau}_i) = -\sum_{i=1}^{n} \frac{E_i \alpha \Delta T_i}{1-\mu} K(t, \overline{\tau_i}) \tag{7.10}$$

where $K(t, \overline{\tau}_i)$ is the relaxation coefficient, $\overline{\tau} = (\tau_{i-1} + \tau_i)/2$.

7.1.4 The Thermal Stresses in Fixed Slab Due to Hydration Heat of Cement

In order to study the variations of thermal stresses in fixed slabs of different thickness, the thermal stresses in fixed slabs with thickness $h = 1, 2, 3, 4, 5$ m are computed in the following.

The adiabatic temperature rise of concrete is

$$\theta = \frac{\theta_0 \tau}{1.00 + \tau} \tag{7.11}$$

The modulus of elasticity is

$$E(\tau) = E_0[1 - \exp(-0.40\tau^{0.34})] \tag{7.12}$$

The relaxation coefficient is

$$K(t,\tau) = 1 - [0.40 + 0.60\exp(-0.62\tau^{0.17})] \times \{1 - \exp[-(0.20 + 0.27\tau^{-0.23})(t-\tau)^{0.36}]\} \tag{7.13}$$

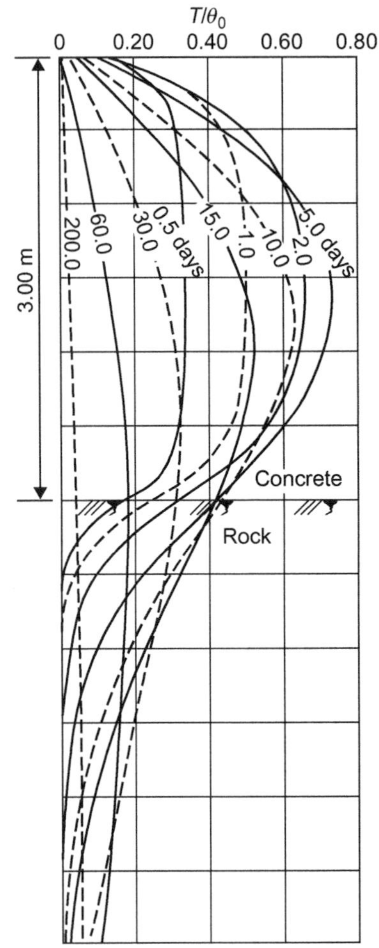

Figure 7.3 Distribution of temperatures in a concrete slab with 3 m thickness.

The unit of time is day. The thermal properties of concrete and rock are the same: the diffusivity $a = 0.10\ \text{m}^2/\text{day}$, the conductivity $\lambda = 220$ kJ/(m d °C), the surface conductance $\beta = 2000$ kJ/(m^2 d °C).

The temperature field is computed by the finite difference method.

The temperature distributions in a fixed slab with thickness 3 m are shown in Figure 7.3. The maximum temperatures appear in the central part of the slab.

The variations of temperatures at the center of fixed slabs with different thickness are shown in Figure 7.4. It is clear that the higher temperature will appear in a thicker slab.

The thermal stresses due to hydration heat of cement at different times in a fixed slab with 3 m thickness are shown in Figure 7.5. The slab is subjected to compressive stress before $\tau = 2$ days, because the expansion induced by temperature rise is

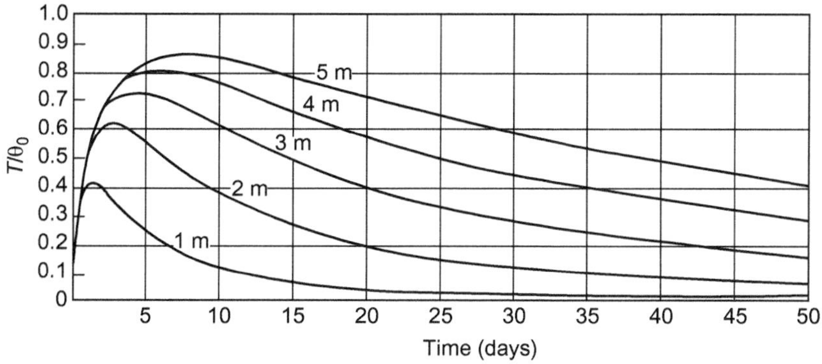

Figure 7.4 Variation of temperatures due to hydration heat of cement at the center of concrete slab of different thickness on rock foundation.

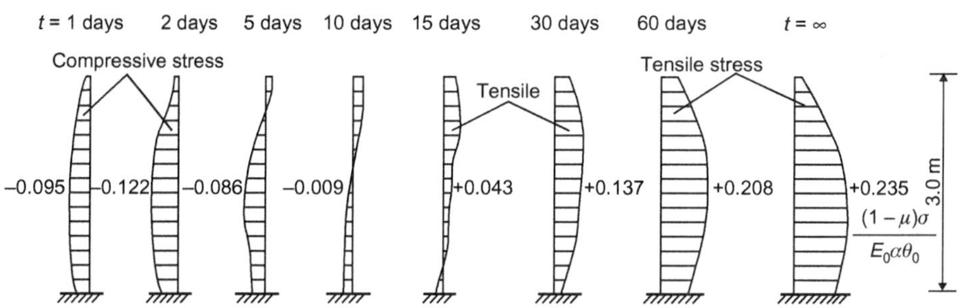

Figure 7.5 Viscoelastic thermal stresses due to hydration heat of cement in a fixed slab of 3 m thickness on rock foundation [unit: $(1-\mu)\sigma/E_0\alpha\theta_0$].

restrained by the foundation. Owing to natural cooling of the slab, the tensile stresses appear near the top surface after 5 days and the whole section of the slab is subjected to tensile stress after 30 days. The maximum tensile stress is 0.235 $E_0\alpha\theta_0/(1-\mu)$ when $t \to \infty$. The compressive stress induced by temperature rise in the early age was small because $E(\tau)$ and $K(t,\tau)$ were small in the early age. On the other hand, $E(\tau)$ and $K(t,\tau)$ were large at late age and when the temperature dropped from the maximum to zero, the tensile stress caused by the temperature change is so big that there will be remarkable residual tensile stress in the slab after complete cooling which may lead to cracking. This is an important peculiarity of thermal stress of massive concrete structures.

The viscoelastic thermal stresses at the centers of fixed slabs due to hydration heat of cement are shown in Figure 7.6. They are compressive stresses in early age and tensile stresses in later age.

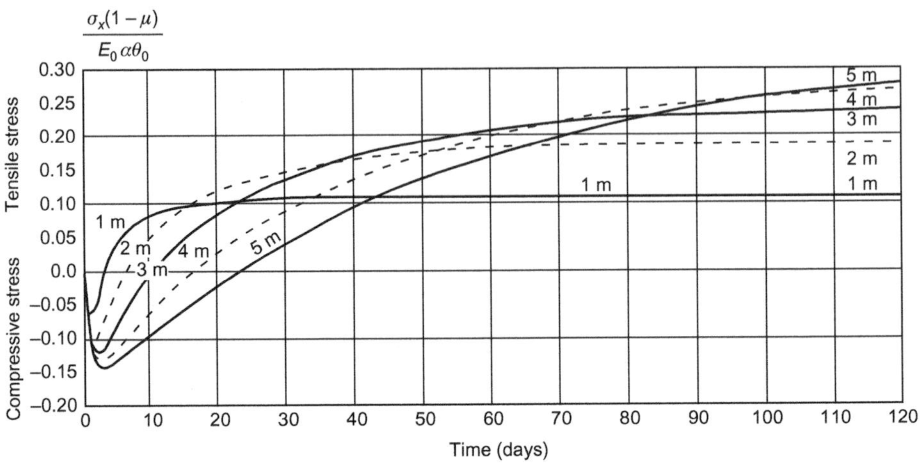

Figure 7.6 The viscoelastic thermal stresses at the center of a fixed slab with different thickness due to hydration heat of cement.

7.2 Method for Computing Thermal Stresses in a Free Slab [7, 8]

7.2.1 Elastic Thermal Stress in a Free Slab When the Modulus of Elasticity is Constant

In an infinite slab shown in Figure 7.7, it is assumed that the temperature varies only in the direction of thickness. As the slab can deform freely in all directions, the normal stress and normal strain are as follows:

$$\left.\begin{aligned} &\sigma_x = \sigma_y = \sigma, \quad \sigma_z = 0 \\ &\varepsilon_x = \varepsilon_y = \varepsilon \end{aligned}\right\} \tag{7.14}$$

As $\sigma_z = 0$, Eq. (7.6) is also applicable to free slab. Substituting Eq. (7.14) into Eq. (7.6), the stress–strain relation for a free slab is derived:

$$\sigma = \frac{E(\varepsilon - \alpha T)}{1 - \mu} \tag{7.15}$$

According to the assumption of plane section, the strain ε should be a linear function of z as follows:

$$\varepsilon = \frac{\mathrm{d}u}{\mathrm{d}x} = \frac{\mathrm{d}v}{\mathrm{d}y} = \alpha(A + Bz) \tag{7.16}$$

in which A and B are constants independent of z. Substituting Eq. (7.16) into Eq. (7.15), we have

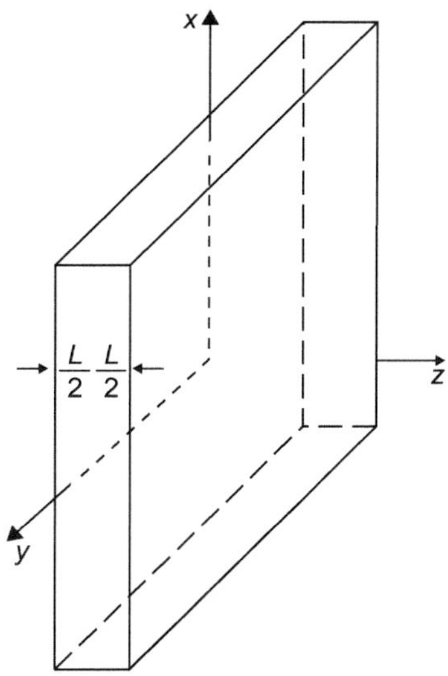

Figure 7.7 A free slab.

$$\sigma = \frac{E\alpha}{1-\mu}(A + Bz - T) \tag{7.17}$$

In a free slab, the axial force and bending moment of any cross section must be equal to zero, namely

$$\left.\begin{aligned} \int_{-L/2}^{L/2} \sigma \, dz &= \int_{-L/2}^{L/2} \frac{E\alpha}{1-\mu}(A + Bz - T)dz = 0 \\ \int_{-L/2}^{L/2} \sigma z \, dz &= \int_{-L/2}^{L/2} \frac{E\alpha}{1-\mu}(A + Bz - T)z \, dz = 0 \end{aligned}\right\} \tag{7.18}$$

As the origin of coordinate z is the center of section, so

$$\int_{-L/2}^{L/2} dz = L, \quad \int_{-L/2}^{L/2} z \, dz = 0, \quad \int_{-L/2}^{L/2} z^2 \, dz = \frac{L^3}{12} \tag{7.19}$$

From Eqs. (7.18) and (7.19), we have

$$\left.\begin{aligned} & A = \frac{1}{L}\int_{-L/2}^{L/2} T \, dz = T_m \\ & B = \frac{12}{L^3} S = \frac{T_d}{L} \\ & T_d = \frac{12S}{L^2}, \quad S = \int_{-L/2}^{L/2} Tz \, dz \end{aligned}\right\} \tag{7.20}$$

where

T_m—mean temperature
T_d—equivalent linear temperature difference
S—moment of temperature
L—thickness of slab.

Substituting Eq. (7.20) into Eq. (7.17), the formula for computing the elastic thermal stress in a free slab is derived as follows:

$$\sigma(z,t) = \frac{E\alpha}{1-\mu}\left[T_m + \frac{T_d}{L}z - T(z,t)\right] \tag{7.21}$$

If the distribution of temperature is symmetrical, i.e., $T(z,t) = T(-z,t)$, then $T_d = 0$ and the stress is given by

$$\sigma(z,t) = \frac{E\alpha}{1-\mu}[T_m - T(z,t)] \tag{7.22}$$

7.2.2 Viscoelastic Thermal Stress in a Free Slab Considering the Influence of Age

Dividing time t into a series of time increments Δt_i, $i = 1,2, \ldots ,n$; as shown in Figure 7.2, the temperature increment in Δt_i is

$$\Delta T_i(z,t_i) = T(z,t_i) - T(z,t_{i-1}) \tag{7.23}$$

and the increment of elastic thermal stress in Δt_i is

$$\Delta\sigma_i^e(z,t_i) = \frac{E_i\alpha}{1-\mu}[\Delta A_i + \Delta B_i z - \Delta T_i(z,t_i)] \tag{7.24}$$

where

$$\left.\begin{aligned}
\Delta A_i &= \Delta T_{mi} = \frac{1}{L}\int_{-L/2}^{L/2}\Delta T_i(z,t_i)\mathrm{d}z\\
\Delta B_i &= \frac{\Delta T_{di}}{L} = \frac{12}{L^3}\Delta S_i\\
\Delta S_i &= \int_{-L/2}^{L/2}\Delta T_i(z,t_i)z\,\mathrm{d}z\\
E_i &= \frac{1}{2}[E(t_i) + E(t_{i-1})]
\end{aligned}\right\} \tag{7.25}$$

Taking account of influence of creep by relaxation coefficient, the viscoelastic thermal stress at time t_n is

$$\sigma(z, t_n) = \sum_{i=1}^{n} \Delta\sigma_i^e(z, t_i) K(t_n, \overline{\tau}_i) \tag{7.26}$$

in which $\overline{\tau}_i = (\tau_i + \tau_{i-1})/2, K(t_n, \overline{\tau}_i)$ is the relaxation coefficient.

Observed results of the bottom slab of a sluice constructed on soft foundation are shown in Figure 7.8. It is clear from the figure that the thermal stresses are close to those in a free slab.

7.3 Thermal Stresses in Free Concrete Slab due to Hydration Heat of Cement

The thermal stresses in free concrete slabs due to hydration heat of cement are computed by Eq. (7.26). The adiabatic temperature rise $\theta(\tau)$, the modulus of elasticity $E(\tau)$, and the relaxation coefficient are expressed by Eqs. (7.11–7.13).

The thicknesses of the slabs are 4 and 10 m.

The stress distributions at different times are shown in Figure 7.9. In the early age, the surface of the slab is subjected to tensile stress and the center to compressive stress. In the late age, the surface of the slab is subjected to compressive stress and the center to tensile stress.

The variations with time of the superficial thermal stresses of two free concrete slabs are shown in Figure 7.10.

7.4 Thermal Stresses in Free Slabs with Periodically Varying Surface Temperature

7.4.1 The Temperature Field

Consider the quasi-steady temperature field in an infinite slab as shown in Figure 7.11.

$$\left.\begin{array}{lll} \text{Equation of heat conduction} & & \dfrac{\partial T}{\partial \tau} = a\dfrac{\partial^2 T}{\partial x^2} \\ \text{Boundary conditions:} & \text{when} & x = L, \quad T = A_1 \cos \omega(\tau - \tau_0) \\ & \text{when} & x = 0, \quad T = A_2 \cos(\tau - \varepsilon - \tau_0) \end{array}\right\} \tag{7.27}$$

in which

$\omega = 2\pi/P$—circular frequency

P—period of variation of temperature

A_1, A_2—amplitudes of variation of surface temperature.

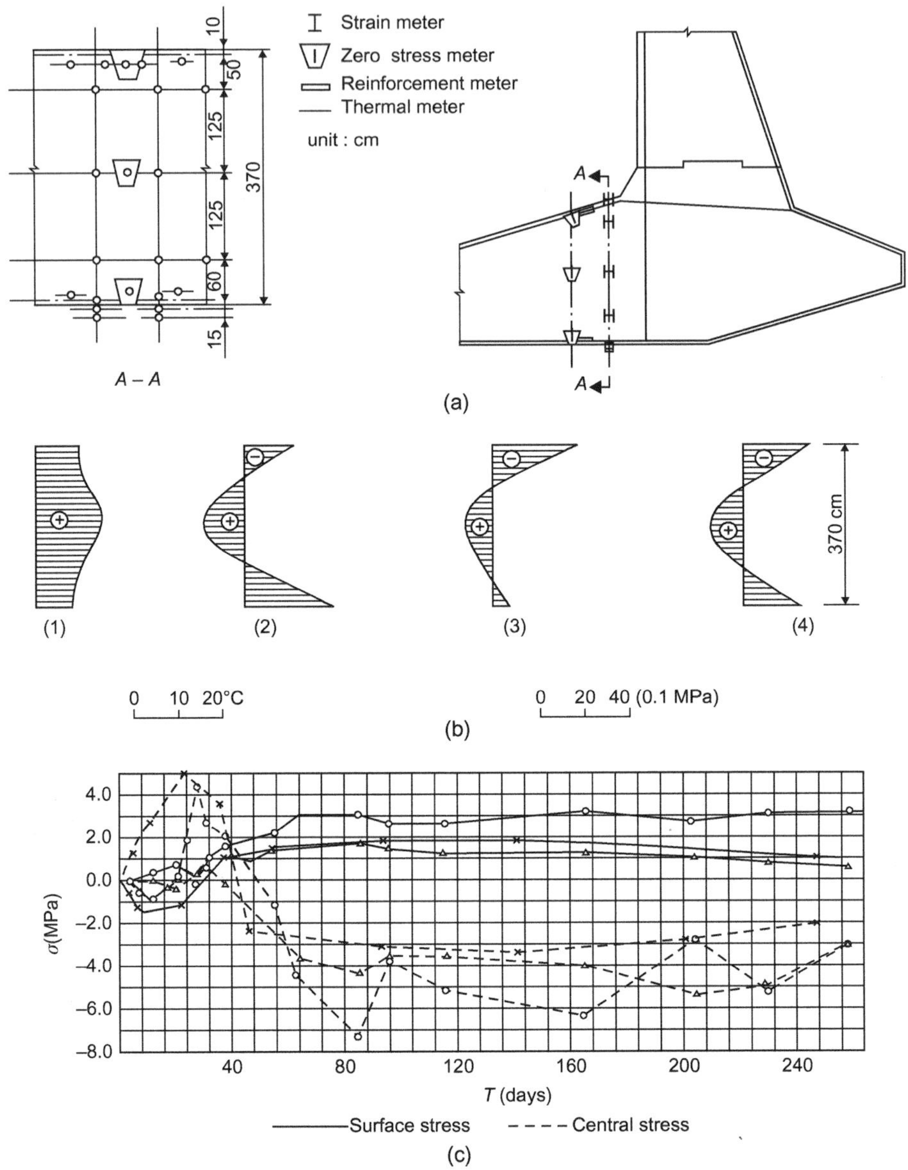

Figure 7.8 Observed results of the bottom slab of a sluice. (1) Observed temperatures, (2) stresses computed by Eq. (7.26) from the observed temperatures, (3) stresses computed from the measured strains, (4) stresses computed by Eq. (7.26) from the computed temperatures. (a) Cross section of observation, (b) temperature and stress distribution, (c) variation of stresses.

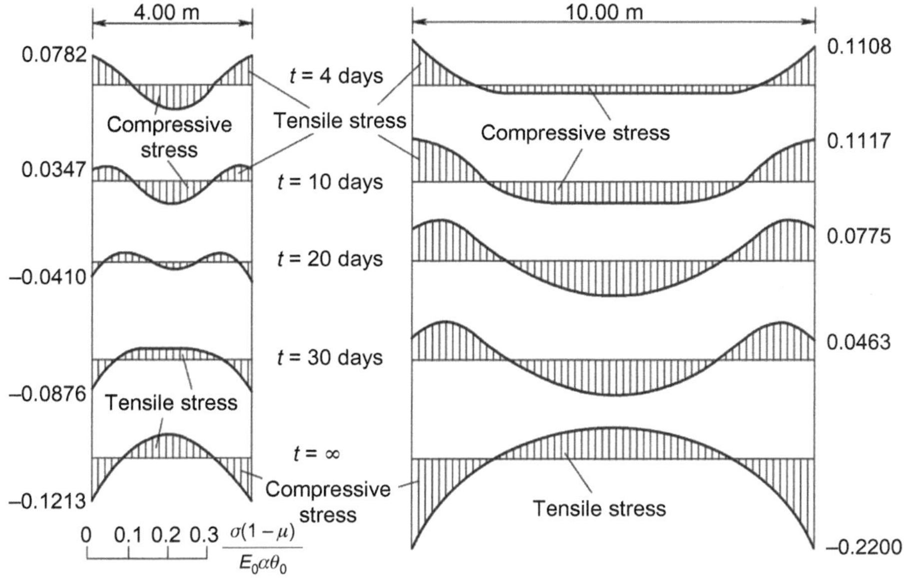

Figure 7.9 Viscoelastic thermal stress in a concrete slab due to hydration heat [unit: $\sigma(1-\mu)/E_0\alpha\theta_0$].

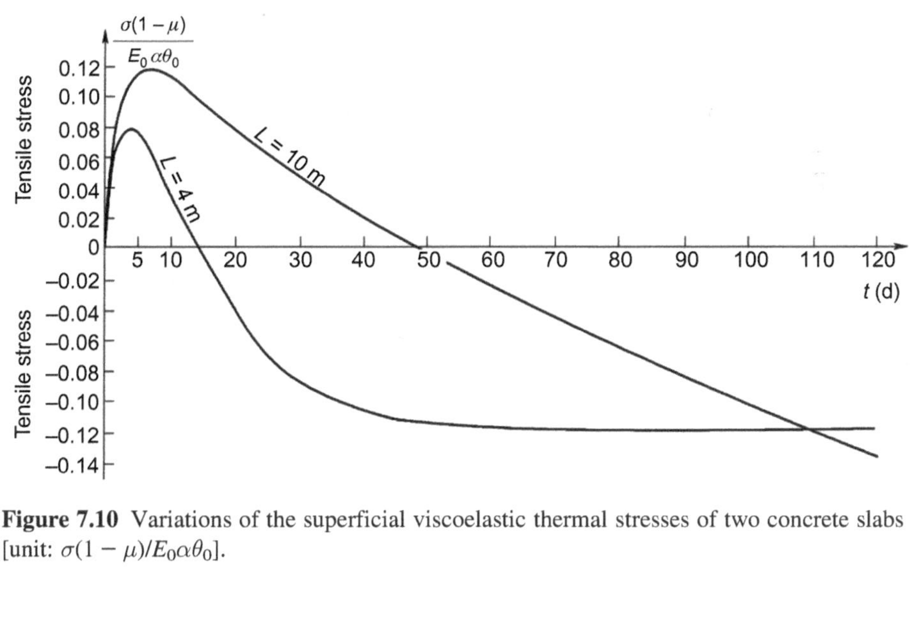

Figure 7.10 Variations of the superficial viscoelastic thermal stresses of two concrete slabs [unit: $\sigma(1-\mu)/E_0\alpha\theta_0$].

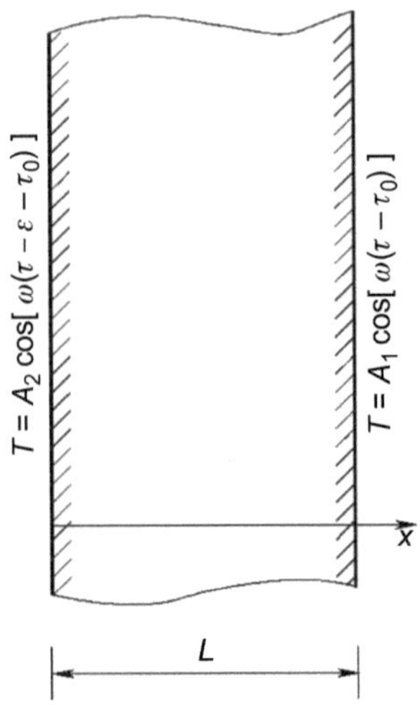

Figure 7.11 A slab.

By method of complex variable, the solution of Eq. (7.27) is

$$T(x,\tau) = A_1 k_1 \cos[\omega(\tau - \tau_0) + \varphi_1] + A_2 k_2 \cos[\omega(\tau - \varepsilon - \tau_0) + \varphi_2] \tag{7.28}$$

$$\left.\begin{aligned}
k_1 &= \sqrt{\frac{\operatorname{ch} 2\zeta - \cos 2\zeta}{\operatorname{ch} 2\zeta_0 - \cos 2\zeta_0}} \\
k_2 &= \sqrt{\frac{\operatorname{ch} 2(\zeta_0 - \zeta) - \cos 2(\zeta_0 - \zeta)}{\operatorname{ch} 2\zeta_0 - \cos 2\zeta_0}} \\
\varphi_1 &= \tan^{-1}\left(\frac{\tan \zeta}{\operatorname{th} \zeta}\right) - \tan\left(\frac{\tan \zeta_0}{\operatorname{th} \zeta_0}\right) \\
\varphi_2 &= \tan^{-1}\left[\frac{\tan(\zeta_0 - \zeta)}{\operatorname{th}(\zeta_0 - \zeta)}\right] - \tan\left(\frac{\tan \zeta_0}{\operatorname{th} \zeta_0}\right) \\
\zeta &= x/\eta, \quad \zeta_0 = L/\eta, \quad \eta = \sqrt{aP/\pi}
\end{aligned}\right\} \tag{7.29}$$

The mean temperature T_m and the equivalent linear temperature difference T_d are given by

$$T_m = \frac{1}{2} k_m [A_1 \cos \omega(\tau - \theta_m - \tau_0) + A_2 \cos \omega(\tau - \varepsilon - \theta_m - \tau_0)] \tag{7.30}$$

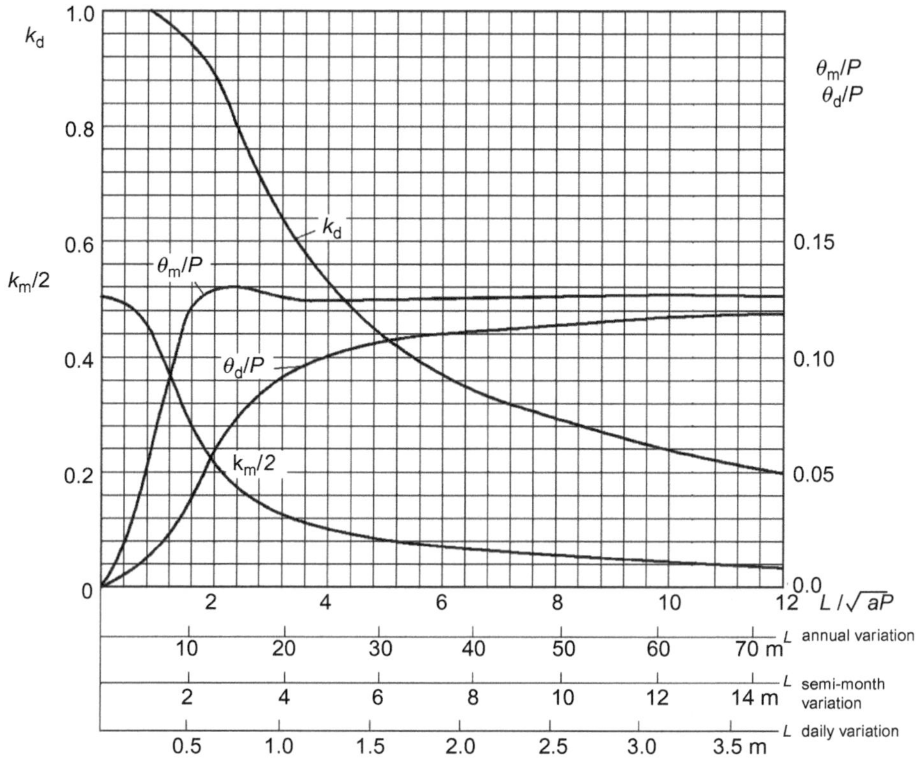

Figure 7.12 Coefficients $k_m/2, k_d, \theta_m/P, \theta_d/P$.

$$T_d = k_d[A_1 \cos \omega(\tau - \theta_d - \tau_0) - A_2 \cos \omega(\tau - \varepsilon - \theta_d - \tau_0)] \tag{7.31}$$

$$\begin{aligned}
k_m &= \frac{1}{\zeta_0}\sqrt{\frac{2(\text{ch}\,\zeta_0 - \cos\zeta_0)}{\text{ch}\,\zeta_0 + \cos\zeta_0}} \\
\theta_m &= \frac{1}{\omega}\left[\frac{\pi}{4} - \tan^{-1}\left(\frac{\sin\zeta_0}{\text{sh}\,\zeta_0}\right)\right] \\
k_d &= \sqrt{a_1^2 + b_1^2}, \quad \theta_d = \frac{1}{\omega}\tan^{-1}(b_1/a_1) \\
a_1 &= \frac{6\sin\omega\theta_m}{k_m\zeta_0^2}, \quad b_1 = \frac{6}{\zeta_0^2}\left(\frac{1}{k_m}\cos\omega\theta_m - 1\right)
\end{aligned} \tag{7.32}$$

The coefficients $k_m/2, k_d, \theta_m/P, \theta_d/P$ are shown in Figure 7.12.

7.4.2 The Viscoelastic Thermal Stresses

The elastic thermal stresses are computed by

$$\sigma(x,\tau)=\frac{E\alpha}{1-\mu}\left[T_{\mathrm{m}}(\tau)+T_{\mathrm{d}}(\tau)\left(\frac{x}{L}-\frac{1}{2}\right)-T(x,\tau)\right] \tag{7.33}$$

From Eq. (6.64), substituting E by ρE and τ by $\tau-\xi$, the formula for computing viscoelastic thermal stress is derived as follows:

$$\sigma(x,\tau)=\frac{\rho E\alpha}{1-\mu}\left[T_{\mathrm{m}}(\tau-\xi)+T_{\mathrm{d}}(\tau-\xi)\left(\frac{x}{L}-\frac{1}{2}\right)-T(x,\tau-\xi)\right] \tag{7.34}$$

where ρ and ξ are given in Eq. (6.63).

Example A free concrete slab with thickness $L=10$ m, diffusivity $a=2.88$ m^2/month, the amplitude of variation of surface temperature at $x=L$ is A_1 and at $x=0$, $A_2=0$. The temperatures in the slab computed by Eq. (7.28) are shown in Figure 7.13. The elastic thermal stresses given by Eq. (7.29) are shown in Figure 7.14.

7.5 Thermal Stress in Free Slab with Third Kind of Boundary Condition and Periodically Varying Air Temperature

An infinite slab is shown in Figure 7.15. On the left side, the surface conductance is β_1 and the air temperature is $T_{\mathrm{a}}=0$; on the right side, the surface conductance is β_2 and the air temperature is $T_{\mathrm{a}}=A_0\cos\omega\tau$. The quasi-steady problem to be solved is

$$\left.\begin{aligned}
&\text{Equation of heat conduction} && && \frac{\partial T}{\partial\tau}=a\frac{\partial^2T}{\partial x^2}\\
&\text{Boundary condition:} && \text{when}\quad x=0 && \lambda\frac{\partial T}{\partial\tau}=\beta_1T\\
& && \text{when}\quad x=L && -\lambda\frac{\partial T}{\partial\tau}=\beta_2(T-A_0\cos\omega\tau)
\end{aligned}\right\} \tag{7.35}$$

The temperature is

$$T(x,\tau)=A_0(g_1\cos\omega\tau+g_2\sin\omega\tau) \tag{7.36}$$

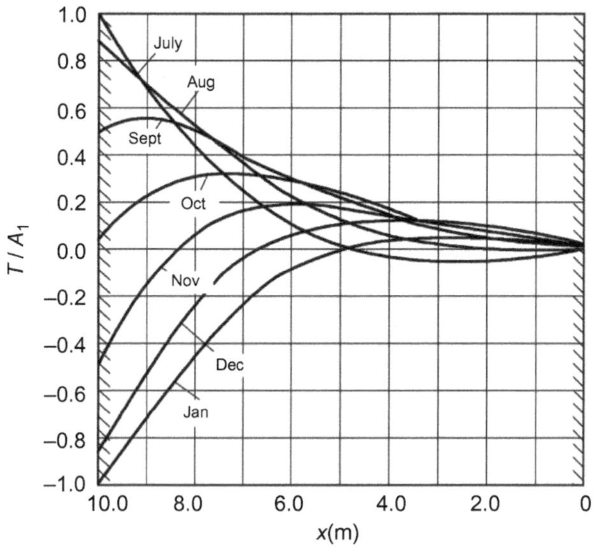

Figure 7.13 Example, $L = 10$ m, temperatures in the slab.

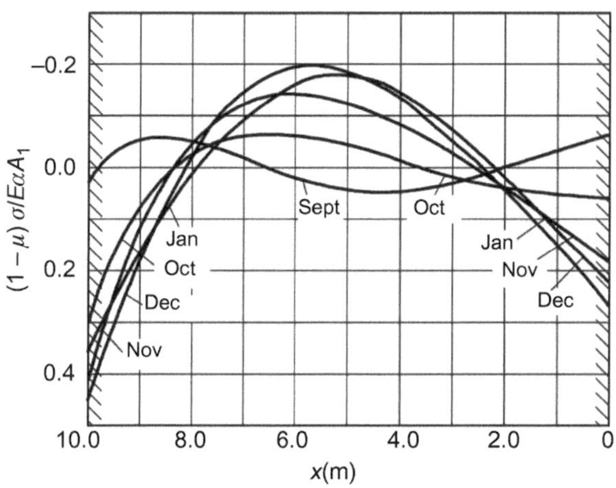

Figure 7.14 Example, $L = 10$ m, elastic thermal stresses in the slab.

The elastic thermal stress is

$$\sigma(x,\tau) = \frac{E\alpha}{1-\mu}\left[T_{\rm m} + \frac{T_{\rm d}}{L}\left(x - \frac{L}{2}\right) - T(x,\tau)\right] \tag{7.37}$$

where

$$T_{\rm m} = (A_0/L)(g_3 \cos \omega\tau + g_4 \sin \omega\tau) \tag{7.38}$$

$$\frac{T_{\rm d}}{L} = \frac{12}{L^3}\left(S_0 - \frac{L^2}{2}T_{\rm m}\right) \tag{7.39}$$

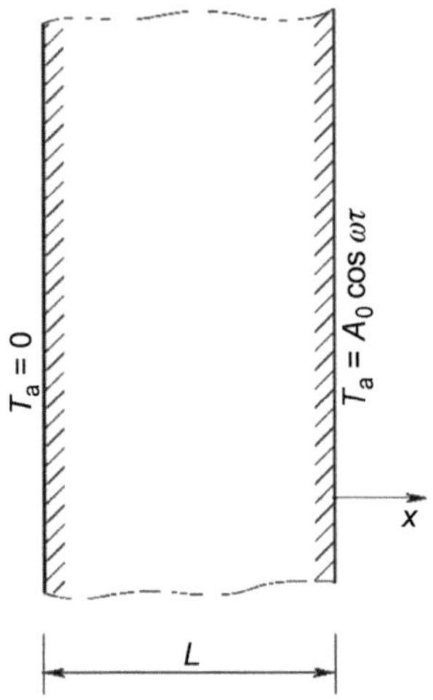

Figure 7.15 A free slab.

$$
\begin{aligned}
&S_0 = A_0(g_5 \cos \omega\tau + g_6 \sin \omega\tau) \qquad (7.40)\\
&g_1 = (a_1b_1 + a_2b_2)/({a^2}_1 + {a^2}_2)\\
&g_2 = (a_2b_1 - a_1b_2)/({a^2}_1 + {a^2}_2)\\
&g_3 = (a_1a_3 + a_2a_4)/({a^2}_1 + {a^2}_2)\\
&g_4 = (a_2a_3 - a_1a_4)/({a^2}_1 + {a^2}_2)\\
&g_5 = (a_1a_6 - a_2a_5)/[2q^2({a^2}_1 + {a^2}_2)]\\
&g_6 = (a_1a_5 + a_2a_6)/[2q^2({a^2}_1 + {a^2}_2)]\\
&a_1 = d_1 - d_2s_4 + s_3(d_3 - d_4), \quad a_2 = d_1s_4 + d_2 + s_3(d_3 + d_4)\\
&a_3 = \frac{1}{2q}(2s_1d_1 + d_3 + d_4 - 1), \quad a_4 = \frac{1}{2q}(2s_1d_2 - d_3 + d_4 + 1)\\
&a_5 = s_1(-2\varsigma_0d_1 - d_3 + d_4 + 1) + \varsigma_0(d_3 - d_4) - d_1\\
&a_6 = s_1(2\varsigma_0d_1 - d_3 - d_4 + 1) + \varsigma_0(d_3 + d_4) - d_2\\
&b_1(\varsigma) = s_1(f_1 - f_2) + f_3, \quad b_2(\varsigma) = s_1(f_1 + f_2) + f_4\\
&f_1(\varsigma) = \text{ch}\,\varsigma \cos \varsigma, \quad f_2(\varsigma) = \text{sh}\,\varsigma \sin \varsigma\\
&f_3(\varsigma) = \text{sh}\,\varsigma \cos \varsigma, \quad f_4(\varsigma) = \text{ch}\,\varsigma \sin \varsigma\\
&d_1 = \text{sh}\,\zeta_0 \cos \zeta_0, \quad d_2 = \text{ch}\,\zeta_0 \sin \zeta_0, \quad d_3 = \text{ch}\,\zeta_0 \cos \zeta_0, \quad d_4 = \text{sh}\,\zeta_0 \sin \zeta_0\\
&s_1 = \lambda q/\beta_1, \quad s_2 = \lambda q/\beta_2, \quad s_3 = s_1 + s_2, \quad s_4 = 2s_1s_2\\
&q = \sqrt{\pi/aP}, \quad \zeta = qx, \quad \zeta_0 = qL \qquad (7.41)
\end{aligned}
$$

The viscoelastic thermal stress is given by

$$\sigma(x,\tau) = \frac{\rho E\alpha}{1-\mu}\left[T_m(\tau-\xi) + T_d(\tau-\xi)\left(\frac{x}{L}-\frac{1}{2}\right) - T(x,\tau-\xi)\right] \qquad (7.42)$$

in which ρ and ξ are given by Eq. (6.63).

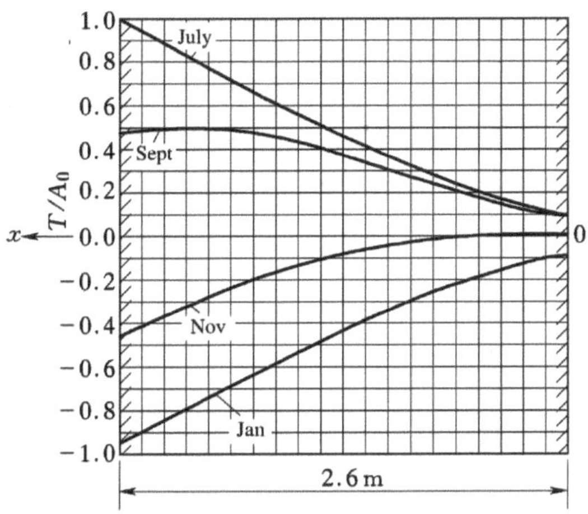

Figure 7.16 Example, distribution of temperatures in the slab with $L = 2.6$ m.

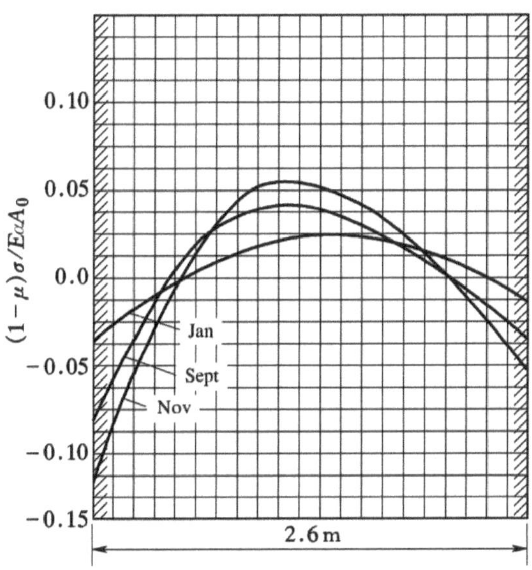

Figure 7.17 Example, elastic thermal stresses in the free slab with $L = 2.60$ m.

Example A free concrete slab with thickness $L = 2.6$ m, diffusivity $a = 0.00217$ m^2/h, conductivity $\lambda = 5.23$ kJ/(m h °C), surface conductance $\beta_1 = 25.1$ kJ/(m^2 h °C), $\beta_2 = 50.2$ kJ/(m^2 h °C). The temperatures in the slab given by Eq. (7.36) are shown in Figure 7.16. The elastic thermal stresses given by Eq. (7.37) are shown in Figure 7.17.

The computation of thermal stress in a free slab with third kind of boundary condition and harmonic air temperature is rather tedious; when the thickness of the slab is bigger than 10 m, the following approximate method can be used in practical engineering analysis: the amplitude of variation of air temperature A_0 is multiplied by a coefficient k, the new amplitude $A_1 = kA_0$ is used as the amplitude of variation of the surface temperature of the slab, k is given by

$$k = \left(1 + \frac{2\lambda}{\beta}\sqrt{\frac{\pi}{aP}} + \frac{2\pi\lambda^2}{aP\beta^2}\right)^{-1/2} \tag{7.43}$$

Coefficient k is given in Table 3.2.

7.6 Thermal Stresses Due to Removing Forms

Owing to the hydration heat of cement, the surface temperature of mass concrete is higher than the air temperature before removing forms. After removing forms, the surfaces of mass concrete come in contact with cold air, and the rapid drop of superficial temperature gives rise to remarkable tensile stress which sometimes may lead to cracking of concrete.

7.6.1 Stresses Due to Removing Forms of Infinite Slab

Consider an infinite slab as shown in Figure 5.8, the thickness is $L = 2l$, the initial temperature of concrete is T_0, the air temperature is T_a, the forms are removed at $\tau = 0$, the temperature in the slab is given by Eq. (5.25). From Eq. (7.21), the elastic stress in the slab is derived as follows:

$$\sigma = \frac{E\alpha(T_0 - T_a)}{1-\mu}(H - G) \tag{7.44}$$

where

$$\left.\begin{aligned} H &= \sum_{n=1}^{\infty} \frac{2\sin\mu_n^2}{\mu_n(\mu_n + \sin\mu_n\cos\mu_n)} e^{-F\mu_n^2} \\ G &= \sum_{n=1}^{\infty} \frac{2\sin\mu_n}{\mu_n + \sin\mu_n\cos\mu_n} \cos\mu_n\frac{x}{l}\cdot e^{-F\mu_n^2} \end{aligned}\right\} \tag{7.45}$$

in which $F = a\tau/l^2$, H represents the mean temperature and G represents the point temperature. $H - G$ of the surface ($x = l$) and the center ($x = 0$) are shown in Figure 7.18, in which $c = \beta l/\lambda$ and $F = a\tau/l^2$ are the Biot's number and the Fourier's number, respectively.

For example, for a concrete slab with thickness $2l = 4.0$ m, $\lambda/\beta = 0.20$ m, $c = \beta l/\lambda = 2.0/0.20 = 10.0$, the variations of the thermal stress $\sigma(1-\mu)/E\alpha(T_0 - T_a)$ of the surface and the center are shown in Figure 7.18 by the curves with $c = 10.0$.

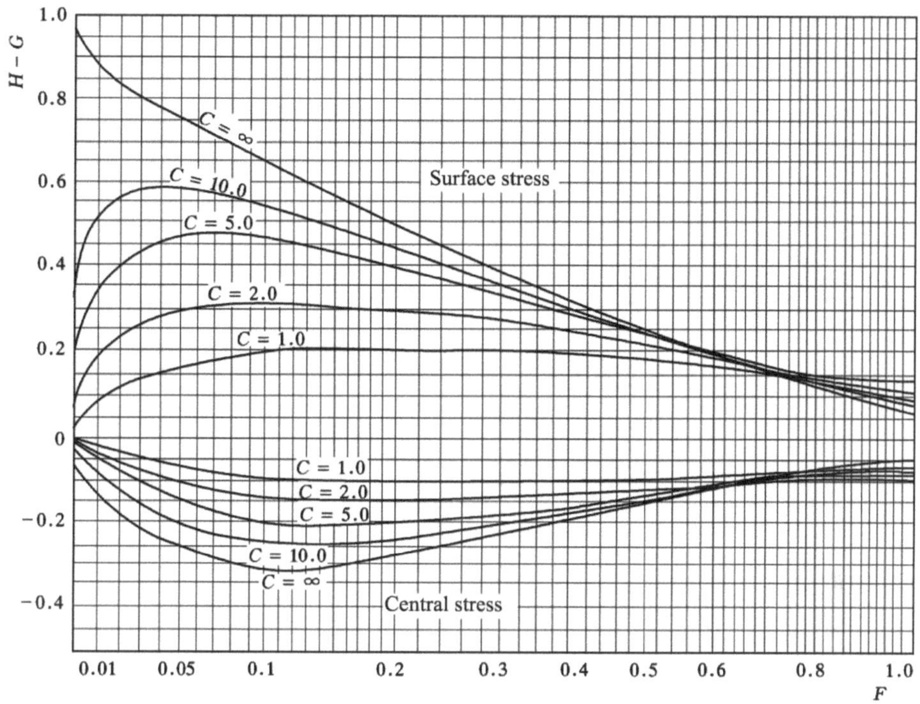

Figure 7.18 Stresses in an infinite concrete slab due to removing forms ($C=\beta l/\lambda$, $F=a\tau/l^2$).

7.6.2 Stresses Due to Removing Forms of Semi-infinite Solid

Consider the semi-infinite solid $x \geq 0$ as shown in Figure 7.19, the initial temperature of concrete is T_0, the air temperature is T_a, the forms are removed when $\tau = 0$. According to Section 5.1 the surface temperature is

$$\left.\begin{array}{l} T(0,\tau) = T_0 + (T_a - T_0)\varphi(\xi) \\ \varphi(\xi) = 1 - e^{\xi^2}(1 - \operatorname{erf} \xi) \\ \xi = \beta\sqrt{a\tau}/\lambda \end{array}\right\} \tag{7.46}$$

The superficial thermal stress is

$$\sigma = -\frac{E\alpha}{1-\mu}[T(0,\tau) - T_0] = \frac{E\alpha}{1-\mu}(T_0 - T_a)\varphi(\xi) \tag{7.47}$$

$\varphi(\xi)$ is shown in Figure 7.20.

Example The semi-infinite solid with $a = 0.10\ \text{m}^2/\text{day}$, $\lambda = 8.0\ \text{kJ/(m h °C)}$, $\beta = 40.0\ \text{kJ/(m}^2\ \text{h °C)}$, initial temperature T_0, air temperature T_a, the forms are removed when $\tau = 0$.

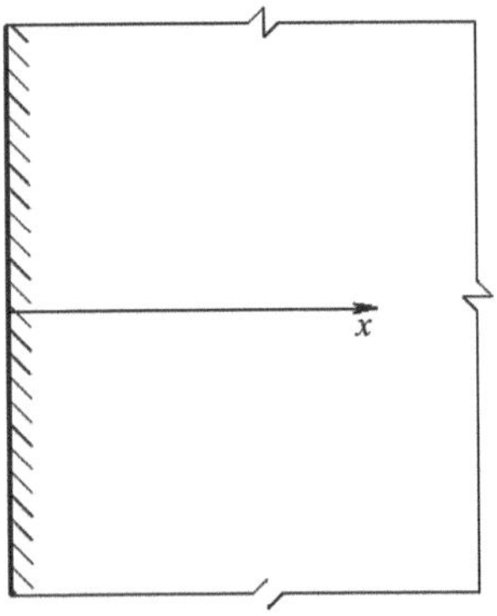

Figure 7.19 Semi-infinite solid.

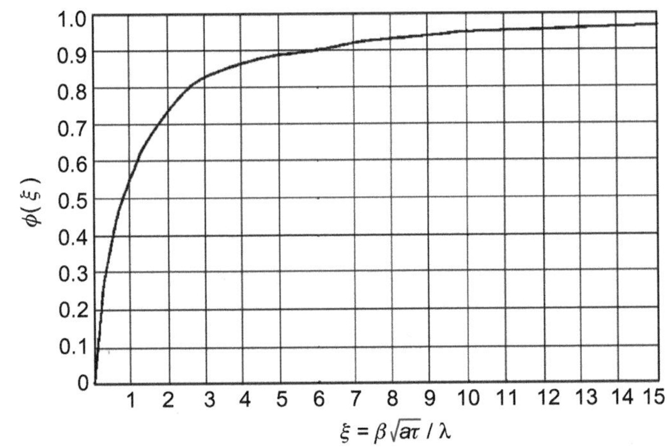

Figure 7.20 Curve $\phi(\xi)$ for semi-infinite solid.

$$\xi = \frac{\sqrt{a\tau}}{\lambda/\beta} = \frac{\sqrt{0.10\tau}}{0.20} = 1.581\sqrt{\tau}$$

The thermal stresses are computed by Eqs. (7.46) and (7.47) and $\sigma(1-\mu)/E\alpha(T_0 - T_a)$ is shown in Figure 7.21. The thermal stress increased rapidly in the first 2 h.

If the concrete surface is covered by foamed plastic plate 1 cm thick with conductivity $\lambda_1 = 0.125$ kJ/(m h °C) immediately after removing forms, the equivalent surface conductance of concrete is

$$\beta = \frac{1}{1/40 + 0.01/0.125} = 9.524 \text{ kJ/(m}^2 \text{ h °C)}, \quad \lambda/\beta = 8.0/9.524 = 0.840 \text{ m},$$

$$\xi = \sqrt{0.10\tau}/0.840 = 0.3765\sqrt{\tau}[0, 1]$$

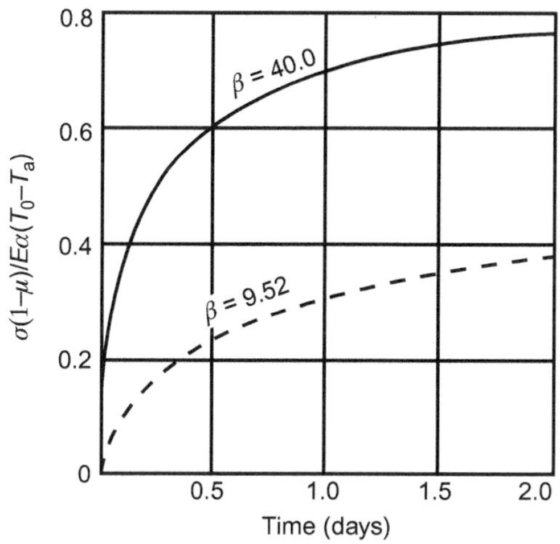

Figure 7.21 Example, stresses due to removing forms of semi-infinite solid.

In this case, the superficial thermal stress is shown by the dotted curve in Figure 7.21. The stress reduced about one half.

7.6.3 Computing Thermal Stress Due to Removing Forms by Finite Element Method

The finite element method (FEM) is suitable for computing the thermal stress due to removing of forms in some complex engineering problems. The effect of removing forms may be considered by the sudden change of surface conductance β. From Eq. (8.78), it is clear that when the temperature field is computed by FEM, only the following two quantities are related to the surface conductance β:

$$g_{ij}^e = \frac{\lambda}{\beta}\int_{\Delta}\int_{C} N_i N_j \, \mathrm{d}s, \quad p_i^e = \frac{\lambda}{\beta}\int_{\Delta}\int_{C} N_i \, \mathrm{d}s \tag{7.48}$$

in which g_{ij}^e and p_i^e are area integrals on the third kind of boundary C. In order to take account of the influence of removing forms on the thermal stress, β_1 and β_2 are used before and after removing forms in the computing ofg_{ij}^e and p_i^e. There is one point for attention: after removing forms, the temperature gradient is large near the surface and the surface temperature drops rapidly, so smaller elements near the surface and smaller time increment $\Delta\tau_i$ must be adopted.

Example A free concrete slab with thickness 6 m, $a = 0.0040$ m²/day, $\lambda = 8.40$ kJ/(m h °C), $\theta(\tau) = 25.0\tau/(1.0 + \tau)$°C, $E(\tau) = 30{,}000[1 - \exp(-0.40\tau^{0.34})]$ MPa, unit creep given by Eq. (6.51).

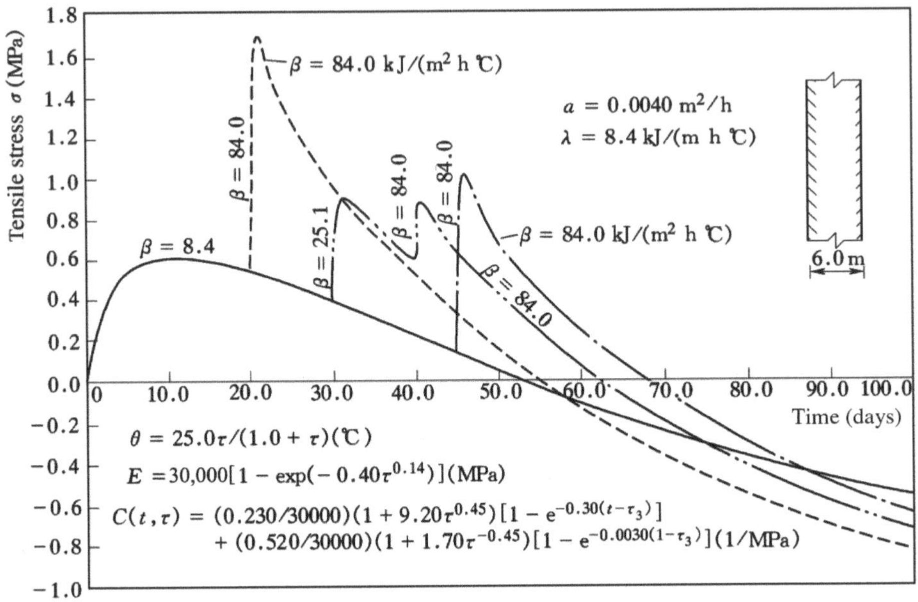

Figure 7.22 Variations of the surface thermal stresses before and after removing forms computed by FEM.

The thermal stresses in the following four cases are computed by FEM:

1. the forms are not removed, $\tau = 0 \sim 100$ days, $\beta = 8.40$ kJ/(m^2 h °C);
2. removing forms at $\tau = 20$ days, $\tau = 1 \sim 20$ days, $\beta = 8.40$ kJ/(m^2 h °C); $\tau = 20 - 100$ days, $\beta = 84.0$ kJ/(m^2 h °C);
3. removing forms at $\tau = 45$ days, $\tau = 0 - 45$ days, $\beta = 8.40$ kJ/(m^2 h °C); $\tau = 45 - 100$ days, $\beta = 84.0$ kJ/(m^2 h °C),
4. $\tau = 0 - 30$ days, $\beta = 8.4$ kJ/(m^2 h °C); when $\tau = 30$ days, the forms are removed and the surface of concrete is covered by insulation layer with $\beta = 25.1$ kJ/(m^2 h °C) which is removed at $\tau = 40$ days, then the surface is in contact with air and $\beta = 84.0$ kJ/(m^2 h °C).

The results of computation are shown in Figure 7.22. It is clear that the tensile stresses due to removing forms are remarkable.

8 Thermal Stresses in Concrete Beams on Elastic Foundation

The method for computing the thermal stresses in concrete beams on elastic foundation proposed by the author in 1977 is discussed in this chapter. This method was adopted in the design specifications of gravity dam, arch dam, and docks in China. The thermal stresses in a beam on elastic foundation are the sum of the self-stresses and the restraint stresses [45,73].

8.1 Self-Thermal Stress in a Beam

A multilayered beam on elastic foundation with length $2l$ is shown in Figure 8.1. The temperature $T(y)$ in the beam varies with y, the modulus of elasticity of the ground is E_f and the modulus of the concrete layers are $E_1, E_2, \ldots$

The beam and the foundation are cut along the plane of contact to make the beam free of restraint of the foundation. Hence, the self-stress in the beam is given by

$$\sigma_x = \frac{E(y)\alpha}{1-\mu}[T_m + \Psi y - T(y)] \tag{8.1}$$

where

$$T_m = \frac{\int E(y)T(y)\mathrm{d}y}{\int E(y)\mathrm{d}y} \tag{8.2}$$

$$\Psi = \frac{\int E(y)T(y)y\,\mathrm{d}y}{\int E(y)y^2\,\mathrm{d}y} \tag{8.3}$$

T_m is the mean temperature with weight $E(y)$, and Ψ is the equivalent linear temperature gradient with weight $E(y)$.

Thermal Stresses and Temperature Control of Mass Concrete. DOI: http://dx.doi.org/10.1016/B978-0-12-407723-2.00008-7

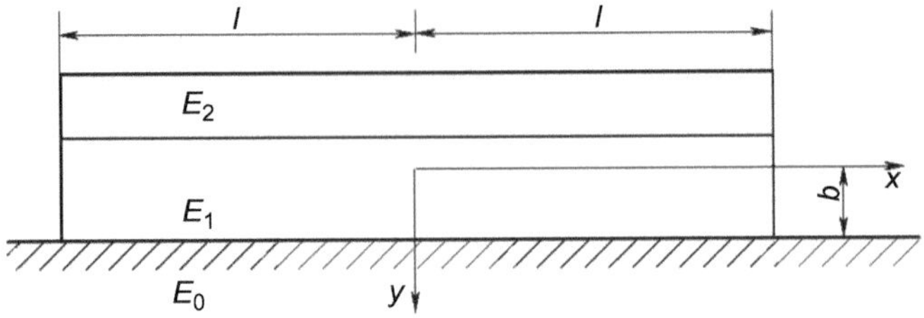

Figure 8.1 Heterogeneous beam on elastic foundation.

Putting the origin of ordinate y on the weighted centroid of the section, the height b of which is determined by

$$\int E(y)y\,\mathrm{d}y = 0 \tag{8.4}$$

If the origin of ordinate y' is put on the bottom plane of the beam, then

$$b = \int E(y')y'\,\mathrm{d}y' / \int E(y')\mathrm{d}y' \tag{8.5}$$

Under the action of temperature and self-stress, the axial displacement w_T and the vertical displacement y_T are respectively

$$w_T = (1+\mu)\alpha T_{\mathrm{m}} x \tag{8.6}$$

$$y_T = \frac{(1+\mu)\alpha\Psi}{2} x^2 \tag{8.7}$$

For a homogeneous beam with height $2h$, width $c = 1$, length $2l$, $b = h$, the self-stress is given by

$$\sigma_x = \frac{E\alpha}{1-\mu}\left[T_{\mathrm{m}} + T_{\mathrm{d}}\frac{y}{2h} - T(y)\right] \tag{8.8}$$

where

$$T_{\mathrm{m}} = \frac{1}{2h}\int_{-h}^{h} T\,\mathrm{d}y \tag{8.9}$$

$$T_{\mathrm{d}} = \frac{3}{2h^2}\int_{-h}^{h} Ty\,\mathrm{d}y \tag{8.10}$$

8.2 Restraint Thermal Stress of Beam on Foundation of Semi-infinite Plane

8.2.1 Nonhomogeneous Beam on Elastic Foundation

Let

$$D = \frac{1}{1-\mu^2}\int E(y)y^2\,\mathrm{d}y \tag{8.11}$$

$$F = \frac{1}{1-\mu^2}\int E(y)\mathrm{d}y \tag{8.12}$$

the forces acting on the beam are shown in Figure 8.2, from the condition of balance of forces, the following equations are derived:

$$\frac{\mathrm{d}M}{\mathrm{d}x} = V - b\tau \tag{8.13}$$

$$\frac{\mathrm{d}V}{\mathrm{d}x} = -p \tag{8.14}$$

$$\frac{\mathrm{d}P}{\mathrm{d}x} = \tau \tag{8.15}$$

where

M—the moment
V—the shearing force
P—the axial force
p—the normal stress on the plane of contact
τ—the shearing stress on the plane of contact
x—the abscissa.

From the stress–strain relation and the hypothesis of plane section of beam, we get

$$M = D\frac{\mathrm{d}^2y}{\mathrm{d}x^2} \tag{8.16}$$

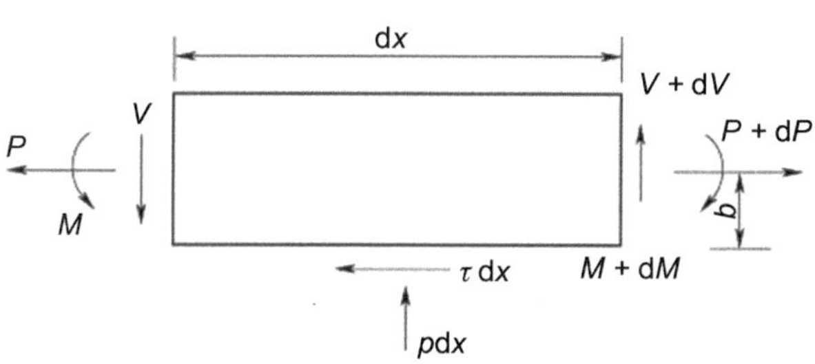

Figure 8.2 The forces acting on the beam.

$$P = F\frac{dw}{dx} \tag{8.17}$$

where

y—the vertical displacement of beam
w—the axial displacement of beam.

Substitution of the above two equations into Eqs. (8.13)–(8.15) yields the following equations:

$$D\frac{d^3y}{dx^3} = V - b\tau \tag{8.18}$$

$$D\frac{d^4y}{dx^4} = -p - b\frac{d\tau}{d\xi} \tag{8.19}$$

$$F\frac{d^2w}{dx^2} = \tau \tag{8.20}$$

For transforming the abscissa into a dimensionless number, let

$$\xi = x/l \tag{8.21}$$

By substituting Eq. (8.21) into Eqs. (8.16)–(8.20), we have

$$M = \frac{D}{l^2}\frac{d^2y}{d\xi^2} \tag{8.22}$$

$$V = \frac{D}{l^3}\frac{d^3y}{d\xi^3} + b\tau \tag{8.23}$$

$$P = \frac{F}{l}\frac{dw}{d\xi} \tag{8.24}$$

$$\frac{D}{l^4}\frac{d^4y}{d\xi^4} + p + \frac{b}{l}\frac{d\tau}{d\xi} = 0 \tag{8.25}$$

$$\frac{F}{l^2}\frac{d^2w}{d\xi^2} - \tau = 0 \tag{8.26}$$

After solution of the simultaneous Eqs. (8.25) and (8.26), we will get $y(\xi)$ and $w(\xi)$ satisfying the boundary conditions. By substituting them into Eqs. (8.22)–(8.24), M, V, and P are obtained.

Due to symmetry, the normal stress $p(\xi)$ on the contact surface is an even function and the shearing stress $\tau(\xi)$ is an odd function.

Let

$$p(\xi) = \frac{A_0 H_0(\xi)}{\sqrt{1-\xi^2}} + \frac{A_2 H_2(\xi)}{\sqrt{1-\xi^2}} + \frac{A_4 H_4(\xi)}{\sqrt{1-\xi^2}} + \ldots \tag{8.27}$$

$$\tau(\xi) = \frac{B_1 H_1(\xi)}{\sqrt{1-\xi^2}} + \frac{B_3 H_3(\xi)}{\sqrt{1-\xi^2}} + \ldots \tag{8.28}$$

where

$$H_n(\xi) = \cos(n \cos^{-1} \xi) \tag{8.29}$$

$H_n(\xi)$ is a Chebyshev polynomial which can be expressed explicitly as follows:

$$\left.\begin{array}{l} H_0(\xi) = 1, \quad H_1(\xi) = \xi, \quad H_2(\xi) = 2\xi^2 - 1, \\ H_3(\xi) = 4\xi^3 - 3\xi, \quad H_4(\xi) = 8\xi^4 - 8\xi^2 + 1, \ldots \end{array}\right\} \tag{8.30}$$

$H_n(\xi)$ is an orthogonal polynomial with weight $(1-\xi^2)^{-1/2}$ as follows:

$$\int_{-1}^{+1} \frac{H_n(\xi) H_m(\xi) \mathrm{d}\xi}{\sqrt{1-\xi^2}} = \begin{cases} 0, & \text{when } n \neq m \\ \pi/2, & \text{when } n = m > 0 \\ \pi, & \text{when } n = m = 0 \end{cases} \tag{8.31}$$

Taking $H_m(\xi) = H_0(\xi) = 1$, from Eqs. (8.27) and (8.31), we have

$$\int_{-1}^{+1} p(\xi) \mathrm{d}\xi = \int_{-1}^{+1} \left[\frac{A_0 H_0(\xi)}{\sqrt{1-\xi^2}} + \frac{A_2 H_2(\xi)}{\sqrt{1-\xi^2}} + \cdots \right] H_0(\xi) \mathrm{d}\xi = A_0 \pi \tag{8.32}$$

Because there is no external load on the beam, by the condition of balance $\int_{-1}^{+1} p(\xi) \mathrm{d}\xi = 0$, from Eq. (8.32) we know that

$$A_0 = 0 \tag{8.33}$$

the other coefficients $A_2, A_4, B_1, B_3, \ldots$ must be determined by the condition of continuity of deformations at the contact surface. Due to the peculiarity of the Chebyshev polynomial, the condition of continuity of deformation is very simple. Experience shows that the computed results are satisfactory even when only one term is taken in $p(\xi)$ and in $\tau(\xi)$, thus, the problem is simplified a great deal. Substituting $p(\xi)$ and $\tau(\xi)$ into Eq. (8.25) and letting $A_0 = 0$, after integration, we get

$$y(\xi)=\frac{l^4}{D}\left\{\frac{C_1\xi^3}{6}+\frac{C_2\xi^2}{2}+C_3\xi+C_4-\frac{A_2}{120}\left[15\xi\sin^{-1}\xi-(2\xi^4-9\xi^2-8)\sqrt{1-\xi^2}\right]\right.$$
$$\left.+\frac{B_1 r}{6}\left[(\xi^2+2)\sqrt{1-\xi^2}+3\xi\sin^{-1}\xi\right]+\cdots\right\} \tag{8.34}$$

where

$$r = b/l \tag{8.35}$$

After differentiating with respect to ξ, we get

$$y'(\xi)=\frac{l_4}{D}\left\{\frac{C_1\xi^2}{2}+C_2\xi+C_3-\frac{A_2}{24}\left[(5-2\xi^2)\xi\sqrt{2-\xi^2}+3\sin^{-1}\xi\right]\right.$$
$$\left.+\frac{B_1 r}{2}\left(\xi\sqrt{1-\xi^2}+\sin^{-1}\xi\right)+\cdots\right\} \tag{8.36}$$

From Eqs. (8.22) and (8.23), we have

$$M = l^2\left[C_1\xi + C_2 - \frac{A_2}{3}(1-\xi^2)\sqrt{1-\xi^2} + B_1 r\sqrt{1-\xi^2} + \cdots\right] \tag{8.37}$$

$$V = l\left[C_1 + A_2\xi\sqrt{1-\xi^2} + \cdots\right] \tag{8.38}$$

The boundary conditions of deflection $y(\xi)$ are as follows:

$$\left.\begin{array}{ll}\text{when } \xi = 0 & y'(\xi)=0,\ V(\xi)=0\\ \text{when } \xi = \pm 1 & M(\xi)=0,\ V(\xi)=0\end{array}\right\} \tag{8.39}$$

from which we get

$$C_1 = C_2 = C_3 = 0$$

One condition in Eq. (8.39) coincides with $A_0 = 0$, so only three coefficients can be determined and the coefficient C_4 remains to be computed. Letting $\xi = 0$ in Eq. (8.34), we have

$$y(0) = \frac{l^4}{D}\left(C_4 - \frac{A_2}{15} + \frac{B_1 r}{3} + \cdots\right) \tag{8.40}$$

In order to eliminate C_4, we take the relative displacement $y^0(\xi) = y(\xi) - y(0)$, namely

$$\left.\begin{aligned} y^0(\xi) &= y(\xi) - y(0) \\ &= \frac{l^4}{D}\left\{ -\frac{A_2}{120}\left[15\xi \sin^{-1}\xi - (2\xi^4 - 9\xi^2 - 8)\sqrt{1-\xi^2} - 8\right]\right. \\ &\quad \left. + \frac{B_1 r}{6}\left[(\xi^2+2)\sqrt{1-\xi^2} + 3\xi \sin^{-1}\xi - 2\right] + \cdots \right\} \\ &\approx \frac{l^4}{D}\left[\xi^2\left(-\frac{A_2}{6} + \frac{B_1 r}{2}\right) + \cdots\right] \end{aligned}\right\} \tag{8.41}$$

The axial displacement $w(\xi)$ must satisfy the following boundary condition:

$$\left.\begin{aligned} &\text{when } \xi = 0, \quad w = 0 \\ &\text{when } \xi = \pm 1, \quad P(\xi) = \frac{F}{l}\frac{dw}{d\xi} = 0 \end{aligned}\right\} \tag{8.42}$$

Substituting Eq. (8.28) into Eq. (8.26), the general solution of $w(\xi)$ is obtained and the coefficients of $w(\xi)$ are determined by the above boundary conditions, finally we get the solution of $w(\xi)$ as follows:

$$w(\xi) = \frac{B_1 l^2}{F}\left(\frac{\xi}{2}\sqrt{1-\xi^2} + \frac{1}{2}\sin^{-1}\xi\right) + \cdots = -\frac{B_1 l^2 \xi}{F} + \cdots \tag{8.43}$$

From the theory of elasticity, the superficial displacements of a semi-infinite half plane due to the action of superficial forces $p(\xi)$ and $\tau(\xi)$ are given by

$$u = -\frac{(1+\mu_f)(1-2\mu_f)l}{2E_f}\left[\int_{-1}^{\xi} p(\eta)d\eta - \int_{\xi}^{+1} p(\eta)d\eta\right]$$
$$-\frac{2(1-\mu_f^2)}{\pi E_f}\int_{-1}^{+1} \tau(\eta)\ln|\xi - \eta| d\eta$$

$$v = -\frac{2(1-\mu_f^2)l}{\pi E_f}\int_{-1}^{+1} p(\eta)\ln|\xi-\eta| d\eta$$
$$+\frac{(1+\mu_f)(1-2\mu_f)l}{2E_f}\left[\int_{-1}^{\xi}\tau(\eta)d\eta - \int_{\xi}^{+1}\tau(\eta)d\eta\right]$$

where E_f and μ_f are respectively the modulus of elasticity and Poisson's ratio of the foundation.

Substituting Eqs. (8.27) and (8.28) into the above two equations, we have

$$\begin{aligned} u(\xi) &= \frac{(1+\mu_f)(1-2\mu_f)l}{E_f}\left[A_2\xi\sqrt{1-\xi^2}+\cdots\right]+\frac{2(1-\mu_f^2)l}{E_f}[B_1\xi+\cdots] \\ &\approx \frac{(1-\mu_f^2)l\xi}{E_f}(sA_2+2B_1)+\cdots \end{aligned} \tag{8.44}$$

$$\begin{aligned} v(\xi) &= \frac{(1-\mu_f^2)lA_2}{E_f}(2\xi^2-1)+\frac{(1+\mu_f)(1-2\mu_f)l}{E_f}\left(-B_1\sqrt{1-\xi^2}\right)+\cdots \\ v^0(\xi) &= v(\xi)-v(0)=\frac{2(1-\mu_f^2)lA_2}{E_f}\xi^2+\frac{(1+\mu_f)(1-2\mu_f)lB_1}{E_f}\left(1-\sqrt{1-\xi^2}\right)+\cdots \\ &\approx \frac{(1-\mu_f^2)l}{E_f}\left(2A_2+\frac{s}{2}B_1\right)\xi^2+\cdots \end{aligned} \tag{8.45}$$

where

$$s=\frac{(1+\mu_f)(1-2\mu_f)}{1-\mu_f^2}=\frac{1-2\mu_f}{1-\mu_f} \tag{8.46}$$

The axial displacement of the beam on the contact surface is $w+w_T-b$ $(\mathrm{d}y/\mathrm{d}x+\mathrm{d}y_T/\mathrm{d}x)$, where the third term is due to the rotation of the section. Assuming that the height of the beam is far smaller than its length, so the effect of rotation may be neglected, then the condition of continuity of displacements on the surface of contact is

$$\left.\begin{aligned} w+w_T&=u \\ y+y_T&=v \end{aligned}\right\} \tag{8.47}$$

Substituting the expressions of w_T, y_T, w, y, u, v into the above equations and comparing the coefficients of both sides, we get

$$\left.\begin{aligned} C_{11}B_1 + C_{12}A_2 &= \beta \\ C_{21}B_1 + C_{22}A_2 &= -\lambda \end{aligned}\right\} \tag{8.48}$$

where

$$\left.\begin{aligned} &C_{11} = 15(kr - s), \quad C_{12} = -5k - 60, \quad C_{21} = -15i - 30, \quad C_{22} = -15s \\ &\beta = \frac{15(1+\mu)\alpha\Psi E_{\mathrm{f}} l}{1-\mu_{\mathrm{f}}^2}, \quad \lambda = \frac{15(1+\mu)\alpha E_{\mathrm{f}} T_{\mathrm{m}}}{1-\mu_{\mathrm{f}}^2} \\ &s = \frac{1-2\mu_{\mathrm{f}}}{1-\mu_{\mathrm{f}}}, \quad i = \frac{E_{\mathrm{f}} l}{(1-\mu_{\mathrm{f}}^2)F}, \quad k = \frac{E_{\mathrm{f}} l^3}{(1-\mu_{\mathrm{f}}^2)D}, \quad r = \frac{b}{l} \end{aligned}\right\} \tag{8.49}$$

Solution of Eq. (8.48) yields

$$B_1 = \frac{C_{22}\beta + C_{12}\lambda}{C_{11}C_{22} - C_{12}C_{21}}, \quad A_2 = \frac{-C_{21}\beta - C_{11}\lambda}{C_{11}C_{22} - C_{12}C_{21}} \tag{8.50}$$

From Eqs. (8.22) and (8.24), we get

$$M(\xi) = l^2\left[B_1 r\sqrt{1-\xi^2} - \frac{A_2}{3}(1-\xi^2)\sqrt{1-\xi^2}\right] \tag{8.51}$$

$$P(\xi) = -B_1 l\sqrt{1-\xi^2} \tag{8.52}$$

We are most interested in the stresses on the central section where the stresses are maximum; letting $\xi = 0$ in the above two equations, we get the moment M_0 and the axial force P_0 on the central section:

$$M_0 = l^2\left(B_1 r - \frac{A_2}{3}\right), \quad P_0 = -B_1 l \tag{8.53}$$

The stresses in the beam due to moment M and axial force P are given by

$$\sigma = \frac{E(y)}{1-\mu^2}\left(\frac{P}{F} - \frac{yM}{D}\right) \tag{8.54}$$

which are the stresses in the beam due to the restraint of foundation; after superposition with the self-stresses given by Eq. (8.1), the elastic thermal stresses in the beam is obtained. For beam of multilayers, different modulus of elasticity of concrete must be adopted in Eq. (8.54) for different layers.

8.2.2 Homogeneous Beam on Elastic Foundation

The computation may be simplified for a homogeneous beam. Let the thickness of the beam be $2h$, the length be $2l$, then

$$b = h, \quad F = \frac{2Eh}{1-\mu^2}, \quad D = \frac{2Eh^3}{3(1-\mu^2)}$$

Let

$$r = \frac{h}{l}, \quad \eta = \frac{E}{E_f}, \quad \mu = \mu_f = \frac{1}{6}$$

and substitute them into Eqs. (8.2), (8.3), (8.50), and (8.53), we get

$$T_m = \frac{1}{2h}\int_{-h}^{+h} T\,dy \tag{8.55}$$

$$\Psi = \frac{3}{2h^3}\int_{-h}^{+h} yT\,dy = \frac{T_d}{2h} \tag{8.56}$$

$$B_1 = \left(\frac{22.5}{\eta r^3} + 180\right)\frac{E\alpha T_m}{(1-\mu)\Delta} + \frac{36E\alpha\Psi h}{(1-\mu)r\Delta} \tag{8.57}$$

$$A_2 = \left(\frac{67.5}{\eta r^2} - 36\right)\frac{E\alpha T_m}{(1-\mu)\Delta} - \left(\frac{22.5}{\eta r^2} + \frac{90}{r}\right)\frac{E\alpha\Psi h}{(1-\mu)\Delta} \tag{8.58}$$

$$\Delta = 331.2\eta + \frac{90}{r} + \frac{54}{r^2} + \frac{45}{r^3} + \frac{11.25}{\eta r^4} \tag{8.59}$$

$$P_0 = -\,lB_1 \tag{8.60}$$

$$M_0 = l_2\left(B_1 r - \frac{A_2}{3}\right) \tag{8.61}$$

The restraint stresses at the top and bottom of the beam are given by

$$\left.\begin{matrix}\sigma_{top}\\ \sigma_{bottom}\end{matrix}\right\} = \frac{P_0}{2h} \pm \frac{3M_0}{2h^2} \tag{8.62}$$

The restraint stress varies linearly between the top and bottom of a homogeneous beam, but the self-stress is generally nonlinear due to the nonlinear distribution of temperature.

Example 1 A homogeneous beam on the rock foundation, $h/l = 0.25$, $E = E_f$, $\mu = \mu_f = 1/6$, there is a uniform temperature drop T in the beam. From Eqs. (8.55)

Table 8.1 The Elastic Thermal Stress $\sigma_x(1-\mu)/E\alpha T$ of a Homogeneous Beam on Rock Foundation ($E = E_f$) due to Uniform Temperature Drop

l/h	Computed Stresses		Stresses by Test	
	Top	**Bottom**	**Top**	**Bottom**
4	− 0.26	− 0.63	− 0.22	− 0.62
8	− 0.59	− 0.68	− 0.57	− 0.68
12.5	− 0.73	− 0.75	− 0.79	− 0.80

and (8.56), $T_m = T$, $\Psi = 0$; from Eq. (8.1), the self-stress is zero. From Eq. (8.59), $\Delta = 7315$; from Eqs. (8.50)–(8.62), we get

$$P_0 = -0.221l\frac{E\alpha T}{1-\mu}, \quad M_0 = 0.00778l^2\frac{E\alpha T}{1-\mu}$$

$$\sigma_{\text{top}} = -0.255\frac{E\alpha T}{1-\mu}, \quad \sigma_{\text{bottom}} = -0.629\frac{E\alpha T}{1-\mu}$$

The stresses given by photo elastic test are

$$\sigma_{\text{top}} = -0.22\left(\frac{E\alpha T}{1-\mu}\right), \quad \sigma_{\text{bottom}} = -0.62\left(\frac{E\alpha T}{1-\mu}\right)$$

which are close to the computed stresses.

The comparison between the stresses given by computation and those given by photo elastic test for different dimensions of beam are shown in Table 8.1. It is clear that all the computed stresses are close to those given by test.

Example 2 As shown in Figure 8.3, a concrete beam of two layers on the rock foundation, the thickness of each layer is $2h$, the length of beam is $2l$, $l/h = 8$, $E_1{:}E_2{:}E_f = 1{:}2{:}3$, there is a uniform temperature drop T in the first layer of the beam, try to compute the elastic thermal stresses in the beam.

From Eq. (8.5), the height b of the weighted centroid of the section is

$$b = \frac{2hE_1 \times 3h + 2hE_2h}{2hE_1 + 2hE_2} = \frac{5}{3}h$$

Putting the origin of the coordinate system on the weighted centroid, the axis y is directed downward, from Eqs. (8.11) and (8.12), we get

$$F = \frac{1}{1-\mu^2}\int E(y)\mathrm{d}y = \frac{2hE_1 + 2hE_2}{1-\mu^2} = \frac{6E_1h}{1-\mu^2}$$

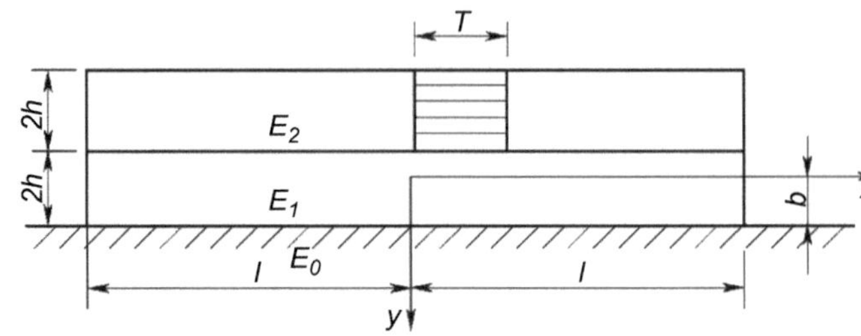

Figure 8.3 A beam of two layers on rock foundation.

$$D = \frac{1}{1-\mu^2}\int E(y)y^2\,\mathrm{d}y = \frac{7.34E_1h^3}{1-\mu^2}$$

From Eqs. (8.2) and (8.3)

$$T_\mathrm{m} = \frac{E_1T \times 2h}{2hE_1 + 2hE_2} = \frac{T}{3}$$

$$\Psi = \frac{E_1T(-(4h/3)) \times 2h}{7.34E_1h^3} = -0.364T/h$$

From Eq. (8.1), the self-thermal stress is

$$\sigma_x = \frac{E(y)\alpha}{1-\mu}\left[\frac{T}{3} - \frac{0.364}{h}Ty - T(y)\right]$$

For the first layer, $E(y) = E_1$, $T(y) = T$ the self-stress is

$$\sigma^\mathrm{I} = \frac{E_1\alpha}{1-\mu}\left(\frac{T}{3} - \frac{0.364}{h}Ty - T\right)$$

$y = -7h/3$ at the top, $y = -h/3$ at the bottom, hence from the above equation, we get

$$\sigma^\mathrm{I}_\mathrm{top} = 0.180\frac{E_1\alpha T}{1-\mu}, \quad \sigma^\mathrm{I}_\mathrm{bottom} = -0.546\frac{E_1\alpha T}{1-\mu}$$

For the second layer, $E_2 = 2E_1$, $T(y) = 0$, the self-thermal stress is

$$\sigma^\mathrm{II} = \frac{E_2\alpha}{1-\mu}\left(\frac{T}{3} - \frac{0.364}{h}Ty\right) = \frac{2E_1\alpha}{1-\mu}\left(\frac{T}{3} - \frac{0.364Ty}{h}\right)$$

Since $y = -h/3$ at the top and $y = +5h/3$ at the bottom of the second layer, so

$$\sigma_{\text{top}}^{\text{II}} = 0.908\frac{E_1\alpha T}{1-\mu}, \quad \sigma_{\text{bottom}}^{\text{II}} = -0.546\frac{E_1\alpha T}{1-\mu}$$

The thermal stresses due to restraint of foundation are computed in the following:

$$r = b/l = 0.209, \quad E_{\text{f}} = 3E_1, \quad \mu = \mu_{\text{f}}$$

From Eq. (8.49),

$$i = 4, \quad k = 209, \quad s = 0.800, \quad C_{11} = 644, \quad C_{12} = -1105, \quad C_{21} = -90,$$
$$C_{22} = -12.00$$
$$\lambda = \frac{15E_1\alpha T}{1-\mu}, \; \beta = -130.5\frac{E_1\alpha T}{1-\mu}$$

From Eq. (8.50),

$$B_1 = 0.1401\frac{E_1\alpha T}{1-\mu}, \quad A_2 = 0.200\frac{E_1\alpha T}{1-\mu}$$

From Eq. (8.53),

$$P_0 = -0.1401\frac{E_1\alpha Tl}{1-\mu}, \quad M_0 = -0.0374\frac{E_1\alpha Tl^2}{1-\mu}$$

From Eq. (8.54), the restraint stress in the first layer is

$$\sigma^{\text{I}} = \frac{E_1\alpha T}{1-\mu}\left(-0.817 + 0.326\frac{y}{h}\right)$$

By substituting the value of y of the top and the bottom of the first layer in the above equation, we get

$$\sigma_{\text{top}}^{\text{I}} = -0.948\frac{E_1\alpha T}{1-\mu}, \quad \sigma_{\text{bottom}}^{\text{I}} = -0.296\frac{E_1\alpha T}{1-\mu}$$

The restraint stresses of the second layer of the beam are as follows:

$$\sigma^{\text{II}} = \frac{E_2\alpha T}{1-\mu}\left(-0.817 + 0.326\frac{y}{h}\right)$$

$$\sigma_{\text{top}}^{\text{II}} = -0.592\frac{E_1\alpha T}{1-\mu}, \quad \sigma_{\text{bottom}}^{\text{II}} = 0.714\frac{E_1\alpha T}{1-\mu}$$

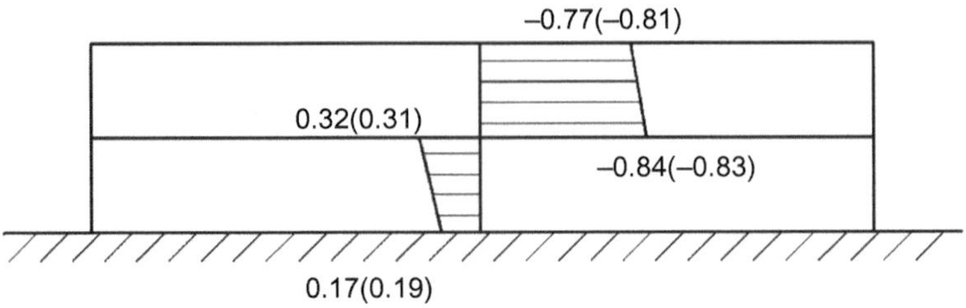

Figure 8.4 The thermal stress $\sigma_x(1-\mu)/E\alpha T$ in a beam of two layers on rock foundation.

Superposition of the restraint stresses and the self-stresses yields the elastic thermal stresses in the beam as shown in Figure 8.4. The numbers in the brackets are the stresses computed by the method of theory of elasticity proposed by the author [7]. It is clear that the results given by two methods are close to each other.

From the two examples, it is evident that the method proposed by the author not only can compute the beam on elastic foundation but also can compute the thermal stresses in concrete blocks with small ratio of height to length. The computation is rather simple and the precision is good, so it is widely applied in practical engineering and adopted in the design specifications of concrete dams and docks in China.

8.3 Restraint Stresses of Beam on Old Concrete Block

When new concrete is poured on an old concrete block which is completely cooled, the thermal stresses in the new concrete may be computed as follows.

The self-stresses of the beam are still given by Eq. (8.1). The foundation is an old concrete block which may be considered as a semi-infinite strip as shown in Figure 8.5, the stress function for which may be taken as follows:

$$\phi(x, y) = Cy \cos \lambda x \cdot e^{-\lambda y}\sqrt{a^2 + b^2} \tag{8.63}$$

the stresses and displacements are given by

$$\sigma_x = C\lambda \cos \lambda x(\lambda y - 2)e^{-\lambda y} \tag{8.64}$$

$$\sigma_y = -Cy\lambda^2 \cos \lambda x e^{-\lambda y} \tag{8.65}$$

$$\tau_{xy} = -C\lambda \sin \lambda x(\lambda y - 1)e^{-\lambda y} \tag{8.66}$$

$$u = \frac{1+\mu_f}{E_f} C \sin \lambda x(\lambda y - 2 + 2\mu_f)e^{-\lambda y} \tag{8.67}$$

where

$$\lambda = \pi/(2l) \tag{8.68}$$

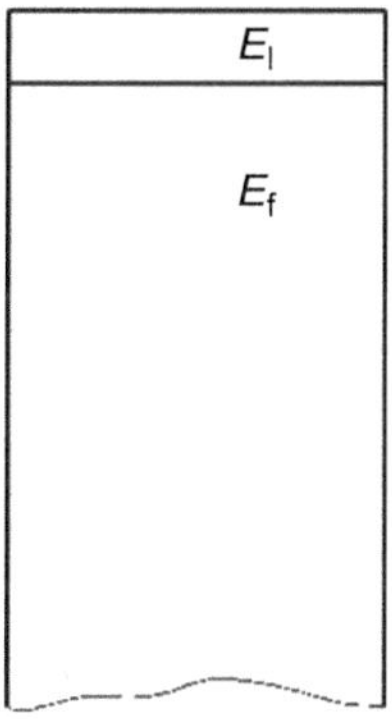

Figure 8.5 Beam on old concrete block.

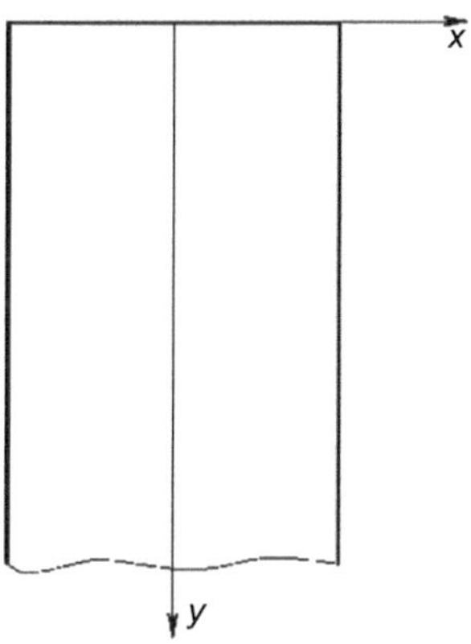

Figure 8.6 The old concrete foundation.

As shown in Figure 8.6, the lateral surface of the strip is free, i.e., when $x = \pm l$, $\sigma_x = 0$, and $\tau_{xy} = 0$, from Eqs. (8.64) and (8.66), when $x = \pm l$, $\sigma_x = 0$, but $\tau_{xy} \neq 0$. Thus, the boundary conditions on the lateral surfaces of the strip are partially satisfied, but the influence of shearing stress of the lateral surfaces on the horizontal displacement of the top surface is small.

On the top of the strip, $y = 0$, from Eqs. (8.65)–(8.67), we have

$$(\tau_{xy})_{y=0} = C\lambda \sin \lambda x, \quad (\sigma_y)_{y=0} = 0 \tag{8.69}$$

$$u_{y=0} = -\frac{2(1-\mu_f^2)}{E_f} C \sin \lambda x \tag{8.70}$$

Let γ_1 be the coefficient of horizontal resistance of the strip foundation, defined by the following equation:

$$\gamma_1 = -\frac{(\tau_{xy})_{y=0}}{u_{y=0}} \tag{8.71}$$

From Eqs. (8.69) and (8.70), we have

$$\gamma_1 = -\frac{\lambda E_f}{2(1-\mu_f^2)} = \frac{\pi E_f}{4(1-\mu_f^2)l} \tag{8.72}$$

The coefficient of semi-infinite plane is greater, the expression of which is given by Eq. (8.109). By comparison of Eqs. (8.72) and (8.109), it is clear that in order to make $\gamma_1 = k_1$, we must take

$$E_f' = -\frac{3.16}{4} E_f \approx 0.80E_f \tag{8.73}$$

For the same modulus of elasticity, the ratio of the rigidity of strip to that of half plane is approximately 0.80. The following approximate method can be used to compute the thermal stresses in beam on old concrete block: let E_f be the modulus of old concrete foundation, the thermal stresses can be computed by the method of beam on semi-infinite plane with $E_f' = 0.80E_f$.

Example A homogeneous beam on old concrete block, $h = 0.20l$, $E = E_f$, there is a uniform temperature T in the beam. Try to compute the thermal stresses.

From Eq. (8.73), $E_f' = 0.80E_f = 0.80E$, $\eta = E/E_f' = 1/0.80 = 1.25$, $r = 0.20$; from Eq. (8.59), $\Delta = 13{,}463$; from Eqs. (8.60) and (8.61), we have

$$P_0 = -0.1803l\frac{E\alpha T}{1-\mu}, \quad M_0 = 0.00357\frac{E\alpha T l^2}{1-\mu}$$

From Eq. (8.62), the horizontal stresses at the top and bottom of beam are as follows:

$$\sigma_x^{\text{top}} = -0.317\frac{E\alpha T}{1-\mu}, \quad \sigma_x^{\text{bottom}} = -0.585\frac{E\alpha T}{1-\mu}$$

The distribution of the stresses is shown in Figure 8.7.

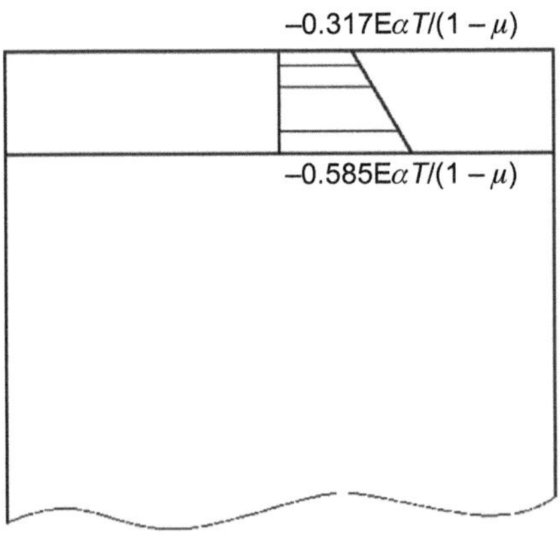

Figure 8.7 The thermal stresses in beam on old concrete block.

8.4 Approximate Analysis of Thermal Stresses in Thin Beam on Half-Plane Foundation

When l/h is large enough, the rotation of the beam may be neglected, only the axial deformation of the beam is considered, the self-stress can be computed by

$$\sigma_1 = \frac{E\alpha}{1-\mu}[T_m - T(y)] \tag{8.74}$$

In the computing of the restraint stress, as the rotation is neglected, only the condition of continuity of horizontal displacement

$$u = w + w_T \tag{8.75}$$

is considered, thus, the restraint stress of the thin beam is

$$\sigma_2 = -\frac{s_1 E\alpha T_m}{1-\mu} \tag{8.76}$$

where

$$s_1 = \frac{1}{1 + (4(1-\mu_f^2)Eh/(1-\mu^2)E_f l)} \tag{8.77}$$

The actual thermal stress in the thin beam on foundation of half plane is

$$\sigma = \sigma_1 + \sigma_2 = \frac{E\alpha}{1-\mu}[(1-s_1)T_m - T(y)] \tag{8.78}$$

Example $E = E_f$, $T(y) = T_m$, $\mu = \mu_f = 1/6$, $l/h = 12.5$, from Eq. (8.77), $s_1 = 0.76$, from Eq. (8.78), the thermal stress in the beam is

$$\sigma = -0.76\frac{E\alpha T_m}{1-\mu}$$

8.5 Thermal Stress on the Lateral Surface of Beam on Elastic Foundation

Sometimes the lateral surfaces of the beam are exposed in the air and superficial cracks may appear due to sudden change of the air temperature. In this case, the temperature of the beam is not only a function of y, but also a function of z, as

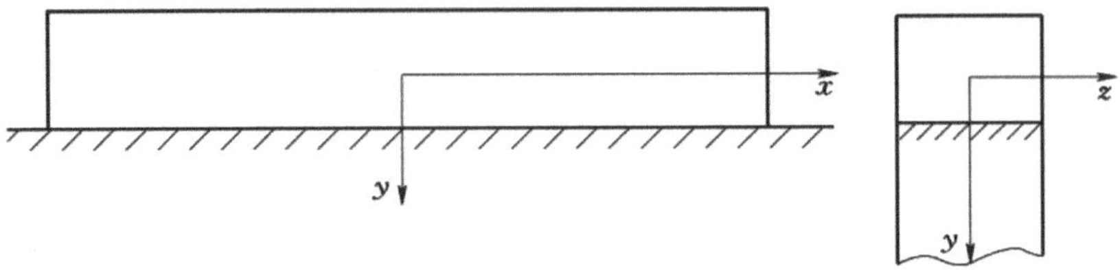

Figure 8.8 Homogeneous beam on elastic foundation.

shown in Figure 8.8, and the foundation is a half space, not a half plane. It is necessary to use the three-dimensional finite element method for precise analysis. If the time of exposure of the lateral surface is not too long, the depth of influence of air temperature is small, hence, the following approximate method may be adopted.

Assuming that the temperature is symmetrical about the y-axis, i.e., $T(y,z) = T(y, -z)$, the self-stress is given by

$$\sigma_x = \frac{E(y)\alpha}{1-\mu}[T_{\mathrm{m}} + \Psi y - T(y,z)] \tag{8.79}$$

$$T_{\mathrm{m}} = \frac{\iint E(y)T(y,z)\mathrm{d}y\,\mathrm{d}z}{\iint E(y)\mathrm{d}y\,\mathrm{d}z} \tag{8.80}$$

$$\Psi = \frac{\iint E(y)T(y,z)y\,\mathrm{d}y\,\mathrm{d}z}{\iint E(y)y^2\,\mathrm{d}y\,\mathrm{d}z} \tag{8.81}$$

Neglecting the influence of lateral foundation, and assuming that the thickness of foundation in the z-direction is equal to the thickness of beam, the restraint stresses of the beam are computed approximately by the method of Section 8.2, namely the beam on half plane, with T_{m} and ψ given by Eqs. (8.80) and (8.81). If the time of exposure of the lateral surface is short, the depth of influence of air temperature is small, T_{m} and ψ may be computed by Eqs. (8.2) and (8.3) which are simpler, but the self-stress must be computed by Eq. (8.79). From comparison of Eqs. (8.79) and (8.1), it is evident that the difference between the stress on the lateral surface and that on the central section of the beam is as follows:

$$\Delta\sigma_x = \sigma_x^{\mathrm{lateral}} - \sigma_x^{\mathrm{central}} = \frac{E(y)\alpha}{1-\mu}[T(y,0) - T(y,z)] \tag{8.82}$$

Thus, the stresses on the lateral surface are given by

$$\sigma_x^{\mathrm{lateral}} = \sigma_x^{\mathrm{central}} + \Delta\sigma_x \tag{8.83}$$

in which $\sigma_x^{\mathrm{central}}$ is given by Eq. (8.54) and $\Delta\sigma_x$ is given by Eq. (8.82).

8.6 Thermal Stresses in Beam on Winkler Foundation [22,45,73]

A strip with unit width is cut from a rectangular plate with uniform thickness, thus, we get a beam with height $2h$, length $2l$, width 1, and temperature $T(y)$ as shown in Figure 8.9. The stresses and internal forces of the beam are given by the following formulas:

$$\left.\begin{aligned}\sigma &= \frac{E}{1-\mu^2}\left[\frac{\mathrm{d}u}{\mathrm{d}x} - (1+\mu)\alpha T\right]\\ N &= \frac{2Eh}{1-\mu^2}\left[\frac{\mathrm{d}u}{\mathrm{d}x} - (1+\mu)\alpha T_{\mathrm{m}}\right]\\ M &= \frac{2Eh^3}{3(1-\mu^2)}\left[\frac{\mathrm{d}^2 v}{\mathrm{d}x^2} - (1+\mu)\alpha\frac{T_{\mathrm{d}}}{2h}\right]\end{aligned}\right\} \tag{8.84}$$

where

u—horizontal displacement
v—vertical displacement
N—axial force
M—bending moment.

The self-stresses are given by Eq. (8.8), and in the following, we shall show how to compute the restraint stresses which are dependent on T_{m} and T_{d}.

8.6.1 Restraint Stress of Beam in Pure Tension

Consider a beam of unit width and height 2h on elastic foundation as shown in Figure 8.9. Neglecting the rotation of the cross section of beam, only axial displacement u is considered. On the contact surface, only the shearing stress q is considered which is expressed by

$$q = k_1 u \tag{8.85}$$

where k_1—the coefficient of horizontal resistance of foundation.

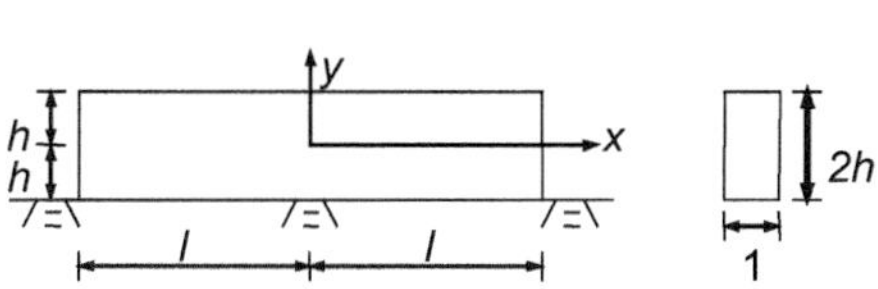

Figure 8.9 Beam of unit width on elastic foundation.

The condition of balance of force of a beam element with length dx is $q = 2h\, \mathrm{d}\sigma/\mathrm{d}x$, from which the following equilibrium equation of the beam is derived as follows:

$$\frac{\mathrm{d}^2 u}{\mathrm{d}x^2} - \lambda^2 u = 0 \tag{8.86}$$

where

$$\lambda = \sqrt{\frac{k_1(1-\mu^2)}{2Eh}} \tag{8.87}$$

Boundary condition: when $x = 0$, $u = 0$; when $x = \pm l$, $\sigma = 0$; hence the solution of Eq. (8.86) is

$$u = \frac{(1+\mu)\alpha T_{\mathrm{m}}}{\lambda} \cdot \frac{\mathrm{sh}\,\lambda x}{\mathrm{ch}\,\lambda l} \tag{8.88}$$

and the stress is

$$\sigma = \frac{E\alpha T_{\mathrm{m}}}{1-\mu}\left(\frac{\mathrm{ch}\,\lambda x}{\mathrm{ch}\,\lambda l} - 1\right) \tag{8.89}$$

The stress at midpoint of beam ($x = 0$) is maximum

$$\sigma_{\mathrm{m}} = -\frac{E\alpha T_{\mathrm{m}}}{1-\mu} g(\lambda l) \tag{8.90}$$

where

$$g(\lambda l) = 1 - \frac{1}{\mathrm{ch}\,\lambda l} \tag{8.91}$$

8.6.2 Restraint Stress of Beam in Pure Bending

Neglecting the axial force and axial displacement, only the bending moment and the rotation of cross section of beam are taken into account. On the contact surface, only the normal stress p is considered which is expressed by

$$p = k_2 v \tag{8.92}$$

where k_2—the coefficient of vertical resistance of the foundation.

The equilibrium equation of the beam is

$$\frac{\mathrm{d}^4 v}{\mathrm{d}\xi^4} + 4v = 0 \tag{8.93}$$

where

$$\xi = \frac{x}{\rho}, \ \rho = \left[\frac{4EI}{(1-\mu^2)k_2}\right]^{1/4} = \left[\frac{8Eh^3}{3(1-\mu^2)k_2}\right]^{1/4} \tag{8.94}$$

Boundary conditions: when $\xi = 0$, $\mathrm{d}v/\mathrm{d}\xi = 0$, shearing force is zero; when $\xi = l/\rho$, both the moment and the shearing force are zero. The solution is as follows:

$$v = A_1 \ \mathrm{ch}\ \xi \cos \xi + A_2 \ \mathrm{sh}\ \xi \sin \xi \tag{8.95}$$

in which

$$\left.\begin{aligned} A_1 &= \frac{4E\alpha T_d h^2}{3(1-\mu^2)k_2\rho^2} \cdot \frac{\mathrm{sh}\ \eta \cos \eta - \mathrm{ch}\ \eta \sin \eta}{\sin 2\eta + \mathrm{sh}\ 2\eta} \\ A_2 &= \frac{4E\alpha T_d h^2}{3(1-\mu^2)k_2\rho^2} \cdot \frac{\mathrm{sh}\ \eta \cos \eta + \mathrm{ch}\ \eta \sin \eta}{\sin 2\eta + \mathrm{sh}\ 2\eta} \\ \eta &= l/\rho \end{aligned}\right\} \tag{8.96}$$

The stress is

$$\sigma = -\frac{2y}{\rho^2}(A_1 \ \mathrm{sh}\ \xi \sin \xi - A_2 \ \mathrm{ch}\ \xi \cos \xi) - \frac{E\alpha y T_d}{2h(1-\mu)} \tag{8.97}$$

Let $\xi = 0$, we get the maximum stress as follows:

$$\sigma_d = -\frac{E\alpha T_d y}{2(1-\mu)h} f(\eta) \tag{8.98}$$

where

$$f(\eta) = 1 - \frac{2(\mathrm{sh}\ \eta \cos \eta + \mathrm{ch}\ \eta \sin \eta)}{\sin 2\eta + \mathrm{sh}\ 2\eta} \tag{8.99}$$

The stresses at the top and bottom of the central section ($\xi = 0$, $y = \pm h$) of beam are as follows:

$$\left.\begin{aligned} &\sigma_{do}^{top} \\ &\sigma_{do}^{bottom} \end{aligned}\right\} = \mp \frac{E\alpha T_d}{2(1-\mu)h} f(\eta) \tag{8.100}$$

$f(\eta)$ and $g(\lambda l)$ are shown in Figure 8.10.

8.6.3 Restraint Stresses of Beam in Bending and Tension

On the contact surface, there are shearing stress q and normal stress p expressed by

$$q = k_1 u, \quad p = k_2 v \tag{8.101}$$

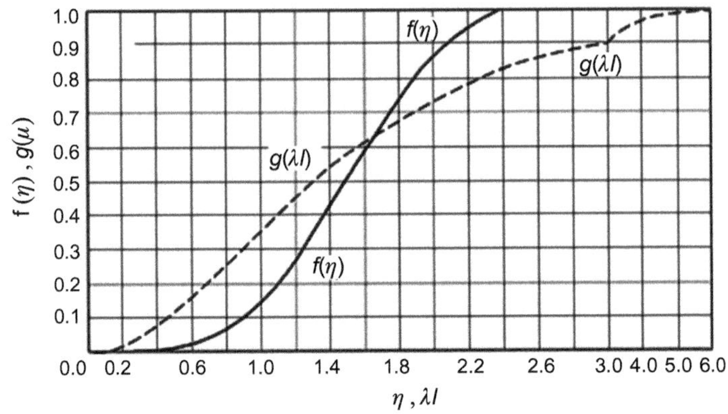

Figure 8.10 Values of $f(\eta)$ and $g(\lambda l)$.

where k_1 and k_2 are respectively the coefficients of horizontal and vertical resistance of foundation. The horizontal displacement of the bottom of beam is $u + h\,\mathrm{d}v/\mathrm{d}x$, assuming that the height of beam h is far smaller than its length l, such that the influence of the rotation of cross section on the axial displacement may be neglected so the horizontal displacement u is still expressed by Eq. (8.88). It has been proved that the vertical displacement v may be expressed by

$$v = c_1\,\mathrm{ch}\frac{x}{\rho}\cos\frac{x}{\rho} + c_2\,\mathrm{sh}\frac{x}{\rho}\sin\frac{x}{\rho} + \gamma\,\mathrm{ch}\,\lambda x \tag{8.102}$$

where

$$\gamma = -\frac{3\lambda^2\alpha T_\mathrm{m}}{Eh\,\mathrm{ch}\,\lambda l \cdot [\lambda^4 + 3k_2/(2Eh^3)]} \tag{8.103}$$

At the ends of the beam ($x = \pm l$), the moment M and the shearing force Q are equal to zero, thus, we have

$$\left.\begin{aligned} &c_1\,\mathrm{sh}\,\eta\sin\eta - c_2\,\mathrm{ch}\,\eta\cos\eta = \frac{2h^2E}{3(1-\mu)k_2\rho^2}(2\gamma h\,\mathrm{ch}\,\lambda l - \alpha T_\mathrm{d}) \\ &c_1(\mathrm{sh}\,\eta\cos\eta + \mathrm{ch}\,\eta\sin\eta) - c_2(\mathrm{sh}\,\eta\cos\eta - \mathrm{ch}\,\eta\sin\eta) \\ &= \frac{4E\gamma h^3\lambda^3}{3(1-\mu^2)k_2\rho}\mathrm{sh}\,\lambda l + \frac{2k_1 h\alpha T_\mathrm{m}}{(1-\mu)k_2\lambda\rho}\mathrm{th}\,\lambda l \end{aligned}\right\} \tag{8.104}$$

The coefficients c_1 and c_2 may be solved from the above equations. The bending moment is given by

$$M = \frac{k_2\rho^2}{2}\left(c_1\,\mathrm{sh}\frac{x}{\rho}\sin\frac{x}{\rho} - c_2\,\mathrm{ch}\frac{x}{\rho}\cos\frac{x}{\rho}\right) - \frac{2Eh^3\gamma\lambda^2}{3(1-\mu^2)}\mathrm{ch}\,\lambda x + \frac{E\alpha h^2 T_\mathrm{d}}{3(1-\mu)} \tag{8.105}$$

On the central section ($x = 0$), the bending moment M_0 and axial force N_0 are as follows:

$$M_0 = -\frac{c_2 k_2 \rho^2}{2} - \frac{2Eh^3\gamma\lambda^2}{3(1-\mu^2)} + \frac{E\alpha h^2 T_\mathrm{d}}{3(1-\mu)}$$

$$N_0 = -\frac{2Eh\alpha T_\mathrm{m}}{1-\mu} g(\lambda l)$$

The restraint stresses on the central section are given by

$$\sigma = -\frac{E\alpha T_\mathrm{m}}{1-\mu} g(\lambda l) \mp \frac{3M_0 y}{2h^3} \tag{8.106}$$

8.6.4 Coefficients of Resistance of Foundation

The above solutions of beam on Winkler foundation were given by Professor G. N. Maslov in 1940, but they have not been applied in engineering due to lack of expressions of coefficients of resistance of foundation which is given by the author as the following.

Assuming that the foundation is an elastic half plane, on the surface of which there are shearing loads $q(x)$ as follows:

$$q(x) = \begin{cases} x, & \text{when } -l \le x \le +l \\ 0, & \text{when } |x| > l \end{cases} \tag{8.107}$$

From the theory of elasticity, the tangential displacement of the surface of foundation is

$$u_0 = -\frac{(1-\mu_\mathrm{f}^2)l^2}{\pi E_\mathrm{f}}\left[\left(1 - \frac{x^2}{l^2}\right)\ln\frac{l+x}{l-x} + \frac{2x}{l}\right] \tag{8.108}$$

which is expressed by solid curve in Figure 8.11. Drawing a dotted straight line close to the curve from which the coefficient of horizontal resistance of the foundation is derived in the following:

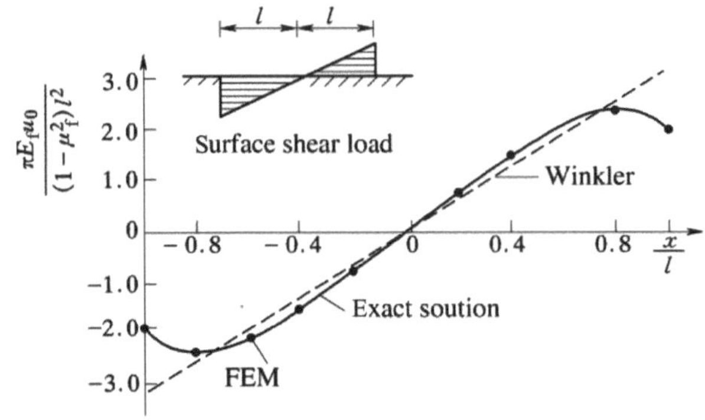

Figure 8.11 The horizontal displacement u_0 of the surface of foundation.

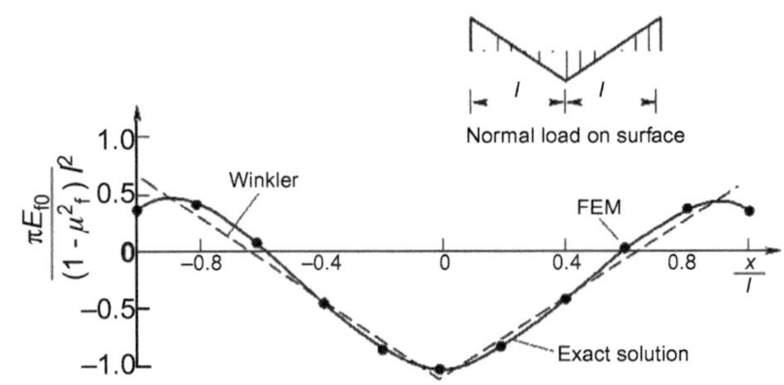

Figure 8.12 The vertical displacement v_0 of the surface of foundation.

$$k_1 = \frac{\pi E_f}{3.16(1-\mu_f^2)l} = \frac{0.9942E_f}{(1-\mu_f^2)l} \tag{8.109}$$

where E_f and μ_f are respectively the modulus of elasticity and Poisson's ratio of the foundation.

If on the surface of foundation, there are normal loadings as follows:

$$p(x) = \begin{cases} \frac{l}{2} - |x|, & \text{when } |x| < l \\ 0, & \text{when } |x| > l \end{cases} \tag{8.110}$$

the vertical displacement v_0 of the surface of foundation will be

$$v_0 = \frac{(1-\mu_f^2)l^2}{\pi E_f}\left[\left(\frac{x}{l} - \frac{x^2}{l^2}\right)\ln\frac{x}{l-x} + \left(\frac{x}{l} + \frac{x^2}{l^2}\right)\ln\frac{l+x}{x} - 1\right] \tag{8.111}$$

which are expressed by the solid curve in Figure 8.12. Drawing a dotted line close to the curve from which the coefficient of vertical resistance of the foundation is obtained as follows:

$$k_2 = \frac{\pi E_f}{1.73(1-\mu_f^2)l} = \frac{1.816E_f}{(1-\mu_f^2)l} \tag{8.112}$$

8.6.5 Approximate Method for Beam on Winkler Foundation

For the exact solution of beam on Winkler foundation, it is necessary to solve the simultaneous Eq. (8.104). For a thin beam with small h/l, we can neglect the influence of axial displacement on the bending moment and the influence of rotation on the axial force; from Eqs. (8.79), (8.90), and (8.98), we get an approximate formula for the stresses on the central cross section ($\xi = 0$) of the beam on elastic foundation as follows:

$$\sigma = \frac{E\alpha}{1-\mu}\left\{[1-g(\lambda l)]T_m + [1-f(\eta)]\cdot\frac{T_d y}{2h} - T(y)\right\} \tag{8.113}$$

where λ, η, $f(\eta)$, $g(\lambda l)$ are given by Eqs. (8.87), (8.91), (8.96), and (8.99).

This is a convenient and useful formula which can be applied widely in engineering.

Let $y = \pm h$, we will get the stresses at the top and bottom of the beam.

8.6.6 Analysis of Effect of Restraint of Soil Foundation

For concrete beam on rock foundation, as the modulus of elasticity of rock is large, both the vertical and horizontal resistance of foundation cannot be neglected, the restraint stresses must be computed by Eq. (8.54). Because the modulus of elasticity of soil is small, the thermal stresses in concrete beam on soil foundation may be simplified in some cases, which will be analyzed in the following.

1. The horizontal restraint

If the horizontal displacement of the beam is fully restrained, the thermal stress is $\sigma_{mo} = -E\alpha T_m/(1-\mu)$, the actual restraint stress σ_m is given by Eq. (8.90), the ratio of σ_m/σ_{mo} is

$$\frac{\sigma_m}{\sigma_{mo}} = -\frac{\sigma_m(1-\mu)}{E\alpha T_m} = 1 - \frac{1}{\text{ch}\,\lambda l} = g(\lambda l) \tag{8.114}$$

where λ, $g(\lambda l)$, and k_1 are given in Eqs. (8.87), (8.91), and (8.109). Let the Poisson's ratio of concrete and foundation be $\mu = 0.167$, $\mu_f = 0.25$, we have

$$\lambda l = \text{ch}^{-1}\left(\frac{1}{1-g}\right) \tag{8.115}$$

Taking $g = 0.05$, from the above equation, the error due to neglect of the horizontal restraint of soil foundation will not exceed 5% if

$$l/h \le l_1/h = 0.202E/E_f \tag{8.116}$$

2. The restraint of rotation of beam

When the rotation of beam is fully restrained, the stresses at the top and bottom of beam are $\sigma_{do} = \mp E\alpha T_d/[2(1-\mu)h]$, when the rotation is partially restrained, the stress is σ_d, the ratio of σ_d/σ_{do} is

$$\frac{\sigma_d}{\sigma_{do}} = -\frac{2\sigma_d(1-\mu)h}{E\alpha T_d} = f(\eta) \tag{8.117}$$

Let $f(\eta) = 0.95$, when

$$\frac{l}{h} > \frac{l_2}{h} = 0.306\left(\frac{E}{E_f}\right)^{1/3} \tag{8.118}$$

the beam may be considered as fully restrained in rotation. Let $f(\eta) = 0.05$, when

$$\frac{l}{h} < \frac{l_3}{h} = 0.763\left(\frac{E}{E_f}\right)^{1/3} \tag{8.119}$$

the restraint of foundation to rotation of beam may be neglected. l_1/h, l_2/h, and l_3/h for different E/E_f are given in Figure 8.13.

Let the modulus of concrete be $E_c = 30{,}000$ MPa, l_1/h, l_2/h, and l_3/h are given in Table 8.2.

For example, modulus of elasticity of concrete $E_c = 30{,}000$ MPa, length of beam $l = 30$ m, height of beam $h = 2.5$ m, modulus of elasticity of foundation $E_f = 50$ MPa, $l/h = 30/2.5 = 12.00$. From Table 8.2, $l_1/h = 121 > 12$, $l_3/h = 6.4 < 12$, so in this case, the horizontal resistance of the foundation may be neglected but the rotational resistance of the foundation must be considered.

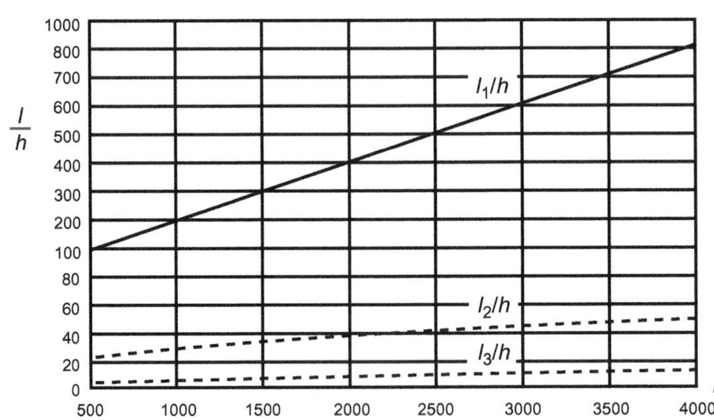

Figure 8.13 l_1/h, l_2/h, and l_3/h for different E/E_f.

Table 8.2 l_1/h, l_2/h, l_3/h ($E_c = 30{,}000$ MPa) for different restraint conditions

Modulus of Elasticity of Foundation, E_f (MPa)	l_1/h for No Horizontal Restraint	l_2/h for Full Rotational Restraint	l_3/h for No Rotational Restraint
10	606	44.1	11.0
20	303	35.0	8.7
30	202	30.6	7.6
40	152	27.8	6.9
50	121	25.8	6.4

Generally speaking, for concrete beam on soil foundation, the horizontal resistance may be neglected but the rotational resistance must be considered in most cases.

8.7 Thermal Stresses in Beams on Elastic Foundation When Modulus of Elasticity of Concrete Varying with Time

In order to consider the variation of modulus of elasticity with age of concrete, the incremental method may be used. In the ith increment of time $\Delta\tau_i = \tau_i - \tau_{i-1}$, the increment of temperature is

$$\Delta T(y, \tau_i) = T(y, \tau_i) - T(y, \tau_{i-1}) \tag{8.120}$$

and the mean modulus of elasticity of concrete is

$$E(\tau_{i-0.5}) = \frac{E(\tau_i) + E(\tau_{i-1})}{2}$$

The elastic stress increment $\Delta\sigma(\tau_i)$ may be computed by the above-mentioned method, the viscoelastic thermal stress at time t may be computed by the following approximate formula:

$$\sigma(t) = \sum \Delta\sigma(\tau_i) K(t, \tau_i) \tag{8.121}$$

where $K(t,\tau_i)$ is the stress relaxation coefficient.

9 Finite Element Method for Computing Temperature Field

Experience shows that the finite difference method is suitable for the solution of one-dimensional (1D) temperature field and the finite element (FEM) method is more suitable for the solution of 2D and 3D temperature field in practical engineering.

9.1 Variational Principle for the Problem of Heat Conduction [9, 35, 39, 40, 44, 46, 53, 70, 76, 78]

9.1.1 Euler's Equation

As shown in Figure 9.1, consider the functional

$$I(T) = \iiint_R F(T, T_x, T_y, T_z)\mathrm{d}x\,\mathrm{d}y\,\mathrm{d}z + \iint_C G(T)\mathrm{d}s \tag{9.1}$$

where $T_x = \partial T/\partial x$, $T_y = \partial T/\partial y$, $T_z = \partial T/\partial z$, and the boundary conditions of temperature T are:

$$\left.\begin{aligned} &\text{On boundary} \quad c' \quad T = T_\mathrm{b} \\ &\text{On boundary} \quad c \quad l_x\frac{\partial T}{\partial x} + l_y\frac{\partial T}{\partial y} + l_z\frac{\partial T}{\partial z} = \frac{\beta}{\lambda}(T - T_\mathrm{a}) \end{aligned}\right\} \tag{9.2}$$

where l_x, l_y, l_z are the direction cosines of the outward normal to the surface and T_a is the air temperature. It has been proved that when the functional $I(T)$ takes the minimum value, namely

$$\delta(I) = 0 \tag{9.3}$$

the following equations are valid:

$$\frac{\partial F}{\partial T} - \frac{\partial}{\partial x}\left(\frac{\partial F}{\partial T_x}\right) - \frac{\partial}{\partial y}\left(\frac{\partial F}{\partial T_y}\right) - \frac{\partial}{\partial z}\left(\frac{\partial F}{\partial T_z}\right) = 0 \tag{9.4}$$

Thermal Stresses and Temperature Control of Mass Concrete. DOI: http://dx.doi.org/10.1016/B978-0-12-407723-2.00009-9

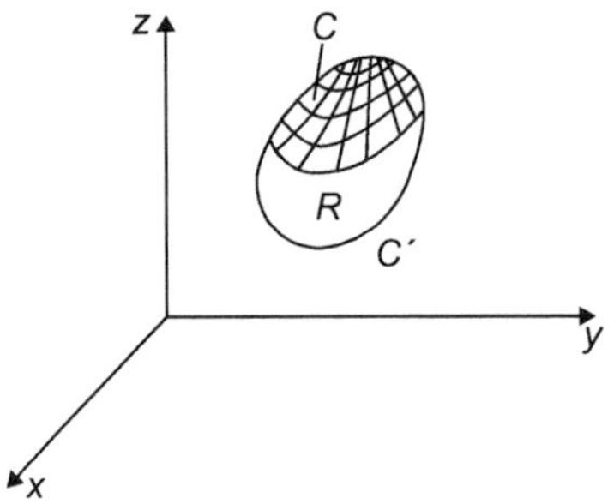

Figure 9.1 A space problem.

$$\frac{\partial G}{\partial T} + l_x \frac{\partial F}{\partial T_x} + l_y \frac{\partial F}{\partial T_y} + l_z \frac{\partial F}{\partial T_z} = 0 \tag{9.5}$$

These are the Euler's equations of the variational problem (Eqs. (9.1)–(9.3)).

9.1.2 Variational Principle of Problem of Heat Conduction

Consider the unsteady temperature $T(x,y,z,t)$ which satisfies the following equations:

In region R

$$\frac{\partial^2 T}{\partial x^2} + \frac{\partial^2 T}{\partial y^2} + \frac{\partial^2 T}{\partial z^2} + \frac{1}{a}\left(\frac{\partial \theta}{\partial \tau} - \frac{\partial T}{\partial \tau}\right) = 0 \tag{9.6}$$

when $\tau = 0$

$$T = T_0(x, y, z) \tag{9.7}$$

$$\text{On surface} \quad c' \quad \text{when } \tau > 0, \ T = T_b \tag{9.8}$$

$$\text{On surface} \quad c \quad \text{when } \tau > 0, \ l_x \frac{\partial T}{\partial x} + l_y \frac{\partial T}{\partial y} + l_z \frac{\partial T}{\partial z} + \frac{\beta}{\lambda}(T - T_a) = 0 \tag{9.9}$$

Now take the function F and G as follows:

$$\left.\begin{aligned} F &= \frac{1}{2}\left[\left(\frac{\partial T}{\partial x}\right)^2 + \left(\frac{\partial T}{\partial y}\right)^2 + \left(\frac{\partial T}{\partial z}\right)^2\right] - \frac{1}{a}\left(\frac{\partial \theta}{\partial \tau} - \frac{\partial T}{\partial \tau}\right) T \\ G &= \frac{\beta}{\lambda}\left(\frac{1}{2}T^2 - T_a T\right) \end{aligned}\right\} \tag{9.10}$$

Substituting F and G into Eq. (9.1), we get

$$I(T) = \iiint_R \left\{ \frac{1}{2}\left[\left(\frac{\partial T}{\partial x}\right)^2 + \left(\frac{\partial T}{\partial y}\right)^2 + \left(\frac{\partial T}{\partial z}\right)^2\right] - \frac{1}{a}\left(\frac{\partial \theta}{\partial \tau} - \frac{\partial T}{\partial \tau}\right) T \right\} \mathrm{d}x\, \mathrm{d}y\, \mathrm{d}z + \iint_C \frac{\beta}{\lambda}\left(\frac{1}{2}T^2 - T_\mathrm{a} T\right) \mathrm{d}s \tag{9.11}$$

From Eq. (9.10), we get the partial derivatives of F and G:

$$\frac{\partial F}{\partial T} = -\frac{1}{a}\left(\frac{\partial \theta}{\partial \tau} - \frac{\partial T}{\partial \tau}\right), \quad \frac{\partial F}{\partial T_x} = \frac{\partial T}{\partial x}, \quad \frac{\partial F}{\partial T_y} = \frac{\partial T}{\partial y}$$

$$\frac{\partial F}{\partial T_z} = \frac{\partial T}{\partial z}, \quad \frac{\partial}{\partial x}\left(\frac{\partial F}{\partial T_x}\right) = \frac{\partial^2 T}{\partial x^2}, \quad \frac{\partial}{\partial y}\left(\frac{\partial F}{\partial T_y}\right) = \frac{\partial^2 T}{\partial y^2}$$

$$\frac{\partial}{\partial z}\left(\frac{\partial F}{\partial T_z}\right) = \frac{\partial^2 T}{\partial z^2}, \quad \frac{\partial G}{\partial T} = \frac{\beta}{\lambda}(T - T_\mathrm{a})$$

Substituting them into Eqs. (9.4) and (9.5), we have
In region R:

$$\frac{\partial F}{\partial T} - \frac{\partial}{\partial x}\left(\frac{\partial F}{\partial T_x}\right) - \frac{\partial}{\partial y}\left(\frac{\partial F}{\partial T_y}\right) - \frac{\partial}{\partial z}\left(\frac{\partial F}{\partial T_z}\right) = -\frac{1}{a}\left(\frac{\partial \theta}{\partial \tau} - \frac{\partial T}{\partial \tau}\right) \tag{9.12}$$

On boundary C:

$$\frac{\partial G}{\partial T} + l_x \frac{\partial F}{\partial T_x} + l_y \frac{\partial F}{\partial T_y} + l_z \frac{\partial F}{\partial T_z} = \frac{\beta}{\lambda}(T - T_\mathrm{a}) + l_x \frac{\partial T}{\partial x} + l_y \frac{\partial T}{\partial y} + l_z \frac{\partial T}{\partial z} = 0 \tag{9.13}$$

Equations (9.12) and (9.13) are identical with Eqs. (9.6) and (9.9). Hence the following conclusion is derived: If the temperature $T(x,y,z,\tau)$ takes the initial temperature T_0 when $\tau = 0$, takes the boundary temperature T_b on the surface c' and enforces the functional $I(T)$ expressed by Eq. (9.11) to take the minimum value, then $T(x,y,z,\tau)$ will satisfy Eq. (9.6) in region R and Eq. (9.9) on surface C, namely, $T(x,y,z,\tau)$ is the unsteady temperature solution that we seek.

9.2 Discretization of Continuous Body

Temperatures are the unknown variables in the solution of temperature field. There are infinite points in a continuous body and every point has a temperature; hence, there are infinite unknown variables in the temperature field of a continuous body which is difficult to be solved by numerical method. In the FEM, the continuous body is substituted by a group of elements which are connected at a finite number of nodes as shown in Figure 9.2. The temperatures in the element are expressed by the temperatures of the nodes with shape functions. Only temperatures of the finite number of nodes are unknown variables, and the original problem with infinite unknown variables is replaced by a new problem with finite unknown variables which can be solved by numerical methods.

9.3 Fundamental Equations for Solving Unsteady Temperature Field by FEM [8, 9]

Consider a 3D unsteady temperature field, the equation of heat conduction, the initial and boundary conditions are given by Eqs. (9.6)–(9.9). According to the variational principle and Eq. (9.10), this heat conduction problem is identical with the minimum value problem of the functional $I(T)$:$T(x, y, z, \tau) = T_0(x, y, z)$ when $\tau = 0$ and $T(x,y,z,\tau) = T_b$ on the first kind of boundary c' and enforce the following functional $I(T)$ to take the minimum value.

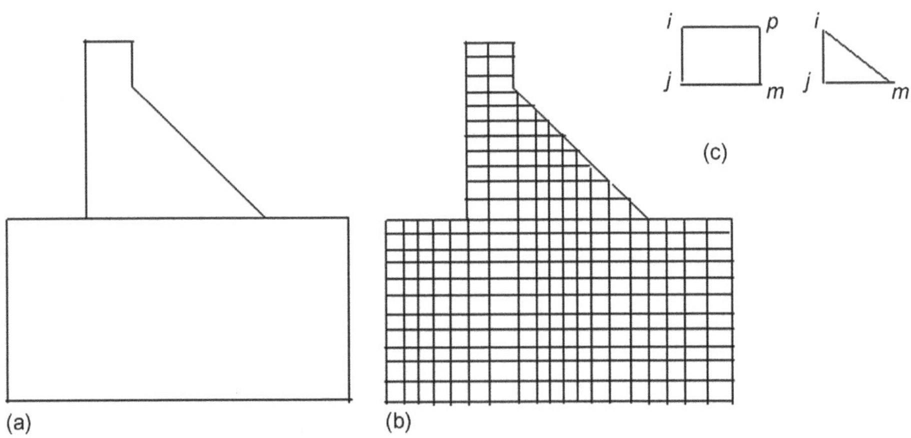

Figure 9.2 Discretization of continuous structure by FEMs: (a) original structure, (b) structure after discretization, and (c) element.

$$I(T) = \iint_R \int \left\{ \frac{1}{2}\left[\left(\frac{\partial T}{\partial x}\right)^2 + \left(\frac{\partial T}{\partial y}\right)^2 + \left(\frac{\partial T}{\partial z}\right)^2\right] - \frac{1}{a}\left(\frac{\partial \theta}{\partial \tau} - \frac{\partial T}{\partial \tau}\right) T \right\} \mathrm{d}x\,\mathrm{d}y\,\mathrm{d}z + \iint_C \frac{\beta}{\lambda}\left(\frac{1}{2}T^2 - T_a T\right) \mathrm{d}s \tag{9.14}$$

The first term on the right part of Eq. (9.14) is the space integral in the region of solution R and the second term is the area integral of the third kind of boundary C'. As shown in Figure 9.2, the solution region R is divided into a finite number of elements. Suppose that the nodes of element e are $i, j, m, \ldots, p$, the temperatures of modes are $T_i(\tau), T_j(\tau), T_m(\tau), \ldots, T_p(\tau)$, and the temperature of any point of the element is expressed by the temperatures of nodes as follows:

$$\begin{aligned} T^e(x, y, z, \tau) &= N_i T_i + N_j T_j + N_m T_m + \cdots + N_p T_p \\ &= [N_i, N_j, N_m \cdots] \begin{Bmatrix} T_i \\ T_j \\ T_m \\ \vdots \end{Bmatrix} = [N]\{T\}^e \end{aligned} \tag{9.15}$$

where the shape function $N_i(\xi,\eta,s)$ is a function of the local coordinates ξ, η, ς and the nodal temperature $T_i(\tau)$ is a function of time.

The rate of change of temperature of any point in the element is

$$\begin{aligned} \frac{\partial T}{\partial \tau} &= N_i \frac{\partial T_i}{\partial \tau} + N_j \frac{\partial T_j}{\partial \tau} + N_m \frac{\partial T_m}{\partial \tau} + \cdots \\ &= [N_i, N_j, N_m + \cdots] \begin{Bmatrix} \dot{T}_i \\ \dot{T}_j \\ \dot{T}_m \\ \cdots \end{Bmatrix} = [N]\{\dot{T}\}^e \end{aligned} \tag{9.16}$$

Considering element e as a subregion ΔR of the solution region R, the value of the functional in the subregion is

$$I^e(T) = \iint_{\Delta R} \int \left\{ \frac{1}{2}\left[\left(\frac{\partial T}{\partial x}\right)^2 + \left(\frac{\partial T}{\partial y}\right)^2 + \left(\frac{\partial T}{\partial z}\right)^2\right] - \frac{1}{a}\left(\frac{\partial \theta}{\partial \tau} - \frac{\partial T}{\partial \tau}\right) T \right\} \mathrm{d}x\,\mathrm{d}y\,\mathrm{d}z + \iint_{\Delta C} \frac{\beta}{\lambda}\left(\frac{1}{2}T^2 - T_a T\right) \mathrm{d}s \tag{9.17}$$

Differentiating Eq. (9.17) in the sign of integral, we obtain

$$\begin{aligned}\frac{\partial I^e}{\partial T_i} &= h_{ii}^e T_i + h_{ij}^e T_j + h_{im}^e T_m + \cdots + r_{ii}^e \frac{\partial T_i}{\partial \tau} + r_{ij}^e \frac{\partial T_j}{\partial \tau} + r_{im}^e \frac{\partial T_m}{\partial \tau} + \cdots \\ &\quad - f_i^e \frac{\partial \theta}{\partial \tau} + g_{ii}^e T_i + g_{ij}^e T_j + g_{im}^e T_m + \cdots - p_i^e T_{\mathrm{a}}\end{aligned} \tag{9.18}$$

where

$$\left.\begin{aligned} h_{ij}^e &= \iint_{\Delta R}\int \left(\frac{\partial N_i}{\partial x}\frac{\partial N_j}{\partial x} + \frac{\partial N_i}{\partial y}\frac{\partial N_j}{\partial y} + \frac{\partial N_i}{\partial z}\frac{\partial N_j}{\partial z} \right) \mathrm{d}x\,\mathrm{d}y\,\mathrm{d}z \\ f_i^e &= \frac{1}{a}\iint_{\Delta R}\int N_i\,\mathrm{d}x\,\mathrm{d}y\,\mathrm{d}z \\ g_{ij}^e &= \frac{\lambda}{\beta}\int_{\Delta C}\int N_i N_j\,\mathrm{d}s \\ p_i^e &= \frac{\lambda}{\beta}\int_{\Delta C}\int N_i\,\mathrm{d}s \end{aligned}\right\} \tag{9.19}$$

where g_{ij}^e and p_i^e are area integrals on the third kind of boundary C.

When the elements are small enough, the original functional may be replaced by the sum of the functional of the elements, namely

$$I(T) \cong \sum_e I^e(T) \tag{9.20}$$

In order to enforce the functional $I(T)$ to take the minimum value, it is necessary that

$$\frac{\partial I}{\partial T_i} \cong \sum_e \frac{\partial I^e}{\partial T_i} = 0 \tag{9.21}$$

Substituting Eq. (9.18) into the above equation, we get

$$\begin{aligned}&\sum_e h_{ii}^e T_i + h_{ij}^e T_j + h_{im}^e T_m + \cdots + r_{ii}^e \frac{\partial T_i}{\partial \tau} + r_{ij}^e \frac{\partial T_j}{\partial \tau} + r_{im}^e \frac{\partial T_m}{\partial \tau} + \cdots \\ &\quad - f_i \frac{\partial \theta}{\partial \tau} + g_{ii}^e T_i + g_{ij}^e T_j + g_{im}^e T_m + \cdots - p_i^e T_{\mathrm{a}} = 0\end{aligned} \tag{9.22}$$

which may be transformed into the following equation by the symbol of matrix:

$$[H]\{T\} + [R]\left\{\frac{\partial T}{\partial \tau}\right\} + \{F\} = 0 \tag{9.23}$$

where

$$\left.\begin{aligned} H_{ij} &= \sum_e (h_{ij}^e + g_{ij}^e) \\ R_{ij} &= \sum_e r_{ij}^e \\ F_i &= \sum_e \left(-f_i \frac{\partial \theta}{\partial \tau} - p_i^e T_{\mathrm{a}} \right) \end{aligned}\right\} \tag{9.24}$$

where $\sum_e$ indicates the sum of all the elements relevant to the node i.

Equation (9.23) is valid at any time τ, so it is valid at $\tau = \tau_n$ and $\tau = \tau_{n+1}$, namely

$$[H]\{T_n\} + [R]\left\{\frac{\partial T}{\partial \tau}\right\}_n + \{F_n\} = 0 \tag{9.25}$$

$$[H]\{T_{n+1}\} + [R]\left\{\frac{\partial T}{\partial \tau}\right\}_{n+1} + \{F_{n+1}\} = 0 \tag{9.26}$$

Now let

$$\Delta T_n = T_{n+1} - T_n = \Delta \tau_n \left[(1-s)\left(\frac{\partial T}{\partial \tau}\right)_n + s\left(\frac{\partial T}{\partial \tau}\right)_{n+1} \right] \tag{9.27}$$

According to the value of s, there are the following cases:

1. For $s = 0$, $\Delta T_n = \Delta \tau_n (\partial T/\partial \tau)_n$, forward difference, explicit solution.
2. For $s = 1$, $\Delta T_n = \Delta \tau_n (\partial T/\partial \tau)_{n+1}$, backward difference, implicit solution.
3. For $s = 1/2$, $\Delta T_n = (1/2)\Delta \tau_n [(\partial T/\partial \tau)_n + (\partial T/\partial \tau)_{n+1}]$, midpoint difference, implicit solution.

Experience shows that the backward difference method ($s = 1$) is the better one in the two implicit methods.

From Eq. (9.27), we have

$$\left\{\frac{\partial T}{\partial \tau}\right\}_{n+1} = \frac{1}{s\,\Delta \tau_n}[\{T_{n+1}\} - \{T_n\}] - \frac{1-s}{s}\left\{\frac{\partial T}{\partial \tau}\right\}_n \tag{9.28}$$

Substituting into Eq. (9.26), we get

$$[H]\{T_{n+1}\} + [R]\left(\frac{1}{s\,\Delta \tau_n}[\{T_{n+1}\} - \{T_n\}] - \frac{1-s}{s}\left\{\frac{\partial T}{\partial \tau}\right\}_n\right) + \{F_{n+1}\} = 0 \tag{9.29}$$

From Eq. (9.25),

$$-[R]\left\{\frac{\partial T}{\partial \tau}\right\}_n = [H]\{T_n\} + \{F_n\}$$

Substituting into Eq. (9.29), the fundamental equation for computing the unsteady temperature field by FEM is derived in the following:

$$\left([H]+\frac{1}{s\Delta\tau_n}[R]\right)\{T_{n+1}\}+\left(\frac{1-s}{s}[H]-\frac{1}{s\Delta\tau_n}[R]\right)\{T_n\}+\frac{1-s}{s}\{F_n\}+\{F_{n+1}\}=0 \tag{9.30}$$

where $\{T_n\}$, $\{F_n\}$, and $\{F_{n+1}\}$ are known and $\{T_{n+1}\}$ are unknown; thus the above equation is a linear equation system about $\{T_{n+1}\}$ and the solution of which will give the temperatures $\{T_{n+1}\}$ of all the nodes at time $\tau = \tau_{n+1}$.

9.4 Two-Dimensional Unsteady Temperature Field, Triangular Elements

In order to give the readers a clear idea, it will be explained in the following how triangular elements are used to compute the 2D unsteady temperature field.

As shown in Figure 9.3, the solution region R is divided into a set of elements each of which has three nodes. Take one element e whose nodes are i, j, and m. The temperature in the element e is expressed in the following (Figure 9.4):

$$T^e(x, y, y, \tau) = N_i(x, y)T_i(\tau) + N_j(x, y)T_j(\tau) + N_m(x, y)T_m(\tau) \tag{9.31}$$

where N_i, N_j, N_m are the following shape functions:

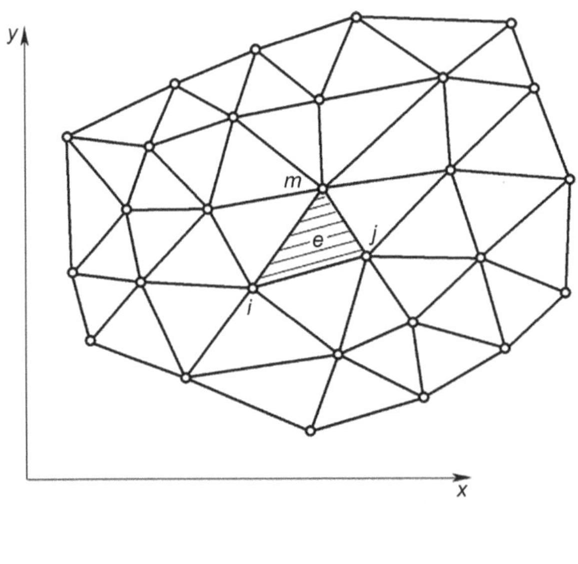

Figure 9.3 Discretization of the solution region.

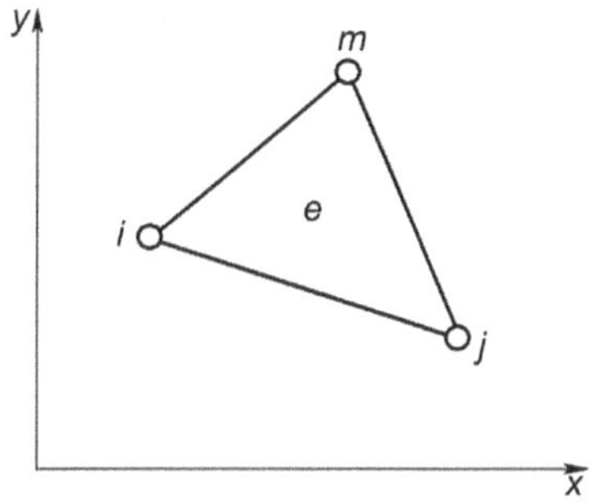

Figure 9.4 Element *e*.

$$\left.\begin{aligned} N_i &= \frac{1}{2A}(a_i + b_i x + c_i y) \\ N_j &= \frac{1}{2A}(a_j + b_j x + c_j y) \\ N_m &= \frac{1}{2A}(a_m + b_m x + c_m y) \end{aligned}\right\} \tag{9.32}$$

where

$$a_i = x_j y_m - x_m y_j, \quad b_i = y_j - y_m, \quad c_i = x_m - x_j \tag{9.33}$$

and A is the area of the triangle ijm

$$A = \frac{1}{2}\begin{vmatrix} 1 & x_i & y_i \\ 1 & x_j & y_j \\ 1 & x_m & y_m \end{vmatrix}$$

Substituting Eq. (9.32) into Eq. (9.19), we get

$$\begin{aligned} &h_{ii}^e = \frac{b_i^2 + c_i^2}{4A}, \quad h_{ij}^e = \frac{b_i b_j + c_i c_j}{4A}, \quad h_{im}^e = \frac{b_i b_m + c_i c_m}{4A} \\ &r_{ii}^e = \frac{A}{6a}, \quad r_{ij}^e = r_{im}^e = \frac{A}{12a} \\ &f_i^e = \frac{A}{3a}, \quad p_i^e = \overline{\beta} L/2 \\ &g_{ii}^e = \overline{\beta} L/3, \quad g_{ij}^e = g_{im}^e = \overline{\beta} L/6, \quad \overline{\beta} = \beta/\lambda \end{aligned} \tag{9.34}$$

Now we have obtained all the coefficients in Eq. (9.24). Let

$$\frac{\partial\theta}{\partial\tau} = \frac{\Delta\theta}{\Delta\tau} \tag{9.35}$$

Then Eq. (9.30) is established and the solution of which will give the nodal temperatures $\{T_{n+1}\}$ for $\tau = \tau_{n+1}$.

9.5 Isoparametric Elements

Today the isoparametric elements are extensively used in practice to raise the efficiency of computing [8, 90, 91, 112–115].

9.5.1 Two-Dimensional Isoparametric Elements

The isoparametric elements are constructed by transformation of coordinates. The 2D linear isoparametric element is shown in Figure 9.5. The element in the local coordinates is a 2×2 square, the origin of the local coordinates (ξ,η) is the center of element and the four boundary elements are $\xi = \pm 1$ and $\eta = \pm 1$. In the Cartesian coordinates, the element is a quadrilateral. The coordinates of any point in the element are expressed by

$$\left.\begin{aligned} x &= \sum N_i x_i = N_1 x_1 + N_2 x_2 + \cdots \\ y &= \sum N_i y_i = N_1 y_1 + N_{2y} x_2 + \cdots \end{aligned}\right\} \tag{9.36}$$

where N_i (ξ,η) is the shape function expressed by local coordinates ξ and η, and (x_i,y_i) are the global coordinates of node i. Equation (9.36) is the coordinate transform formula for a 2D isoparametric element.

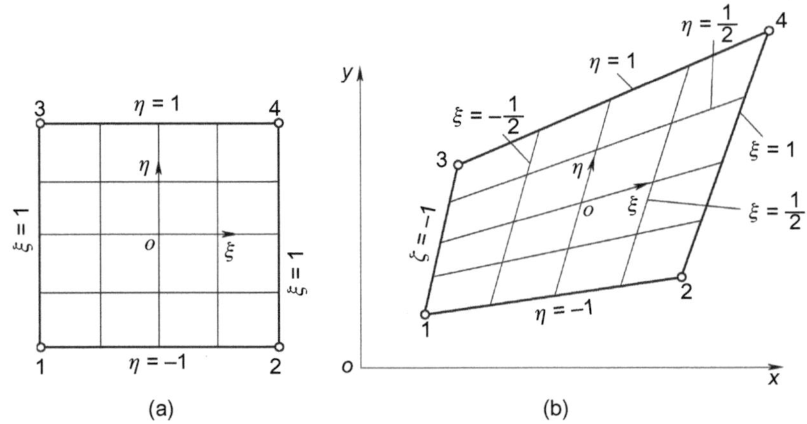

Figure 9.5 Two-dimensional linear isoparametric element: (a) local coordinates and (b) Cartesian coordinates.

The shape functions for a 2D linear isoparametric element are

$$\begin{aligned} N_1 &= \frac{1}{4}(1-\xi)(1-\eta) \\ N_2 &= \frac{1}{4}(1+\xi)(1-\eta) \\ N_3 &= \frac{1}{4}(1-\xi)(1+\eta) \\ N_4 &= \frac{1}{4}(1+\xi)(1+\eta) \end{aligned} \tag{9.37}$$

The temperature at any point of the element in the Cartesian coordinate system is

$$T = \sum N_i T_i = N_1 T_1 + N_2 T_2 + \cdots \tag{9.38}$$

where $N_i(\xi,\eta)$ is the shape function expressed by local coordinates (ξ,η) and T_i is the temperature at node i in the Cartesian coordinate.

The 2D quadratic isoparametric element with eight nodes is shown in Figure 9.6.

The coordinates are transformed by Eq. (9.36) with shape functions as follows:

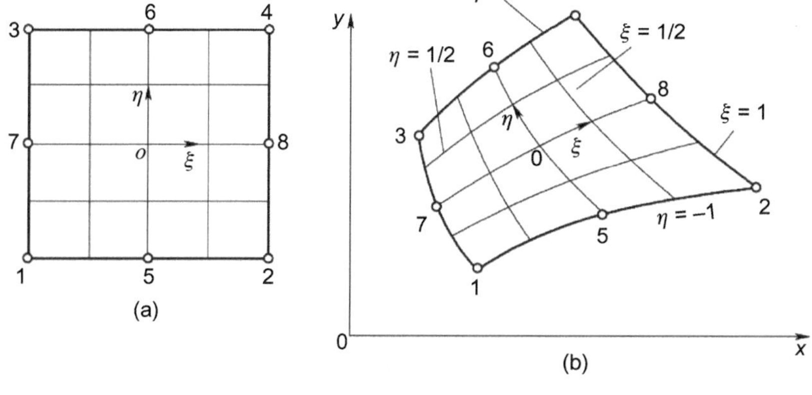

Figure 9.6 Two-dimensional quadratic isoparametric element: (a) local coordinates and (b) Cartesian coordinates.

$$\left.\begin{array}{ll} \text{Corner point} & N_i = \frac{1}{4}(1+\xi_0)(1+\eta_0)(\xi_0+\eta_0-1), \quad i=1,2,3,4 \\ \text{Midpoint of side} & N_i = \frac{1}{2}(1-\xi^2)(1+\eta_0), \quad i=5,6 \\ & N_i = \frac{1}{2}(1-\eta^2)(1+\xi_0), \quad i=7,8 \\ & \xi_0 = \xi_i\xi, \eta_0 = \eta_i\eta \end{array}\right\} \tag{9.39}$$

The shape functions are quadratic on the four sides of the element.

9.5.2 Three-Dimensional Isoparametric Elements

A 3D quadratic isoparametric element with 20 nodes is shown in Figure 9.7. The coordinates of any point of the element in the Cartesian coordinates are

$$\left.\begin{array}{l} x = \sum N_i x_i = N_1 x_1 + N_2 x_2 + \cdots \\ y = \sum N_i y_i = N_1 y_1 + N_2 y_2 + \cdots \\ z = \sum N_i z_i = N_1 z_1 + N_2 z_2 + \cdots \end{array}\right\} \tag{9.40}$$

where x_i, y_i, z_i are the Cartesian coordinates of node i and $N_i(\xi,\eta,\zeta)$ is the shape function expressed by local coordinates ξ,η,ζ as follows:

$$\left.\begin{array}{ll} \text{Corner point} & N_i = \frac{1}{8}(1+\xi_0)(1+\eta_0)(1+\zeta_0)(\xi_0+\eta_0+\zeta_0-2) \\ & (\xi_i = 0, \ \eta_i = \pm 1, \ \zeta_i = \pm 1) \\ \text{Midpoint of side} & N_i = \frac{1}{4}(1-\xi^2)(1+\eta_0)(1+\zeta_0) \end{array}\right\} \tag{9.41}$$

The temperature in the element in the Cartesian coordinates is still expressed by Eq. (9.38), but the shape functions in which are expressed by ξ,η,ζ as Eq. (9.41).

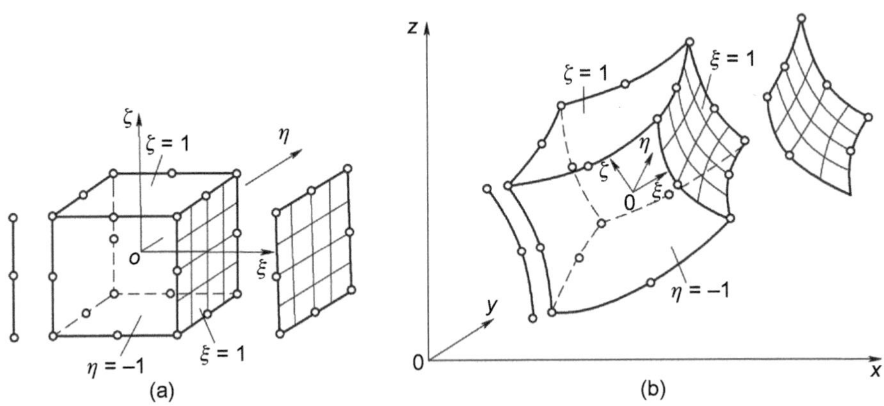

Figure 9.7 Three-dimensional quadratic isoparametric element: (a) local coordinates and (b) Cartesian coordinates.

9.6 Computing Examples of Unsteady Temperature Field

Example 1 Cooling of rectangular prism A rectangular prism of infinite length, the cross section is 10 m × 4 m, the initial temperature is $T_0 = 20°C$, the sides AD and BC are insulated, the sides AB and CD are kept at $T_b = 10°C$, the diffusivity $a = 0.10\ m^2/day$, the net of computation is shown in Figure 9.8, the results computed by FEM are shown in Figure 9.9. The points indicating the computed results all lie on the curves indicating the theoretical solutions.

Example 2 Cooling of concrete dam A concrete gravity dam is shown in Figure 9.10(a), the height of dam is $H = 65$ m, the width of the base of dam is $L = 49$ m, the initial temperature of the dam is 20°C and that of the rock foundation is 0°C, and the boundary temperature is 0°C. The process of cooling is computed by FEM and the computed results are shown in Figure 9.10. It is clear from

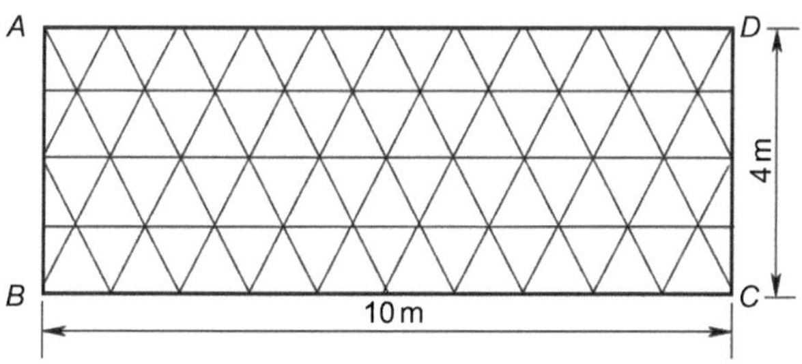

Figure 9.8 Example of net of computation.

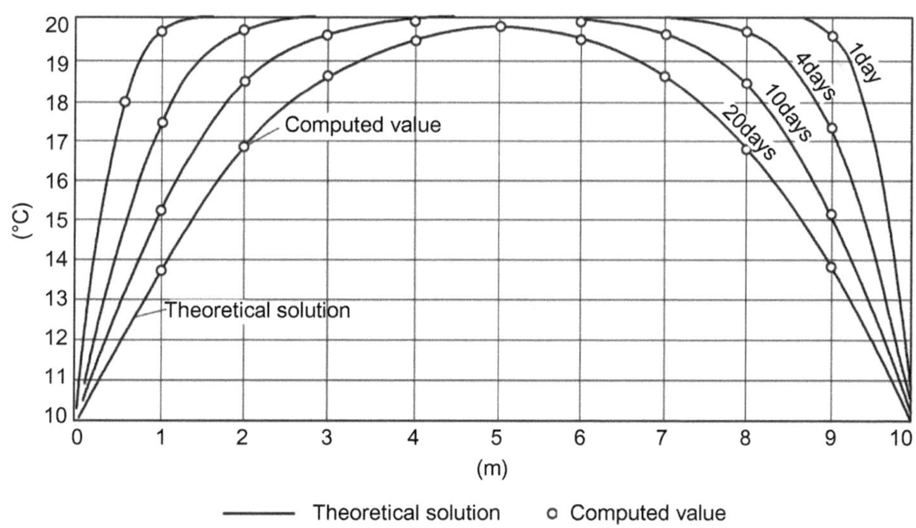

Figure 9.9 Example 1: Comparison of the results given by FEM and the theoretical solution.

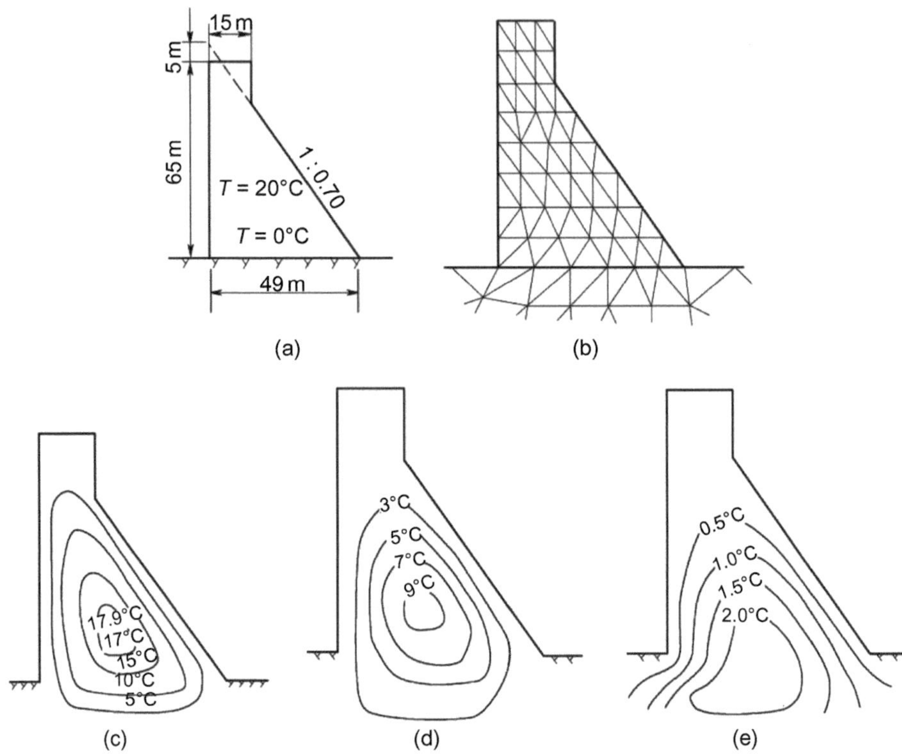

Figure 9.10 Example 2: Cooling of concrete dam: (a) dimensions of dam and initial temperature T_0, (b) net of computing (only part of net is shown for the foundation), (c) temperature distribution after cooling of 1 year, (d) temperature distribution after cooling of 3 years, and (e) temperature distribution after cooling of 8 years.

Figure 9.10(e) that the temperature in the interior of the dam is still 2°C after cooling of 8 years.

10 Finite Element Method for Computing the Viscoelastic Thermal Stresses of Massive Concrete Structures

The thermal stresses of massive concrete structures depend on the process of construction. Sometimes the geometrical shape of a structure is irregular, so it is difficult to compute them by theoretical methods; generally the finite element method (FEM) is used.

10.1 FEM for Computing Elastic Thermal Stresses

As shown in Figure 10.1, a two-dimensional continuous body is divided into a finite number of elements which are connected at n nodes. Two displacements of each node are unknown variables; thus there are $2n$ unknowns for the whole structure. There are three displacements for each node and $3n$ unknowns for the whole structure for a spatial problem [8, 9, 35–40].

10.1.1 Displacements of an Element

For a spatial element shown in Figure 10.2, the displacements of any node i are

$$\{\delta_i\} = \begin{Bmatrix} u_i \\ v_i \\ w_i \end{Bmatrix}$$

where u_i, v_i, w_i are the displacements in the x, y, z directions, respectively. The vector constructed by the displacements of all the nodes of element e is indicated by $\{\delta\}^e$ in the following:

$$\{\delta\}^e = [\delta_1 \delta_2 \delta_3 \cdots]^{\mathrm{T}} = [u_1 v_1 w_1 u_2 v_2 w_2 \cdots]^{\mathrm{T}} \tag{10.1}$$

Thermal Stresses and Temperature Control of Mass Concrete. DOI: http://dx.doi.org/10.1016/B978-0-12-407723-2.00010-5

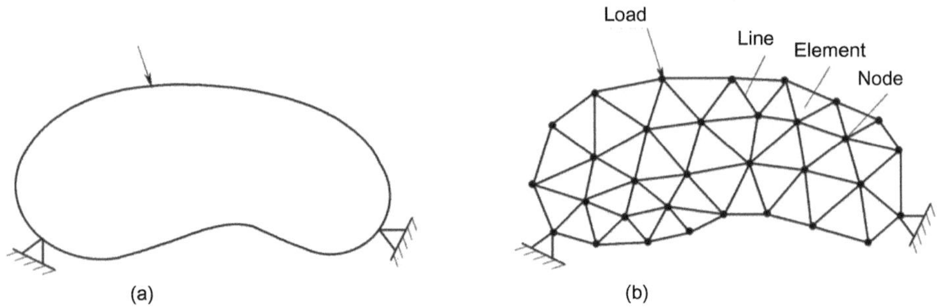

Figure 10.1 Discretion of plane problem: (a) continuous body and (b) model of finite element.

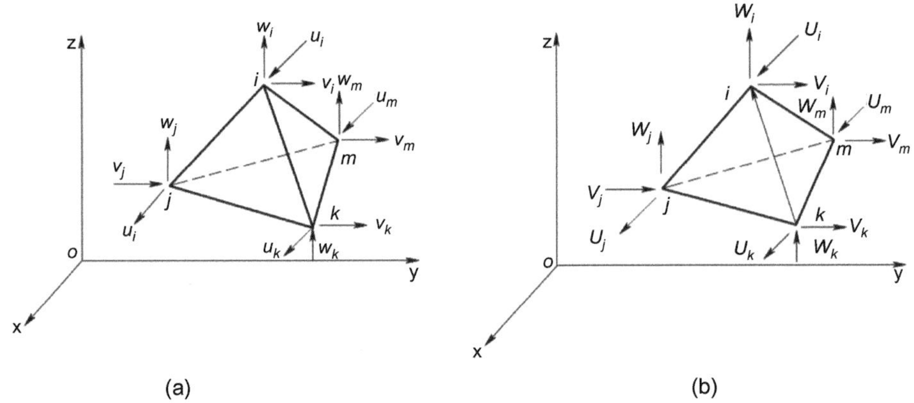

Figure 10.2 Spatial element: (a) nodal displacements and (b) nodal force.

The displacements of any point in the element are expressed by shape functions and nodal displacements as follows:

$$\left.\begin{aligned} u &= \sum N_i u_i = N_1 u_1 + N_2 u_2 + \cdots \\ v &= \sum N_i v_i = N_1 v_1 + N_2 v_2 + \cdots \\ w &= \sum N_i w_i = N_1 w_1 + N_{2w} u_2 + \cdots \end{aligned}\right\} \tag{10.2}$$

or expressed by matrices

$$\{r\} = \begin{Bmatrix} u \\ v \\ w \end{Bmatrix} = \begin{bmatrix} N_1 & 0 & 0 & N_2 & 0 & 0 & \cdots \\ 0 & N_1 & 0 & 0 & N_2 & 0 & \cdots \\ 0 & 0 & N_1 & 0 & 0 & N_2 & \cdots \end{bmatrix} \begin{Bmatrix} u_1 \\ v_1 \\ w_1 \\ u_2 \\ v_2 \\ \vdots \end{Bmatrix} = [N_1 I, N_2 I, \cdots]\{\delta\}^e = [N]\{\delta\}^e \tag{10.3}$$

where

$[N]$—the shape function matrix
I—a 3×3 unit matrix.

10.1.2 Strains of an Element

For the spatial problem, there are six strain components at any point which may be expressed in the following:

$$\{\varepsilon\} = \begin{Bmatrix} \varepsilon_x \\ \varepsilon_y \\ \varepsilon_z \\ \gamma_{xy} \\ \gamma_{yz} \\ \gamma_{zx} \end{Bmatrix} = \begin{Bmatrix} \frac{\partial u}{\partial x} \\ \frac{\partial v}{\partial y} \\ \frac{\partial w}{\partial z} \\ \frac{\partial u}{\partial y} + \frac{\partial v}{\partial x} \\ \frac{\partial v}{\partial z} + \frac{\partial w}{\partial y} \\ \frac{\partial w}{\partial x} + \frac{\partial u}{\partial z} \end{Bmatrix} \tag{10.4}$$

Substituting Eq. (10.2) into Eq. (10.4), we have

$$\{\varepsilon\} = \begin{bmatrix} \frac{\partial N_1}{\partial x} & 0 & 0 & \frac{\partial N_2}{\partial x} & 0 & 0 & \dots \\ 0 & \frac{\partial N_1}{\partial y} & 0 & 0 & \frac{\partial N_2}{\partial y} & 0 & \dots \\ 0 & 0 & \frac{\partial N_1}{\partial z} & 0 & 0 & \frac{\partial N_2}{\partial z} & \dots \\ \frac{\partial N_1}{\partial y} & \frac{\partial N_1}{\partial x} & 0 & \frac{\partial N_2}{\partial y} & \frac{\partial N_2}{\partial x} & 0 & \dots \\ 0 & \frac{\partial N_1}{\partial z} & \frac{\partial N_1}{\partial y} & 0 & \frac{\partial N_2}{\partial z} & \frac{\partial N_2}{\partial y} & \dots \\ \frac{\partial N_1}{\partial z} & 0 & \frac{\partial N_1}{\partial x} & \frac{\partial N_2}{\partial z} & 0 & \frac{\partial N_2}{\partial x} & \dots \end{bmatrix} \begin{Bmatrix} u_1 \\ v_1 \\ w_1 \\ u_2 \\ v_2 \\ w_2 \\ \vdots \end{Bmatrix} = [B_1 B_2 \dots]\{\delta\}^e = [B]\{\delta\}^e \tag{10.5}$$

where

$$[B_i] = \begin{bmatrix} \frac{\partial N_i}{\partial x} & 0 & 0 \\ 0 & \frac{\partial N_i}{\partial y} & 0 \\ 0 & 0 & \frac{\partial N_i}{\partial z} \\ \frac{\partial N_i}{\partial y} & \frac{\partial N_i}{\partial x} & 0 \\ 0 & \frac{\partial N_i}{\partial z} & \frac{\partial N_i}{\partial y} \\ \frac{\partial N_i}{\partial z} & 0 & \frac{\partial N_i}{\partial x} \end{bmatrix} \tag{10.6}$$

10.1.3 Stresses of an Element

There are six stress components at any point in an element:

$$\{\sigma\} = \begin{bmatrix} \sigma_x & \sigma_y & \sigma_z & \tau_{xy} & \tau_{yz} & \tau_{zx} \end{bmatrix}^{\mathrm{T}} \tag{10.7}$$

According to the generalized Hooke's law, the stress–strain relationship is as follows:

$$\{\sigma\} = [D](\{\varepsilon\} - \{\varepsilon_0\}) + \{\sigma_0\} \tag{10.8}$$

where $\{\varepsilon_0\}$ is the initial strain and $\{\sigma_0\}$ the initial stress. The elasticity matrix for isotropic elastic body is expressed in the following:

$$[D] = \frac{E(1-\mu)}{(1+\mu)(1-2\mu)} \begin{bmatrix} 1 & \frac{\mu}{1-\mu} & \frac{\mu}{1-\mu} & 0 & 0 & 0 \\ & 1 & \frac{\mu}{1-\mu} & 0 & 0 & 0 \\ & & 1 & 0 & 0 & 0 \\ \text{symmetrical} & & & \frac{1-2\mu}{2(1-\mu)} & 0 & 0 \\ & & & & \frac{1-2\mu}{2(1-\mu)} & 0 \\ & & & & & \frac{1-2\mu}{2(1-\mu)} \end{bmatrix} \tag{10.9}$$

10.1.4 Nodal Forces and Stiffness Matrix of an Element

Nodal forces are concentrated forces acting on the nodes which are equivalent statically to the boundary stresses and distributed loads on the element. As shown in Figure 10.2(b), the nodal forces at node i are expressed by

$$\{F_i\} = \begin{Bmatrix} U_i \\ V_i \\ W_i \end{Bmatrix}$$

where U_i, V_i, W_i are the nodal force acting at node i in the x, y, z directions, respectively. The vector consisting of all the nodal forces of the element e is indicated by $\{F\}^e$ as follows:

$$\{F\}^e = [F_1 F_2 \cdots]^{\mathrm{T}} = [U_1 V_1 W_1 U_2 V_2 W_2 \cdots]^{\mathrm{T}} \tag{10.10}$$

By the principle of virtual displacement, the element nodal forces may be computed as follows:

$$\{F\}^e = \iiint [B]^{\mathrm{T}}\{\sigma\}\mathrm{d}x\,\mathrm{d}y\,\mathrm{d}z \tag{10.11}$$

If there is no initial stress, the stresses are expressed by

$$\{\sigma\} = [D]\{\varepsilon\} = [D][B]\{\delta\}^e$$

Substituting the above equation into Eq. (10.11), we get

$$\{F\}^e = \iiint [B]^{\mathrm{T}}[D][B]\mathrm{d}x\,\mathrm{d}y\,\mathrm{d}z\{\delta\}^e$$

Let

$$[k]^e = \iiint [B]^{\mathrm{T}}[D][B]\mathrm{d}x\,\mathrm{d}y\,\mathrm{d}z \tag{10.12}$$

then

$$\{F\}^e = [k]^e\{\delta\}^e \tag{10.13}$$

which establishes the relation between the nodal forces and the nodal displacements. The matrix $[k]^e$ is called the element stiffness matrix.

10.1.5 Nodal Loads

All the distributed loads are replaced by equivalent nodal loads. Let $\{P_i\}$ be the equivalent nodal loads at node i,

$$\{P_i\} = \begin{Bmatrix} X_i \\ Y_i \\ Z_i \end{Bmatrix}$$

where X_i, Y_i, Z_i are the concentrated loads at node i in the x, y, z directions, respectively.

Let $\{P\}^e$ be the vector consisting of all the nodal loads:

$$\{P\}^e = [P_1 P_2 \cdots]^{\mathrm{T}} = [X_1 Y_1 Z_1 X_2 Y_2 Z_2 \cdots]^{\mathrm{T}} \tag{10.14}$$

The formulas for various nodal loads derived by the principle of virtual displacement are given in the following:

1. Distributed body force

   If the body force in a unit volume is

$$\{q\} = \begin{Bmatrix} q_x \\ q_y \\ q_z \end{Bmatrix}$$

   The equivalent nodal load produced by $\{q\}$ are

$$\{P\}_q^e = \iiint [N]^{\mathrm{T}} \{q\} \mathrm{d}x\, \mathrm{d}y\, \mathrm{d}z \tag{10.15}$$

2. Distributed surface force

   If the element e keeps to the side of the structure and the surface forces $\{p\}$ are acting on the surface of the element:

$$\{p\} = \begin{Bmatrix} p_x \\ p_y \\ p_z \end{Bmatrix}$$

   The equivalent nodal load induced by $\{p\}$ is

$$\{P\}_p^e = \int [N]^{\mathrm{T}} \{p\} dS \tag{10.16}$$

3. Initial strain and initial stress

   If there are initial strain $\{\varepsilon_0\}$ and initial stress $\{\sigma_0\}$ in the element, the equivalent nodal loads produced by them are

$$\{P\}_{\varepsilon_0}^e = \iiint [B]^{\mathrm{T}} [D] \{\varepsilon_0\} \mathrm{d}x\, \mathrm{d}y\, \mathrm{d}z \tag{10.17}$$

$$\{P\}_{\sigma_0}^e = -\iiint [B]^{\mathrm{T}} [\sigma_0] \mathrm{d}x\, \mathrm{d}y\, \mathrm{d}z \tag{10.18}$$

10.1.6 Equilibrium Equation of Nodes and the Global Stiffness Matrix

For any node i cut from the body, the equilibrium equations in the x, y, z directions must be satisfied as follows:

$$\sum_e U_i = X_i, \quad \sum_e V_i = Y_i, \quad \sum_e W_i = Z_i \tag{10.19}$$

where U_i, V_i, W_i are nodal forces and X_i, Y_i, Z_i are nodal loads. The above equations may be expressed by matrices as follows:

$$\sum_e \{F_i\} = \{P_i\}$$

Substituting $\{F\}^e = [k]^e\{\delta\}^e$, we obtain the nodal equilibrium equations expressed by nodal displacements in the following:

$$[K]\{\delta\} = \{P\} \tag{10.20}$$

where $[K]$ is the global stiffness matrix, whose elements K_{ij} are computed by

$$K_{rs} = \sum_e k_{ij}^e \tag{10.21}$$

where the symbol $\sum_e$ means the sum of all the coefficients k_{ij}^e relevant to the element "e."

10.1.7 Collection of FEM Formulas

For the convenience of the reader, the FEM formulas are collected in the following:

$$\{\delta\}^e = [u_1 v_1 w_1 u_2 v_2 w_2 \cdots]^{\mathrm{T}} \tag{10.22}$$

$$\{\varepsilon\} = [\varepsilon_x \varepsilon_y \varepsilon_z \gamma_{xy} \gamma_{yz} \gamma_{zx}]^{\mathrm{T}} \tag{10.23}$$

$$\{\sigma\} = [\sigma_x \sigma_y \sigma_z \tau_{xy} \tau_{yz} \tau_{zx}]^{\mathrm{T}} \tag{10.24}$$

$$\{r\} = \begin{Bmatrix} u \\ v \\ w \end{Bmatrix} = \begin{bmatrix} N_1 & 0 & 0 & N_2 & 0 & 0 & \ldots \\ 0 & N_1 & 0 & 0 & N_2 & 0 & \ldots \\ 0 & 0 & N_1 & 0 & 0 & N_2 & \ldots \end{bmatrix} \begin{Bmatrix} u_1 \\ v_1 \\ w_1 \\ u_2 \\ \vdots \end{Bmatrix} = [N]\{\delta\}^e \tag{10.25}$$

$$\{\varepsilon\} = [B]\{\delta\}^e \tag{10.26}$$

$$\{\sigma\} = [D](\{\varepsilon\} - \{\varepsilon_0\}) + \{\sigma_0\} \tag{10.27}$$

$$\{F\}^e = \iiint [B]^{\mathrm{T}}\{\sigma\}\mathrm{d}x\,\mathrm{d}y\,\mathrm{d}z \tag{10.28}$$

$$\{F\}^e = [k]^e\{\delta\}^e \tag{10.29}$$

$$\{k\}^e = \iiint [B]^{\mathrm{T}}[D][B]\mathrm{d}x\,\mathrm{d}y\,\mathrm{d}z \tag{10.30}$$

$$\{P\}_q^e = \iiint [N]^{\mathrm{T}}\{q\}\mathrm{d}x\,\mathrm{d}y\,\mathrm{d}z \tag{10.31}$$

$$\{P\}_p^e = \int_S [N]^{\mathrm{T}}\{p\}\mathrm{d}s \tag{10.32}$$

$$\{P\}_{\varepsilon_0}^e = \iiint [B]^{\mathrm{T}}[D]\{\varepsilon_0\}\mathrm{d}x\,\mathrm{d}y\,\mathrm{d}z \tag{10.33}$$

$$\{P\}_{\sigma_0}^e = -\iiint [B]^{\mathrm{T}}[\sigma_0]\mathrm{d}x\,\mathrm{d}y\,\mathrm{d}z \tag{10.34}$$

$$[K]\{\delta\} = \{P\} \tag{10.35}$$

10.2 Implicit Method for Solving Viscoelastic Stress–Strain Equation of Mass Concrete

Because the creep strain of concrete depends on the history of stress which must be recorded in the process of computation, a huge volume of memory is required. How to reduce the volume of memory is important in computing viscoelastic thermal stress of mass concrete by FEM. Zienkiewicz had proposed an explicit method with equal time interval, the writer had proposed an implicit method with unequal time interval which is more efficient and will be described in the following.

10.2.1 Computing Increment of Strain

In the explicit calculation, the stress is assumed to be constant in $\Delta\tau$, the stress–time curve has the shape of steps, as shown in Figure 10.3(a), and the error

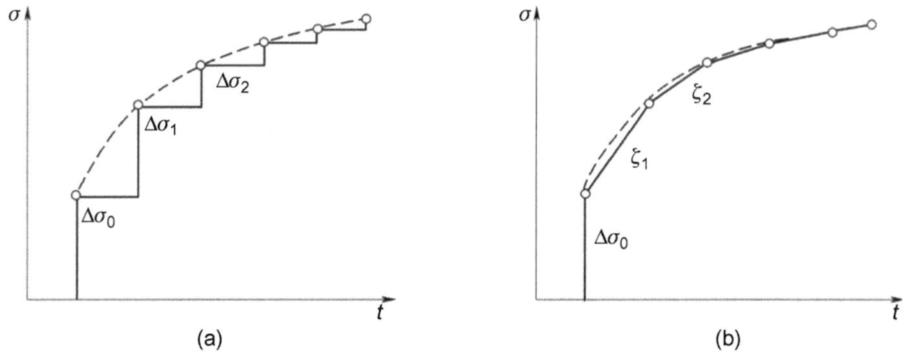

Figure 10.3 History of stress: (a) explicit method and (b) implicit method.

of computation is rather big. In the implicit calculation, it is assumed that the stress varies linearly in $\Delta\tau$, namely

$$\frac{\mathrm{d}\sigma}{\mathrm{d}\tau} = \xi_i = \text{constant}$$

in $\Delta\tau_i$, as shown in Figure 10.3(b), the $\sigma - \tau$ curve is a broken line.

From Eq. (2.23), the strain at time t is

$$\varepsilon(t) = \varepsilon^{\mathrm{e}}(t) + \varepsilon^{\mathrm{c}}(t) \tag{10.36}$$

where

$$\varepsilon^{\mathrm{e}}(t) = \frac{\sigma(\tau_0)}{E(\tau_0)} + \int_{\tau_0}^{t} \frac{1}{E(\tau)}\frac{\mathrm{d}\sigma}{\mathrm{d}\tau}\mathrm{d}\tau \tag{10.37}$$

$$\varepsilon^{\mathrm{c}}(t) = \sigma(\tau_0)C(t,\tau_0) + \int_{\tau_0}^{t} C(t,\tau)\frac{\mathrm{d}\sigma}{\mathrm{d}\tau}\mathrm{d}\tau \tag{10.38}$$

in which $\varepsilon^{\mathrm{e}}(t)$ is the elastic strain and $\varepsilon^{\mathrm{c}}(t)$ is the creep strain.

Dividing the time τ into a series of unequal time increments as shown in Figure 10.4, the increment of elastic strain in $\Delta\tau_i$ is given by

$$\Delta\varepsilon_n^{\mathrm{e}} = \varepsilon^{\mathrm{e}}(t_n) - \varepsilon^{\mathrm{e}}(t_{n-1}) = \int_{\tau_{n-1}}^{\tau_n} \frac{1}{E(\tau)}\frac{\mathrm{d}\sigma}{\mathrm{d}\tau}\mathrm{d}\tau = \left(\frac{\mathrm{d}\sigma}{\mathrm{d}\tau}\right)_n \int_{\tau_{n-1}}^{\tau_n} \frac{1}{E(\tau)}\mathrm{d}\tau$$

By the mean value theorem of integration, we get

$$\Delta\varepsilon_n^{\mathrm{e}} = \frac{1}{E(\overline{\tau}_n)}\left(\frac{\mathrm{d}\sigma}{\mathrm{d}\tau}\right)_n (\tau_n - \tau_{n-1}) = \frac{\Delta\sigma_n}{E(\overline{\tau}_n)} \tag{10.39}$$

From τ_0 to t, the creep strain is

$$\begin{aligned}\varepsilon^c(t) &= \Delta\sigma_0 C(t,\tau_0) + \sum_i \int_{\tau_{n-1}}^{\tau_n} C(t,\tau)\frac{d\sigma}{d\tau}d\tau \\ &= \Delta\sigma_0 C(t,\tau_0) + \sum_i \left(\frac{d\sigma}{d\tau}\right) \int_{\tau_{n-1}}^{\tau_n} C(t,\tau)d\tau\end{aligned}$$

By the mean value theorem, we have

$$\varepsilon^c(t) = \Delta\sigma_0 C(t,\tau_0) + \sum_i C(t,\overline{\tau}_i)\left(\frac{d\sigma}{d\tau}\right)_i \Delta\tau_i = \Delta\sigma_0 C(t,\tau_0) + \sum_i C(t,\overline{\tau}_i)\Delta\sigma_i \tag{10.40}$$

Let the unit creep be expressed by

$$C(t,\tau) = \Psi(\tau)[1 - e^{-r(t-\tau)}] \tag{10.41}$$

Substituting into Eq. (10.40), we get

$$\varepsilon^c(t) = \Delta\sigma_0\Psi(\tau_0)[1 - e^{-r(t-\tau_0)}] + \sum_i \Delta\sigma_i\Psi(\overline{\tau}_i)[1 - e^{-r(t-\overline{\tau}_i)}] \tag{10.42}$$

As shown in Figure 10.4, take three adjacent instants of time τ_{n-1}, τ_n, τ_{n+1}, with

$$\Delta\tau_n = \tau_n - \tau_{n-1}, \quad \Delta\tau_{n+1} = \tau_{n+1} - \tau_n$$

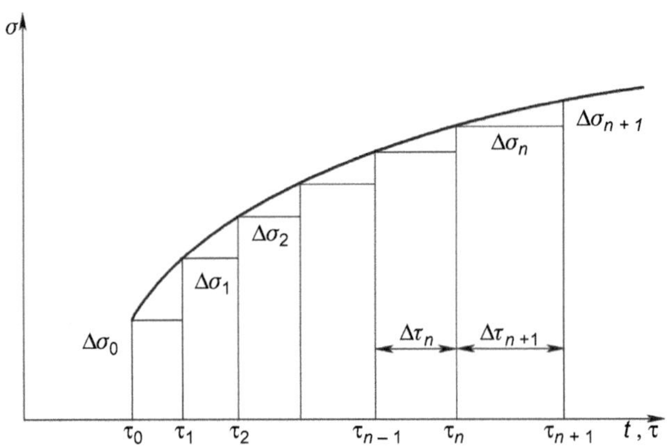

Figure 10.4 Incremental method.

From Eq. (10.42), the creep strains at the three adjacent instants of time are

$$\begin{aligned}\varepsilon^{c}(t_{n-1}) = &\Delta\sigma_0\Psi(\tau_0)[1-e^{-r(\tau_n-\Delta\tau_n-\tau_0)}]\\ &+\Delta\sigma_1\Psi(\overline{\tau}_1)[1-e^{-r(\tau_n-\Delta\tau_n-\overline{\tau}_1)}]+\cdots\\ &+\Delta\sigma_{n-1}\Psi(\overline{\tau}_{n-1})[1-e^{-r(\tau_n-\Delta\tau_n-\overline{\tau}_{n-1})}]\end{aligned} \tag{10.43}$$

$$\begin{aligned}\varepsilon^{c}(t_n) = &\Delta\sigma_0\Psi(\tau_0)[1-e^{-r(t_n-\tau_0)}]+\Delta\sigma_1\Psi(\overline{\tau}_1)[1-e^{-r(t_n-\overline{\tau}_1)}]+\cdots\\ &+\Delta\sigma_{n-1}\Psi(\overline{\tau}_{n-1})[1-e^{-r(t_n-\overline{\tau}_{n-1})}]+\Delta\sigma_n\Psi(\overline{\tau}_n)[1-e^{-r(t_n-\overline{\tau}_n)}]\end{aligned} \tag{10.44}$$

$$\begin{aligned}\varepsilon^{c}(t_{n+1}) = &\Delta\sigma_0\Psi(\tau_0)[1-e^{-r(t_n+\Delta\tau_{n+1}-\tau_0)}]+\Delta\sigma_1\Psi(\overline{\tau}_1)[1-e^{-r(\tau_n+\Delta\tau_{n+1}-\overline{\tau}_1)}]+\cdots\\ &+\Delta\sigma_{n-1}\Psi(\overline{\tau}_{n-1})[1-e^{-r(t_n+\Delta\tau_{n+1}-\overline{\tau}_{n-1})}]+\Delta\sigma_n\Psi(\overline{\tau}_n)[1-e^{-r(t_n+\Delta\tau_{n+1}-\overline{\tau}_n)}]\\ &+\Delta\sigma_{n+1}\Psi(\overline{\tau}_{n+1})[1-e^{-r(t_n+\Delta\tau_{n+1}-\overline{\tau}_{n+1})}]\end{aligned}$$

10.45

From Eq. (10.44) minus Eq. (10.43), we get

$$\begin{aligned}\Delta\varepsilon_n^{c} &= \varepsilon^{c}(t_n)-\varepsilon^{c}(t_{n-1})\\ &= \Delta\sigma_0\Psi(\tau_0)[e^{-r(t_n-\Delta\tau_n-\tau_0)}-e^{-r(t_n-\tau_0)}]\\ &\quad+\Delta\sigma_1\Psi(\overline{\tau}_1)[e^{-r(t_n-\Delta\tau_n-\overline{\tau}_1)}-e^{-r(t_n-\overline{\tau}_1)}]\\ &\quad\vdots\\ &\quad+\Delta\sigma_{n-1}\Psi(\overline{\tau}_{n-1})[e^{-r(t_n-\Delta\tau_n-\overline{\tau}_{n-1})}-e^{-r(t_n-\overline{\tau}_{n-1})}]\\ &\quad+\Delta\sigma_n\Psi(\overline{\tau}_n)[1-e^{-r(t_n-\overline{\tau}_n)}]\\ &= (1-e^{-r\Delta\tau_n})[\Delta\sigma_0\Psi(\tau_0)e^{-r(t_n-\Delta\tau_n-\tau_0)}\\ &\quad+\Delta\sigma_1\Psi(\overline{\tau}_1)e^{-r(t_n-\Delta\tau_n-\overline{\tau}_1)}+\cdots\\ &\quad+\Delta\sigma_{n-1}\Psi(\overline{\tau}_{n-1})e^{-r(t_n-\Delta\tau_n-\overline{\tau}_{n-1})}]\\ &\quad+\Delta\sigma_n\Psi(\overline{\tau}_n)[1-e^{-r(t_n-\overline{\tau}_n)}]\end{aligned} \tag{10.46}$$

Similarly

$$\begin{aligned}\Delta\varepsilon_{n+1}^{c} &= \varepsilon^{c}(t_{n+1})-\varepsilon^{c}(t_n)\\ &= (1-e^{-r\Delta\tau_{n+1}})[\Delta\sigma_0\Psi(\tau_0)e^{-r(t_n-\tau_0)}+\Delta\sigma_1\Psi(\overline{\tau}_1)e^{-r(t_n-\overline{\tau}_1)}+\cdots\\ &\quad+\Delta\sigma_{n-1}\Psi(\overline{\tau}_{n-1})e^{-r(t_n-\overline{\tau}_{n-1})}+\Delta\sigma_n\Psi(\overline{\tau}_n)e^{-r(t_n-\overline{\tau}_n)}]\\ &\quad+\Delta\sigma_{n-1}\Psi(\overline{\tau}_{n-1})[1-e^{-r(t_n+\Delta\tau_{n+1}+\overline{\tau}_{n+1})}]\end{aligned} \tag{10.47}$$

Comparison of Eqs. (10.46) and (10.47) yields a set of recurrence formulas as follows:

$$\left.\begin{aligned}\Delta\varepsilon_{n+1}^{c} &= (1-e^{-r\Delta\tau_{n+1}})\omega_{n+1}+\Delta\sigma_{n+1}C(t_{n+1},\overline{\tau}_n)\\ \omega_{n+1} &= \omega_n e^{-r\Delta\tau_n}+\Delta\sigma_n\Psi(\overline{\tau}_n)e^{-0.5r\Delta\tau_n}\\ \omega_1 &= \Delta\sigma_0\Psi(\tau_0)\end{aligned}\right\} \tag{10.48}$$

which may be rewritten in the following:

$$\left.\begin{aligned}\Delta\varepsilon_n^{\mathrm{c}} &= \varepsilon^{\mathrm{c}}(t^n) - \varepsilon^{\mathrm{c}}(t_{n-1})\\ &= (1 - \mathrm{e}^{-r\Delta\tau_n})\omega_n + \Delta\sigma_n C(t_n, \overline{\tau}_n)\\ \omega_n &= \omega_{n-1}\,\mathrm{e}^{-r\Delta\tau_{n-1}} + \Delta\sigma_{n-1}\Psi(\overline{\tau}_{n-1})\Psi\,\mathrm{e}^{-0.5r\Delta\tau_{n-1}}\\ \omega_1 &= \Delta\sigma_0\Psi(\tau_0)\end{aligned}\right\} \tag{10.49}$$

If the unit creep is expressed by

$$C(t,\tau) = \sum_{s=1}\Psi_s(\tau)[1 - \mathrm{e}^{-r_s(t-\tau)}] \tag{10.50}$$

then the incremet of creep strain is given by

$$\begin{aligned}\Delta\varepsilon_n^{\mathrm{c}} &= \varepsilon^{\mathrm{c}}(t_n) - \varepsilon^{\mathrm{c}}(t_{n-1})\\ &= \sum_s (1 - \mathrm{e}^{-r_s\Delta\tau_n})\omega_{sn} + \Delta\sigma_n C(t_n - \overline{\tau}_n)\\ &= \eta_s + \Delta\sigma_n C(t_n, \overline{\tau}_n)\end{aligned} \tag{10.51a}$$

$$\eta_n = \sum_s (1 - \mathrm{e}^{-r_s\Delta\tau_n})\omega_{sn} \tag{10.51b}$$

$$\omega_{sn} = \omega_{s,n-1}\,\mathrm{e}^{-r_s\Delta\tau_{n-1}} + \Delta\sigma_{n-1}\Psi_s(\overline{\tau}_{n-1})\mathrm{e}^{-0.5r_s\Delta\tau_{n-1}} \tag{10.51c}$$

$$\omega_{s1} = \Delta\sigma_0\Psi_s(\tau_0) \tag{10.51d}$$

Using the above recurrence formulas to compute the viscoelastic thermal stresses of massive concrete structure, it is not necessary to record the history of stress, only ω_{sn} is required to store; thus the volume of memory is reduced a great deal.

10.2.2 Relationship Between Stress Increment and Strain Increment for One-Directional Stress

Besides the elastic strain and creep strain induced by stress, there are thermal strain, shrinkage strain, and autogenous strain in concrete. Generally the strain of concrete under the action of one-directional stress may be expressed by

$$\varepsilon(t) = \varepsilon^{\mathrm{e}}(t) + \varepsilon^{\mathrm{c}}(t) + \varepsilon^{\mathrm{T}}(t) + \varepsilon^{0}(t) + \varepsilon^{\mathrm{s}}(t) \tag{10.52}$$

where

$\varepsilon^{\mathrm{e}}(t)$—elastic strain
$\varepsilon^{\mathrm{c}}(t)$—creep strain
$\varepsilon^{\mathrm{T}}(t)$—free thermal strain
$\varepsilon^{0}(t)$—autogenous strain
$\varepsilon^{\mathrm{s}}(t)$—shrinkage strain.

The strain increment in $\Delta\tau_n$ is

$$\Delta\varepsilon_n = \varepsilon(t_n) - \varepsilon(t_{n-1}) = \Delta\varepsilon_n^{\mathrm{e}} + \Delta\varepsilon_n^{\mathrm{c}} + \Delta\varepsilon_n^{\mathrm{T}} + \Delta\varepsilon_n^{0} + \Delta\varepsilon_n^{\mathrm{s}} \tag{10.53}$$

Substituting Eqs. (10.39) and (10.36) into Eq. (10.53), we have

$$\Delta\varepsilon_n = \frac{\Delta\sigma_n}{E(\overline{\tau}_n)} + \eta_n + \Delta\sigma_n C(t_n, \overline{\tau}_n) + \Delta\varepsilon_n^{\mathrm{T}} + \Delta\varepsilon_n^{0} + \Delta\varepsilon_n^{\mathrm{s}} \tag{10.54}$$

Hence the formula for stress increment $\Delta\sigma_n$ is derived

$$\Delta\sigma_n = \overline{E}_n(\Delta\varepsilon_n - \eta_n - \Delta\varepsilon_n^{\mathrm{T}} - \Delta\varepsilon_n^{0} - \Delta\varepsilon_n^{\mathrm{s}}) \tag{10.55}$$

where

$$\overline{E}_n = \frac{E(\overline{\tau}_n)}{1 + E(\overline{\tau}_n)C(t_n, \overline{\tau}_n)} \tag{10.56}$$

η_n is given by Eq. (10.51b), $\Delta\varepsilon_n^{\mathrm{T}} = \alpha\Delta T_n$, $\Delta\varepsilon_n^{0} = \varepsilon^0(t_n) - \varepsilon^0(t_{n-1})$, $\Delta\varepsilon_n^{\mathrm{s}} = \varepsilon^{\mathrm{s}}(t_n) - \varepsilon^{\mathrm{s}}(t_{n-1})$, $\overline{\tau}_n = (\tau_{n-1} + \tau_n)/2$.

10.2.3 Relationship Between Stress Increment and Strain Increment for Complex Stress State

For plane problem

$$\{\varepsilon\} = [\varepsilon_x, \varepsilon_y, \gamma_{xy}]^{\mathrm{T}} \tag{10.57}$$

$$\{\sigma\} = [\sigma_x, \sigma_y, \tau_{xy}]^{\mathrm{T}} \tag{10.58}$$

For spatial problem

$$\{\varepsilon\} = [\varepsilon_x, \varepsilon_y, \varepsilon_z, \gamma_{xy}, \gamma_{yz}, \gamma_{zx}]^{\mathrm{T}} \tag{10.59}$$

$$\{\sigma\} = [\sigma_x, \sigma_y, \sigma_z, \tau_{xy}, \tau_{yz}, \tau_{zx}]^{\mathrm{T}} \tag{10.60}$$

From Eq. (9-2-4), the vector of elastic strain increment is given by

$$\{\Delta\varepsilon_n^{\mathrm{e}}\} = \frac{1}{E(\overline{\tau}_n)}[Q][\Delta\sigma_n] \tag{10.61}$$

where $[Q]$ is given by Eqs. (10.65)–(10.67).

From Eqs. (10.51a)–(10.51d) the vector of creep strain increment is given by

$$\{\Delta\varepsilon_n^c\} = \{\eta_n\} + C(t_n, \overline{\tau}_n)[Q]\{\Delta\sigma_n\} \tag{10.62}$$

where

$$\{\eta_n\} = \sum_s (1 - e^{-r_s\Delta\tau_n})\{\omega_{sn}\} \tag{10.63}$$

$$\{\omega_{sn}\} = \{\omega_{s,n-1}\}e^{-r_s\Delta\tau_{n-1}} + [Q]\{\Delta\sigma_{n-1}\}\Psi_s(\overline{\tau}_{n-1})e^{-0.5r_s\Delta\tau_{n-1}} \tag{10.64}$$

For plane stress problem

$$[Q] = \begin{bmatrix} 1 & -\mu & 0 \\ -\mu & 1 & 0 \\ 0 & 0 & 2(1+\mu) \end{bmatrix},\ [Q]^{-1} = \frac{1}{1-\mu^2}\begin{bmatrix} 1 & \mu & 0 \\ \mu & 1 & 0 \\ 0 & 0 & (1-\mu)/2 \end{bmatrix},\ \{\Delta\varepsilon_n^T\} = \begin{Bmatrix} \alpha\Delta T_n \\ \alpha\Delta T_n \\ 0 \end{Bmatrix} \tag{10.65}$$

For plane strain problem

$$[Q] = (1+\mu)\begin{bmatrix} 1-\mu & -\mu & 0 \\ -\mu & 1-\mu & 0 \\ 0 & 0 & 2 \end{bmatrix},\ [Q]^{-1} = \frac{1-\mu}{(1+\mu)(1-2\mu)}\begin{bmatrix} 1 & \dfrac{\mu}{1-\mu} & 0 \\ \dfrac{\mu}{1-\mu} & 1 & 0 \\ 0 & 0 & \dfrac{1-2\mu}{2(1-\mu)} \end{bmatrix}$$

$$\{\Delta\varepsilon_n^T\} = [(1+\mu)\alpha\Delta T_n,\ \ (1+\mu)\alpha\Delta T_n, 0]^T \tag{10.66}$$

For spatial problem

$$[Q] = \begin{bmatrix} 1 & -\mu & -\mu & 0 & 0 & 0 \\ & 1 & -\mu & 0 & 0 & 0 \\ & & 1 & 0 & 0 & 0 \\ & \text{symmetrical} & & 2(1+\mu) & 0 & 0 \\ & & & & 2(1+\mu) & 0 \\ & & & & & 2(1+\mu) \end{bmatrix}$$

$$[Q]^{-1} = \frac{1-\mu}{(1+\mu)(1-2\mu)} \begin{bmatrix} 1 & \frac{\mu}{1-\mu} & \frac{\mu}{1-\mu} & 0 & 0 & 0 \\ & 1 & \frac{\mu}{1-\mu} & 0 & 0 & 0 \\ & & 1 & 0 & 0 & 0 \\ & \text{symmetrical} & & \frac{1-2\mu}{2(1-\mu)} & 0 & 0 \\ & & & & \frac{1-2\mu}{2(1-\mu)} & 0 \\ & & & & & \frac{1-2\mu}{2(1-\mu)} \end{bmatrix}$$

$$\{\Delta\varepsilon_n^{\mathrm{T}}\} = [\alpha\Delta T_n, \ \ \alpha\Delta T_n, \ \ \alpha\Delta T_n, \ \ 0, \ \ 0, \ \ 0]^{\mathrm{T}} \tag{10.67}$$

Similar to Eq. (10.53), the strain increment vector is

$$\{\Delta\varepsilon_n\} = \{\Delta\varepsilon_n^{\mathrm{e}}\} + \{\Delta\varepsilon_n^{\mathrm{c}}\} + \{\Delta\varepsilon_n^{\mathrm{T}}\} + \{\Delta\varepsilon_n^{0}\} + \{\Delta\varepsilon_n^{\mathrm{s}}\} \tag{10.68}$$

Substituting Eqs. (10.61) and (10.62) into the above equation, the relation between the stress increment vector and the strain increment vector for complex stress state is derived:

$$\{\Delta\sigma_n\} = [\overline{D}_n](\{\Delta\varepsilon_n\} - \{\eta_n\} - \{\Delta\varepsilon_n^{\mathrm{T}}\} - \{\Delta\varepsilon_n^{0}\} - \{\Delta\varepsilon_n^{\mathrm{s}}\}) \tag{10.69}$$

where

$$[\overline{D}_n] = \overline{E}_n[Q]^{-1} \tag{10.70}$$

where $\overline{E}_n$ is given by Eq. (10.56) and $[\overline{D}_n]$ is the elasticity matrix.

10.3 Viscoelastic Thermal Stress Analysis of Concrete Structure

The creep compliance of concrete is

$$J(t,\tau) = \frac{1}{E(\tau)} + C(t,\tau) \tag{10.71}$$

The modulus of elasticity is expressed by one of the two equations:

$$E(\tau) = E_0(1 - \mathrm{e}^{-a\tau^b}) \tag{10.72}$$

$$E(\tau) = \frac{E_0\tau}{q+\tau} \tag{10.73}$$

The unit creep $C(t, \tau)$ is expressed by

$$C(t,\tau) = \sum_{s=1}^{m} \Psi_s(\tau)[1 - \mathrm{e}^{-r_s(t-\tau)}] \tag{10.74}$$

$$\text{in which}\quad \left.\begin{aligned} &\Psi_s(\tau) = f_s + g_s\tau^{-p}, && \text{when } s = 1 \text{ to } m-1 \\ &\Psi_s(\tau) = D\mathrm{e}^{-r_s\tau}, && \text{when } s = m \end{aligned}\right\} \tag{10.75}$$

where E_0, a, b, q, f_s, g_s, p_s, D, r_s are material constants.

Because the modulus of elasticity and unit creep both vary with time, incremental method is used and the time τ is divided into a series of time increments $\Delta\tau_1, \Delta\tau_2, \ldots, \Delta\tau_n$.

The strain increment in $\Delta\tau_n$ is

$$\{\Delta\varepsilon_n\} = \{\varepsilon(t_n)\} - \{\varepsilon(t_{n-1})\} = \{\Delta\varepsilon_n^{\mathrm{e}}\} + \{\Delta\varepsilon_n^{\mathrm{c}}\} + \{\Delta\varepsilon_n^{\mathrm{T}}\} + \{\Delta\varepsilon_n^{0}\} + \{\Delta\varepsilon_n^{\mathrm{s}}\} \tag{10.76}$$

By implicit method, the elastic strain increment is computed by

$$\{\Delta\varepsilon_n^{\mathrm{e}}\} = \frac{1}{E(\overline{\tau}_n)}[Q]\{\Delta\sigma_n\} \tag{10.77}$$

From Eqs. (10.51(a–d)), the creep strain increment is computed by

$$\{\Delta\varepsilon_n^{\mathrm{c}}\} = \{\eta_n\} + C(t, \overline{\tau}_n)[Q]\{\Delta\sigma_n\} \tag{10.78}$$

where

$$\{\eta_n\} = \sum_s (1 - \mathrm{e}^{-r_s\Delta\tau_n})\{\omega_{sn}\} \tag{10.79}$$

$$\{\omega_{sn}\} = \{\omega_{s,n-1}\}\mathrm{e}^{-r_s\Delta\tau_{n-1}} + [Q]\{\Delta\sigma_{n-1}\}\Psi_s(\overline{\tau}_{n-1})\mathrm{e}^{-0.5r_s\Delta\tau_{n-1}} \tag{10.80}$$

The relation between stress increment and strain increment is

$$\{\Delta\sigma_n\} = [\overline{D}_n](\{\Delta\varepsilon_n\} - \{\eta_n\} - \{\Delta\varepsilon_n^{\mathrm{T}}\} - \{\Delta\varepsilon_n^{0}\} - \{\Delta\varepsilon_n^{\mathrm{s}}\}) \tag{10.81}$$

where

$$[\overline{D}_n] = \overline{E}_n[Q]^{-1} \tag{10.82}$$

$$\overline{E}_n = \frac{E(\overline{\tau}_n)}{1 + E(\overline{\tau}_n)C(t_n, \overline{\tau}_n)} \tag{10.83}$$

From Eq. (10.6), the element nodal forces increment is given by

$$\{\Delta F\}^{\mathrm{e}} = \iiint [B]^{\mathrm{T}}\{\Delta\sigma\}\mathrm{d}x\,\mathrm{d}y\,\mathrm{d}z \tag{10.84}$$

Substituting $\{\Delta\sigma_n\}$ expressed by Eq. (10.55) into the above equation, we have

$$\{\Delta F\}^{\mathrm{e}} = [k]^{\mathrm{e}}\{\delta_n\}^{\mathrm{e}} - \int\int\int [B]^T[\overline{D}_n](\{\eta_n\} + \{\Delta\varepsilon_n^{\mathrm{T}}\} + \{\Delta\varepsilon_n^0\} + \Delta\varepsilon_n^{\mathrm{s}})\mathrm{d}x\,\mathrm{d}y\,\mathrm{d}z \tag{10.85}$$

where $[k]^e$ is the element stiffness matrix given by

$$[k]^e = \iiint [B]^{\mathrm{T}}[\overline{D}_n][B]\mathrm{d}x\,\mathrm{d}y\,\mathrm{d}z \tag{10.86}$$

The third term on the right part of Eq. (10.85) represents the nodal forces induced by the creep strain, thermal strain, autogenous strain, and shrinkage strain; after changing their sign, we get the element nodal loads increment as follows

$$\{\Delta P_n\}_e^{\mathrm{c}} = \iiint [B]^{\mathrm{T}}[\overline{D}_n]\{\eta_n\}\mathrm{d}x\,\mathrm{d}y\,\mathrm{d}z \tag{10.87}$$

$$\{\Delta P_n\}_e^{\mathrm{T}} = \iiint [B]^{\mathrm{T}}[\overline{D}_n]\{\Delta\varepsilon_n^{\mathrm{T}}\}\mathrm{d}x\,\mathrm{d}y\,\mathrm{d}z \tag{10.88}$$

$$\{\Delta P_n\}_e^0 = \iiint [B]^{\mathrm{T}}[\overline{D}_n]\{\Delta\varepsilon_n^0\}\mathrm{d}x\,\mathrm{d}y\,\mathrm{d}z \tag{10.89}$$

$$\{\Delta P_n\}_e^{\mathrm{s}} = \iiint [B]^{\mathrm{T}}[\overline{D}_n]\{\Delta\varepsilon_n^{\mathrm{s}}\}\mathrm{d}x\,\mathrm{d}y\,\mathrm{d}z \tag{10.90}$$

where

$\{P_n\}_e^{\mathrm{c}}$—nodal load increment due to creep strain
$\{\Delta P_n\}_e^{\mathrm{T}}$—nodal load increment due to temperature change

$\{\Delta P_n\}_e^0$—nodal load increment due to autogenous strain
$\{\Delta P_n\}_e^s$—nodal load increment due to shrinkage strain.

Collection of all the nodal forces and nodal loads yields the global equilibrium equation as follows:

$$[K]\{\Delta\delta_n\} = \{\Delta P_n\}^{\mathrm{L}} + \{\Delta P_n\}^{\mathrm{C}} + \{\Delta P_n\}^{\mathrm{T}} + \{\Delta P_n\}^{0} + \{\Delta P_n\}^{\mathrm{S}} \tag{10.91}$$

where $\{\Delta P_n\}^{\mathrm{L}}$ is the increment of nodal load due to external load and $[K]$ is the global stiffness matrix, whose elements are given by

$$K_{rs} = \sum_e k_{ij}^e \tag{10.92}$$

Solution of equilibrium equation (10.91) yields the nodal displacement increments $\{\Delta\delta_n\}$, then the stress increment $\{\Delta\sigma_n\}$ may be computed by Eq. (10.55) and the stresses at time t_n are given by

$$\{\sigma_n\} = \{\Delta\sigma_1\} + \{\Delta\sigma_2\} + \cdots + \{\Delta\sigma_n\} = \sum\{\Delta\sigma_n\} \tag{10.93}$$

10.4 Compound Element

The dimensions of element should be suited to the temperature gradients and stress gradients of the structure. As shown in Figure 10.5, in the region R_1 of new concrete, there must be several layers of elements in each lift of concrete; in the region R_2, there may be one layer of element in one lift of concrete; and in the region R_3 of old concrete, several lifts of concrete may be combined to one layer of element which is a compound element to reduce the number of unknowns [8].

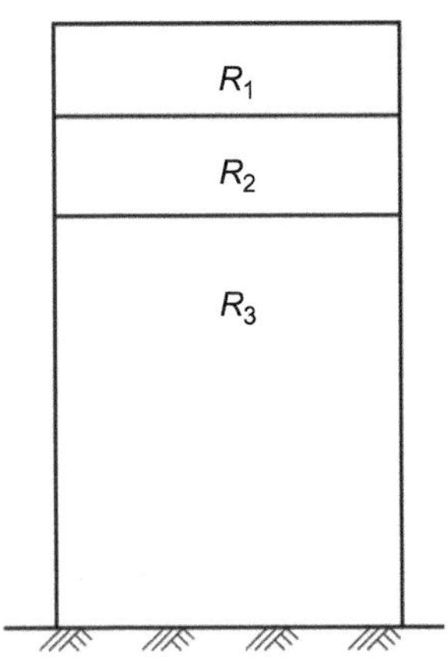

Figure 10.5 Dam block.

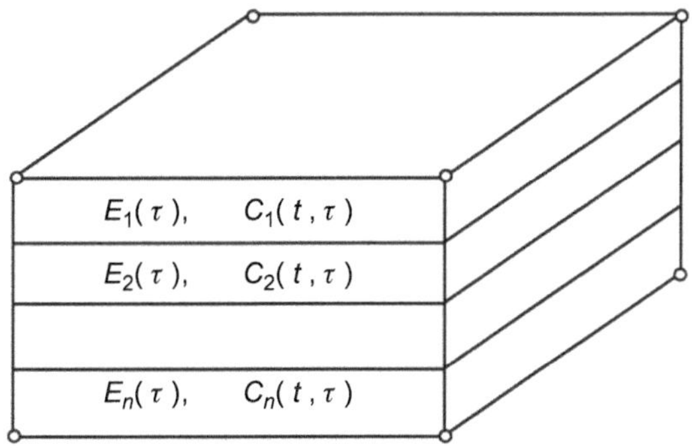

Figure 10.6 Compound element.

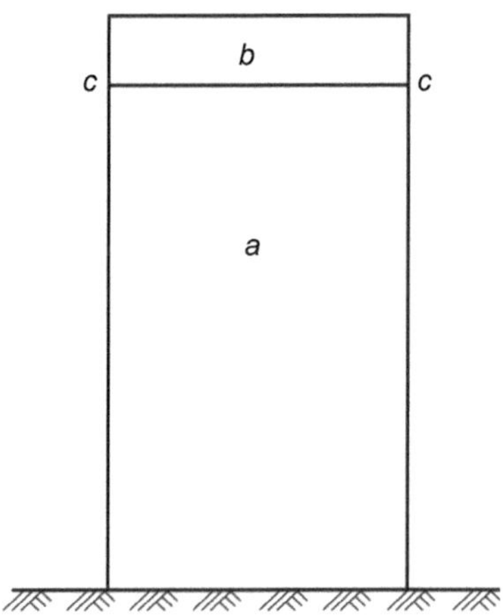

Figure 10.7 Different $\Delta\tau$ in different region ($\Delta\tau_a \neq \Delta\tau_b$).

As shown in Figure 10.6, in a compound element, each lift of concrete will have different modulus of elasticity $E_i(\tau)$, unit creep $C_i(t, \tau)$, and adiabatic temperature rise $\theta_i(\tau)$.

10.5 Method of Different Time Increments in Different Regions

Generally the same time increment $\Delta\tau_i$ is used in different regions of structure in computing temperature field and stress field. As shown in Figure 10.7, a new method using different time increments $\Delta\tau_i$ in different regions is proposed which is more efficient [7].

11 Stresses due to Change of Air Temperature and Superficial Thermal Insulation

Experience shows that most cracks in mass concrete structures are originally superficial cracks, but some of them may become larger and deeper cracks later which will reduce the safety and durability of the structure. Thermal insulation is the most efficient measure for preventing superficial cracks of mass concrete structures.

11.1 Superficial Thermal Stress due to Linear Variation of Air Temperature During Cold Wave

The variation of air temperature T_a during a cold wave is expressed in the following:

$$\left.\begin{array}{lll} \text{When} & 0 \le \tau \le Q & T_a = k\tau \\ \text{When} & Q \le \tau \le 2Q & T_a = k\tau - 2k(\tau - Q) \end{array}\right\} \tag{11.1}$$

where

coefficient $k = A/Q$
A—the maximum drop of air temperature
Q—the duration of drop of air temperature.

According to the principle of superposition, it is necessary to compute the superficial temperature of concrete $f(\tau)$ under the action of air temperature $T_a = k\tau$ only, when $\tau > Q$ and the superficial temperature of concrete is $f(\tau) - 2f(\tau - Q)$.

Consider an infinite slab with thickness $2R$ and the initial and boundary conditions are as follows:

$$\left.\begin{array}{lll} \text{When} & \tau = 0,\ 0 \le x \le R & T = 0 \\ \text{When} & \tau > 0,\ x = 0 & \dfrac{\partial T}{\partial x} = 0 \\ \text{When} & \tau > 0,\ x = R & \lambda \dfrac{\partial T}{\partial x} - \beta(k\tau - T) = 0 \end{array}\right\} \tag{11.2}$$

Thermal Stresses and Temperature Control of Mass Concrete. DOI: http://dx.doi.org/10.1016/B978-0-12-407723-2.00011-7

The solution is

$$T(x,\tau) = k\tau - \frac{k}{2a}\left[R^2\left(1+\frac{2\lambda}{\beta R}\right) - x^2\right] + \frac{kR^2}{a}\sum_{n=1}^{\infty}\frac{A_n}{\mu_n^2}\cos\left(\frac{\mu_n x}{R}\right)\exp\left(-\mu_n^2\frac{a\tau}{R^2}\right) \tag{11.3}$$

The temperature is minimum and the tensile stress is maximum on the surface of concrete. Let $x = R$ in Eq. (11.3), we get the superficial temperature of concrete in the following:

$$T(R,\tau) = k\tau - \frac{\lambda kR}{\beta a} + \frac{kR^2}{a}\sum_{n=1}^{\infty}\frac{A_n}{\mu_n^2}\cos\mu_n\exp\left(-\mu_n^2\frac{a\tau}{R^2}\right) \tag{11.4}$$

$$A_n = 2\sin\mu_n/(\mu_n + \sin\mu_n\cos\mu_n)$$

where

λ—the conductivity
a—the diffusivity
μ_n—root of the characteristic equation $\cot\mu_n - \mu_n\lambda/(\beta R) = 0$
β—surface conductance.

The superficial temperature computed by Eq. (11.4) for $Q = 1$ day is shown in Figure 11.1. The minimum superficial temperatures during a cold wave are shown in Figure 11.2.

The depth of influence of a cold wave is so small that the thermal deformation is perfectly restrained and the elastic thermal stress is $\sigma = E\alpha T(\tau)/(1-\mu)$ and the viscoelastic thermal stress $\sigma^*(t)$ is given by

$$\frac{(1-\mu)\sigma^*(t)}{E\alpha T_0} = \frac{1}{T_0}\sum K(t,\tau)\Delta T(\tau) \tag{11.5}$$

where

$K(t,\tau)$—the relaxation coefficient of concrete
T_0—the maximum drop of air temperature in a cold wave.

Generally the thickness of a concrete dam is 5–50 m, but the influence of thickness on the superficial temperature is very small if the thickness is greater than 5 m; thus an infinite slab with thickness 10 m and $a = 0.0040\ \text{m}^2/\text{h}$ is analyzed, the maximum superficial viscoelastic thermal stress for different thermal insulation is shown in Figure 11.3 from which it is clear that the superficial thermal stress drops rapidly with the increase of λ/β.

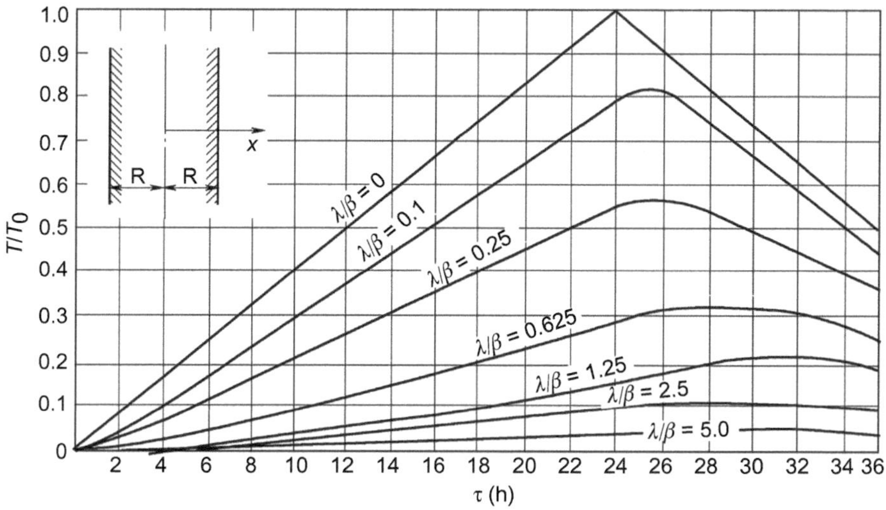

Figure 11.1 The variation of superficial temperature for $Q = 1$ day.

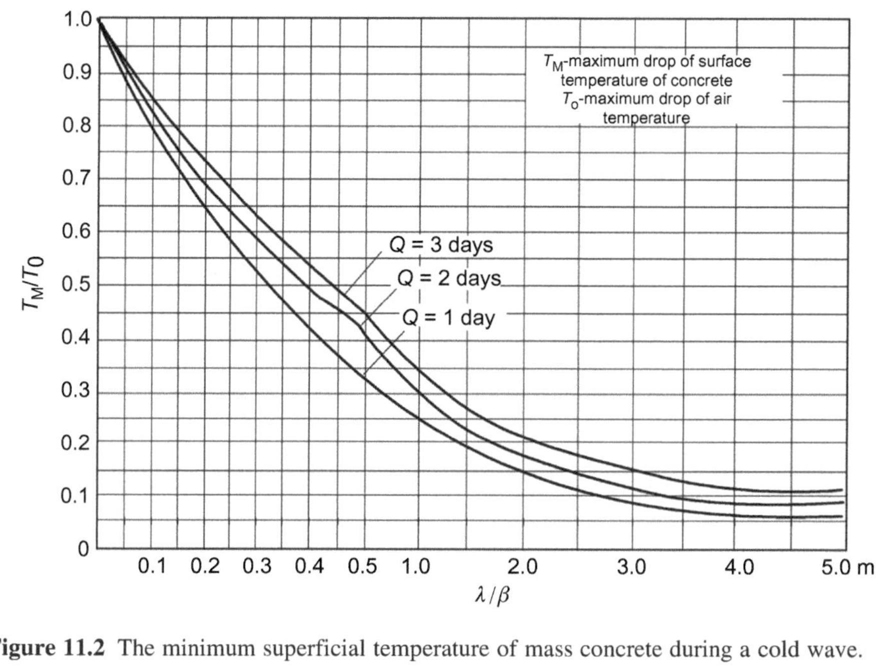

Figure 11.2 The minimum superficial temperature of mass concrete during a cold wave.

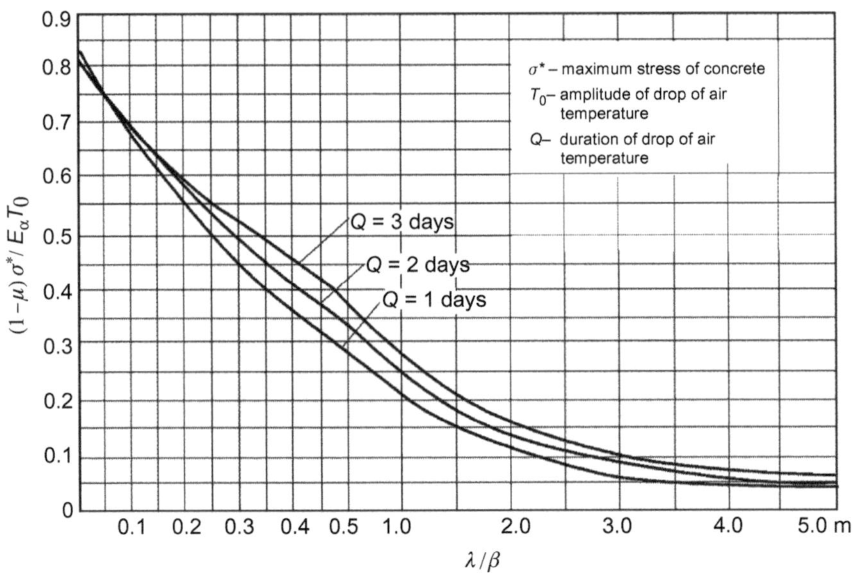

Figure 11.3 Influence of superficial thermal insulation on the maximum viscoelastic superficial thermal stress of mass concrete during a cold wave.

11.2 Superficial Thermal Insulation, Harmonic Variation of Air Temperature, One-Dimensional Heat Flow

11.2.1 Superficial Thermal Insulation, Daily Variation of Air Temperature, One-Dimensional Heat Flow

The daily variation of air temperature is expressed by

$$T_a = A \sin\left(\frac{2\pi\tau}{P}\right) \tag{11.6}$$

where

T_a—air temperature
A—amplitude of variation
P—period of variation ($P = 1$ day).

The depth of influence of daily variation of air temperature is less than 1 m and the thickness of concrete dam is more than 5 m; thus the temperature field of semi-infinite solid is analyzed. As the air temperature varies periodically, we consider a quasi-steady temperature field. The temperature T should satisfy the following equations:

Equation of heat conduction

$$\frac{\partial T}{\partial \tau} = a\frac{\partial^2 T}{\partial x^2} \tag{11.7}$$

Boundary conditions

$$\left.\begin{array}{ll}\text{When} \quad x=0, \tau>0, & \lambda \dfrac{\partial T}{\partial x}=\beta(T_{\mathrm{a}}-T) \\ \text{When} \quad x=\infty, & T=0\end{array}\right\} \tag{11.8}$$

The solution of Eqs. (11.7) and (11.8) is

$$T=fA\,\mathrm{e}^{-px}\sin\left[\frac{2\pi(\tau-m)}{P}-px\right] \tag{11.9}$$

where

$$\begin{aligned} f&=\frac{1}{\sqrt{1+2u+2u^2}}, \quad u=\frac{\lambda}{\beta}\sqrt{\frac{\pi}{aP}} \\ m&=\frac{P}{2\pi}\tan^{-1}\left(\frac{1}{1+1/u}\right), \quad p=\sqrt{\frac{\pi}{aP}} \end{aligned} \tag{11.10}$$

As the temperature variation is maximum on the surface of concrete, letting $x=0$ in Eq. (11.9), we get the surface temperature of concrete as follows:

$$T=fA\sin[2\pi(\tau-m)/P] \tag{11.11}$$

The depth of variation of temperature is very small so the mass concrete structure may be analyzed as it is a semi-infinite solid. The maximum viscoelastic superficial thermal stress is given by

$$\sigma=f\rho E(\tau)\alpha A/(1-\mu) \tag{11.12}$$

where ρ is the relaxation coefficient considering the effect of creep, for daily variation of air temperature $\rho \cong 0.90$.

The modulus of elasticity is given by

$$E(\tau)=E_0[1-\exp(1-0.40\tau^{0.34})] \tag{11.13}$$

where E_0 is the final modulus of elasticity, approximately $E_0=1.20E(90)$, $E_0=1.05E(365)$, or $E_0=1.45E(28)$.

If the allowable tensile stress of concrete is σ_a, the stress induced by external load, hydration heat, initial temperature difference, etc. is σ_0, the thermal stress due to cold wave given by Eq. (11.12) should satisfy the following equation:

$$\sigma+\sigma_0 \leq \sigma_a=R_t/k \tag{11.14}$$

where

R_t—the tensile strength of concrete
k—the coefficient of safety.

Taking the equal sign" = "in the above equation and substituting Eq. (11.12) in it, we get

$$\sigma = \sigma_a - \sigma_0 = f\rho E(\tau)aA/(1-\mu) \tag{11.15}$$

Substituting f of Eq. (11.10) into Eq. (11.15), we have

$$1 + 2u + 2u^2 = b^2 \tag{11.16}$$

where

$$b = \frac{1}{f} = \frac{\rho E(\tau)aA}{(1-\mu)(\sigma_a - \sigma_0)} \tag{11.17}$$

From Eq. (11.16), we obtain

$$u = 0.50\left(\sqrt{2b^2 - 1} - 1\right) \tag{11.18}$$

The capability of superficial thermal insulation is indicated by the surface conductance β. After computing of u by Eq. (11.18), from Eq. (11.10), the required surface conductance of concrete is given by

$$\beta = \frac{\lambda}{u}\sqrt{\frac{\pi}{aP}} \tag{11.19}$$

β given by the above equation can ensure that the thermal stress will not exceed the allowable tensile stress of concrete.

The variation of the diffusivity of concrete is small, generally we may take $a = 0.10\ \text{m}^2/\text{day}$. For daily variation, $P = 1$ day, so we get the required surface conductance as follows:

$$\beta = 5.605\lambda/u \tag{11.19a}$$

The relation between surface conductance β and the thickness h of the superficial thermal insulation layer is

$$\beta = \frac{1}{(h/\lambda_s + 1/\beta_0)} \tag{11.20}$$

where

h—thickness of insulation layer
λ_s—conductivity of insulation layer, see Table 2.1

β_0—surface conductance between air and the outside of the insulation layer, generally $\beta_0 = 40-80$ kJ/(m^2 h °C).

From Eq. (11.20), the thickness of the surface insulation layer is given by

$$h = \lambda_s \left(\frac{1}{\beta} - \frac{1}{\beta_0} \right) \tag{11.21}$$

If the result of computation is $\beta \geq \beta_0$ or $h \leq 0$, it means that the superficial thermal insulation is not required.

Example 1 The amplitude of variation of air temperature $A = 8°C$, modulus of elasticity of concrete at 90 days $E(90) = 24{,}000$ MPa, Poisson's ratio $\mu = 0.16$, $\alpha = 1 \times 10^{-5}°C^{-1}$, $\lambda = 9.21$ kJ/(m h °C), $a = 0.10$ m^2/day, $\beta_0 = 83.7$ kJ/(m^2 h °C). Try to compute the thermal stress induced by daily variation of air temperature at $\tau = 5$ days. Let $\sigma_0 = 0.2$ MPa, allowable stress $\sigma_a = 1.0$ MPa. Try to check whether the superficial thermal insulation is required.

First, consider the case of no superficial thermal insulation, take $\beta = \beta_0 = 83.7$ kJ/(m^2 h °C), from Eqs. (11.10), (11.12), and (11.13), $\sigma = 0.712 + 0.200 = 0.912$ MPa, which is less than the allowable tensile stress, so the superficial thermal insulation is not necessary.

11.2.2 Superficial Thermal Insulation for Cold Wave, One-Dimensional Heat Flow

Assuming that the cold wave begins at $\tau = \tau_1$, as shown in Figure 11.4, the air temperature during the cold wave is expressed by

$$T_a = -A \sin\left[\frac{\pi(\tau - \tau_1)}{2Q}\right], \quad \tau \geq \tau_1 \tag{11.22}$$

where

Q—duration of decrease of air temperature in cold wave
A—amplitude of decrease of air temperature in cold wave.

For a semi-infinite solid, the temperature field must satisfy the equation of heat conduction (11.7), the boundary conditions (11.8), and the initial condition:

When $\tau = 0, \quad T = 0$

The theoretical solution is too complex for practical application, so the following approximate formula for the surface temperature of mass concrete during cold wave is proposed by the writer:

$$T = -f_1 A \sin\left[\frac{\pi(\tau - \tau_1 - \Delta)}{2P}\right] \tag{11.23}$$

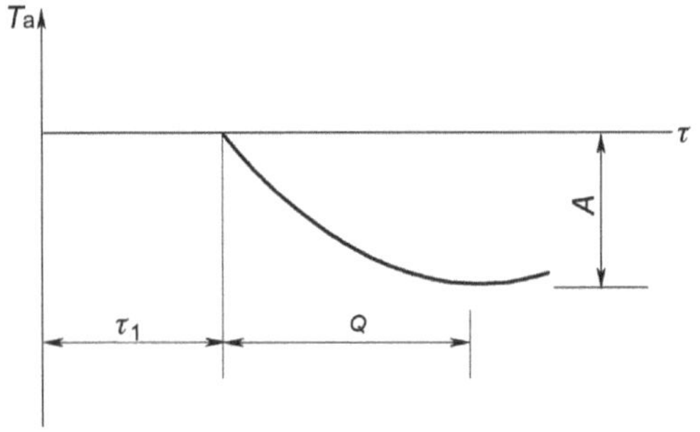

Figure 11.4 Variation of air temperature in a cold wave.

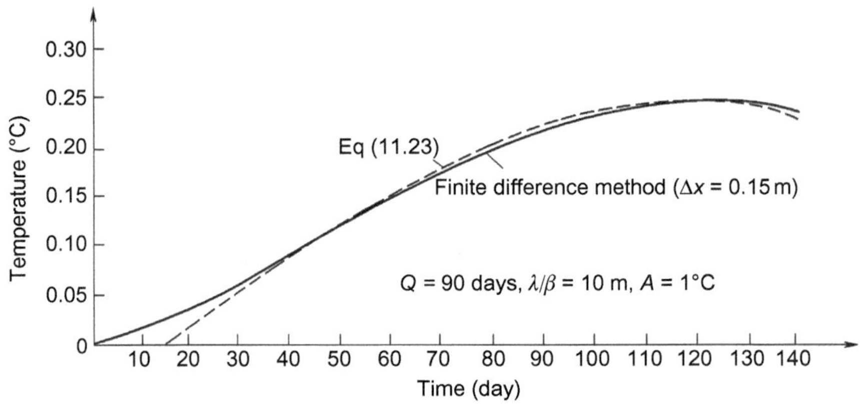

Figure 11.5 Surface temperature of mass concrete during a cold wave.

$$
\begin{aligned}
& f_1 = 1/\sqrt{1 + 1.85u + 1.12u^2}, \quad \Delta = 0.4gQ \\
& P = Q + \Delta, \quad g = \frac{2}{\pi}\tan^{-1}\left[\frac{1}{1 + 1/u}\right] \\
& u = \frac{\lambda}{2\beta}\sqrt{\frac{\pi}{Qa}} = \frac{2.802\lambda}{\beta\sqrt{Q}} \quad (\text{when } a = 0.10\ \text{m}^2/\text{day})
\end{aligned}
\tag{11.24}
$$

In order to check the precision of Eq. (11.23), the following case is analyzed by the finite difference method: $Q = 3$ days, $\lambda/\beta = 4.4$ m, $A = 1°\text{C}$, $\Delta x = 0.03$ m, the temperatures obtained by the finite difference method are shown in Figure 11.5 by a solid curve and the temperatures given by Eq. (11.23) are shown by a dotted curve. It is clear from Figure 11.5 that the temperatures given by two methods agree very well in the region of big temperature drop which we are interested in.

The viscoelastic thermal stress on the surface of mass concrete is

$$
\sigma^*(t) = \sum K(t,\tau)E(\tau)\alpha\,\Delta T(\tau)/(1-\mu) \tag{11.25}
$$

For the conventional mass concrete with $E(\tau)$ given by Eq. (6.50) and $K(t,\tau)$ given by (6.52), the maximum viscoelastic superficial thermal stress during a cold wave is

$$\sigma = f_1 \rho_1 E(\tau_m)\alpha A/(1-\mu) \tag{11.26}$$

$$\rho_1 = \frac{0.830 + 0.051\tau_m}{1 + 0.051\tau_m}\exp\left[-0.095(P-1)^{0.60}\right], \quad P = 1 - 8 \text{ days} \tag{11.27}$$

where

ρ_1—the relaxation coefficient considering creep of concrete
τ_m—the mean age of concrete during the time of drop of air temperature, given by

$$\tau_m = \tau_1 + \Delta + \frac{1}{2}P \tag{11.28}$$

From Eq. (11.24), we have $1 + 1.85u + 1.12u^2 = b^2 = 1/f_1^2$.

In order to determine the required superficial thermal insulation, it is necessary to compute u by

$$u = 0.9449\sqrt{b^2 - 0.2360} - 0.8259 \tag{11.29}$$

where

$$b = \frac{1}{f_1} = \frac{\rho_1 E(\tau_m)\alpha A}{(1-\mu)(\sigma_a - \sigma_0)} \tag{11.30}$$

The surface conductance β for resisting a cold wave is given by

$$\beta = \frac{\lambda}{2u}\sqrt{\frac{\pi}{aQ}} = \frac{2.802\lambda}{u\sqrt{Q}} \quad (\text{for } a = 0.10 \text{ m}^2/\text{day}) \tag{11.31}$$

Example 2 The cold wave begins at $\tau_1 = 5$ days, $Q = 3$ days, $A = 11.9°\text{C}$, $E(90) =$ 24 GPa, $\mu = 0.16$, $\alpha = 1 \times 10^{-5}°\text{C}^{-1}$, $a = 0.10$ m²/day, $\beta_0 = 83.7$ kJ/(m² h °C), $\lambda_s = 0.1256$ kJ/(m h °C), the initial stress $\sigma_0 = 0.30$ MPa, the allowable tensile stress $\sigma_a = 1.10$ MPa. Try to compute the required superficial thermal insulation for resisting a cold wave.

In order to compute Δ, assuming $\beta = 6.28$ kJ/(m² h °C), from Eq. (11.24), get $u = 2.37$, $g = 0.390$, $\Delta = 0.468$ day, $P = 3.468$ day. From Eqs. (11.28), (11.13), and (11.27), get $\tau_m = 7.203$ days, $E(\tau_m) = 15{,}633$ MPa, $\rho_1 = 0.7437$. From Eq. (11.30), $b = 2.232$; from Eq. (11.29), $u = 1.232$; from Eq. (11.31), we get the required surface conductance $\beta = 12.09$ kJ/(m² h °C). Based on $\beta = 12.09$ kJ/(m² h °C), compute once again, get the required surface conductance $\beta = 12.12$ kJ/(m^2 h °C) which is very close to $\beta = 12.09$ kJ/(m^2 h °C) given by the first time computation. It is clear that the assumed value of β has some influence on Δ, but its influence on the final value of required β is very small, and the error is only 0.25%.

If foamed plastic plate with $\lambda_s = 0.1256$ kJ/(m h °C) is used as the insulation layer, from Eq. (11.21), we get the required thickness $h = 0.886$ cm.

Check: substituting $\beta = 12.12$ kJ/(m^2 h °C) in Eq. (11.31), we get $u = 1.2297$; from Eq. (11.24), $f_1 = 0.4486$; from Eq. (11.26), $\sigma = 0.800$ MPa, $\sigma + \sigma_0 = 0.800 + 0.300 = 1.10$ MPa which is equal to the expected allowable tensile stress.

11.2.3 Superficial Thermal Insulation, Temperature Drop in Winter, One-Dimensional Heat Flow

The placing of concrete generally is stopped in winter in the cold region. In this case, the surface of concrete must be insulated carefully to prevent cracking. A detailed method of computation will be given in Section 11.5 and an approximate method of computation proposed by the writer will be described in the following, which is adopted in the design specifications of concrete dams in China.

The monthly mean air temperature in the winter may still be expressed by Eq. (11.22). Assuming that the concrete is placed at $\tau_1 = 0$ and A is the drop of air temperature from the time of placing concrete to the time of minimum air temperature (generally the middle of January), the surface temperature of concrete can still be computed by Eq. (11.23), the comparison of which with the results computed by the finite difference method is shown in Figure 11.6. It is apparent that the temperatures computed by two different methods are close to each other.

The maximum viscoelastic thermal stress on the surface of concrete is given by

$$\sigma = rf_1\rho_2 E(\tau_m)\alpha A/(1-\mu) \tag{11.32}$$

$$\rho_2 = \frac{0.830 + 0.051\tau_m}{1.00 + 0.051\tau_m}\exp(-0.104P^{0.35}), \quad P \geq 20 \text{ days} \tag{11.33}$$

$$\tau_m = \Delta + P/2 \tag{11.34}$$

and f_1 and u are given by Eqs. (11.24) and (11.18) and b is computed by

$$b = \frac{1}{f_1} = \frac{r\rho_2 E(\tau_m)\alpha A}{(1-\mu)(\sigma_a - \sigma_0)} \tag{11.35}$$

where r is the restraint coefficient shown in Figure 11.7. The required superficial thermal insulation for preventing cracking in winter is given by Eqs. (11.29), (11.31), and (11.21).

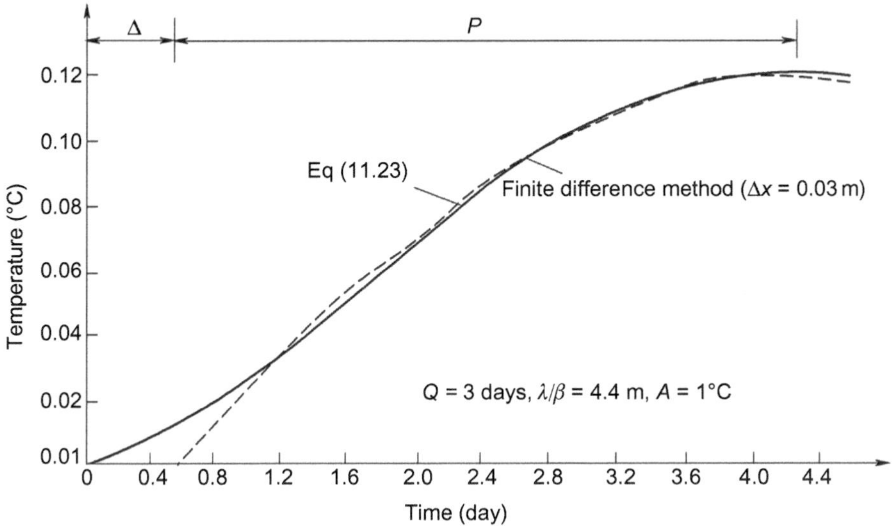

Figure 11.6 Surface temperature of mass concrete during the winter.

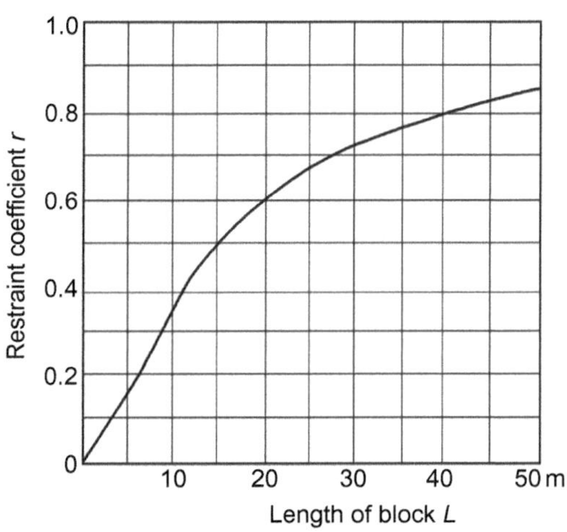

Figure 11.7 Restraint coefficient *r*.

Example 3 The placing of concrete is stopped at mid-October and the air temperature drops to mid-January. The duration of drop of air temperature is $Q = 90$ days. The maximum drop of air temperature is $A = 22°C$. The length of dam block is $L = 40$ m, the allowable stress $\sigma_a = 1.10$ MPa, the initial stress $\sigma_0 = 0.2$ MPa, $\lambda_s = 0.140$ kJ/(m h °C), the rest is the same as example 2. Try to compute the required superficial thermal insulation.

In order to compute Δ, first assuming $\beta = 2.51$ kJ/(m^2 h °C); from Eq. (11.24), $u = 1.083$, $g = 0.3052$, $\Delta = 10.99$ days, $P = 100.99$ days; from Figure 11.7, $r = 0.80$;

from Eqs. (11.28)–(11.34), $\tau_m = 61.48$ days, $\rho_2 = 0.5684$, $E(\tau_m) = 23{,}115$ MPa, $b = 3.08$, $u = 2.047$. From Eq. (11.31), we get the required $\beta = 1.329$ kJ/(m^2 h °C). The conductivity of foamed plastics $\lambda_s = 0.140$ kJ/(m^2 h °C); from Eq. (11.21), the required thickness is $h = 10.36$ cm.

11.3 Superficial Thermal Insulation, Harmonic Variation of Air Temperature, Two-Dimensional Heat Flow [22, 36, 41, 42]

Experience shows that superficial cracks frequently appear on the corners of massive concrete structures where the concrete temperature drops more rapidly than elsewhere during a cold wave.

11.3.1 Two-Dimensional Heat Flow, Thermal Insulation for Daily Variation of Air Temperature

As shown in Figure 11.8, the faces *AC* and *BC* are covered by an insulation layer with thickness h and equivalent surface conductance β. The variation of air temperature is still expressed by Eq. (11.6). Because the depth of influence of daily variation is less than 1.0 m, at any point 1.0 m beyond corner *C*, the superficial temperature may be computed by the one-dimensional equation (11.11). Due to two-dimensional heat flow, it is suggested to compute the temperature at the corner point *C* by the following approximate formula:

$$T_c = f_c A \sin\left[\frac{2\pi(\tau - m)}{P}\right] \tag{11.36}$$

where m is given by Eq. (11.10). Refer to the product theorem in the theory of heat conduction, it is suggested to compute the coefficient f_c in Eq. (11.36) by

$$f_c = 1 - (1 - f)^2 \tag{11.37}$$

where f is given by Eq. (11.10).

The maximum viscoelastic thermal stress at corner *C* is given by

$$\sigma_c = f_c \rho E(\tau) \alpha A \tag{11.38}$$

For daily variation of temperature, the relaxation coefficient $\rho \cong 0.90$.

Assuming that the initial stress is σ_0, the allowable tensile stress is σ_a, hence

$$\sigma_c + \sigma_0 \leq \sigma_a \tag{11.39}$$

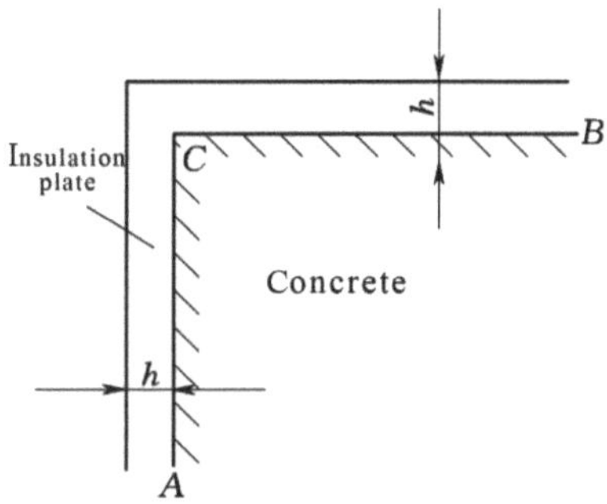

Figure 11.8 The corner of a concrete dam.

Taking equal sign in the above equation, we get $\sigma_c = \sigma a - \sigma_0$, by substituting in Eq. (11.38), we have

$$c = f_c = \frac{\sigma_a - \sigma_0}{\rho E(\tau)\alpha A} \tag{11.40}$$

Substituting $b = 1/f$ into Eq. (11.37), we get $c = 1 - (1 - 1/b)^2$, thus

$$b = \frac{1}{1 - \sqrt{1 - c}} \tag{11.41}$$

By substituting b in Eq. (11.16), u will be given by Eq. (11.18). Thus, it is clear that after computing b and c by Eqs. (11.40) and (11.41), the required capability of superficial thermal insulation near corner C may be determined by Eqs. (11.18) and (11.19) and the thickness of the insulation layer is given by Eq. (11.21).

Example 1 The basic data is the same as example 1 in Section 11.2; try to compute the required capability of superficial thermal insulation at the corner of the dam block.

From Eq. (11.40), $c = 0.7730$; from Eq. (11.41), $b = 1.910$; from Eq. (11.18), $u = 0.7546$; from Eq. (11.19), the required surface conductance is $\beta = 68.41$ kJ/(m^2 h °C).

11.3.2 Two-Dimensional Heat Flow, Thermal Insulation for Cold Wave

The variation of air temperature during a cold wave is expressed by Eq. (11.22). Since the depth of influence of a cold wave is about 1.5 m, the surface temperature of concrete 1.5 m beyond the corner may still be computed by Eq. (11.13). The temperature of the corner C is suggested to be given by

$$T_c = -f_{c1} A \sin\left[\frac{\pi(\tau - \tau_1 - \Delta)}{2P}\right] \tag{11.42}$$

where Δ and P are given by Eq. (11.24) and referring to the theorem of product in the theory of heat conduction, the coefficient f_{c1} is computed by

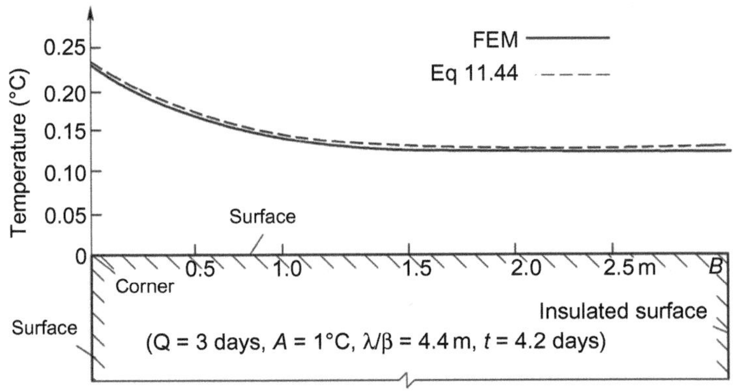

Figure 11.9 The surface temperature near the corner of a dam block, two-dimensional heat flow.

$$f_{c1} = 1 - (1 - f_1)^2 \tag{11.43}$$

where f_1 is given by Eq. (11.24).

Based on Eq. (11.9), it is suggested to compute the temperatures along the two sides of the corner by the following approximate formula:

$$T_b = A[f_1 + (f_{c1} - f_1)e^{-qx}], \quad q = \sqrt{\frac{\pi}{4Qa}} \tag{11.44}$$

where x is the distance between the reference point and the corner C. The dotted lines shown in Figure 11.9 are the temperatures given by Eq. (11.44) which are close to the temperatures computed by the finite element method (FEM).

It will be shown in the following how to estimate the depth of influence of two-dimensional heat flow. Let

$$(T_b - T_s)/T_s = \varepsilon$$

where

T_b—the surface temperature due to two-dimensional heat flow
T_s—the surface temperature due to one-dimensional heat flow.

From Eq. (11.44), we get the depth of influence of two-dimensional heat flow as follows:

$$x = \frac{1}{q}\ln\left(\frac{f_{c1} - f_1}{\varepsilon f_1}\right), \quad q = \sqrt{\frac{\pi}{4Qa}} \tag{11.45}$$

For example, if $Q = 3$ days, $\lambda/\beta = 4.40$ m, $a = 0.10$ m^2/day, $\varepsilon = 0.10$; from Eq. (11.45), we get the depth of influence $x = 1.34$ m.

The maximum viscoelastic thermal stress at the corner C during a cold wave is

$$\sigma_c = \rho_1 f_{c1} E(\tau_m)\alpha A \tag{11.46}$$

where ρ_1 is the relaxation coefficient given by Eq. (11.27). After computing c by

$$c = \frac{\sigma_a - \sigma_0}{\rho_1 E(\tau_m)\alpha A} \tag{11.47}$$

compute b by Eq. (11.41), u by Eq. (11.29), we will get β by Eq. (11.31), and h by Eq. (11.21).

Example 2 The basic data are the same as example 2 in Section 11.2; try to compute the required superficial thermal insulation at the corner during a cold wave.

From Eq. (11.47), $c = 0.533$; from Eq. (11.41), $b = 3.156$; from Eq. (11.29), $u = 2.121$; from Eq. (11.31), the required surface conductance at the corner is $\beta = 7.03$ kJ/(m^2 h °C). If foamed plastic plate is used, from Eq. (11.21), the required thickness is $h = 1.637$ cm.

From example 2 in Section 11.2, the required thickness of insulation layer for one-dimensional heat flow is 0.886 cm, which is 54% of the required thickness at the corner.

Summarizing the results of the preceding two examples, the required thickness of insulation plate is 1.637 cm in the range 1.5 m near the corner and 0.886 cm beyond 1.5 m.

If the superficial thermal insulation layers on the two sides of the corner are of different kinds of materials, there are different surface conductances β_1 and β_2 on the two sides. From Eq. (11.24), we will get two different coefficients f_1 and f_2. In this case, it is suggested to compute coefficient f_{c1} by

$$f_{c1} = 1 - (1 - f_1)(1 - f_2) \tag{11.48}$$

In order to check the precision of the above equation, a 6 m × 6 m concrete block with $\lambda/\beta_1 = 1.10$ m, $\lambda/\beta_2 = 4.40$ m, and $Q = 3$ days is computed by FEM. The results given by two different methods are shown in the following:

By the FEM

$$f_1 = 0.355, \quad f_2 = 0.118, \quad f_{c1} = 0.425$$

By Eqs. (11.24) and (11.48)

$$f_1 = 0.357, \quad f_2 = 0.119, \quad f_{c1} = 0.433$$

It is apparent that the results given by two different methods are very close to each other, and the error of f_{c1} is 1.88%.

11.3.3 Two-Dimensional Heat Flow, the Superficial Thermal Insulation During Winter

The temperature at the corner during winter can still be computed by Eq. (11.42). As the time of cooling is long, the effect of creep is substantial, and the thermal stress is computed by the following formula:

$$\sigma_c = r\rho_2 f_{c1} E(\tau_m)\alpha A \tag{11.49}$$

where r is the restraint coefficient shown in Figure 11.7, the relaxation coefficient ρ_2 is given by Eq. (11.33), and the coefficient c is computed by

$$c = \frac{\sigma_a - \sigma_0}{r\rho_2 E(\tau_m)\alpha A} \tag{11.50}$$

Compute b by Eq. (11.41), u by Eq. (11.29), then the required surface conductance β is given by Eq. (11.31) and the thickness by Eq. (11.21).

Example 3 The basic data are the same as example 3 given in Section 11.2; try to compute the required surface conductance β near the corner to prevent cracking in winter.

From Figure 11.7, $r = 0.80$; from Eq. (11.50), $c = 0.3894$; from Eq. (11.41), $b = 4.575$; from Eq. (11.29), $u = 3.472$; from Eq. (11.31), the required $\beta = 0.7836$ kJ/(m^2 h °C). If foamed plastic plate is used, the required thickness is $h = 17.69$ cm which is 171% of the required thickness for one-dimensional heat flow. Taking $\varepsilon = 0.15$, from Eq. (11.45), the depth of influence of two-dimensional heat flow is $x = 5.58$ m. For example, the thickness of the thermal insulation plate is 17.69 cm in the range of 5.6 m near the corner and 10.36 cm beyond 5.6 m.

When determining β, Eq. (11.14) is used, i.e., $\sigma + \sigma_0 = \sigma_a$, in which the initial stress σ_0 is also related to the value of β, thus the method of iteration should be used, generally two to three iterations are required.

11.4 Thermal Stresses in Concrete Block During Winter and Supercritical Thermal Insulation [41, 42, 66, 71, 72, 89]

11.4.1 Superficial Thermal Stresses During Winter

A series of concrete blocks with height H are shown in Figure 11.10. The placing of concrete is stopped in the winter. One block with height H, width L, and length S is taken from Figure 11.10 and shown in Figure 11.11. The top, upstream, and downstream surface of the block are exposed in the air and the two lateral surfaces are

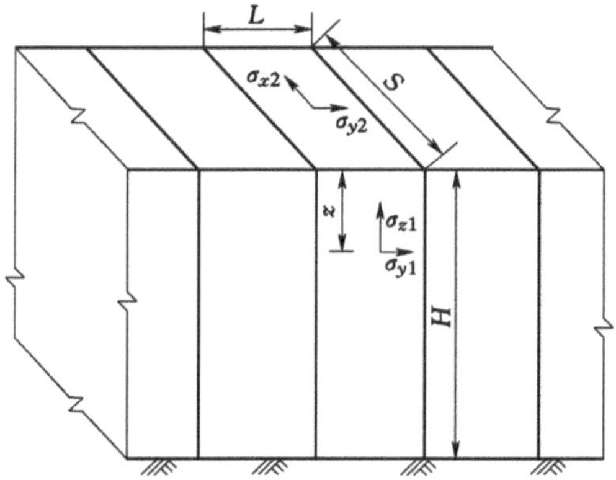

Figure 11.10 A series of concrete dam blocks.

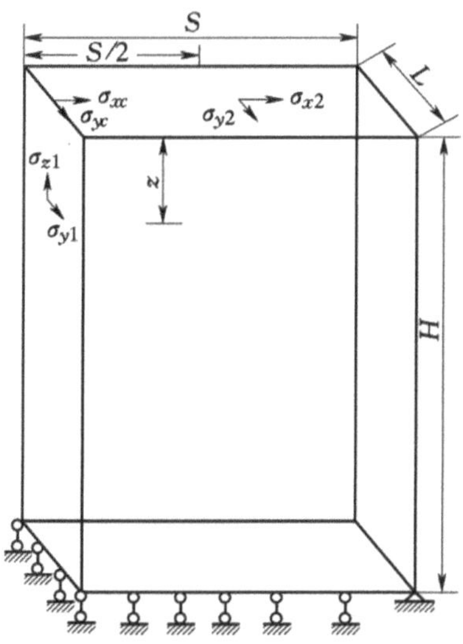

Figure 11.11 A single dam block.

thermally insulated. Take the x axis in the direction of flow of water, the y axis in the direction of dam axis, and the z axis in the vertical direction. As shown in Figure 11.12, the air temperature T_a is expressed by

$$T_a = -A \sin\left(\frac{\pi\tau}{2Q}\right) \tag{11.51}$$

where A is the amplitude of variation, °C; τ is the time; Q is the duration of drop of temperature, generally taking $Q = 90$ days. Because the depth of influence of a cold wave is about 1.5 m and the length of dam block S is usually more than 20 m, the temperature field may be analyzed as it is a semi-infinite solid. The solution strictly satisfying the initial condition is too complex for practical application, so an approximate solution of temperature is given in the following. Assuming that

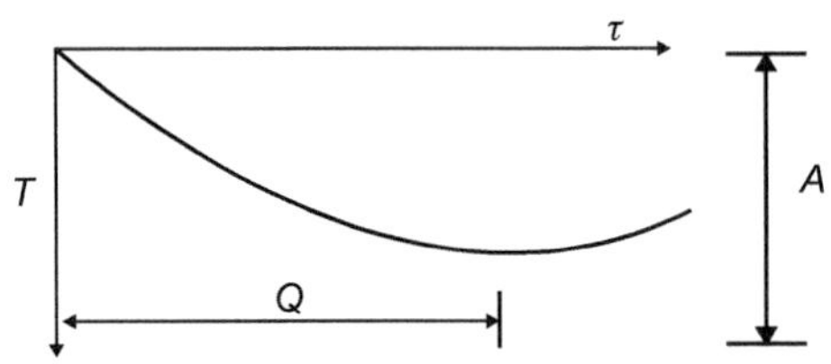

Figure 11.12 Air temperature in the winter.

the initial temperature is zero, under the action of the air temperature expressed by Eq. (11.67), the surface temperature of the dam block is given by

$$T = -f_1 A \sin\left[\frac{\pi(\tau - \Delta)}{2P}\right] \tag{11.52}$$

where

$$\left.\begin{aligned} &f_1 = 1/\sqrt{1 + 1.85\mu + 1.12u^2}, \quad \Delta = 0.4gQ \\ &P = Q + \Delta, \quad g = \frac{2}{\pi}\tan^{-1}\left[\frac{1}{1 + 1/u}\right] \\ &u = \frac{\lambda}{2\beta}\sqrt{\frac{\pi}{Qa}} = \frac{2.802\lambda}{\beta\sqrt{Q}} \quad (\text{when a} = 0.10\ \text{m}^2/\text{day}) \end{aligned}\right\} \tag{11.53}$$

where

a—the diffusivity
λ—the conductivity
β—the surface conductance; when the surface is covered by an insulation layer, β is given by Eq. (11.20).

The temperature of the corner is given by

$$T_c = -f_c A \sin\left[\frac{\pi(\tau - \Delta)}{2P}\right] \tag{11.54}$$

$$f_c = 1 - (1 - f_1)^2 \tag{11.55}$$

In order to get an approximate formula for the viscoelastic thermal stress in the block, the following analysis is conducted. For 16 concrete blocks with different dimensions (height $H = 300$ m; thickness $L = 10$ m, 20 m, 30 m, 50 m; length $S = 20$ m, 40 m, 60 m, 100 m, see Figure 11.11), the initial temperature is zero, the air temperature is expressed by Eq. (11.51), the temperature field and the viscoelastic thermal stresses in these blocks are computed by FEM. The results of these computations provide a basis for the approximate formulas. The stresses interested in engineering practice are the horizontal stress σ_{y1} in the direction of dam axis and the vertical stress σ_{z1} on the midpoint of the upstream face, the horizontal stress σ_{x2} in the direction of water flow and the horizontal stress σ_{y2} in the direction of dam axis at the center of the top surface. These stresses may be computed as follows:

$$\left.\begin{aligned}\sigma_{y1} &= R_{y1}f_1\rho_2E(\tau_m)\alpha A/(1-\mu)\\ \sigma_{z1} &= R_{z1}f_1\rho_2E(\tau_m)\alpha A/(1-\mu)\\ \sigma_{x2} &= R_{x2}f_1\rho_2E(\tau_m)\alpha A/(1-\mu)\\ \sigma_{y2} &= R_{y2}f_1\rho_2E(\tau_m)\alpha A/(1-\mu)\end{aligned}\right\} \tag{11.56}$$

$$\rho_2 = \frac{0.830 + 0.051\tau_m}{1.00 + 0.051\tau_m}\exp(-0.104P^{0.35}) \tag{11.57}$$

$$\tau_m = \Delta + P/2 \tag{11.58}$$

where ρ_2 is the relaxation coefficient, R_{y1}, R_{z1}, R_{x2}, and R_{y2} are the restraint coefficients of σ_{y1}, σ_{z1}, σ_{x2}, and σ_{y2} as shown in Figure 11.13 which are computed by FEM.

It is evident from Figure 11.13 that the restraint coefficients are dependent on the surface conductance β of concrete. This is due to the fact that the restraint coefficients are related to the depth of variation of temperature. The smaller the value of β, the smaller the depth of variation of temperature, the bigger the restraint coefficient.

The stress σ_{yc} at the corner of the midpoint on the upstream face is given by

$$\sigma_{yc} = f_cR_{yc}\rho_2E(\tau_m)\alpha A \tag{11.59}$$

11.4.2 Computation of Superficial Thermal Insulation

The thickness of the superficial thermal insulation layer is determined by the thermal stress as follows:

$$\sigma_x + \sigma_{x0} \leq [\sigma_x] \tag{11.60}$$

$$\sigma_y + \sigma_{y0} \leq [\sigma_y] \tag{11.61}$$

$$\sigma_z + \sigma_{z0} - \gamma z \leq [\sigma_z] \tag{11.62}$$

where σ_x, σ_y, σ_z are the stresses induced by the variation of air temperature in the winter; σ_{x0}, σ_{y0}, and σ_{z0} are the initial stresses; γ is the density of concrete, γz is the compressive stress due to weight of concrete, z is the height of the block. $[\sigma_x]$, $[\sigma_y]$, and $[\sigma_z]$ are the allowable stresses computed by

$$[\sigma_x] = [\sigma_y] = \frac{E\varepsilon_p}{K_1} \quad \text{or} \quad \frac{R_t}{K_2} \tag{11.63}$$

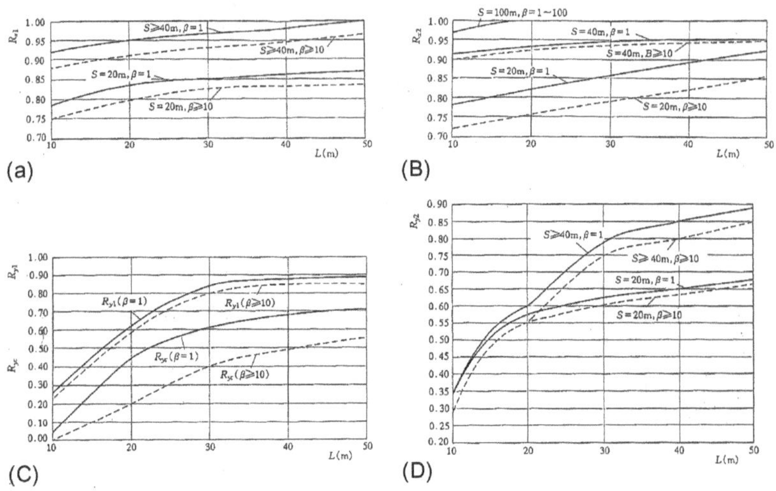

Figure 11.13 The restraint coefficient of thermal stress in dam blocks during winter (unit of β: kJ/(m^2 h° C)).

$$[\sigma_z] = \frac{\eta R_t}{K_2} \tag{11.64}$$

where E is the modulus of elasticity, ε_p is the extensibility of concrete, R_t is the tensile strength, η is the reduction coefficient of tensile strength of the horizontal construction joint, generally $\eta = 0.5\text{—}0.7$; K_1 and K_2 are the coefficients of safety, it is suggested to take $K_1 = 1.6\text{—}2.2$, $K_2 = 1.4\text{—}1.9$.

Example Thermal stresses in winter of concrete block with thickness $L = 20$ m, width $s = 40$ m, density $\gamma = 24.5$ kN/m^3, $z = 20$ m, $\lambda = 10$ kJ/(m h °C), $a = 0.10$ m^2/day, $\beta_0 = 80$ kJ/(m^2 h °C), $\alpha = 1 \times 10^{-5}$(1/°C), $E(\tau) = 30{,}000(1 - \exp(-0.40\tau^{0.34}))$ MPa, $\mu = 0.167$, $\varepsilon_p(90) = 0.80 \times 10^{-4}$, $R_t(90) = 1.80$ MPa, reduction coefficient of horizontal joint $\eta = 0.60$, initial stress $\sigma_{x0} = \sigma_{y0} = \sigma_{z0} = 0.20$ MPa. Amplitude and duration of air temperature drop $A = 22$°C, $Q = 90$ days. The surface of block is covered by foamed plastic plate with thickness $h = 10$ cm and $\lambda_s = 0.14$ kJ/(m h °C).

Take

$$K_1 = 1.8, \quad [\sigma_x] = [\sigma_y] = 1.12 \text{ MPa}$$

$$K_2 = 1.6, \quad \eta = 0.60, \quad [\sigma_z] = 0.68 \text{ MPa}$$

The allowable thermal stresses are

$$[\sigma_x] - [\sigma_{x0}] = 1.12 - 0.20 = 0.92\ \text{MPa},$$

$$[\sigma_z] + \gamma z - \sigma_{z0} = 0.68 + 0.49 - 0.20 = 0.97\ \text{MPa}$$

The equivalent surface conductance is

$$\beta = \frac{1}{1/80 + 0.10/0.14} = 1.376\ \text{kJ}/(\text{m}^2\ \text{h}\ ^\circ\text{C})$$

The thermal stresses are

$$\sigma_{x2} = 1.04\ \text{MPa},\quad \sigma_{y2} = 0.675\ \text{MPa},\quad \sigma_{z1} = 1.056\ \text{MPa},\quad \sigma_{y1} = 0.685\ \text{MPa},$$
$$\sigma_{yc} = 0.648\ \text{MPa}$$

σ_{x2} and σ_{z1} exceed the allowable stresses 0.92 MPa and 0.97.

11.4.3 Determining the Thickness of Superficial Thermal Insulation Plate

Let

$$b = \frac{R_{bi}\rho_2 E(\tau_m)\alpha A}{([\sigma_x] - \sigma_{x0})(1-\mu)} \tag{11.65a}$$

or

$$b = \frac{R_z\rho_2 E(\tau_m)\alpha A}{([\sigma_z] + \gamma z - \sigma_{z0})(1-\mu)} \tag{11.65b}$$

or

$$b = \frac{R_{yc}\rho_2 E(\tau_m)\alpha A}{[\sigma_{yc}] - \sigma_{yc0}} \tag{11.66}$$

The computing process is as follows:

1. Assume a preliminary value of h and compute the corresponding β by Eq. (11.20); (2) get the restraint coefficients from Figure 11.13 and compute b by Eqs. (11.65a) or (11.65b); (3) compute u by Eq. (11.29) and β by Eq. (11.31); (4) compute the required thickness h by Eq. (11.21).

Example 2 The basic data are the same as example 1. Try to compute the required thickness of thermal insulation plate. From example 1, it is evident that the control stresses for this example are σ_{x2} and σ_{z1}.

I. h_1 determined by σ_{x2}

From example 1, $R_{x2} = 0.925$, $\rho_2 = 0.565$, $E(\tau_m) = 23{,}940$ MPa, $\alpha = 1 \times 10^{-5}{}^\circ C^{-1}$, $A = 22^\circ C$, $[\sigma_x] = 1.12$ MPa, $\sigma_{x0} = 0.20$ MPa.

From Eq. (11.65a):

$$b = \frac{0.925 \times 0.565 \times 23940 \times 10^{-5} \times 22}{(1.12 - 0.20)(1 - 0.167)} = 3.59$$

From Eq. (11.29)

$$u = 0.9449\sqrt{3.59^2 - 0.236} - 0.8259 = 2.535$$

From Eq. (11.31)

$$\beta = \frac{10}{2 \times 2.535}\sqrt{\frac{\pi}{0.10 \times 90}} = 1.165$$

From Eq. (11.21), the required thickness is

$$h = 0.14\left(\frac{1}{1.165} - \frac{1}{80}\right) = 0.1184 \text{ m} = 11.84 \text{ cm}$$

II. h_2 determined by σ_{z1}

By the similar computation, we get $h_2 = 11.24$ cm.

III. The required thickness is

$h = \max(h_1, h_2) = \max(11.84, 11.24) = 11.84$ cm.

11.5 Comprehensive Analysis of Effect of Superficial Thermal Insulation for Variation of Air Temperature

The effect of superficial thermal insulation may be indicated by A_0/A, where A_0 is the amplitude of variation of surface temperature of concrete and A is the amplitude of variation of air temperature. Assuming that the variation of air temperature is expressed by Eqs. (11.6) and (11.22), the conductivity $\lambda = 10.0$ kJ/(m h °C), the diffusivity $a = 0.10$ m^2/day, the surface conductance between air and the insulation plate is $\beta_0 = 80.0$ kJ/(m^2 h °C); from Eqs. (11.10) and (11.24), the effect of thermal insulation may be computed by the following formulas:

For annual and daily variation

$$A_0/A = [1 + 112/\beta\sqrt{P} + 6272/\beta^2 P]^{-1/2} \tag{11.67}$$

For cold wave

$$A_0/A = [1+51.8/\beta\sqrt{Q}+878/\beta^2 Q]^{-1/2} \tag{11.68}$$

Thickness of insulation plate

$$h = \lambda_s(1/\beta - 1/\beta_0) \tag{11.69}$$

where

P—period of variation
Q—duration of drop of air temperature in cold wave
λ_s—conductivity of the insulation plate.

Assuming $\lambda = 10.0$ kJ/(m^2 h °C) and $a = 0.10$ m^2/day, the results computed by Eqs. (11.67)–(11.69) are given in Table 11.1 and Figure 11.14.

11.6 The Necessity of Long Time Thermal Insulation for Important Concrete Surface

In the construction of a concrete dam, some cracks often appear after a severe cold wave. Hence people only pay attention to superficial thermal insulation in the early age of concrete and overlook it in the late age. DL/T 5144-2001 "Specifications for hydraulic concrete construction" stipulates that "the concrete of age within 28 days should be covered by insulation layer before sudden drop of air temperature" [13] which presents possibly a false impression that it is not necessary to protect the concrete after the age of 28 days. Practical experience shows that cracks may appear in mass concrete of age over 28 days in the first or second winter after construction.

For example, a lot of cracks appeared on the upstream and downstream face of a gravity dam as shown in Figure 11.15. These cracks appeared not within the 28-day age but in the first and second winter after construction of the dam. The annual mean air temperature at the damsite is 17.3°C, the amplitude of annual variation of air temperature is 11.3°C, and the mean daily air temperature drop is 13.3°C in 3 days in a maximum cold wave. The tensile strength of concrete is 3.32 MPa, taking safety coefficient $k = 1.50$, the allowable tensile stress is $[\sigma_x] = 2.21$ MPa, let the reduction coefficient of horizontal construction joint be $\eta = 0.60$, the allowable vertical tensile stress is $[\sigma_y] = 1.33$ MPa. According to the situation of practical construction process, the simulation thermal stresses are computed by FEM.

The results of computation are given in Table 11.2. The maximum tensile stresses appear in the first and second winter after construction. In the case of no thermal insulation, the maximum horizontal tensile stress $\sigma_x = 4.2$ MPa, the maximum vertical tensile stress $\sigma_y = 2.6$ MPa, all exceed the allowable tensile stress which explains the cause of cracking.

In order to prevent cracking, two schemes for superficial thermal insulation plates are computed: (1) scheme 5 cm + 2 cm, the upstream and downstream faces are covered with foamed polystyrene plate 5 cm thick, the horizontal construction joint and the lateral vertical surface are covered with foamed polythene quilt 2 cm

Table 11.1 Effect of Superficial Thermal Insulation A_0/A and Thickness of Insulation Plate h

Equivalent Surface Conductance β (kJ/(m^2 h °C))	λ/β	Thickness of Plastic Plate h(cm)				Effect of Insulation A_0/A				
		EPS $\lambda_s = 0.148$	XPS $\lambda_s = 0.108$	Polythene $\lambda_s = 0.160$	Polyurethane $\lambda_s = 0.095$	Daily Variation	Semi Month Variation	Annual Variation	2-Day Cold Wave	4-Day Cold Wave
80.0	0.125	0	0	0	0	0.544	0.837	0.964	0.809	0.858
40.0	0.25	0.19	0.14	0.20	0.12	0.360	0.710	0.930	0.676	0.749
20.0	0.50	0.56	0.41	0.60	0.36	0.212	0.535	0.865	0.504	0.593
10.0	1.00	1.30	0.95	1.40	0.83	0.116	0.352	0.754	0.332	0.416
7.0	1.43	1.93	1.41	2.09	1.24	0.083	0.271	0.676	0.257	0.330
5.0	2.00	2.78	2.03	3.00	1.78	0.060	0.206	0.591	0.197	0.259
4.0	2.50	3.52	2.56	3.80	2.26	0.049	0.171	0.532	0.163	0.217
3.0	3.33	4.75	3.47	5.13	3.05	0.037	0.132	0.453	0.127	0.171
2.0	5.00	7.22	5.27	7.80	4.63	0.025	0.091	0.349	0.088	0.121
1.0	10.00	14.62	10.67	15.80	9.38	0.0125	0.047	0.204	0.0458	0.0637
0.5	20.00	29.42	21.47	31.80	18.88	0.0063	0.024	0.111	0.0234	0.0328

Note: A_0—amplitude of variation of temperature of concrete surface; A—amplitude of variation of air temperature; h—thickness of insulation plate; λ_s—conductivity of insulation plate.

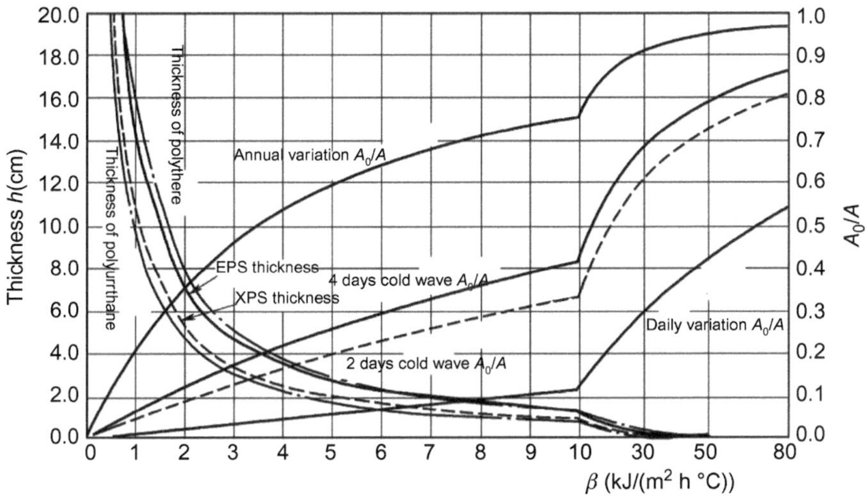

Figure 11.14 Relation between β, h, and A_0/A. (β—equivalent surface conductance; h—thickness of insulation plate; A_0—amplitude of variation of temperature of concrete surface; A—amplitude of variation of air temperature.)

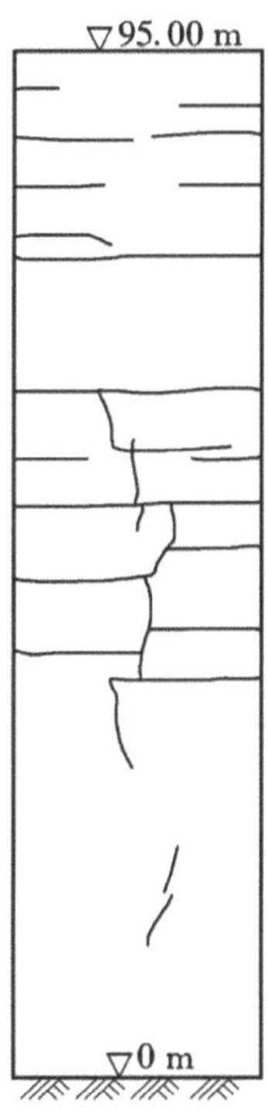

Figure 11.15 Cracks on the upstream face of a gravity dam.

Table 11.2 The Maximum Tensile Stress on the Upstream Face of a Gravity Dam in Winter (MPa)

Thickness of polystyrene plate on upstream and downstream face	0	5 cm	3 cm	Allowable stress (late age)
Thickness of polythene quilt on horizontal surface	0	2 cm	1 cm	
Horizontal tensile stress	4.2	1.6	1.9	2.2
Vertical tensile stress	2.6	−0.1	0.1	1.33

thick; (2) scheme 3 cm + 1 cm, the corresponding thicknesses are changed to 3 and 1 cm. The results of computation are given in Table 11.2, which show that the thermal stresses are reduced remarkably by thermal insulation and the maximum stresses are less than the allowable stresses. No crack appeared in the later 5,000,000 m^3 concrete in this gravity dam.

DL/T 5144-2001 "Specifications for hydraulic concrete construction" stipulates that the time of curing of concrete should not be less than 28 days. Practical experiences of dam construction show that 28 days is too short for curing of mass concrete.

11.7 Materials for Superficial Thermal Insulation

11.7.1 Foamed Polystyrene Plate

Polystyrene, polythene, and polyurethane are now used in the superficial thermal insulation of mass concrete structures. Their properties are given in Table 11.3.

Generally EPS is used in the construction of concrete dams for superficial thermal insulation. The outer face of EPS is painted with acrylic acid cement for water proofing.

A lot of horizontal cracks appeared in the upstream face in the early stage of construction of Jingshuitang arch dam. Since January 1985, all the upstream face is covered by foamed polystyrene plates whose dimensions are 1.0 m × 1.5 m × 0.02 m and no cracks appeared thereafter.

11.7.2 Foamed Polythene Wadded Quilt

In the construction of Three Gorges dam, the foamed polythene wadded quilts are used to cover the horizontal construction joints and the vertical surfaces with keys. Two layers of foamed polythene of thickness 1 cm are wadded in a canvas bag to form a 1.5 m × 2.0 m quilt, the outer face of which is painted with waterproofing glue.

Table 11.3 Properties of Foamed Plastics

Kinds	Density (kg/m^3)	Conductivity (kJ/(m h °C))	Water Adsorption (%)	Compressive Strength (kPa)	Tensile Strength (kPa)
Expansive type polystyrene (EPS)	15–30	0.148	2–6	60–280	130–340
Compressive type polystyrene (XPS)	42–44	0.108	1	200	500
Polythene	22–40	0.160	2	33	190
Polyurethane	35–55	0.080–0.108	1	150–300	500

11.7.3 Polyurethane Foamed Coating

The polyurethane and the foaming agent sprayed by a nozzle on the concrete surface will form a foamed layer whose conductivity is $\lambda \leqq 0.11$ kJ/(m h °C). If it is used for long time insulation, it must be covered by a protective coating.

11.7.4 Compound Permanent Insulation Plate

For long time thermal insulation of mass concrete, it is suggested to use the compound permanent insulation plate whose structure is

(Foamed plastic plate) + (protective layer)

The foamed plastic plate generally is made of polystyrene and the protective layer may be one of the following types (refer to Figure 11.16):

Type A: polymer layer
Type B: cement and sand reinforced by steel wire
Type C_1: reinforced concrete
Type C_2: reinforced concrete fixed by steel bolt.

11.7.5 Permanent Thermal Insulation and Anti-Seepage Plate

In order to possess the effect of anti-seepage as well as thermal insulation, an anti-seepage curtain may be added between the surface of concrete and the foamed plastic plate.

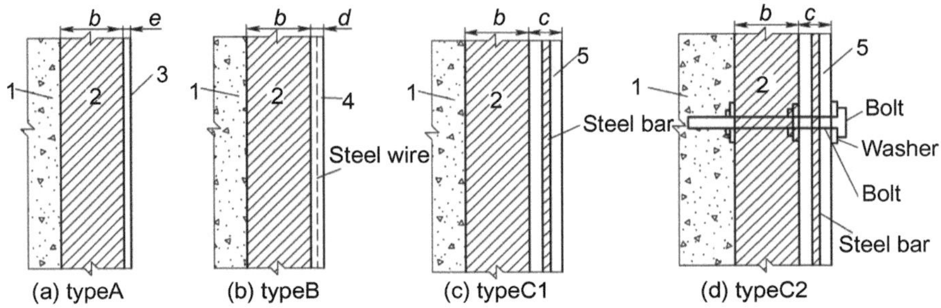

Figure 11.16 Compound permanent insulation plate: (a) type A, (b) type B, (c) type C1, and (d) type C2. 1—dam; 2—foamed polystyrene plate; 3—polymer protective coating; 4—cement and sand reinforced by steel wire; 5—reinforced concrete.

Table 11.4 Thermal Properties of Sand Layer

Material	Conductivity λ (kJ/(m h °C))	Specific Heat *c* (kJ/(kg °C))	Diffusivity *a* (m²/h)	Density ρ (kg/m³)
Dry sand	1.17	0.80	0.00098	1500
Wet sand	4.06	2.09	0.00118	1650

11.7.6 Straw Bag

The conductivity of a dry straw bag is about 0.20 kJ/(m h °C) and that of a wet straw bag is about 0.60 kJ/(m h °C). The dry straw bag burns easily and the wet straw bag is corruptible. In the past, straw bags had been extensively used for the thermal insulation of mass concrete, but due to the development of the plastics industry, the application of straw bags had reduced in recent years.

11.7.7 Sand Layer

The thermal properties of sand are given in Table 11.4. If the construction stopped in the winter, the top surface of the dam may be covered by two sand layers: the upper layer is dry sand and the lower layer is wet sand. The air temperature is expressed by Eq. (11.22).

The minimum temperature of the concrete surface is $T_1 = -f_1 A$, so

$$T_1/A = -f_1 = -1/\sqrt{1 + 1.85u + 1.12u^2} \tag{11.70}$$

and

Table 11.5 Effect of Thermal Insulation of Sand Layer in Winter ($-T_1/A$)

Thickness of sand (m)	0.50	1.00	1.50	2.00
Dry sand	0.436	0.277	0.202	0.160
Wet sand	0.725	0.573	0.472	0.401

$$u = \frac{1}{2}\sqrt{\frac{\pi}{aQ}}\left(\frac{\lambda}{\beta_0} + \frac{\lambda}{\lambda_{s1}}h_1 + \frac{\lambda}{\lambda_{s1}}h_2\right) \tag{11.71}$$

where

Q—time of temperature drop
a—diffusivity of concrete
λ_{s1}, λ_{s2}—conductivity of dry and wet sand
h_1, h_2—thickness of dry and wet sand
β_0—surface conductance of top surface of sand.

Example Effect of thermal insulation of sand layer in the winter. The thermal properties for concrete are $\lambda = 10.0$ kJ/(m h °C), $a = 0.10$ m^2/day. The top of concrete is covered by one layer of dry or wet sand, $Q = 90$ days. The effect of thermal insulation indicated by T_1/A is given in Table 11.5.

11.7.8 Requirements of Thermal Insulation for Different Concrete Surfaces

The requirements of thermal insulation for different parts of surfaces of mass concrete are different as given in Table 11.6.

Table 11.6 Requirements of Thermal Insulation for Different Parts of Concrete Dam

Type of Dam				Common Region			Cold Region		
				Important Dam	Common Dam	Small Dam	Important Dam	Common Dam	Small Dam
Conventional concrete dam	Arch dam	Upstream face	Dam heel	*A*	*A*	*C*	*A*	*A*	*B*
			The rest	*B*	*B*	*C*	*B*	*B*	*B*
		Downstream face	Tensile region	*B*	*B*	*C*	*B*	*B*	*B*
			The rest	*B*	*C*	*D*	*B*	*B*	*C*
	Gravity dam	Upstream face		*B*	*B*	*C*	*B*	*B*	*B*
		Downstream face		*B*	*C*	*D*	*B*	*B*	*C*
RCC	Arch dam	Upstream face		*A*	*A*	*A*	*A*	*A*	*A*
		Downstream face		*B*	*C*	*D*	*B*	*B*	*B*
	Gravity dam	Upstream face		*A*	*A*	*A*	*A*	*A*	*A*
		Downstream face		*B*	*C*	*D*	*B*	*B*	*C*

Note: *A*—permanent thermal insulation and seepage proof; *B*—permanent thermal insulation; *C*—long time thermal insulation in construction period; *D*—short time thermal insulation in construction period.

12 Thermal Stresses in Massive Concrete Blocks

For the convenience of construction and prevention of cracking, in the period of construction, massive concrete structures generally are divided into many blocks by contraction joints which are grouted after cooling of the structure. Before grouting of joints, the thermal stresses in the structure are actually the thermal stresses in the concrete blocks. After grouting of joints, the concrete blocks are combined into a monolithic structure. Thus, the thermal stresses in the period of operation are the stresses of the global structure.

Strictly speaking, the stresses in the concrete structure are accumulated from the beginning of placing concrete to the operation of the structure: The stress analysis simulating the process of construction will be described later in Chapter 16. The thermal stresses in concrete blocks induced by change of temperatures are introduced in the following.

The thermal stresses of concrete blocks are influenced by the following factors: (1) the modulus of deformation of the foundation, (2) the shape and dimensions of the block, (3) the temperature differences and temperature gradients of concrete.

Generally speaking, the modulus of elasticity of concrete $E_c = 30 - 40$ GPa and the modulus of deformation of rock foundation $E_R = 5 - 30$ GPa. For important projects, E_c and E_R must be determined by tests at the damsite.

12.1 Thermal Stresses of Concrete Block on Elastic Foundation due to Uniform Cooling

12.1.1 Thermal Stresses of Block on Horizontal Foundation

As shown in Figure 12.1, a concrete block with height H, length L, and modulus of elasticity E lies on rock foundation with modulus of elasticity E_R. Under the action of uniform cooling, the horizontal stress on the vertical central cross-section of the block is expressed by

$$\sigma_x = -\frac{RE\alpha T}{1-\mu} \tag{12.1}$$

Thermal Stresses and Temperature Control of Mass Concrete. DOI: http://dx.doi.org/10.1016/B978-0-12-407723-2.00012-9

where

T—the temperature difference
R—the restraint coefficient which is dependent on E/E_R and H/L.

By means of stress function, Schleeh had computed the thermal stresses of a concrete block on a rigid foundation due to uniform cooling [92]. The restraint coefficients R of horizontal stresses σ_x of the central section are shown in Figure 12.2.

By means of stress function and relaxation method, Zienkiewicz had computed the thermal stresses in a concrete block on elastic foundation due to uniform cooling ($H = L$, $E_c = E_R$). The results of computing are shown in Figure 12.3 [89].

By means of the finite element method (FEM), the writer and Song Jingting had computed the thermal stresses in a concrete block on elastic foundation with different E_c/E_R due to uniform cooling. The results of computing are shown in Figure 12.4. It is evident that the influence of the restraint of rock foundation is very small when $y \geq 0.40L$ [8].

For a concrete block with $H = L$, the relation between the maximum restraint coefficient R and the ratio E_c/E_R may be expressed by

$$R = \exp\left[-0.58\left(\frac{E_c}{E_R}\right)^{0.60}\right] \tag{12.2}$$

where

E_c—modulus of elasticity of concrete
E_R—modulus of elasticity of rock foundation.

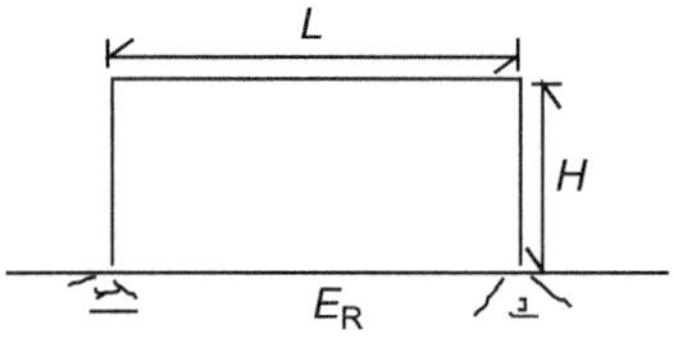

Figure 12.1 Concrete block on rock foundation.

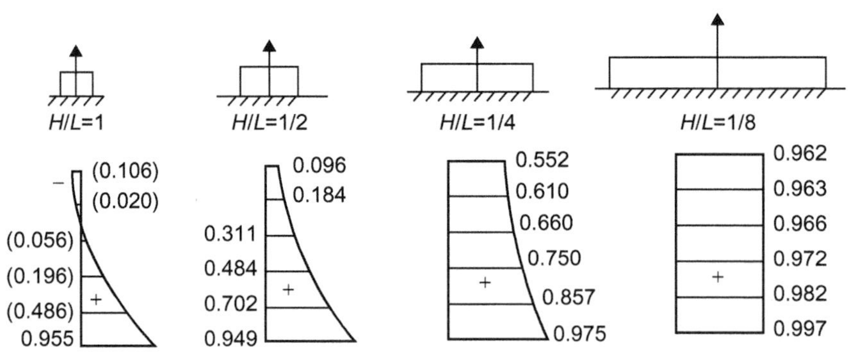

Figure 12.2 The restraint coefficient R of thermal stress in a concrete block on rigid foundation due to uniform cooling.

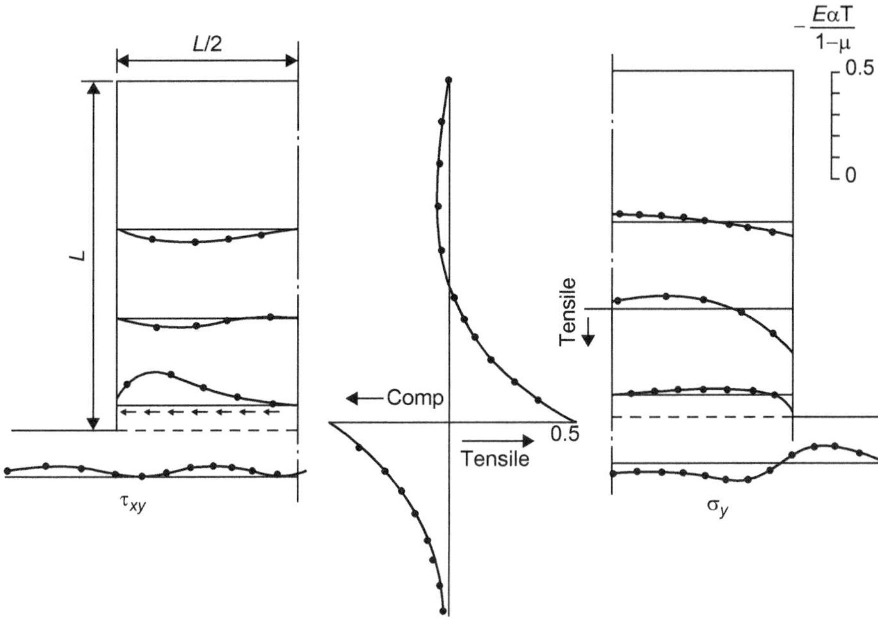

Figure 12.3 Thermal stresses of a concrete block on elastic foundation due to uniform cooling ($E = E_R$).

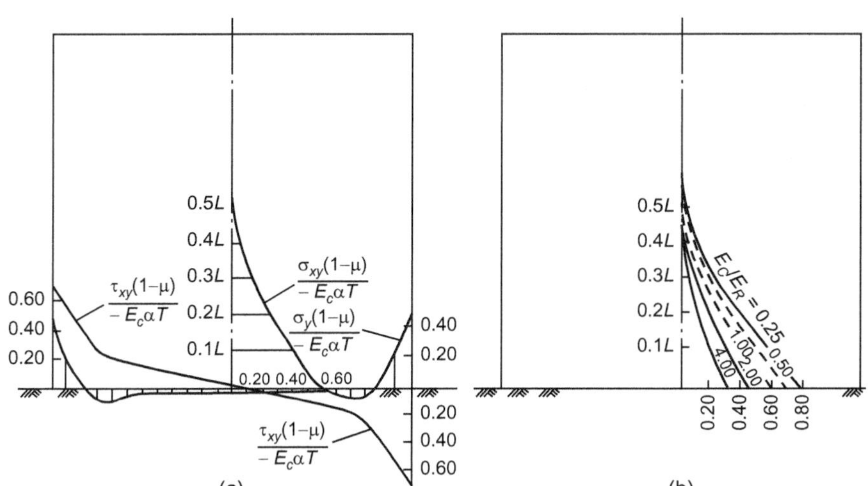

Figure 12.4 Thermal stresses of a concrete block on elastic foundation: (a) distribution of stress when $E_c = E_R$ and (b) $R = \sigma_x(1-\mu)/E_c\alpha T$ for different E_c/E_γ.

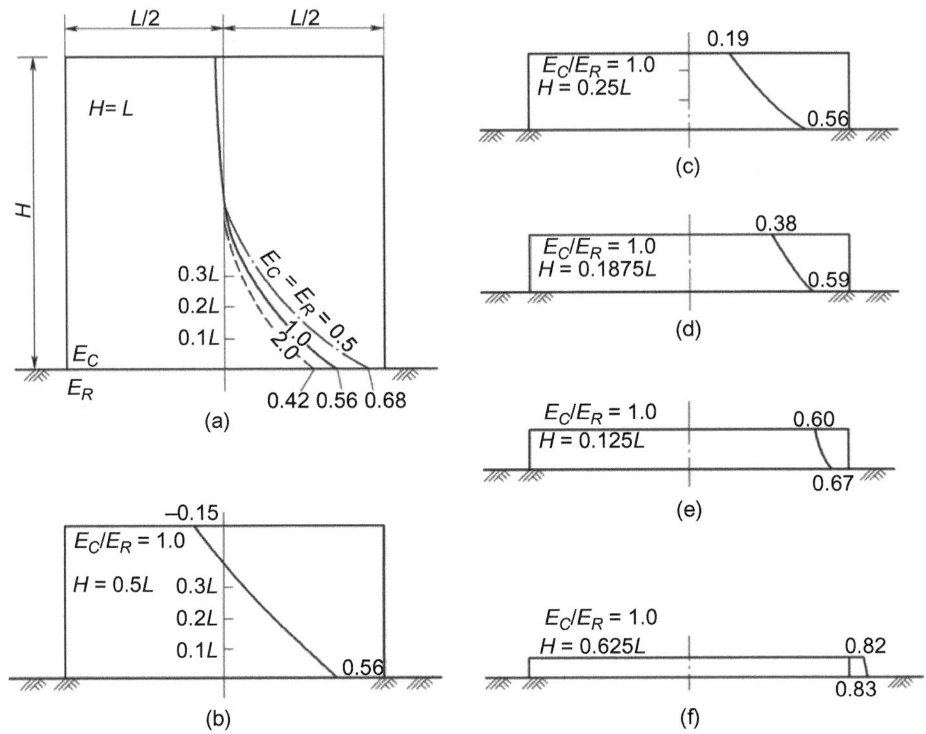

Figure 12.5 The restraint coefficients of concrete blocks with different *H*/*L* on elastic foundation, $R = -(1-\mu)\sigma/E\alpha\Delta T$.

The thermal stresses of concrete blocks of different shapes ($H/L = 1.0$, 0.5, 0.25, 0.1875, 0.125, 0.0625) on elastic foundation ($E_R = E_c$) are shown in Figure 12.5 [7].

12.1.2 Danger of Cracking of Thin Block with Long Time of Cooling

From the above-mentioned results of computing, it is clear that large and deep cracks are liable to develop in thin blocks with long time of cooling. This is due to the following reasons:

1. The restraint coefficients are large in a thin and long block.
2. The rate of cooling is rapid in a thin block.
3. The whole section is subjected to tensile stress in the late age, once a cold wave appears and the superficial fissure induced is liable to become a large and deep crack.

12.1.3 Concrete Block on Inclined Foundation

The viscoelastic thermal stresses in the concrete block on an inclined foundation due to uniform unit temperature difference of Tongfeng arch dam computed by

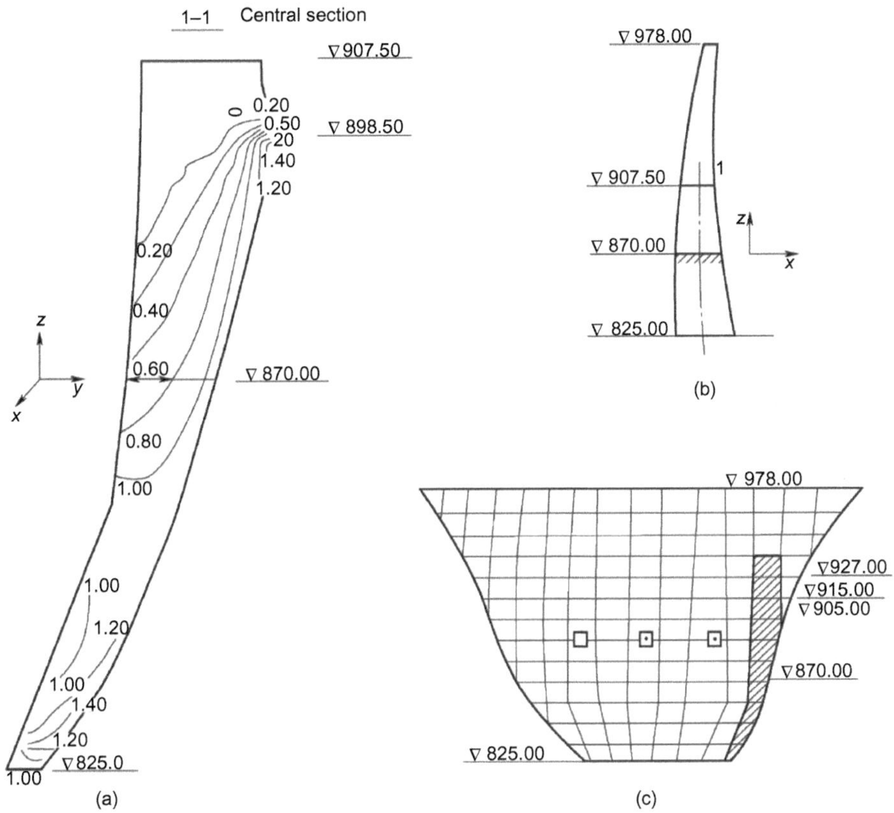

Figure 12.6 Viscoelastic stresses of a concrete block on inclined foundation due to uniform temperature difference in Tongfeng arch dam ($\alpha = 6 \times 10^{-6}$°C^{-1}, $E_R = 20$ GPa, $E_c(90) = 41.2$ GPa): (a) iso-stress curve, (b) cross-section and (c) downstream elevation.

Wang Guobing by FEM are shown in Figure 12.6. $E_R = 20$ GPa, $E_c(90) = 41.2$ GPa, $\alpha = 6 \times 10^{-6}$°C^{-1}. Although E_R and α are rather small, tensile stresses over 1 MPa appeared in a large area.

12.2 Influence Lines of Thermal Stress in Concrete Block

A concrete block with height H and length L on rock foundation is shown in Figure 12.7(a). The temperature $T(y)$ in the block varies only in the direction y. The horizontal thermal stress on the central cross-section is given by

$$\sigma_x(y) = -\frac{E\alpha T(y)}{1-\mu} + \frac{E\alpha}{1-\mu}\int_0^H T(\xi)A_y'(\xi)\mathrm{d}\xi \tag{12.3}$$

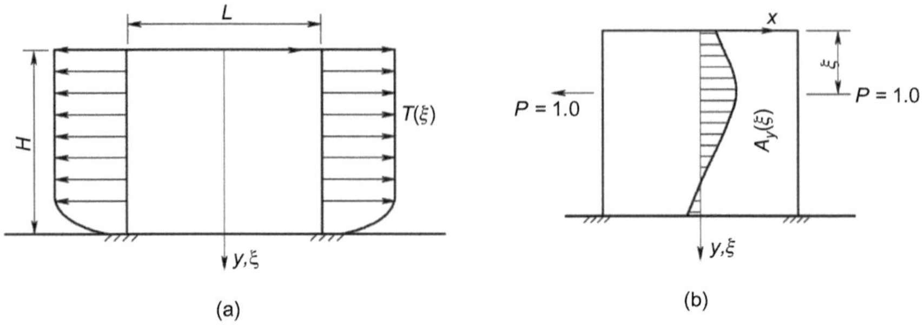

Figure 12.7 Method of influence lines.

where $A'_y(\xi)$—influence coefficient of stress, namely, the horizontal stress at point y of the central cross-section when a pair of unit concentrated force $P = 1.0$ acting at two points $y = \xi$ on the two sides.

Substituting $d\xi$ by $\Delta\xi$, Eq. (12.3) may be transformed into the following equation:

$$\left.\begin{aligned} \sigma_x(y) &= -\frac{E\alpha T(y)}{1-\mu} + \sum_{i=1}^{n} A'_y(\xi)P_i(\xi) \\ P_i(\xi) &= \frac{E\alpha}{1-\mu}T(\xi_i)\Delta\xi_i \end{aligned}\right\} \tag{12.4}$$

where ξ_i is the ordinate of the point of application of $P_i(\xi)$ which is a concentrated force acting on the sides of the block with unit thickness. The dimension of $P_i(\xi)$ is t/m, that of σ_x is MPa, thus the dimension of $A'_y(\xi)$ is m^{-1}.

Let

$$A_y(\xi) = A'_y(\xi)L \tag{12.5}$$

where $A'_y(\xi)$ is a dimensionless number and L is the length of the concrete block. The value of $A_y(\xi)$ depends on E_c/E_R and H/L. Substituting Eq. (12.5) into Eq. (12.4), we obtain

$$\begin{aligned} \sigma_x(y) &= -\frac{E\alpha T(y)}{1-\mu} + \frac{1}{L}\sum_{i=1}^{n} A_y(\xi)P_i(\xi) \\ &= \frac{E\alpha}{1-\mu}\left[-T(y) + \frac{1}{L}\sum_{i=1}^{n} A_y(\xi)T(\xi_i)\Delta\xi_i\right] \end{aligned} \tag{12.6}$$

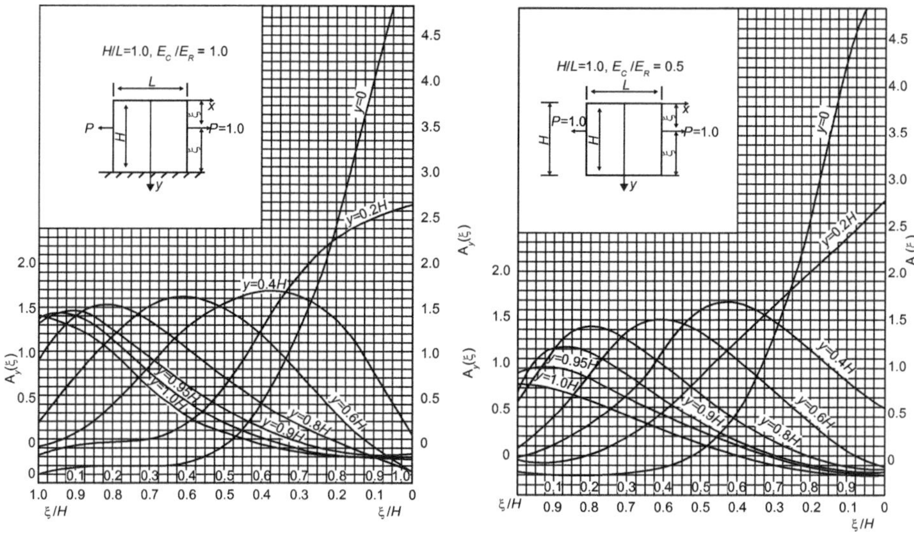

Figure 12.8 Influence coefficient $A_y(\xi) = A'_y(\xi)L$ of a concrete block on elastic foundation: (a) $E_c/E_R = 1$ and (b) $E_c/E_R = 0.5$, $H = L$.

The influence coefficient $A_y(\xi) = A'_y(\xi)L$ of a concrete block on elastic foundation, $H = L$, $E_c/E_R = 1$ and 0.5 are shown in Figure 12.8 [7].

Some examples of computing are given in the following.

Example 1 A concrete block on rock foundation, placed every 4 days a lift of thickness 1.5 m, the adiabatic temperature rise $\theta = 20(1 - e^{-0.34\tau})$°C. By the finite difference method, the temperature field is computed by the envelop which is $T_k(y)$. The stresses induced by $T_k(y)$ is computed by method of influence lines and shown in Figure 12.9. The maximum stress is $-0.45\ E\alpha T_r/(1-\mu)$ which is 26% less than $-0.61E\alpha T_r/(1-\mu)$ computed by restraint coefficient. From this example, it is clear that the distribution of temperature has a great influence upon thermal stress.

Example 2 Local cooling of concrete block. The pipe cooling before grouting of joints is carried out stage by stage. Generally the first stage pipe cooling will induce the maximum thermal stress because the concrete in the cooled region is subjected to restraint of rock foundation as well as the upper uncooled concrete. Let $L = H = 30$ m, height of local cooling $\Delta H = 10$ and 18 m, $E_C = E_2 = 24$ GPa, $\alpha = 1 \times 10^{-5}$°C^{-1}. The temperature differences and the stresses computed by influence lines are shown in Figure 12.10.

Example 3 Thermal stresses due to stepwise temperature difference are shown in Figure 12.11.

Example 4 Influence of temperature gradient on thermal stresses in a concrete block on rock foundation is shown in Figure 12.12.

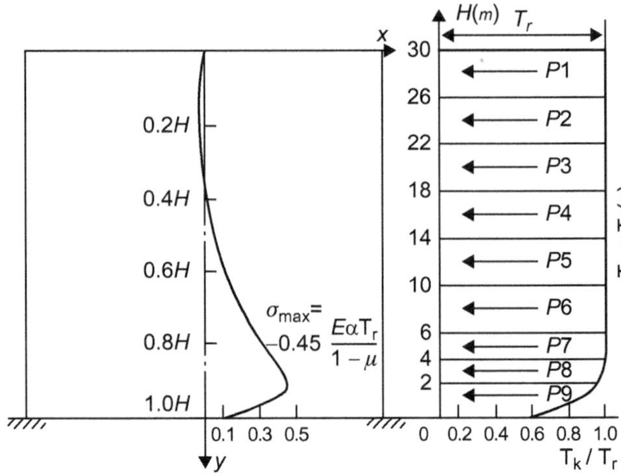

Figure 12.9 Example 1: (unit of stress: $-E\alpha T_r/(1-\mu)$).

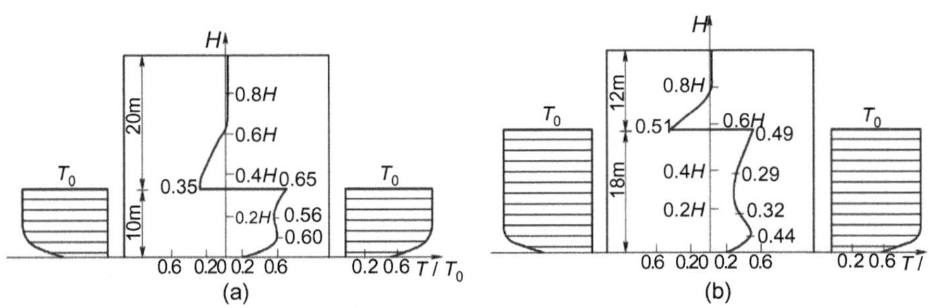

Figure 12.10 Thermal stresses in a concrete block due to local cooling ($-\sigma(1-\mu)/E\alpha T_0$): (a) $\Delta H = 10$ m and (b) $\Delta H = 18$ m.

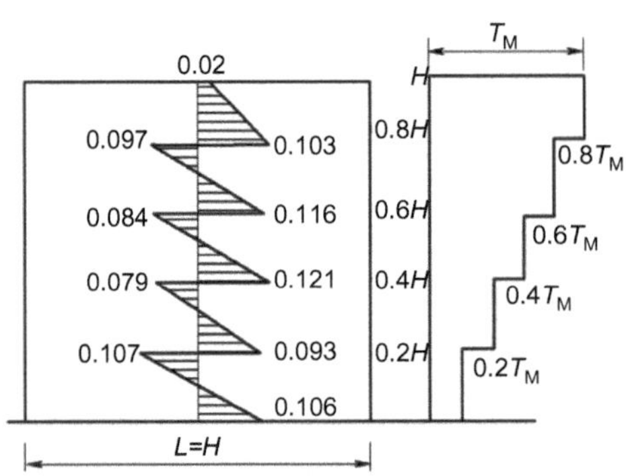

Figure 12.11 Thermal stresses in a concrete block due to stepwise temperature difference (unit of stress: $-E\alpha T_M/(1-\mu)$).

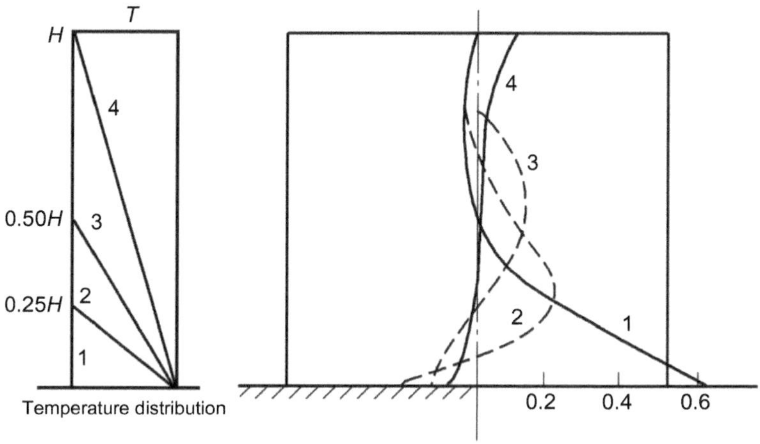

Figure 12.12 Influence of temperature gradient on thermal stress (stress unit: $-E\alpha T_M/(1-\mu)$).

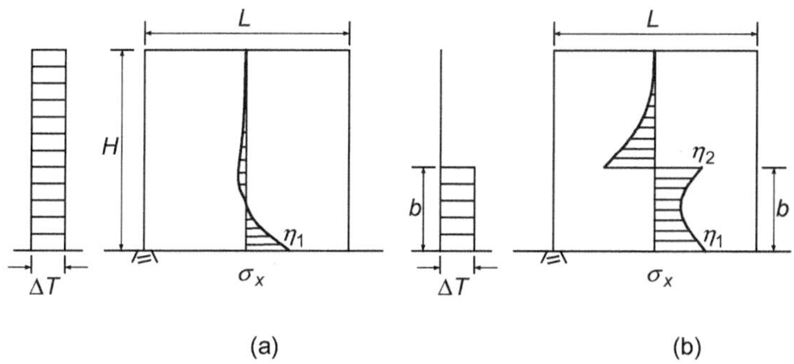

Figure 12.13 Comparison of (a) overall cooling and (b) local cooling.

12.3 Influence of Height of Cooling Region on Thermal Stresses

12.3.1 Influence of Height of Cooling Region on Elastic Thermal Stresses

It is necessary to lower the temperature of the concrete to the steady temperature of the dam before the grouting of contraction joints by pipe cooling. Generally the height of the cooling region is less than the height of the block, i.e., $b < H$, as shown in Figure 12.13.

As shown in Figure 12.13, the difference between overall cooling and local cooling is remarkable. In overall cooling, the tensile stresses appear only in the region of foundation restraint. In local cooling, the tensile stresses appear not only in the region of foundation restraint but also in the upper part of the cooling region.

Consider a concrete block on rock foundation with height H, width L, $H = 3L$, $E = E_R$. When $y = 0$–b, $\Delta T = 1°C$; when $y = b$–H, $\Delta T = 0$. The thermal stresses are

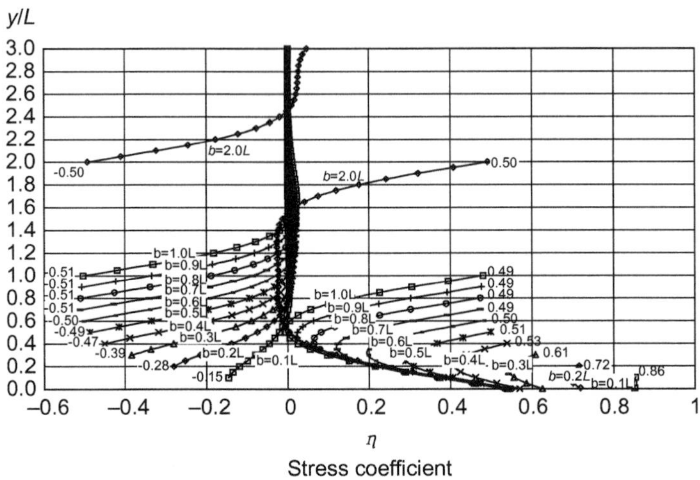

Figure 12.14 Stress coefficient $\eta = (1-\mu)\sigma_x/E\alpha\Delta T$ in local cooling of concrete block on rock foundation ($H = 3L$; $E = E_R$; $y = 0$–b, $\Delta T = 1$, $y > b$, $\Delta T = 0$).

computed by FEM. The stress coefficients $\eta = (1-\mu)\sigma_x/E\alpha\Delta T$ of the central cross-section are shown in Figure 12.14 and Table 12.1. It is evident that the ratio *b*/*H* has remarkable influence on the thermal stress and an important conclusion is derived: the ratio *b*/*H* should not be less than 0.40 in local cooling, i.e., it is suggested to control

$$b/H \geq 0.40 \tag{12.7}$$

Practically pipe cooling is conducted step by step. The second cooling and the third cooling will be conducted after the first cooling. Assuming that *b* is the height of the first cooling region and *c* the height of the second cooling region, the influence of *c*/*L* on the thermal stress will be studied in the following.

Consider three cases: (1) $c/L = 0.10$, $b/L = 0.50$, $(b+c)/L = 0.60$, from Table 12.1, $\eta_2 = 0.50$; (2) $c/L = 0.20$, $b/L = 0.50$, $(b+c)/L = 0.70$, $\eta_2 = 0.49$; (3) $c/L = 0.50$, $b/L = 0.50$, $(b+c)/L = 1.00$, $\eta_2 = 0.49$. Hence, when $b/L = 0.50$, c/L practically has no influence on the thermal stress.

As shown in Figure 12.15(a), the concrete above the cooling region is subjected to compressive stress, when $b = 0.5L$, $y/L = 0.5, 0.6, 0.7$, $\eta = -0.50, -0.32, -0.18$. As shown in Figure 12.15(b), in the second cooling, if the concrete between $y = 0.5L$ and $y = 0.6L$ is cooled, $\eta = 0.82$; if the cooling region is $y = 0.5L$ to $y = 0.7L$, $\eta = 0.67$. The comprehensive stresses of first and second cooling are shown in Figure 12.15(c), when $b + c = 0.6L$, $\eta = 0.82 - 0.32 = 0.50$; when $b + c = 0.7L$, $\eta = 0.67 - 0.18 = 0.49$, The two stress coefficients η are close to each other.

Hence, for a high concrete block on rock foundation with uniform initial temperature, the height *b* of first cooling region has great influence on the thermal stresses, but the height increment *c* of the second cooling region practically has no influence on the thermal stress. Of course the practical situation is complex, *c*/*L* should not be too small.

Table 12.1 Stress Coefficients η of Local Cooling of Block

b/L	**0.1**	**0.2**	**0.3**	**0.4**	**0.5**	**0.6**	**0.7**	**0.8**	**0.9**	**1.0**	**2.0**
η_2 at top of cooling region	0.85	0.72	0.61	0.53	0.51	0.50	0.49	0.49	0.49	0.49	0.50
η_1 at base of cooling region	0.86	0.73	0.64	0.57	0.56	0.56	0.56	0.56	0.56	0.56	0.56

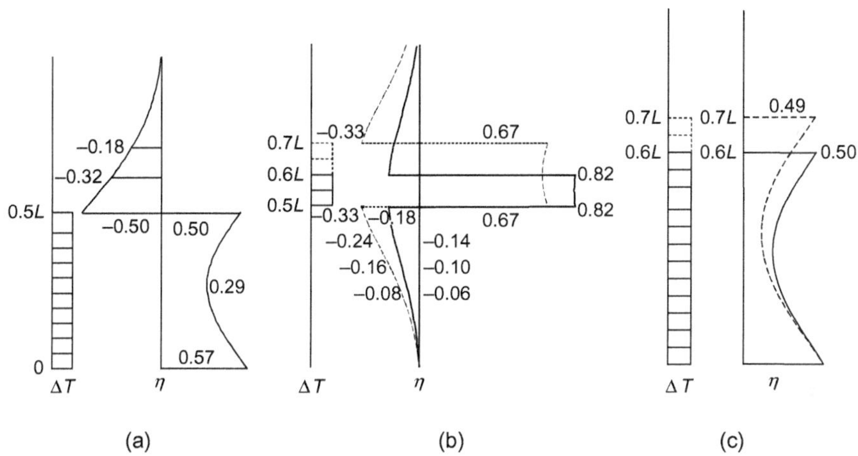

Figure 12.15 Influence of the increment of height of cooling region on thermal stress: (a) first cooling, $b = 5L$, (b) second cooling, $c = 0.1L$, $0.2L$ and (c) comprehensive stress.

12.3.2 Influence of Height of Cooling Region on the Viscoelastic Thermal Stresses

Consider a concrete block on rock foundation with height $H = 60$ m and width $L = 60$ m. The modulus of elasticity $E(\tau)$ and unit creep $C(t, \tau)$ are given by

$$E(\tau) = 35,000[1 - \exp(-0.40\tau^{0.34})] \text{ MPa} \tag{12.8}$$

$$\begin{aligned} C(t, \tau) = {} & 6.50 \times 10^{-6}(1 + 9.20/\tau^{0.45})[1 - e^{0.30(t-\tau)}] \\ & + 14.80 \times 10^{-6}(1 + 1.70/\tau^{0.45})[1 - e^{0.0050(t-\tau)}] \end{aligned} \tag{12.9}$$

$\alpha = 1 \times 10^{-5}$°C^{-1}, $a = 0.10$ m^2/day, $\beta = 70$ kJ/(m^2 h °C), $\mu = 0.167$, allowable tensile stress $(\sigma_t) = 2.10\tau/(7.0 + \tau)$ MPa. For the rock foundation, $E_R = 35{,}000$ MPa, $\mu = 0.25$, no creep.

The height of the first cooling region is b, the initial temperature of concrete and rock is $T_0 = 30$°C, the steady temperature $T_f = 10$°C, the temperature of cooling water $T_w = 8$°C. The pipe cooling begins at $\tau = 90$ days and ends when the temperature of concrete drops to the steady temperature. The spacing of cooling pipe is 1.5 m × 1.5 m, $b/L = 0.10$, 0.20, 0.30, 0.40. The viscoelastic thermal stresses are computed by FEM. The stresses on the central section are shown in Figure 12.16. For $b/L = 0.40$, 0.30, 0.20, 0.10, the maximum stresses are $\sigma_x = 2.75$, 3.05, 3.51, 3.90 MPa, respectively.

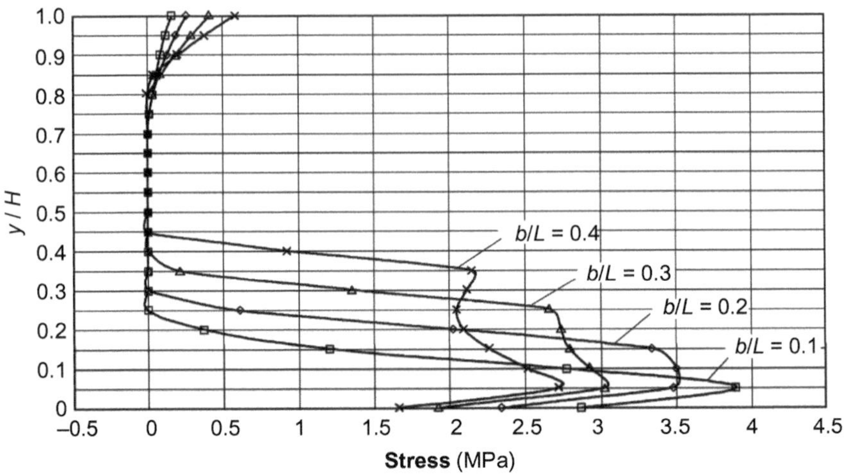

Figure 12.16 The final viscoelastic thermal stress on the central section of the block after cooling.

By means of the stress coefficients in Figure 12.14, neglecting the effect of creep, the maximum elastic stresses are $\sigma_x = 5.15$, 5.62, 6.3, and 7.33 MPa.

It is clear that: (1) the smaller the ratio *b*/*L*, the bigger the stress; (2) because the block is very long ($L = 60$ m), for the four cases computed, the maximum tensile stresses all exceed the allowable stress.

12.4 Influence of Height of Cooling Region on Opening of Contraction Joints

Two concrete blocks standing side by side on rock foundation are shown in Figure 12.17. The height of cooling region is b. There is a uniform temperature difference ΔT in the cooling region. For the concrete, the modulus of elasticity is E_c, Poisson's ratio $\mu = 0.167$, coefficient of thermal expansion $\alpha = 1 \times 10^{-5}{}^\circ C^{-1}$. For the rock foundation, the modulus of elasticity $E_R = E_c$, $\mu_R = 0.25$, $\alpha_R = \alpha$. Computed by FEM as a plane strain problem.

The opening of contraction joint is expressed by

$$\delta = \rho \alpha L \Delta T \tag{12.10}$$

where

δ—the opening of joint
ρ—coefficient of opening of joint
α—coefficient of thermal expansion
L—length of block
ΔT—temperature difference.

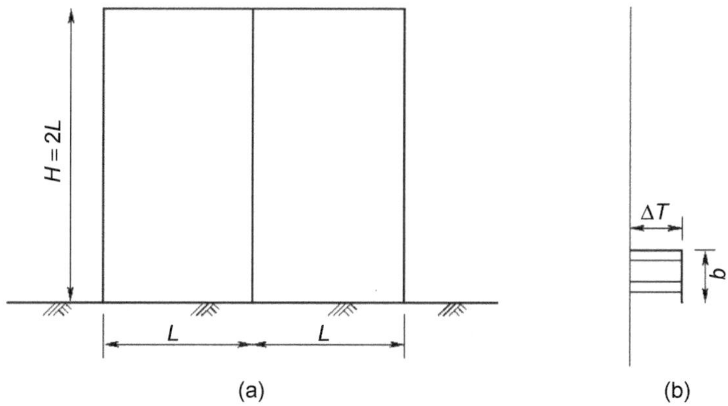

Figure 12.17 Concrete blocks standing side by side: (a) concrete blocks and (b) temperature difference.

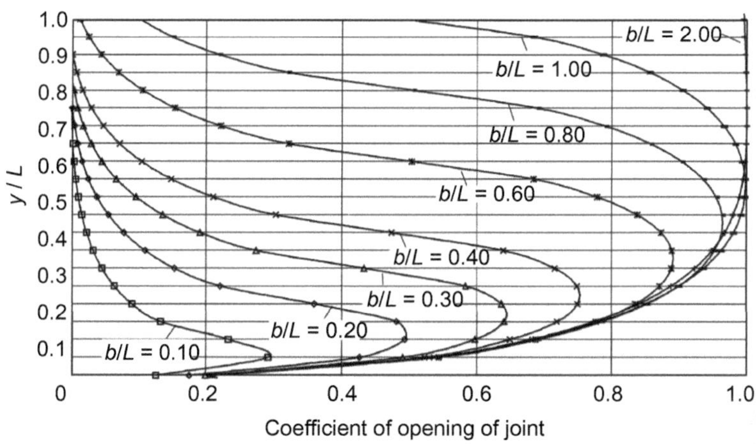

Figure 12.18 The coefficient ρ of opening of joint.

The coefficient of opening of joint ρ is shown in Figure 12.18. It is evident that the ratio b/L is of vital importance to the opening of joint. If $b/L \leq 0.60$, the joint cannot open fully. Here b is the height of the cooling region before the grouting of joint which is generally greater than the height of the first cooling region.

12.5 Two Kinds of Temperature Difference Between Upper and Lower Parts of Block

In the specifications for design of concrete dams, the temperature difference between the upper and lower parts of a concrete block is defined as the difference between the maximum temperatures of the upper and lower parts of the block.

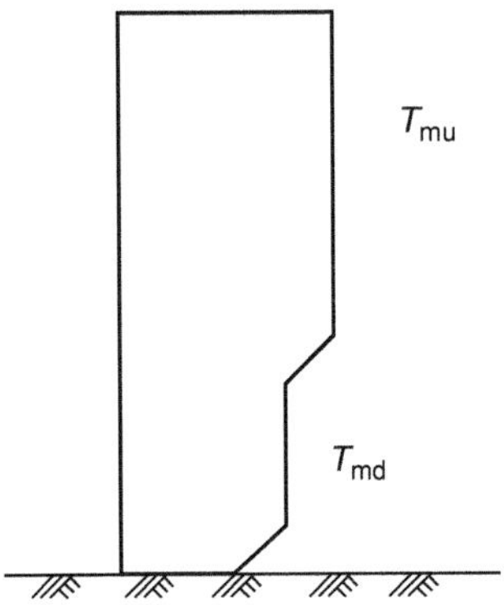

Figure 12.19 Final temperature difference between the upper and lower parts of a concrete block.

This is incomplete. Actually there are two kinds of temperature difference between the upper and lower parts of the block as follows.

1. Final temperature difference between the upper and lower parts of the block

As shown in Figure 12.19, the maximum temperature of concrete in the period of construction is T_m; due to natural or artificial cooling the temperatures finally drop to the steady temperature T_f. The temperature difference of concrete is $\Delta T_f = T_m - T_f$ which is different in the upper and lower parts. In the upper part, $\Delta T_u = T_{mu} - T_{fu}$, where T_{mu} and T_{fu} are the maximum temperature and the steady temperature of the upper part, respectively. In the lower part, $\Delta T_d = T_{md} - T_{fd}$, where T_{md} and T_{fd} are the maximum and steady temperatures of the lower part of the block. The final temperature difference between the upper and lower parts of the block is given by

$$\Delta T_1 = \Delta T_u - \Delta T_d = (T_{mu} - T_{fu}) - (T_{md} - T_{fd})$$

Since the steady temperature T_f varies slowly in the vertical direction, $T_{fu} - T_{fd} \cong 0$, thus the final temperature difference between the upper and lower parts of the block is

$$\Delta T_1 = T_{mu} - T_{md} \tag{12.11}$$

Because there is no pipe cooling before joint grouting, the temperature difference between the upper and lower parts of RCC gravity dam is computed by Eq. (12.11).

2. Transient temperature difference between the upper and lower parts of the block

If the joints are to be grouted, the temperature of the dam must be lowered to steady temperature T_f before grouting of joints by pipe cooling. As shown in Figure 12.20(b), after local cooling, the temperature in the lower part is steady temperature T_f, but the temperature in the upper part is still T_{mu}, thus the transient temperature difference between the upper and lower parts of the block is

$$\Delta T_2 = T_{mu} - T_f \tag{12.12}$$

Generally speaking, the maximum temperature of the lower part is higher than the steady temperature T_f, i.e., $T_{md} > T_f$, so from Eqs. (12.11) and (12.12), we have

$$\Delta T_2 > \Delta T_1 \tag{12.13}$$

Thus, we come to the conclusion that the transient temperature difference ΔT_2 between the upper and lower parts of the block generally is bigger than the final

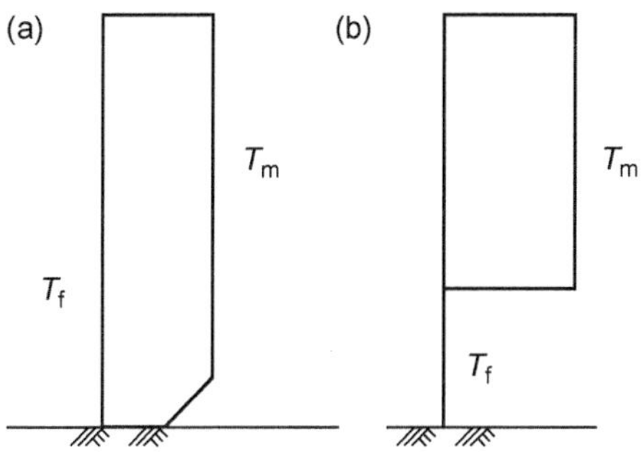

Figure 12.20 Transient temperature difference between upper and lower parts of a concrete block: (a) initial temperature and (b) after local cooling.

temperature difference ΔT_1 between the upper and lower parts of the block. This explains why sometimes large cracks appeared in the upper region of dam block which is out of restraint of rock foundation.

12.6 Two Principles for Temperature Control and the Allowable Temperature Differences of Mass Concrete on Rock Foundation

12.6.1 Stresses due to Stepwise Temperature Difference

Consider a concrete block on rock foundation with height H and length L, $H = L$. As shown in Figure 12.21, the temperature in the rock is 0°C, the distribution of temperature in the block is uniform in the horizontal direction and stepwise in the vertical direction, namely

$$\left.\begin{aligned} \text{When } y &= 0 \sim b_1;\ b_1 \sim b_2,\ b_2 \sim b_3, b_3 \sim H \\ T &= T_0;\ T_0 + T_1,\ T_0 + T_1 + T_2,\ T_0 + T_1 + T_2 + T_3 \end{aligned}\right\} \tag{12.14}$$

Assuming that the modulus of elasticity of rock foundation is E_R, the Poisson's ratio is $\mu_R = 0.25$, the temperature is 0°C; the modulus of elasticity of concrete is $E = E_R$, the coefficient $\eta_{bi}(y)$ of horizontal stress at point y of the central section due to the unit temperature difference shown in Figure 12.22 are computed by FEM and shown in Figure 12.23.

For the temperature difference shown in Figure 12.21, the horizontal stress σ_x on the central section may be computed by

$$\sigma_x(y) = \frac{E\alpha}{1-\mu}\left[T_0\eta_0(y) + T_1\eta_{b1}(y) + T_2\eta_{b2}(y) + T_3\eta_{b3}(y)\right] \tag{12.15}$$

where $\eta_{bi}(y)$ is the stress coefficient of point y on the central section when $b/L = b_i$, e.g., $\eta_{0.15}(y)$ is the stress coefficient when $b/L = 0.15$. $\eta_{bi}(y)$ is shown in Figure 12.23 and Table 12.2 in which there are two coefficients for $y = b_i$ because the stress σ_x is discontinuous at this point.

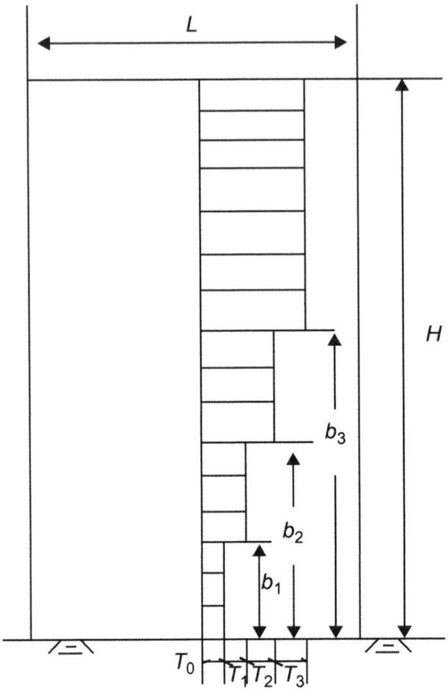

Figure 12.21 Distribution of temperature.

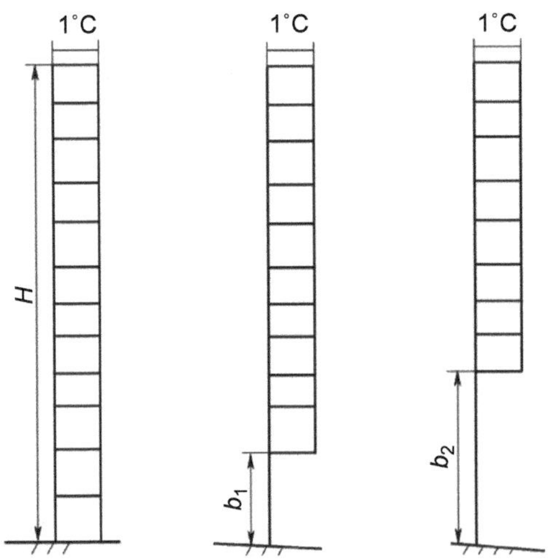

Figure 12.22 Unit temperature difference: (a) uniform temperature difference, (b) nonuniform temperature difference 1 and (d) nonuniform temperature difference 1.

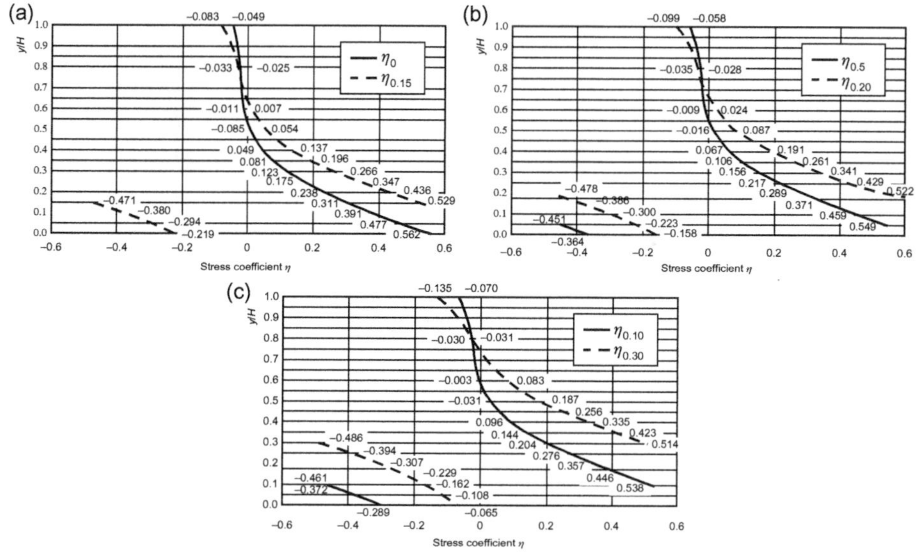

Figure 12.23 Stress coefficient $\eta_{b/H}(y)$ of central section of concrete block due to unit temperature difference: (a) η_0, $\eta_{0.15}$, (b) $\eta_{0.05}$, $\eta_{0.20}$ and (c) $\eta_{0.10}$, $\eta_{0.30}$.

Table 12.2 Stress Coefficient η due to Unit Temperature Difference

y/L	*b/L*					
	0	**0.05**	**0.10**	**0.15**	**0.20**	**0.30**
0.00	0.562	−0.364	− 0.289	− 0.219	− 0.158	− 0.065
0.05	0.477	− 0.451,0.49	− 0.372	− 0.294	− 0.223	− 0.108
0.10	0.391	0.459	− 0.462, 0.538	− 0.380	− 0.300	− 0.162
0.15	0.311	0.371	0.446	− 0.47, 0.53	− 0.386	− 0.229
0.20	0.238	0.289	0.357	0.436	− 0.478, 0.522	− 0.307
0.25	0.175	0.217	0.276	0.347	0.429	− 0.394
0.30	0.123	0.156	0.204	0.266	0.341	− 0.486, 0.514
0.35	0.0811	0.106	0.144	0.196	0.261	0.423
0.40	0.049	0.067	0.096	0.127	0.191	0.335
0.45	0.026	0.038	0.059	0.091	0.123	0.256
0.50	0.008	0.016	0.031	0.054	0.087	0.187
0.60	− 0.011	− 0.009	0.011	0.007	0.024	0.083
0.70	− 0.020	− 0.021	− 0.021	− 0.018	− 0.012	0.016
0.80	− 0.025	− 0.028	− 0.031	− 0.034	− 0.035	− 0.030
1.00	− 0.049	− 0.058	− 0.070	− 0.083	− 0.099	− 0.125

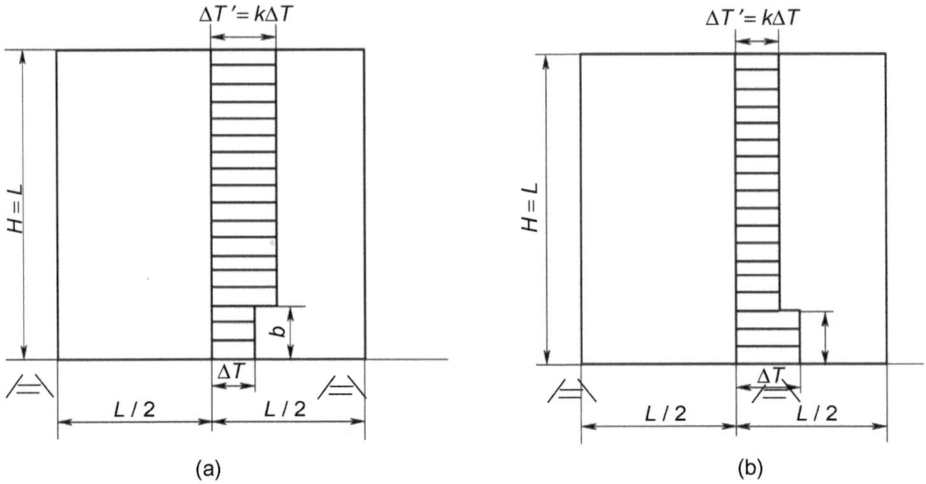

Figure 12.24 Second-order stepwise temperature difference in concrete block on rock foundation: (a) positive stepwise temperature difference ($k > 1.0$) and (b) negative stepwise temperature difference ($k < 0$).

12.6.2 Positive Stepwise Temperature Difference and the First Principle About the Control of Temperature Difference of Concrete on Rock Foundation

Consider a concrete block on rock foundation as shown in Figure 12.24, the temperature in the rock is zero and the temperatures in the concrete block are

$$\left.\begin{array}{lll}\text{When} & y = 0 \sim b & T = \Delta T \\ \text{When} & y = b \sim H & \Delta T = \Delta T' = k\,\Delta T\end{array}\right\} \tag{12.16}$$

The temperature difference in Eq. (12.16) may be changed to the sum of the following two temperature differences:

$$\left.\begin{array}{ll}\text{When} & y = 0 \sim H, T = \Delta T \\ \text{When} & y = b \sim H, T = (k-1)\Delta T\end{array}\right\} \tag{12.17}$$

The stresses in the block induced by the above temperature difference is

$$\sigma_x(y) = \frac{E\alpha}{1-\mu}\left[\eta_0(y)\Delta T + \eta_b(y)(k-1)\Delta T\right] \tag{12.18}$$

The stress coefficients $\eta = \sigma_x(y)(1-\mu)/E\alpha\Delta T$ computed by Eq. (12.18) for $k \geq 1.0$ are shown in Figure 12.25.

In the following we will explain how to choose a suitable value of k to get a favorable stress coefficient η.

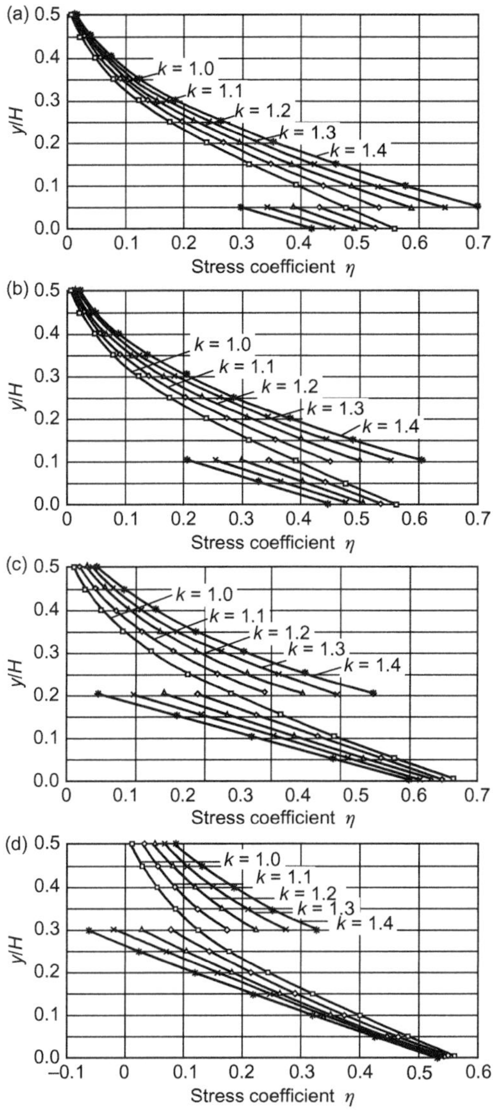

Figure 12.25 Stress coefficients for positive stepwise temperature difference: (a) $b/L = 0.05$, (b) $b/L = 0.10$, (c) $b/L = 0.20$, and (d) $b/L = 0.30$.

The maximum η may appear either at $y = 0$ or at $y = b$, from the condition $\sigma_x(0) = \sigma_x(y)$, we have

$$\eta_0(0) + (k-1)\eta_b(0) = \eta_0(b) + (k-1)\eta_b(b) \tag{12.19}$$

Hence, a critical value k_{cr} is derived in the following:

$$k_{cr} = 1 + \frac{\eta_0(0) - \eta_0(b)}{\eta_b(b) - \eta_b(0)} \tag{12.20}$$

Table 12.3 k_{cr} and $\eta = (1-\mu)\sigma/E\alpha\,\Delta T$

b/L		0	0.05	0.10	0.15	0.20	0.30
Critical value	k_{cr}	1.00	1.093	1.206	1.336	1.477	1.758
	η	0.562	0.528	0.502	0.488	0.487	0.512
Suggested value	k	1.00	1.10	1.20	1.30	1.30	1.30
	η	0.562	0.532	0.504	0.496	0.515	0.542

Table 12.4 Suggested Allowable Temperature Differences ΔT (°C) of Concrete on Rock Foundation

Height of Low Temperature *b*	Height of Temperature Control *y*	*L*(m)				
		<17	17–20	20–30	30–40	>40
0	0.00–0.40*L*	26–25	25–22	22–19	19–16	16–14
0.05*L*	0.00–0.05*L*	26–25	25–22	22–19	19–16	16–14
	0.05–0.40*L*	29–28	28–24	24–21	21–18	18–15
0.10 *L*	0.00–0.10*L*	26–25	25–22	22–19	19–16	16–14
	0.10–0.40*L*	33–31	31–28	28–26	24–20	20–18
0.20 *L*	0.00–0.20*L*	26–25	25–22	22–19	19–16	16–14
	0.20–0.40*L*	34–33	33–29	29–25	25–21	21–19

when $k = k_{cr}$, the thermal stress will have the minimum value σ_{min}; if $k < k_{cr}$, $\sigma_x(0) > \sigma_{min}$; if $k > k_{cr}$, $\sigma_x(b) > \sigma_{min}$, k_{cr} computed by Eq. (12.20) and the corresponding stress coefficient $\eta = (1-\mu)\sigma/Ex\,\Delta T$ are given in Table 12.3. In order to reduce the thermal stress, it is suggested to take $k = k_{cr}$, but if k is too big, the temperature difference between the surface and the interior of the block will be large; hence, it would be better if k does not exceed 1.30.

The values adopted in the allowable temperature differences in the design specifications of a concrete dam are $b/L = 0.20$ and $k = 1.1-1.2$, if b/L is reduced from 0.20 to 0.05–0.10 and k is increased from 1.1–1.2 to 1.30, the thermal stresses will reduce and it will be favorable for construction. Hence the following principle is derived.

The first principle about the control of temperature of concrete on rock foundation: for the positive stepwise temperature difference, reducing the height of region of low temperature and suitable increase of the temperature difference between the upper and lower parts of the block will reduce the maximum thermal stress and are favorable for the construction.

The allowable temperature differences in concrete block on rock foundation stipulated in the design specifications of concrete gravity and arch dam are given in Table 12.4.

Based on the above-mentioned first principle, theoretical analysis and practical experiences, it is suggested to adopt the allowable temperature differences ΔT(°C) of a concrete block on rock foundation as given in Table 12.4, which are more convenient for the construction and the coefficients of safety are 6–9% higher than those in the design specification [15] and [16].

12.6.3 Negative Stepwise Temperature Difference and the Second Principle About the Control of Temperature Difference of Concrete on Rock Foundation

The negative stepwise temperature difference shown in Figure 12.24(b) is unfavorable to thermal stress. In the process of cooling, the lower part of concrete is subjected to the restraint of the upper part of concrete as well as the rock foundation and the tensile stress will be large. The stress coefficients $\eta=(1-\mu)\sigma_x/E\alpha\,\Delta T$ for negative temperature difference are shown in Figure 12.26 which are larger than those for positive temperature difference. Based on the analysis of the relation between b/L, k, and η, we derived the following principle.

The second principle about the temperature control of concrete on rock foundation: for the negative stepwise temperature difference, it is not only necessary to reduce the maximum temperature in the strong restraint region (lower part of concrete block), but also necessary to avoid too low temperature in the weak restraint region (upper part of concrete block).

There are no allowable temperature differences for negative stepwise temperature difference of mass concrete on rock foundation in the current literature, but negative stepwise temperature differences appear frequently in practical engineering. According to the above-mentioned second principle, the analysis of thermal stresses and practical experiences, the suggested allowable temperature differences are given in Table 12.5, where y is the height of temperature control.

For positive temperature difference, the allowable temperature differences given in Table 12.5 are the upper limits of temperature. For the negative temperature difference, the allowable temperature differences given in Table 12.6 are the upper limits of temperature difference for the lower part of the block and the lower limits of temperature difference for the upper part of the block, namely

$$\left.\begin{array}{lll}\text{When} & y=0\sim b & \Delta T\le[\Delta T_{\mathrm{a}}]\\ \text{When} & y=b\sim L & \Delta T\ge[\Delta T_{\mathrm{b}}]\end{array}\right\}\qquad(12.21)$$

where

$[\Delta T_{\mathrm{a}}]$—the allowable temperature difference in the lower part of the block
$[\Delta T_{\mathrm{b}}]$—in the upper part of the block.

12.6.4 Stresses due to Multi-Stepwise Temperature Difference

Consider a concrete block on rock foundation shown in Figure 12.27. The elastic thermal stresses due to multi-stepwise temperature difference are computed by FEM. There are four schemes A, B, C, D as shown in Figure 12.28 and Table 12.6, and the computed results are shown in Figure 12.29. The maximum stress coefficients of scheme A, B, C, D are 0.475, 0.580, 0.511, 0.460, respectively. The stress coefficient η of scheme D is the minimum among them.

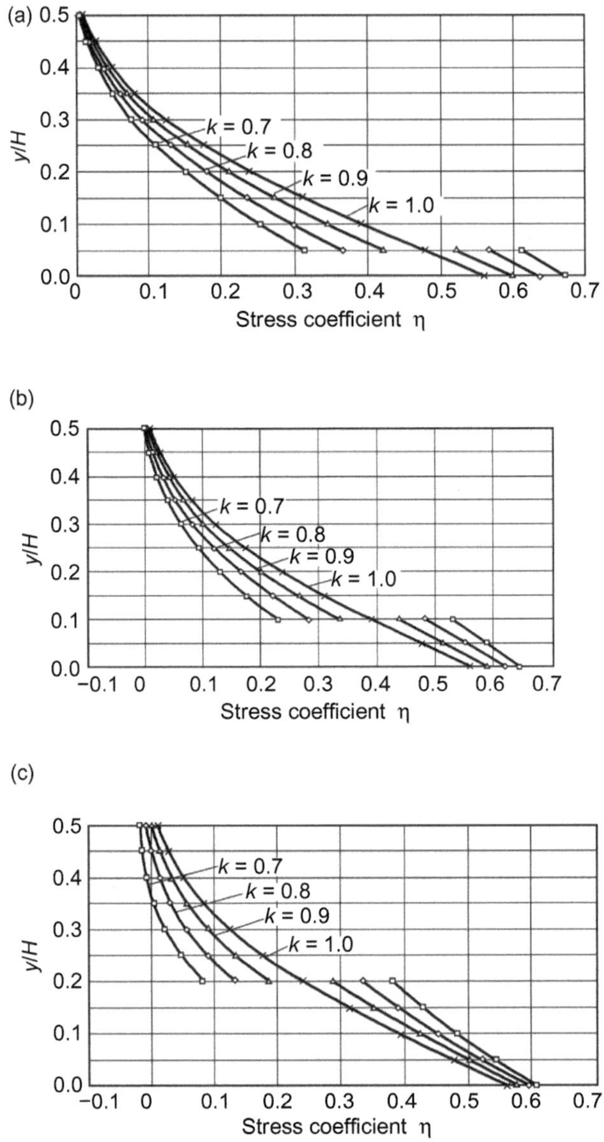

Figure 12.26 Stress coefficients for negative stepwise temperature difference: (a) $b/L = 0.05$, (b) $b/L = 0.10$, and (c) $b/L = 0.20$.

12.6.5 Viscoelastic Thermal Stresses Simulating Process of Construction of Multilayer Concrete Block on Rock Foundation

The practical temperature differences and stresses are very complex, but the above-mentioned two principles about temperature control are still suitable, which will be explained by two examples in the following.

Table 12.5 Allowable Temperature Difference for Negative Stepwise Temperature Difference of Concrete Block on Rock Foundation (°C)

Ratio of ΔT k	b/L	y/L	ΔT	L(m)					η
				< 17	17–20	20–30	30–40	> 40	
0.90	0.05	0.00–0.05	$\Delta T_a \leq$	24	23	21	18	15	0.598
		0.05–0.40	$\Delta T_b \geq$	22	21	19	16	14	
	0.10	0.00–0.10	$\Delta T_a \leq$	25	24	21	18	15	0.591
		0.10–0.40	$\Delta T_b \geq$	22	21	19	16	12	
	0.20	0.00–0.20	$\Delta T_a \leq$	25	24	21	18	16	0.578
		0.20–0.40	$\Delta T_b \geq$	23	22	19	17	14	
0.80	0.05	0.00–0.05	$\Delta T_a \leq$	23	22	19	17	14	0.635
		0.05–0.40	$\Delta T_b \geq$	18	17	15	14	11	
	0.10	0.00–0.10	$\Delta T_a \leq$	24	23	20	17	14	0.620
		0.10–0.40	$\Delta T_b \geq$	19	18	16	14	12	
	0.20	0.00–0.20	$\Delta T_a \leq$	25	24	21	18	15	0.594
		0.20–0.40	$\Delta T_b \geq$	20	19	17	14	12	
0.70	0.05	0.00–0.05	$\Delta T_a \leq$	22	21	18	16	12	0.671
		0.05–0.40	$\Delta T_b \geq$	15	15	12	11	9	
	0.10	0.00–0.10	$\Delta T_a \leq$	23	22	19	16	14	0.649
		0.10–0.40	$\Delta T_b \geq$	16	15	12	12	10	
	0.20	0.00–0.20	$\Delta T_a \leq$	24	23	20	17	15	0.609
		0.20–0.40	$\Delta T_b \geq$	17	16	14	12	10	

Table 12.6 Computed Schemes

Scheme	*A*	*B*	*C*	*D*	Scheme	*A*	*B*	*C*	*D*
b/L	0.05	0.05	0.10	0.10	T_2	1.20	1.40	1.20	1.40
T_0	1.00	1.00	1.00	1.00	T_3	1.30	1.60	1.30	1.60
T_1	1.10	1.20	1.10	1.20	Max. η	0.475	0.580	0.511	0.460

Consider a multilayered concrete block on rock foundation, length $L = 60$ m, height $H = 60$ m, thickness of lift 3.0 m, time interval 7 days, diffusivity of concrete $a = 0.10$ m^2/day, surface conductance $\beta = 70$ kJ/(m^2 h °C), Poisson's ratio $\mu = 0.167$, modulus of elasticity $E = 35{,}000(1 - \exp(-0.40\tau^{0.34}))$ MPa, unit creep expressed by Eq. (12.9), coefficient of thermal expansion $\alpha = 1 \times 10^{-5}$°C^{-1}, allowable tensile stress $[\sigma_t] = 2.10\tau/(7.0 + \tau)$ MPa, adiabatic temperature rise $\theta(\tau) = 25.0\tau/(1.70 + \tau)$°C, unit of time is day. There is no creep in the rock foundation, $E_R = 35{,}000$ MPa, $\mu_R = 0.25$. The initial temperature of concrete and rock foundation is $T_0 = 25$°C, air temperature $T_a = 25.0$°C.

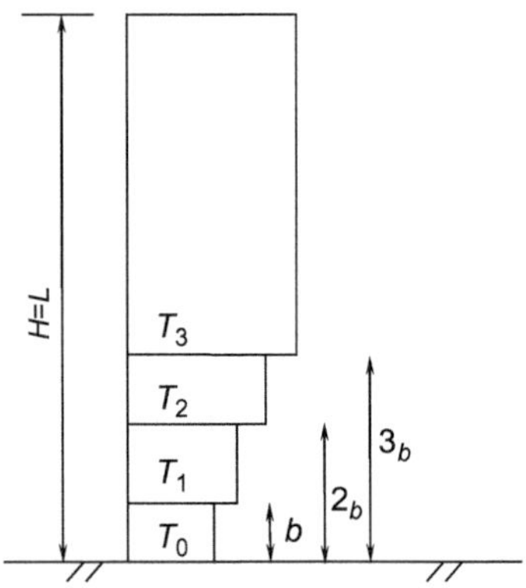

Figure 12.27 Multi-stepwise temperature difference.

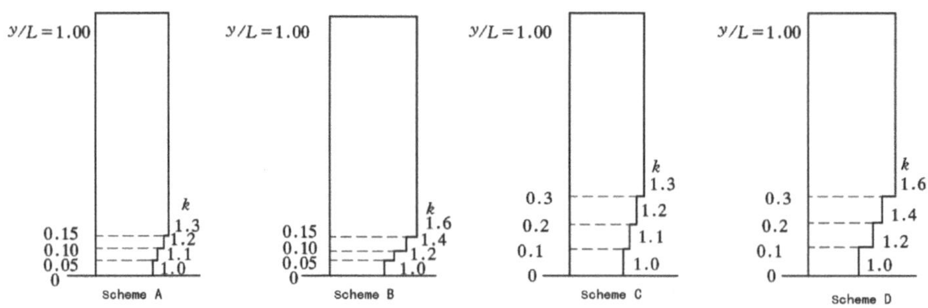

Figure 12.28 Schemes for computing.

Cooling Scheme 1 Spacing of cooling pipe: elevation $y = 0\text{–}30$ m, 1.0 m × 0.50 m; $y = 30\text{–}60$ m, 1.50 m × 1.50 m. There are three stages of cooling: First stage of cooling, $\tau = 0\text{–}20$ days, water temperature $T_w = 20°C$; second stage of cooling, $\tau = 150\text{–}200$ days, $T_w = 15°C$; third stage of cooling, cooling begins from $\tau = 240$ days, $T_w = 11.5°C$, cooling stops when the temperature drops to $T_f = 12.0°C$.

Cooling Scheme 2 Spacing of cooling pipe; $y = 0\text{–}6$ m, 1.0 m × 0.50 m; $y = 6$ m–60 m, 1.5 m × 1.5 m. The time of cooling and water temperatures are the same as scheme 1.

The envelopes of the temperatures and stresses on the central section of the block for the two schemes are shown in Figures 12.30 and 12.31. The height of dense cooling pipe (1.0 m × 0.50 m) in scheme 1 is 30 m and in scheme 2 is reduced to 6 m, but the maximum tensile stress reduced from 1.90 MPa of scheme 1 to 1.82 MPa of scheme 2. It is apparent that scheme 2 is better than scheme 1.

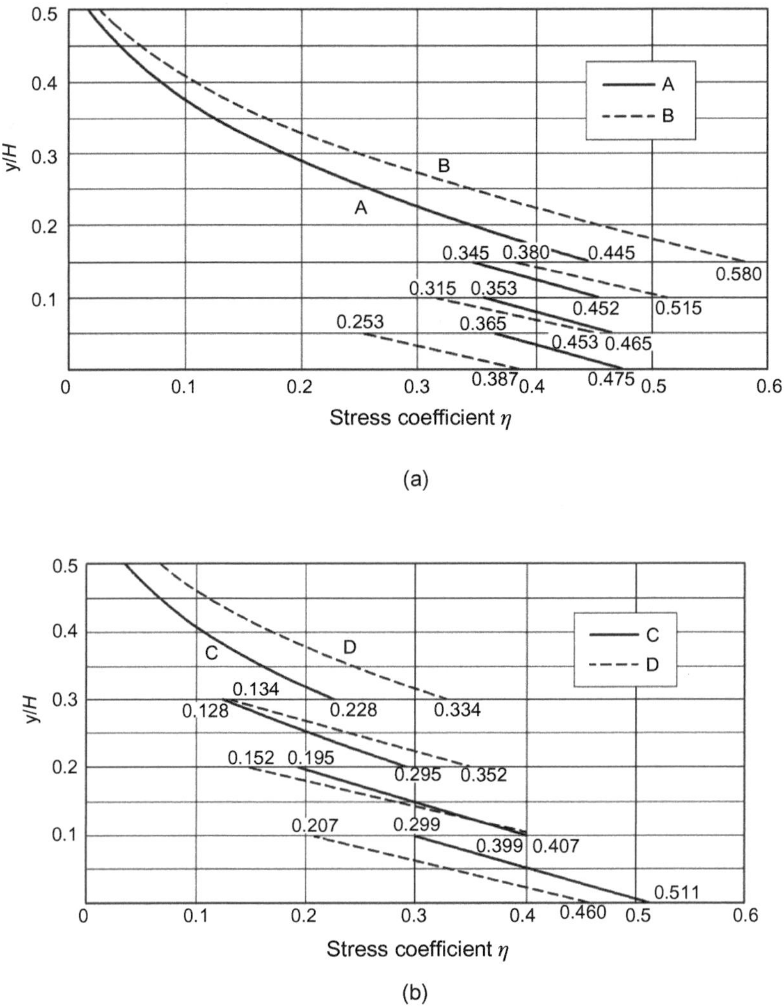

Figure 12.29 Stress coefficients $\eta = (1-\mu)\sigma_x/E\alpha T$ for multi-stepwise temperature difference: (a) scheme A, B and (b) scheme C, D.

12.7 Approximate Formula for Thermal Stress in Concrete Block on Rock Foundation in Construction Period

The maximum thermal stress in a concrete block on rock foundation in the construction period may be computed by

$$\sigma_x = -\frac{KRE\alpha(T_p - T_f)}{1-\mu} - \frac{Kk_r AE\alpha T_r}{1-\mu} \tag{12.22}$$

where

T_p—placing temperature of concrete
T_f—steady temperature of dam
T_r—temperature rise of concrete due to hydration heat of cement
K—the coefficient of relaxation to consider the effect of creep, generally $K = 0.5-0.7$ for conventional concrete, $K = 0.6-0.8$ for RCC
R—restraint coefficient of foundation, given by Eq. (12.2)
A—stress coefficient for temperature rise due to hydration heat of cement, see Figure 12.32. and Eq. (12.23)
k_r—the reduction coefficient to consider the compressive stress induced by the temperature rise in early age due to hydration heat of cement, approximately, $k_r = 0.70-0.85$, because the problem is complex, generally assume $k_r = 1.0$.

The stress coefficient A may also be computed by the writer's formula:

$$A = \exp\left[-0.50\left(\frac{E}{E_R}\right)^{0.75}\right]\{1 - 0.49\exp[-0.055(L-10.0)^{0.85}]\}, \quad L \geq 10\text{ m} \tag{12.23}$$

12.8 Influence of Length of Concrete Block on the Thermal Stress

12.8.1 Influence of Length of Concrete Block on the Thermal Stress due to Temperature Difference Between the Upper and Lower Parts

A concrete dam is generally constructed in the period of several years. The difference of temperature between the upper and lower parts of the dam block is primarily due to the seasonal variation of air temperature. In order to study the influence of length of concrete block on the thermal stress, a concrete block of Three

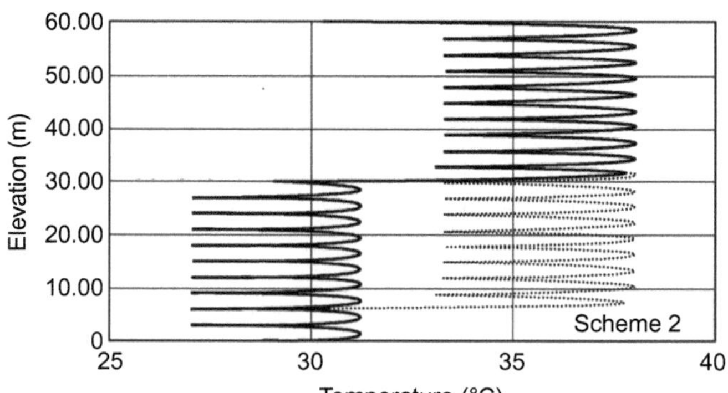

Figure 12.30 Envelopes of temperatures on the central section of the block.

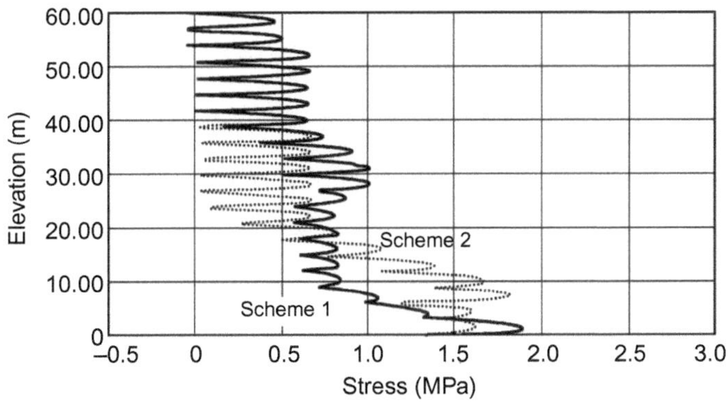

Figure 12.31 Envelopes of the horizontal stresses on the central section of the block.

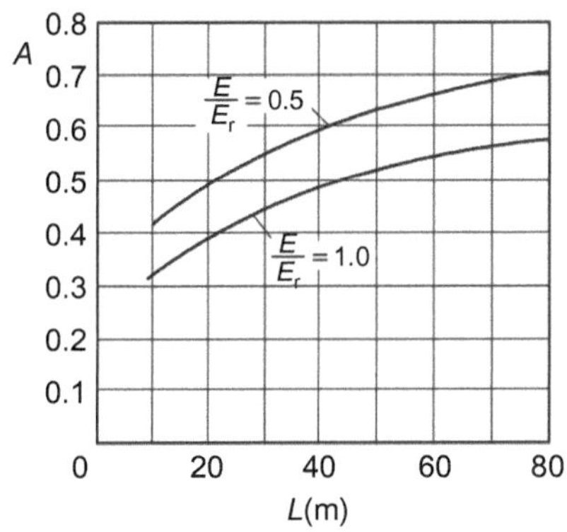

Figure 12.32 Stress coefficient A for temperature rise due to hydration heat of cement.

Georges dam with height $H = 158$ m and length of block $L = 20$, 40, 80, 120 m is computed by FEM. The elevation of the top of the block is 185 m and the elevation of the surface of rock foundation is 27 m. The temperature difference is computed according to the process of construction and then the thermal stresses are computed. The computed results are shown in Figure 12.33 from which it is clear that the influence of length of block on the stress is remarkable. For a concrete block with $L = 20$ m, the stress is very small except in the region near the rock foundation, but for a long concrete block, there are remarkable thermal stresses due to difference of temperature between the upper and lower parts of the block, as well as temperature difference of concrete above rock foundation.

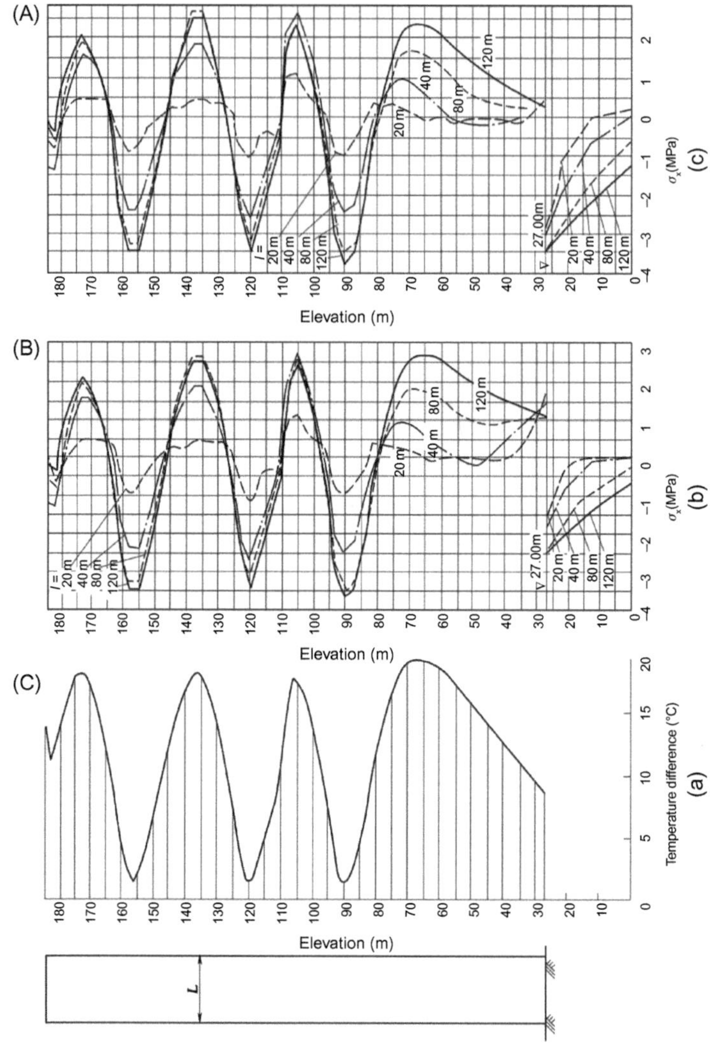

Figure 12.33 Thermal stress in concrete blocks of different length: (a) concrete block and temperature difference, (b) stress without self-weight, and (c) stress with self-weight.

12.8.2 Influence of Joint Spacing on the Thermal Stress due to Annual Variation of Temperature

The annual variation of the surface temperature of a concrete dam may be expressed by

$$T(\tau) = A_0 \cos\frac{2\pi\tau}{P} \tag{12.24}$$

where

A_0—the amplitude of annual variation of surface temperature
$P(=1\mathrm{a})$—period of temperature variation.

The maximum horizontal thermal stress of the surface is given by

$$\sigma_{x0} = -\frac{E\alpha A_0 RK}{1-\mu} \tag{12.25}$$

where

R—the stress coefficient given by Figure 12.34
K—the relaxation coefficient due to creep of concrete, generally $k = 0.60{-}0.70$.

It may be seen from Figure 12.34 that the stress coefficient R is closely related to joint spacing L.

12.9 Danger of Cracking due to Over-precooling of Concrete

Precooling of concrete may effectively reduce the placing temperature, thus it is an important measure of temperature control, but precooling should be suitable, over-precooling is unfavorable for the prevention of cracks [27].

Firstly, in order to avoid the negative stepwise temperature difference as shown in Figure 12.24(b), it is necessary to avoid over-precooling of concrete of the upper part.

Secondly, precooling of new concrete is actually a cold attack on the old concrete. Assuming that at the time of placing new concrete, the surface temperature of the old

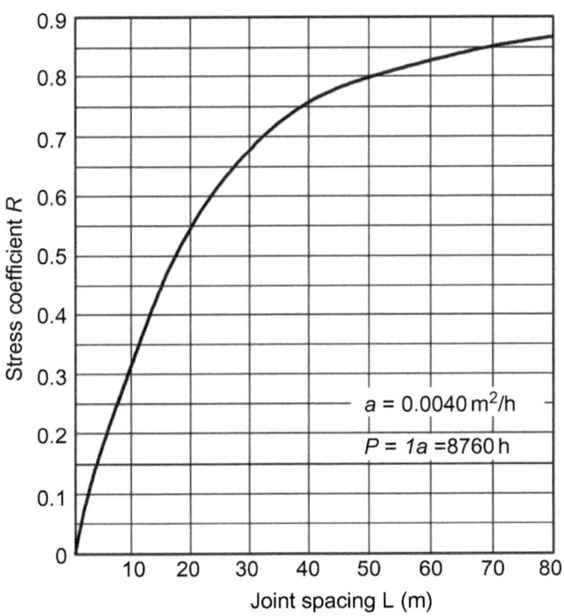

Figure 12.34 Stress coefficient R for thermal stress at surface of gravity dam due to annual variation of temperature.

concrete is T_1 and the placing temperature of the new concrete is T_2, the surface temperature of the old concrete will drop to $(T_1 + T_2)/2$ immediately after placing of new concrete with $\Delta T = T_1 - (T_1 + T_2)/2 = (T_1 - T_2)/2$; because the temperature drops very quickly, the coefficient of relaxation and coefficient of restraint are both close to 1.00, thus the stress increment induced by ΔT is approximately

$$\Delta\sigma(\tau) = \frac{E(\tau)\alpha}{1-\mu} \cdot \frac{T_1 - T_2}{2} \tag{12.26}$$

where

$E(\tau)$—the modulus of elasticity of old concrete
T_1—the surface temperature of old concrete at the time of placing new concrete
T_2—placing temperature of new concrete.

In order to prevent cracking of old concrete, it is necessary that

$$\Delta\sigma(\tau) + \sigma_0(\tau) \le R_t/k$$

Hence, the allowable temperature difference for placing of new concrete is

$$T_1 - T_2 = \frac{2(1-\mu)}{E(\tau)\alpha}\left[R_t(\tau)/k - \sigma_0(\tau)\right] \tag{12.27}$$

where

$R_t(\tau)$—the tensile strength of old concrete
$\sigma_0(\tau)$—the initial stress of old concrete
k—coefficient of safety.

Example Assuming that $E(\tau) = 35{,}000(1 - \exp(-0.44\tau^{0.29}))$ MPa, $R_t(\tau) = 3.30[1 - \exp(-0.33\tau^{0.37})]$ MPa, $\alpha = 1 \times 10^{-5}\,^\circ\mathrm{C}^{-1}$; when $\tau = 0$–15 days, $\sigma_0 = -0.10$ MPa; when $\tau = 16$–30 days, $\sigma_0 = 0$; the allowable placing temperature difference $T_1 - T_2$ computed by Eq. (12.27) is given in Table 12.7. For example, taking $k = 1.1$, the allowable placing temperature difference $T_1 - T_2 \leqq 14^\circ$C; if the

Table 12.7 Example of Allowable Placing Temperature Difference $T_1 - T_2$ (°C)

Age of concrete τ (day)		5	7	14	28
Initial stress σ_0 (MPa)		−0.10	−0.10	−0.10	0
Coefficient of safety k	1.0	14.98	15.24	15.78	15.53
	1.1	12.70	12.94	14.42	14.12
	1.3	11.74	11.93	12.32	11.95
	1.5	10.30	10.46	10.78	10.36
	1.8	8.74	8.86	9.11	8.63

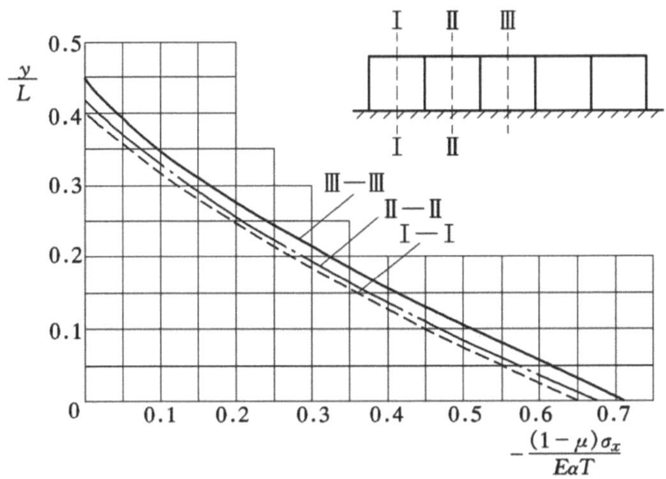

Figure 12.35 Thermal stresses in five concrete blocks standing side by side on rock foundation.

placing temperature of new concrete is $T_2 = 8°C$, the surface temperature of old concrete should be $T_1 \leq 22°C$, which may be a problem in summer.

The measures for control of placing temperature difference (1) reduce the surface temperature T_1, e.g., by spraying water on the surface of old concrete before placing the new concrete, placing the new concrete at night, and pipe cooling of old concrete near the surface with superficial thermal insulation.

12.10 Thermal Stresses in Concrete Blocks Standing Side by Side

The elastic thermal stresses in five concrete blocks standing side by side on rock foundation due to uniform cooling are analyzed by photoelastic model test [6]. The stress coefficients of the central section of the blocks are shown in Figure 12.35 which are greater than the stress coefficient of a single concrete block. Among the five blocks, the thermal stress in the central block is bigger than those in the lateral blocks.

12.11 Equivalent Temperature Rise due to Self-Weight of Concrete

The lateral expansive strain due to action of self-weight of concrete is equivalent to temperature rise which may offset part of the shrinkage strain. Let the block be cut on the surface of foundation, the stresses in the block are $\sigma_x = 0$ and $\sigma_y = -\gamma H$,

Table 12.8 Equivalent Temperature Rise T_g due to Weight of Concrete

H(m)	50	100	150	200	250	300
T_g(°C)	1.36	2.72	4.08	5.44	6.80	8.16

where γ is the density of concrete and H is the height of the block. Due to influence of Poisson's ratio μ, the horizontal lateral strain is

$$\varepsilon_x = \mu\gamma H\left(\frac{1}{E} + C\right)$$

which is equal to the strain induced by temperature rise

$$T_g = \frac{\mu\gamma H}{\alpha}\left(\frac{1}{E} + C\right) \tag{12.28}$$

where

T_g—equivalent temperature rise due to weight of concrete
γ—density of concrete
H—height of concrete
E—modulus of elasticity of concrete
C—unit creep of concrete.

For example, let $\mu = 0.17$, $\gamma = 2.4\ \text{t/m}^3$, $E = 30$ GPa, $C = 1/E$, $\alpha = 1 \times 10^{-5}\,°\text{C}^{-1}$, from Eq. (12.28), we get

$$T_g = 0.0272H(°\text{C}) \tag{12.29}$$

which gives the equivalent temperature rise T_g due to weight of concrete as given in Table 12.8. It is evident that T_g is appreciable when $H \geq 100$ m.

13 Thermal Stresses in Concrete Gravity Dams*

13.1 Thermal Stresses in Gravity Dams due to Restraint of Foundation

As shown in Figure 13.1, T_d is the temperature of a gravity dam without longitudinal joint or the temperature at the time of joint grouting of a gravity dam with longitudinal joints; T_g is the temperature of the rock foundation before the construction of dam; T_f is the final steady temperature of the dam and foundation; the temperature difference shown in Figure 13.1(d) may be divided into two parts as follows:

1. $T_d - T_g$, the difference between the temperature of the dam and that of the foundation which will induce stresses in the dam.
2. $T_g - T_f$, the difference of the temperature of the foundation before construction of the dam and the steady temperature, as $\nabla^2(T_g - T_f) = 0$, so $T_g - T_f$ does not induce stress in the dam.

The stresses in a gravity dam induced by uniform cooling of the dam body are shown in Figure 13.2.

For a gravity dam without longitudinal joint, the temperature in the interior of the dam decreases very slowly because there is no artificial cooling for grouting of the joint. The process of construction is already completed when the dam is cooled to the final steady temperature; thus the stresses due to weight of concrete, water pressure, and temperature change must be superposed. The stresses in a gravity dam without longitudinal joint are computed by the finite element method (FEM) for the following loading cases:

1. Temperature change of the dam body $\Delta T = -11°C$
2. Weight of concrete
3. Low water head (water surface elevation 135 m)
4. Temperature change + weight of concrete
5. Temperature change + weight of concrete + low water head (135 m)
6. Temperature change + weight of concrete + moderate water head (155 m)
7. Temperature change + weight of concrete + high water head (175 m).

The horizontal stresses σ_x of the above-mentioned seven loading cases in this dam are shown in Figure 13.3. It is evident that the tensile stress will be reduced

* The thermal stresses in concrete gravity dams are also explained in Refs [48-50, 54, 55, 102 and 104]

Thermal Stresses and Temperature Control of Mass Concrete. DOI: http://dx.doi.org/10.1016/B978-0-12-407723-2.00013-0

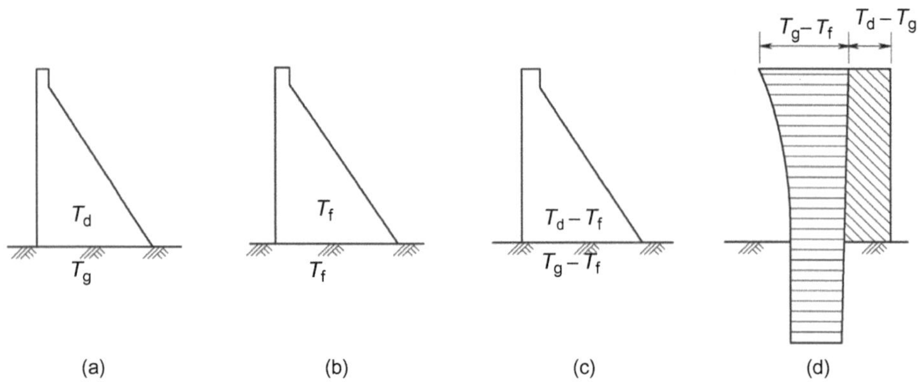

Figure 13.1 Temperature in gravity dam: (a) initial temperature, (b) final steady temperature, (c) temperature difference, and (d) schematic diagram of temperature difference.

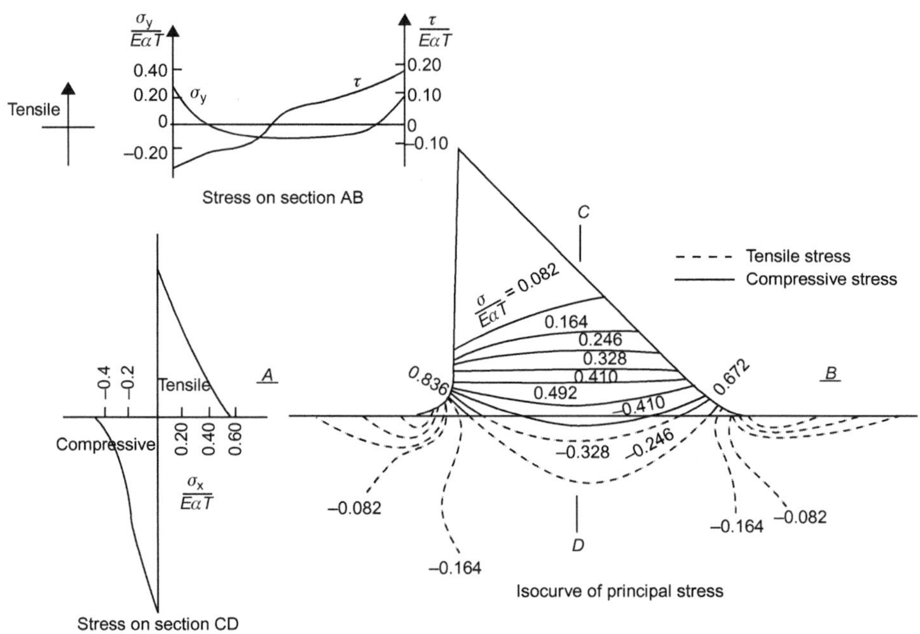

Figure 13.2 The stresses in a gravity dam due to uniform cooling of the dam body.

more than 0.50 MPa by the effect of self-weight and water pressure even the low water elevation 135 m is considered. The vertical stresses σ_y on the horizontal cross section of the dam 3 m above the rock foundation are shown in Figure 13.4.

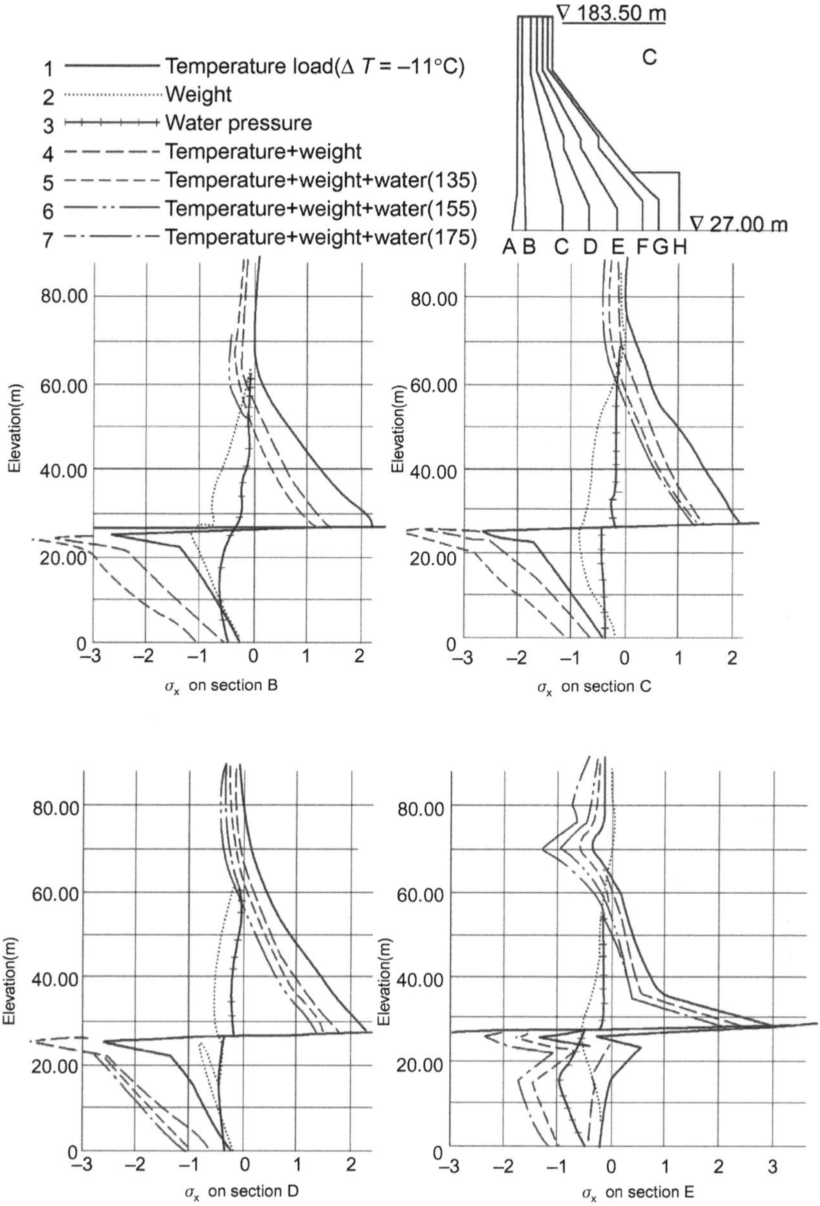

Figure 13.3 Horizontal stress σ_x in a gravity dam due to temperature, self-weight, and water pressure (positive is tensile stress) (MPa).

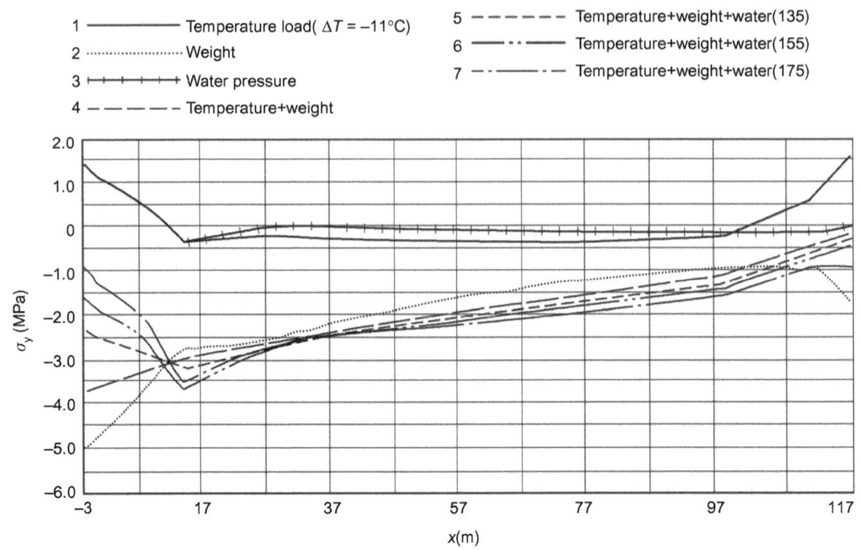

Figure 13.4 The vertical stress σ_y on the horizontal cross section of the dam 3 m above the rock foundation due to weight of concrete, water pressure, and temperature change $\Delta T = -11^\circ C$ of the dam body.

13.2 Influence of Longitudinal Joints on Thermal Stress in Gravity Dam

The influence of longitudinal joints on the thermal stresses in a gravity dam is remarkable and is displayed in the following respects:

1. Influence of the temperature difference between the surface and the interior of the dam.
   For a gravity dam with longitudinal joints, the dam must be cooled to the final steady temperature by artificial cooling before the grouting of joints; thereafter, the temperature difference between the surface and the interior of the dam is small and the surface of the dam is subjected to compressive stress or small tensile stress. For a gravity dam without longitudinal joints, there is no artificial cooling for joint grouting, the internal temperature of the dam is high and decreases very slowly, the surface temperature is low, so tensile stress will appear on the surface, particularly in the winter, and the tensile stress may be so large as to give rise to cracks, as shown in Figure 13.5.
2. Influence of the length of dam block.
   For a gravity dam with longitudinal joints, the length of concrete blocks is 20–40 m, while the length of the base of a gravity dam without longitudinal joints may be 100–200 m or more which is unfavorable to thermal stress.
3. Influence of loading condition.
   For a gravity dam with longitudinal joints, the dam is cooled before grouting of joints, so the thermal stress is computed independently. For a gravity dam without longitudinal joints, the process of construction of the dam is already completed when the dam is cooled to the final steady temperature; thus the stresses due to temperature change must

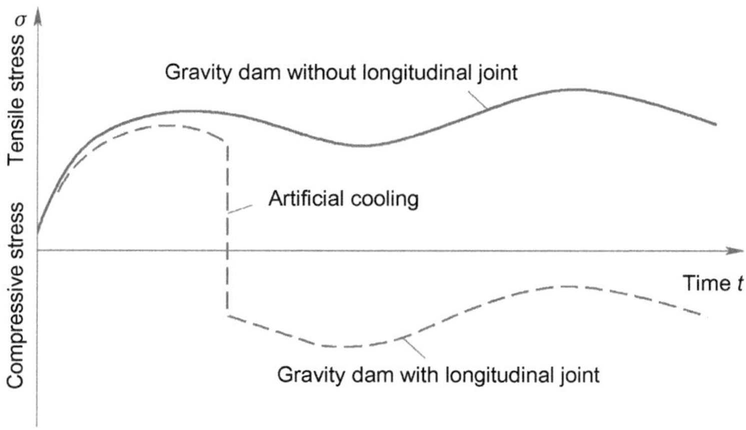

Figure 13.5 Superficial stress due to temperature difference between surface and interior of gravity dam with or without longitudinal joint.

be superimposed with the stresses due to weight of concrete and water pressure which will reduce the tensile stress and is favorable to prevention of cracking.

Thus, the stresses in a gravity dam without longitudinal joints are quite different from those in a gravity dam with longitudinal joints.

13.3 The Temperatures and Stresses in a Gravity Dam without Longitudinal Joint

Generally the dam is constructed in 1.5–2.3 m lifts with time interval 5–15 days, the period of construction may be several years, so the process of construction has appreciable influence on the temperatures and stresses of the dam. In order to consider these influences, the computation must simulate the process of construction by the FEM.

The Longtan roller compacted concrete (RCC) gravity dam with height 192 m is shown in Figure 13.6. A layer of conventional concrete with thickness 6 m is placed on the rock surface and then the RCC is placed. The temperature field and stress field are computed by SAPTIS—a three-dimensional finite element program for simulation computation compiled by Prof. Zhang Guoxin. The influence of climate condition, temperature control, and process of construction are considered in the computation, and the results are shown in Figures 13.7 and 13.8.

13.4 Gravity Dam with Longitudinal Crack

As shown in Figure 13.9, a longitudinal crack appeared in the Norfork dam. The stresses in the dam with and without crack are computed by FEM and the results

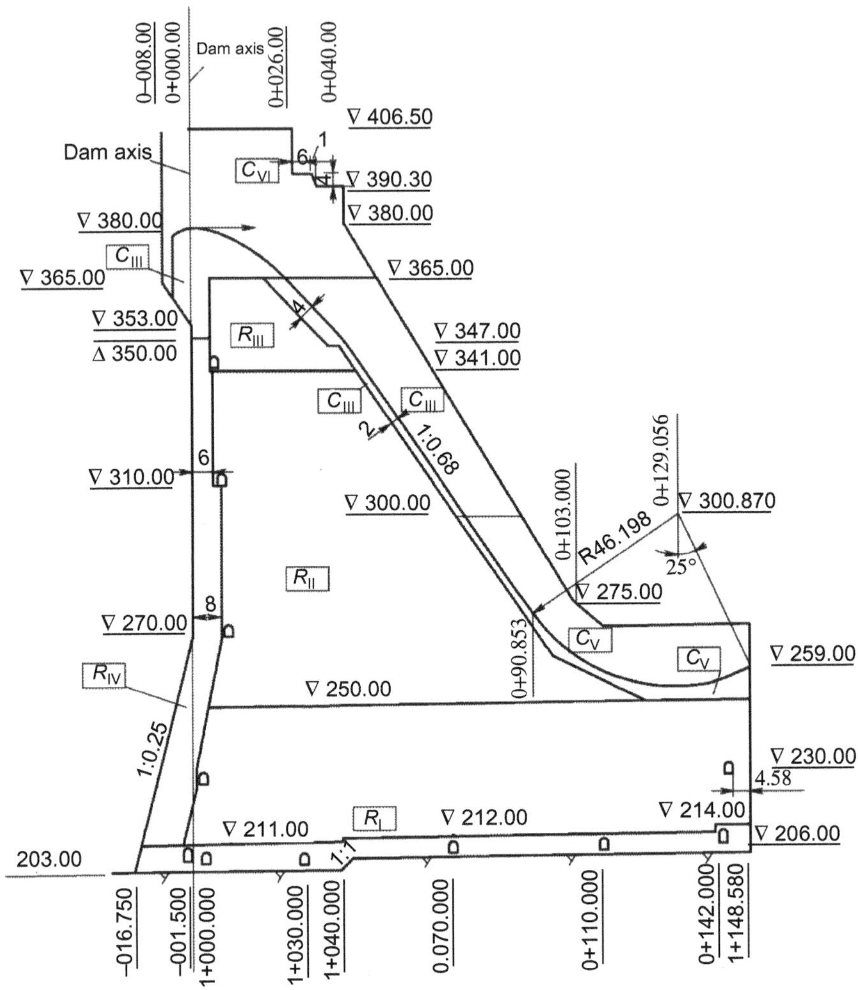

Figure 13.6 Longtan RCC gravity dam, the overflowing section.

are shown in Figures 13.10 and 13.11. It is apparent that the influence of longitudinal crack on the stresses in the dam is remarkable [101].

13.5 Deep Crack on the Upstream Face of Gravity Dam

There are deep cracks on the upstream face of some gravity dams which are very harmful to the structure. The deep crack on the upstream face of the Dworshak dam is shown in Figure 13.12. In order to prevent this type of crack, it is necessary to put a long-time superficial thermal insulation layer, as foamed plastic plate, on the upstream face of the dam.

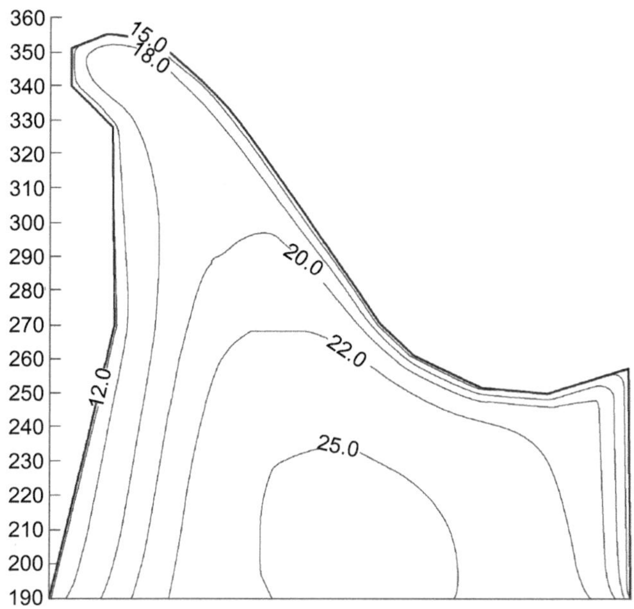

Figure 13.7 Iso-temperature of Longtan RCC gravity dam at 20 years after completion of dam construction.

13.6 Opening of Longitudinal Joint of Gravity Dam in the Period of Operation

Before the grouting of longitudinal joints, generally the temperature of the dam is reduced to the steady temperature T_f by artificial cooling. The actual temperature $T(x,y,z,t)$ of the dam in the period of operation is a function of position (x,y,z) and time t. The steady temperature $T_f(x,y,z)$ is the annual mean value of $T(x,y,z,t)$ in the time domain. The actual temperature $T(x,y,z,t)$ may be divided into two parts as follows:

$$T(x,y,z,t) = T_f + T_q \tag{13.1}$$

$$T_q = T(x,y,z,t) - T_f \tag{13.2}$$

where T_f is the steady temperature and T_q is the quasi-steady temperature, as shown in Figure 13.13.

As the longitudinal joints are grouted at the steady temperature T_f, so T_f will not induce the opening of joints. The opening of joints is induced by the quasi-steady temperature $T_q(x,y,z,t)$ which varies sinusoidally with time.

A computed example is shown in Figures 13.14 and 13.15.

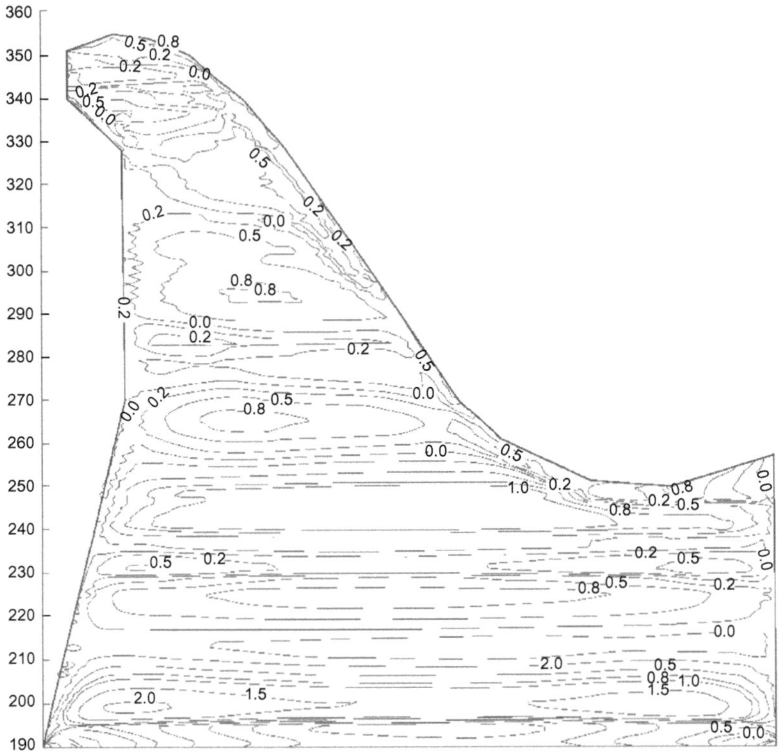

Figure 13.8 Iso-lines of horizontal stress in the river direction at 20 years after completion of dam construction.

13.7 Thermal Stresses of Gravity Dams in Severe Cold Region

13.7.1 Peculiarity of Thermal Stresses of Gravity Dam in Severe Cold Region

The climatic condition is harsh for construction of gravity dams without longitudinal joint in severe cold region. As the air temperature is too low, concrete cannot be poured in about half a year in the winter. The air temperature in the summer is rather high; the temperature of the concrete poured in summer is high. Thus the temperature difference between the upper and lower parts of the dam block is large. The air temperature is very low and the time of low temperature is long in the winter, so there are big temperature differences between surface and interior of the concrete poured in summer. It is necessary to adopt a special superficial thermal insulation layer to prevent cracking of concrete.

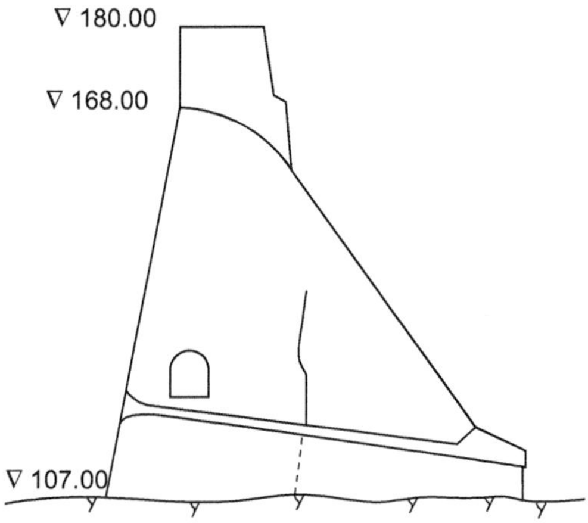

Figure 13.9 Crack in Norfork dam (m).

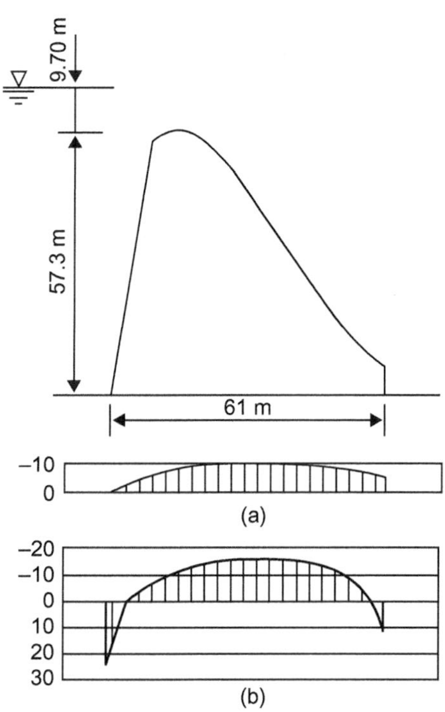

Figure 13.10 Vertical stress σ_y in the dam without crack: (a) water pressure + weight and (b) water pressure + weight + $\Delta T = 12-20°C$.

13.7.2 Horizontal Cracks and Upstream Face Cracks

For gravity dams in a severe cold region, under the simultaneous action of the big temperature difference between the upper and the lower parts of the dam and the temperature difference between the surface and the interior of the dam, there may be large vertical tensile stress near the surface of lift of concrete stopped in

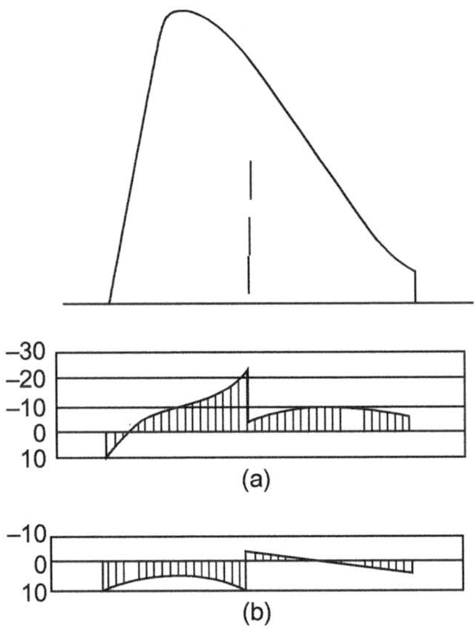

Figure 13.11 Stress in the dam with crack, weight + water pressure (0.1 MPa): (a) vertical stress σ_y and (b) shearing stress τ_{xy}.

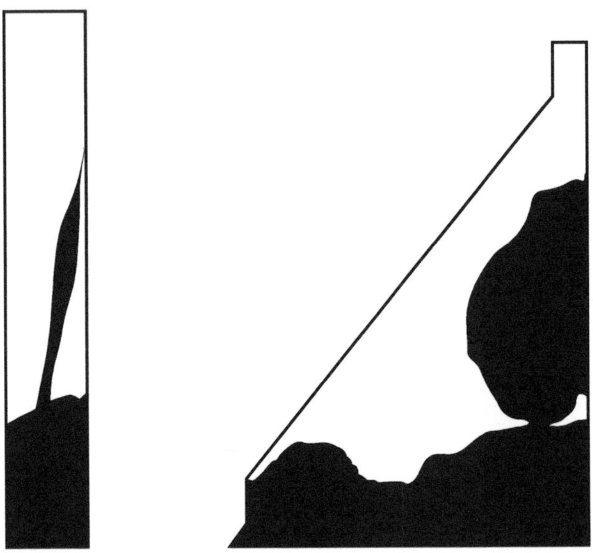

Figure 13.12 The plastic model showing the deep crack on the upstream face of Dworshak dam.

previous winters and large horizontal tensile stresses on the upstream and downstream faces of the dam which may lead to horizontal and vertical cracks.

The Guanyinge RCC gravity dam with maximum height 82 m is located in Liaoning province in China. At the damsite, the annual mean temperature of air is 4.77°C, the maximum monthly mean air temperature is 23.3°C, there are 5 months in which the monthly air temperature is below 0°C, and the pouring of concrete was stopped. The dam was constructed from May 1990 to September 1995, and the pouring of concrete was stopped for 5 months in the winter every year. The

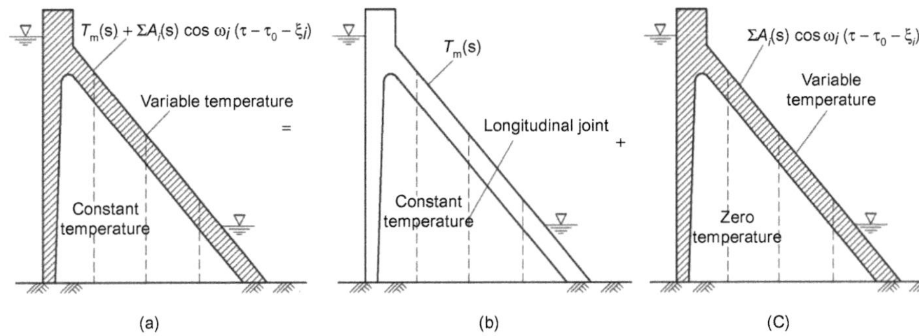

Figure 13.13 Temperature field T_d of gravity dam in the operation period: (a) temperature field T_d in the operation period, (b) steady temperature, and (c) quasi-steady temperature field T_q.

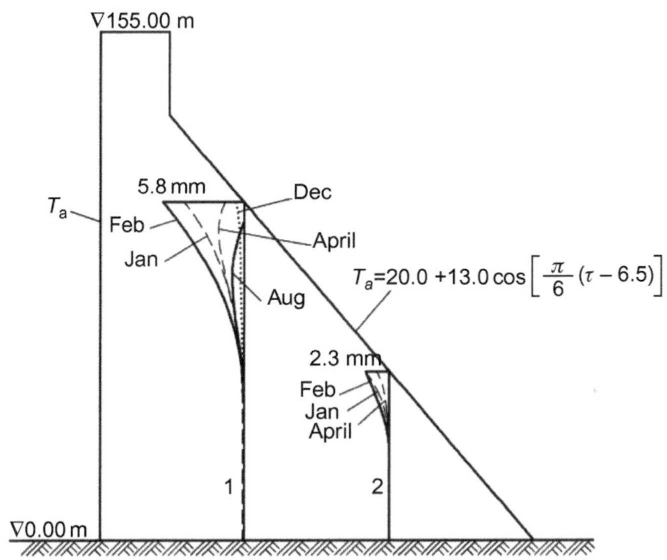

Figure 13.14 Example, opening of longitudinal joints (mm) for empty reservoir.

upstream face and the downstream face are insulated by foamed polystyrene plate of thickness 3–7 and 5 cm, respectively. The horizontal surface of the concrete stopped in winter is insulated by three layers of straw mat with a polystyrene cloth on the top.

The actual temperature difference between the upper and the lower parts of the dam is 13.25–20.65°C and the temperature difference between the surface and the interior of the dam is 15.0–29.0°C. The simulating computation by FEM showed that the vertical stresses on the upstream and downstream faces near the horizontal

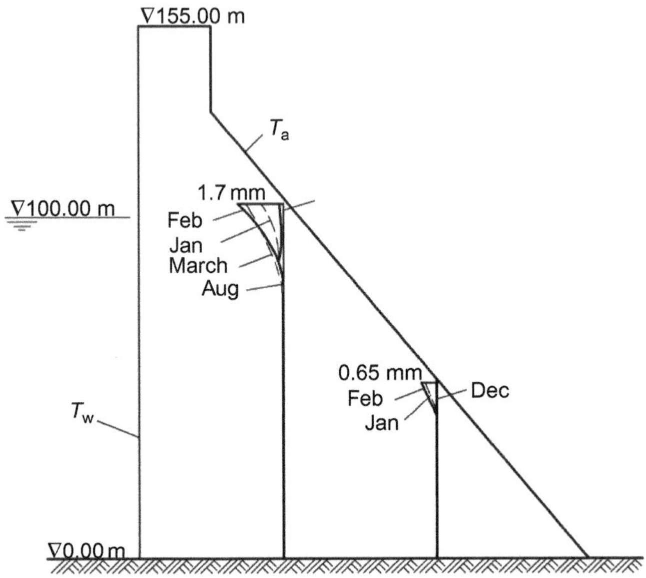

Figure 13.15 Example, opening of longitudinal joints (mm) for water head 100 m.

joints at which the placing of concrete was stopped in the winter were 2.5–4.0 MPa, which exceed the tensile strength of concrete a great deal. As shown in Figure 13.16, on the upstream surface of Guanyinge dam appeared 53 cracks of which 51 were horizontal cracks, most of them were near the horizontal joints at which the pouring of concrete was stopped in the winter, The width of crack was 0.5–1.2 mm, the depth 3.0–6.0 m, and the length of crack was nearly equal to the length of the dam block. On the downstream surface, there appeared 79 cracks in which 60 were horizontal cracks.

13.7.3 Measures for Preventing Cracking of Gravity Dam in Severe Cold Region

1. Strengthen superficial thermal insulation

   Superficial thermal insulation is the most important measure for preventing cracking of gravity dams. The simulating computation by FEM shows that if the upstream and downstream surfaces of Guanyinge dam are insulated by foamed polystyrene plate of thickness 17 cm, the maximum thermal tensile stresses in the winter are reduced to 1.80 and 1.30 MPa, respectively, which will be smaller due to the effect of weight of concrete and no crack will appear.

2. Pipe heating

   It is suggested to set water pipes in the range of 2–3 m in the concrete below the horizontal surface at which pouring of concrete is stopped in winter. At 10–20 days before resuming the placing of concrete next year, 15–20°C water is used to increase the

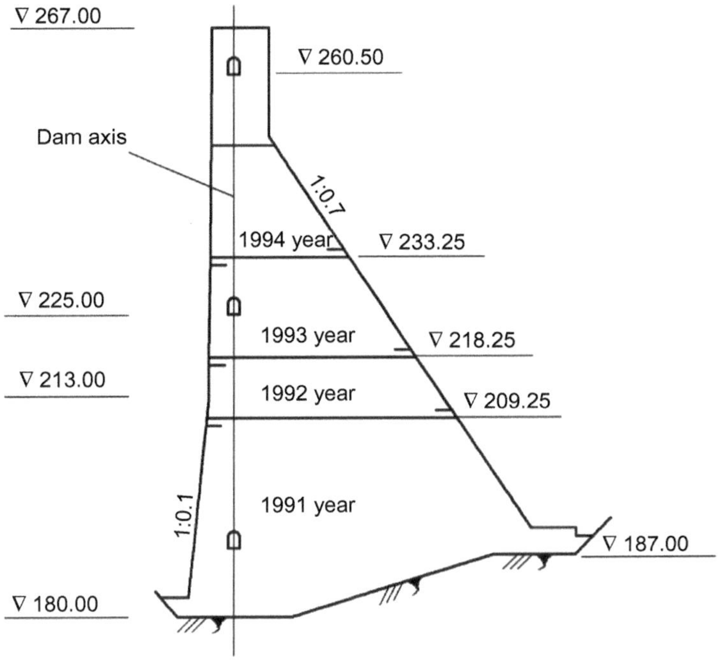

Figure 13.16 The horizontal cracks of Guanyinge RCC gravity dam.

temperature of old concrete to reduce the temperature difference between the upper and the lower parts of concrete.

3. Pipe cooling in the new concrete.
4. Preset artificial cracks.

In the Baishi RCC gravity dam with height 50.3 m, horizontal artificial cracks with water stop are preset on the upstream face (depth 1.0 m) and the downstream face (depth 3.0 m) near the horizontal surface at which concrete is stopped in winter to relieve the tensile stresses as shown in Figure 13.17. In the operation period of this dam, the preset cracks opened in the winter and no crack appeared.

13.8 Thermal Stresses due to Heightening of Gravity Dam

In the heightening of a gravity dam, the old part of the dam is fully cooled, while the temperature in the new concrete will be higher due to hydration heat of cement and high placing temperature of concrete in the hot season. There will be a temperature difference between the old and new concrete which will induce stresses both in the new and old concrete [65, 77, 100, 103].

The author had proposed two methods for computing the stresses due to heightening of a gravity dam, the first one is based on the formula in strength of materials and the second one is based on the theory of elasticity [62]. The first one will be introduced in the following.

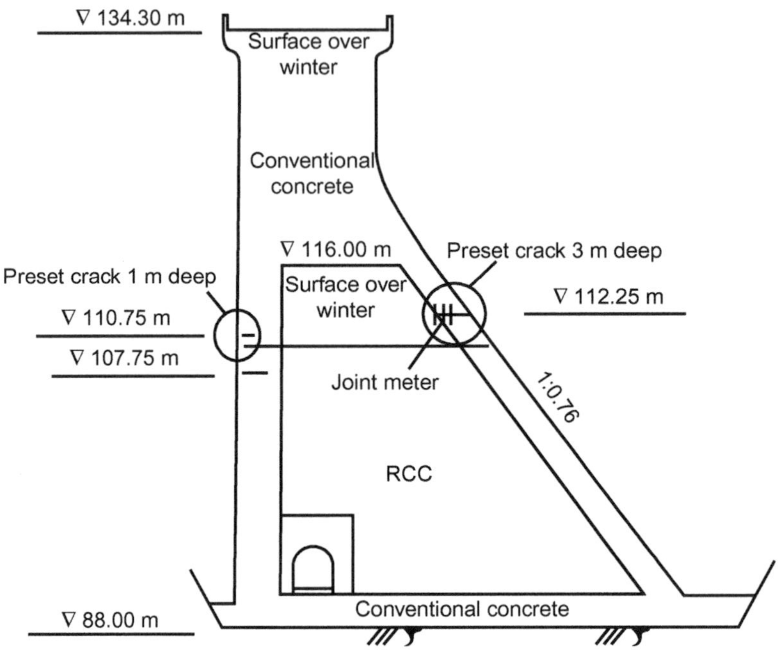

Figure 13.17 Preset artificial cracks in Baishi RCC gravity dam.

As shown in Figure 13.18, E_1 and E_2 are the modulus of elasticity of the old and new concrete respectively and b_1 and b_2 are the width of the old and new concrete. If the dam is fully restrained in the direction CD of the downstream face of the old dam and is free in the perpendicular direction, the initial stress in the dam will be

$$\left.\begin{aligned} \sigma_x^0 &= -E\alpha T \sin^2\beta \\ \sigma_y^0 &= -E\alpha T \cos^2\beta \\ \tau_{xy}^0 &= -E\alpha T \sin\beta\cos\beta \end{aligned}\right\} \tag{13.3}$$

The normal force N and moment M produced by the above initial stress are

$$N = -\cos^2\beta \int E\alpha T\,\mathrm{d}x, \quad M = -\cos^2\beta \int E\alpha Tx\,\mathrm{d}x \tag{13.4}$$

According to the hypothesis of plane section in the theory of strength of materials, the vertical strain on section 1-1 is a linear function of x as follows:

$$\varepsilon_y = A + Bx \tag{13.5}$$

where A and B are constants. The stress will be

$$\sigma_y = E_i\varepsilon_y = E_i(A + Bx) \tag{13.6}$$

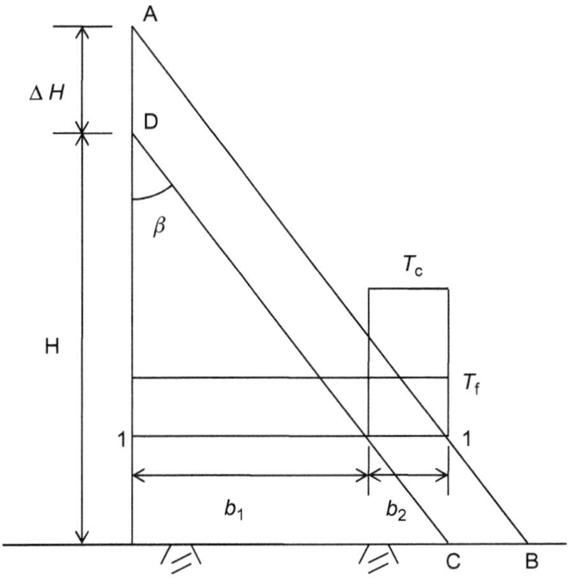

Figure 13.18 Example 1: Heightening of a gravity dam.

where $E_i = E_1$ for the old concrete and $E_i = E_2$ for the new concrete. The normal force N and the moment M will be

$$\begin{aligned} N &= \int \sigma_y \, dx = A\int E \, dx + B\int Ex \, dx \\ M &= \int \sigma_y x \, dx = A\int Ex \, dx + B\int Ex^2 \, dx \end{aligned} \tag{13.7}$$

Put the origin of coordinate at the centroid of the cross section with weight E, then

$$\int Ex \, dx = 0 \tag{13.8}$$

From Eqs. (13.6–13.9), we have

$$A = \frac{N}{D}, \quad B = \frac{M}{F} \tag{13.9}$$

where

$$\left.\begin{aligned} D &= \int E \, dx = E_1 b_1 + E_2 b_2 \\ F &= \int Ex^2 \, dx = \frac{E_1}{3}\left[(b_1 - b_0)^3 + b_0^3\right] + \frac{E_2}{3}\left[(b - b_0)^3 - (b_1 - b_0)^3\right] \end{aligned}\right\} \tag{13.10}$$

Now let the boundary of the dam be free from external restraint, the resulting normal stress on section 1-1 due to removal of N and M will be

$$\sigma_y = -E\left(\frac{N}{D} + \frac{M}{F}x\right) \tag{13.11}$$

Superposing the above stress with the initial stress given by Eq. (13.3), finally we get the stress due to heightening of the dam as follows:

$$\sigma_y = -E\left(\frac{N}{D} + \frac{M}{F}x\right) - E\alpha T \cos^2 \beta \tag{13.12}$$

From Eq. (13.8), the distance from the upstream face to the E weighted centroid O of the section is

$$b_0 = \frac{E_1 b_1^2/2 + E_2 b_2(b_1 + b_2/2)}{E_1 b_1 + E_2 b_2} \tag{13.13}$$

If $E_1 = E_2$, then

$$\begin{aligned} b_0 &= (b_1 + b_2)/2 = b/2 \\ D &= Eb, \ F = Eb^3/12 \end{aligned} \tag{13.14}$$

Example 1 As shown in Figure 13.19, the height of the old dam is 38.0 m, the base is $b_1 = 25.5$ m, the increment of dam height is $\Delta H = 5.0$ m, $b_2 = 3.36$ m, the uniform temperature difference in the new concrete is T_0, and $E_1 = E_2$, try to compute the thermal stress on the base section of the dam.

Now $b = b_1 + b_2 = 28.86$ m, $b_0 = b/2 = 14.43$ m, $\beta = 33.9°$, $\cos^2 \beta = 0.689$. From Eq. (13.4)

$$\begin{aligned} N &= -3.36 \times 0.689 E\alpha T_0 = -2.315 E\alpha T_0 \\ M &= N(b - b_0 - b_2/2) = -2.31 E\alpha T_0 (14.43 - 3.36/2) = -29.52 E\alpha T_0 \end{aligned}$$

From Eq. (13.12),

$$\begin{aligned} \sigma_y &= E\alpha T_0 \left(\frac{2.315}{28.86} + \frac{12 \times 29.52}{28.86^3}x\right) - 0.689 E\alpha T \\ &= E\alpha T_0 (0.0802 + 0.014736x) - 0.689 E\alpha T \end{aligned} \tag{13.15}$$

On the upstream face, $x = -14.43$ m, $T = 0$, from Eq. (13.15),

$$\sigma_y = E\alpha T_0 (0.0802 + 0.014736 \times (-14.43)) = -0.1324 E\alpha T_0$$

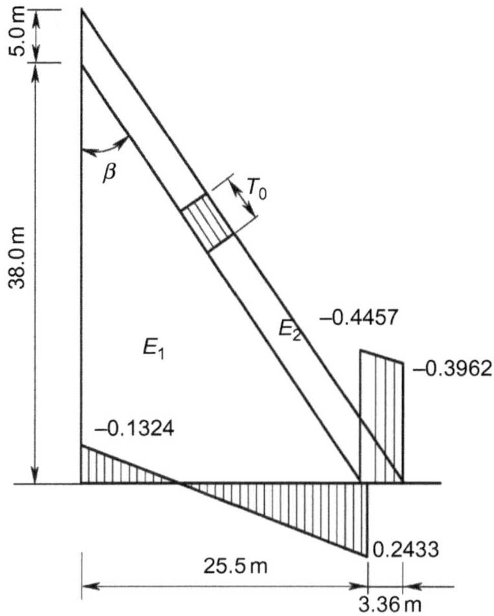

Figure 13.19 Example 1: Heightening of a gravity dam, $\sigma_y/E\alpha T_0$.

On the downstream face of old dam, $x = 14.43 - 3.36 = 11.07$ m, $T = 0$, from Eq. (13.15)

$$\sigma_y = E\alpha T_0(0.0802 + 0.014736 \times 11.07) = 0.2433T$$

On the upstream face of new concrete, $x = 11.07$, $T = T_0$, from Eq. (13.15)

$$\sigma_y = E\alpha T_0(0.0802 + 0.014736 \times 11.07) - 0.689E\alpha T_0 = -0.4457E\alpha T_0$$

On the downstream face of new concrete, $x = 14.43$, $T = T_0$, from Eq. (13.15)

$$\sigma_y = E\alpha T_0(0.0802 + 0.014736 \times 14.43) - 0.689E\alpha T_0 = -0.3962E\alpha T_0$$

The distribution of stress σ_y is shown in Figure 13.19.

Example 2 For the same dam, if $E_1 = 1.5E_2$, from Eqs. (13.13) and (13.10),

$$b_0 = \frac{1.5 \times 1/2 \times 25.5^2 + 3.36(25.5 + 3.36/2)}{1.5 \times 25.5 + 3.36} = 13.915 \text{ m}$$

$$D = 41.61E_2, \quad F = 2719.2E_2$$

$$\sigma_y = E\alpha T_0(0.05563 + 0.01129x) - E\alpha T \cos^2 \beta$$

For the old concrete, $E = 1.5E_2$, $T_0 = 0$, and for the new concrete, $E = E_2$, $T = T_0$.

The vertical stress at the upstream face, the downstream face of the old concrete and new concrete are respectively:

$$\sigma_{y1} = -0.1015E_1\alpha T_0 = -0.1523E_2\alpha T_0$$
$$\sigma_{y2} = 0.1864E_1\alpha T_0 = 0.2797E_2\alpha T_0$$
$$\sigma_{y3} = -0.5025E_2\alpha T_0$$
$$\sigma_{y4} = -0.4646E_2\alpha T_0$$

13.9 Technical Measures to Reduce the Thermal Stress due to Heightening of Gravity Dam

From Figure 13.18, it is evident that the thermal stresses due to heightening of a gravity dam are induced by the temperature difference

$$\Delta T = T_c - T_f \tag{13.16}$$

where T_f is the final steady temperature and T_c is the closing temperature which is the mean temperature of the new concrete at the time of closing of dam top or at the time when the height of new concrete increases to the height of the old dam block. There will be no thermal stress if $\Delta T = 0$; thus the criterion of temperature control for heightening of a gravity dam is

$$T_c \leq T_f \tag{13.17}$$

Referring to Figure 13.20, in the process of heightening a gravity dam, the new concrete is poured by lifts with thickness of 1.5–3.0 m. The variations of

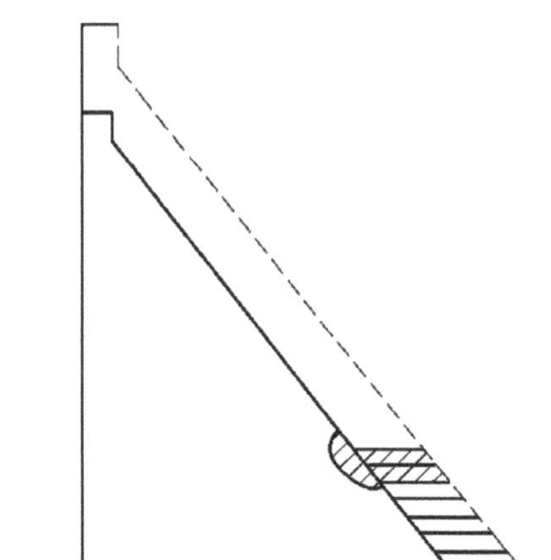

Figure 13.20 Heightening of a gravity dam.

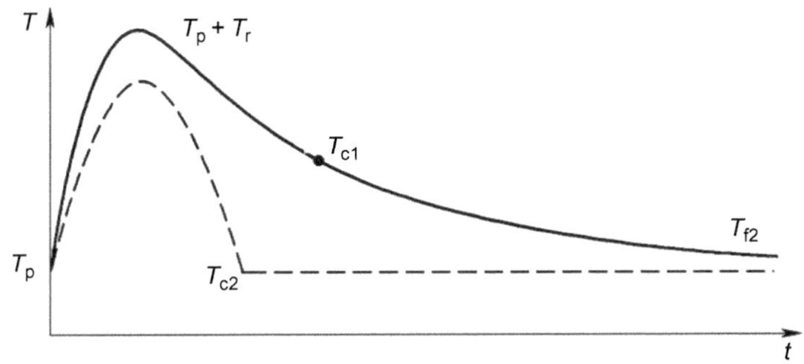

Figure 13.21 The variation of temperature in the new concrete.

temperature in the new concrete are shown in Figure 13.21. In the case of natural cooling, the temperature decreases very slowly and the closing temperature T_{c1} will be higher than the steady temperature T_f. But in the case of artificial cooling, primarily pipe cooling, the temperature in the new concrete may be reduced to $T_{c2} \leq T_f$ in 30–60 days. The time required for cooling may be controlled by adjusting the spacing of pipes and the temperature of cooling water. Too fast cooling of dam concrete is unfavorable. It is suggested that the time of cooling is controlled as follows:

$$t = (3 - 5)\Delta t \qquad (13.18)$$

where t is the time of cooling and Δt is the time interval between concrete lifts. In this case, the temperature gradient in the vertical direction is small and the stresses induced by the difference between the maximum temperature and the closing temperature will be also small and confined in the vicinity of the new concrete and will not influence the stress of the dam as a whole.

14 Thermal Stresses in Concrete Arch Dams

The thermal stresses in the arch dam before the grouting of transverse joints are computed as concrete blocks; see Chapter 12 and the stresses in the dam after the grouting of joints must be computed as a whole body which will be described in this chapter [47, 60, 63, 67, 72, 74, 75, 88, 99, 105].

14.1 Introduction

14.1.1 Self-Thermal Stresses of Arch Dam

Taking the coordinate system (x,y,z) as shown in Figure 14.1, the self-induced thermal stresses are computed in the following:

$$\begin{aligned} \sigma_y = \sigma_z &= \frac{E\alpha}{1-\mu}\left(T_{\mathrm{m}} + T_{\mathrm{d}}\frac{x}{L} - T\right) = \frac{E\alpha T_n}{1-\mu} \\ T_{\mathrm{n}} &= T_{\mathrm{m}} + T_{\mathrm{d}}\frac{x}{L} - T \end{aligned} \tag{14.1}$$

where

$$\left.\begin{aligned} T_{\mathrm{m}} &= \frac{1}{L}\int_{-L/2}^{L/2} T\,\mathrm{d}x \\ T_{\mathrm{d}} &= \frac{12}{L^2}\int_{-L/2}^{L/2} xT\,\mathrm{d}x \end{aligned}\right\} \tag{14.2}$$

where

T_{m}—mean temperature of the section
T_{d}—equivalent linear temperature difference
T_{n}—nonlinear temperature difference, namely, the difference between the actual temperature and the equivalent linear temperature.

The temperature distribution is shown in Figure 14.2.

Thermal Stresses and Temperature Control of Mass Concrete. DOI: http://dx.doi.org/10.1016/B978-0-12-407723-2.00014-2

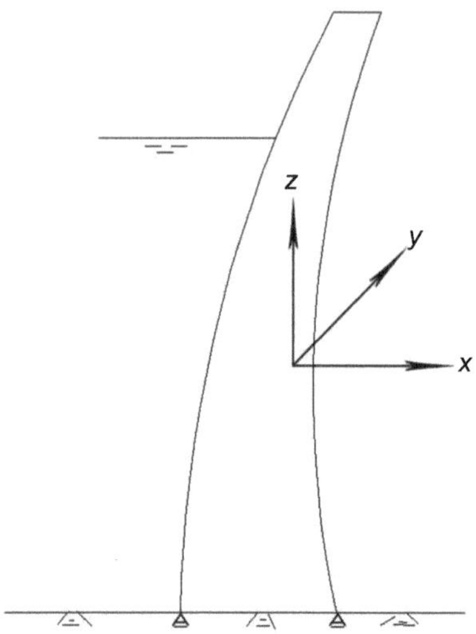

Figure 14.1 Arch dam.

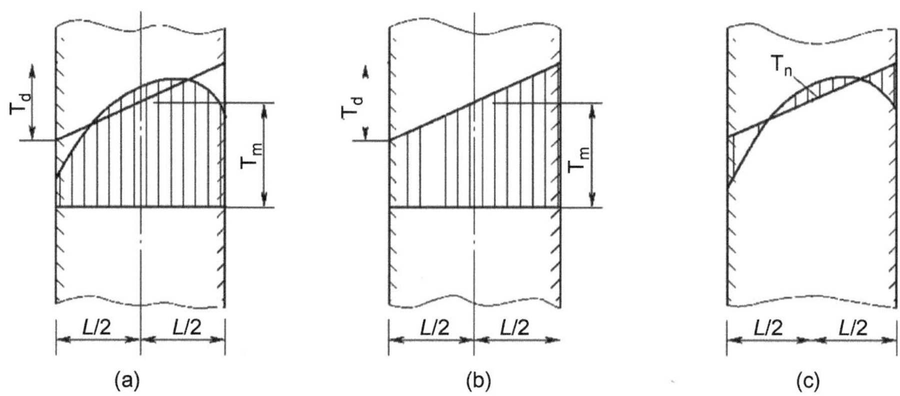

Figure 14.2 Temperature distribution in arch dam: (a) actual temperature, (b) equivalent temperature, and (c) difference between actual and equivalent temperature.

14.1.2 Three Characteristic Temperature Fields in Arch Dam

There are three important characteristic temperature fields in a concrete arch dam:

1. Grouting temperature $T_0(x)$

   Grouting temperature $T_0(x)$ is the temperature of the dam when the transverse joints are grouted. The mean temperature of $T_0(x)$ is T_{m0} and the equivalent linear temperature difference of $T_0(x)$ is T_{d0} which can be computed by Eq. (14.2).
2. Mean temperature of point x in the operation period $T_1(x)$

   $T_1(x)$ is the annual mean temperature at point x in the operation period. In the direction x of thickness, the mean temperature of $T_1(x)$ is T_{m1} and the equivalent temperature difference of $T_1(x)$ is T_{d1}. Generally the arch dam may be considered as a plate, the annual mean temperature of point is a linear function of x, so T_{m1} and T_{d1} may be computed as follows:

$$\left.\begin{aligned} T_{m1} &= \frac{1}{2}(T_{um} + T_{dm}) \\ T_{d1} &= T_{dm} - T_{um} \end{aligned}\right\} \tag{14.3}$$

where

T_{um}—annual mean temperature of the upstream face, generally it is equal to the annual mean temperature of water in the reservoir

T_{dm}—annual mean temperature of the downstream face, generally it is equal to the annual mean air temperature plus the influence of sunshine

3. Temperature variation $T_2(x, \tau)$ in operation period

$T_2(x, \tau)$ is the temperature variation at point x due to the temperature change of air and water. In the direction of thickness, the mean temperature of $T_2(x, \tau)$ is T_{m2} and the equivalent linear temperature difference is T_{d2}, of course both T_{m2} and T_{d2} are functions of time.

14.1.3 Temperature Loading on Arch Dams

According to the three characteristic temperature fields, it is suggested by the writer to compute the temperature loading on arch dams by the following equation (this suggestion is adopted in the design specifications of concrete arch dams in China) [36,43,64,75]:

$$\left.\begin{aligned} T_m &= T_{m1} + T_{m2} - T_{m0} \\ T_d &= T_{d1} + T_{d2} - T_{d0} \end{aligned}\right\} \tag{14.4}$$

where

T_m, T_d—temperature loadings on arch dam

T_{m0}, T_{d0}—the mean grouting temperature and the equivalent linear grouting temperature difference, respectively

T_{m1}, T_{d1}—the mean temperature and the equivalent linear temperature difference of the annual mean temperature $T_1(x)$ in the direction of thickness, respectively

T_{m2}, T_{d2}—the mean temperature and the equivalent linear temperature difference of temperature variation $T_2(x, \tau)$ in the direction of thickness.

From Eq. (14.4), it is apparent that the temperature loadings of an arch dam consist of two parts: the first part is $T_{m1} - T_{m0}$ and $T_{d1} - T_{d0}$ which do not vary with time and the second part is T_{m2} and T_{d2} which vary sinusoidally with time.

14.2 Temperature Loading on Arch Dam for Constant Water Level

For constant water level, T_{m1} and T_{d1} are given by Eq. (14.3). The methods for computing T_{m2} and T_{d2} are introduced in the following.

14.2.1 Formulas for T_{m2} and T_{d2}

The boundary condition for the temperature field of an arch dam is shown in Figure 14.3, the thickness of the computed section is L, the depth of water is y, and the downstream face is in contact with air. The temperature of the downstream face is expressed by

$$T_d = T_{dm} + A_d \cos \omega(\tau - \tau_0) \tag{14.5}$$

where

T_d—temperature of downstream face of the dam
T_{dm}—annual mean temperature of downstream face, which is equal to the annual mean air temperature plus the influence of sunshine
A_d—amplitude of annual variation of temperature of downstream face which is equal to the amplitude of annual variation of air temperature plus the influence of sunshine (about 1–2°C)
τ—time, in month
τ_0—the time for maximum air temperature, generally $\tau_0 = 6.5$ month
$\omega(=2\pi/P)$—circular frequency
$P(=12$ month)—period of variation of temperature.

For the upstream face of the dam, the temperature above the water level is equal to the air temperature and may be computed in the same way as Eq. (14.5); the temperature below the water level is equal to the temperature of water and may be expressed by the following equation:

$$T_u = T_{um} + A_u \cos \omega(\tau - \tau_0 - \varepsilon) \tag{14.6}$$

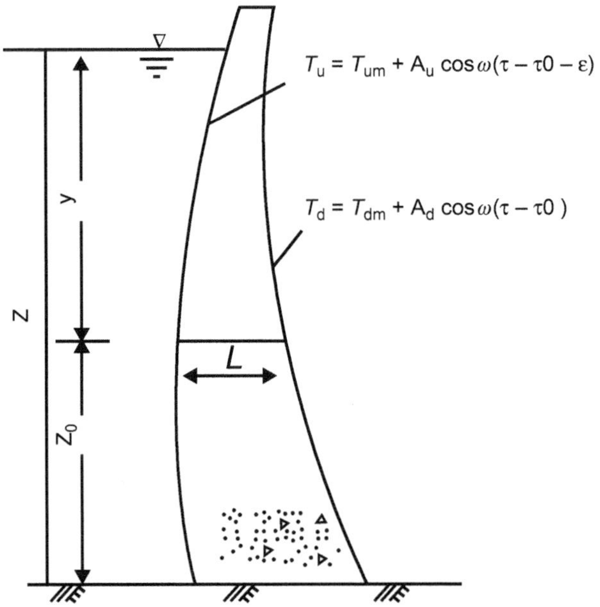

Figure 14.3 Boundary condition for temperature field of arch dam.

where

T_u—temperature of the upstream face of dam
T_{um}—annual mean temperature of upstream face
A_u—amplitude of annual variation of temperature of upstream face
ε—phase difference, in month, the difference between the time for the maximum temperature of upstream face and that of downstream face.

From the above boundary condition and Eq. (5.11), T_{m2} and T_{d2} are given by the following formulas:

$$\left.\begin{aligned} T_{m2} &= \frac{\rho_1}{2}[A_D \cos\omega(\tau-\tau_0-\theta_1) + A_U \cos\omega(\tau-\tau_0-\varepsilon-\theta_1)] \\ T_{d2} &= \rho_2[A_D \cos\omega(\tau-\tau_0-\theta_2) - A_U \cos\omega(\tau-\tau_0-\varepsilon-\theta_2)] \end{aligned}\right\} \tag{14.7}$$

$$\left.\begin{aligned} \rho_1 &= \frac{1}{\eta}\sqrt{\frac{2(\text{ch}\,\eta-\cos\eta)}{\text{ch}\,\eta+\cos\eta}} \\ \rho_2 &= \sqrt{a_1^2+b_1^2} \\ \theta_1 &= \frac{1}{\omega}\left[\frac{\pi}{4}-\tan^{-1}\left(\frac{\sin\eta}{\text{sh}\,\eta}\right)\right] \\ \theta_2 &= \frac{1}{\omega}\tan^{-1}(b_1/a_1) \\ a_1 &= \frac{6}{\rho_1\eta^2}\sin\omega\theta_1 \\ b_1 &= \frac{6}{\eta^2}\left(\frac{1}{\rho_1}\cos\omega\theta_1-1\right) \\ \eta &= \sqrt{\frac{\pi}{aP}}L,\quad \omega=\frac{2\pi}{P} \end{aligned}\right\} \tag{14.8}$$

The values of A_v, A_D, and ε are given in Section 2.5.

14.2.2 Physical Meaning of the Equivalent Linear Temperature

As shown in Figure 14.2, the equivalent linear temperature is

$$T_e(x) = T_m + \frac{T_d}{L}\cdot x \tag{14.9}$$

$T_e(x)$ is not the true temperature and is a fictitious temperature the mechanical effect of which is equal to the true temperature $T(x)$. When the structure is free of external restraint, the deformations induced by the equivalent temperature $T_e(x)$ are equal to those induced by the true temperature $T(x)$. If the structure is fully restrained, the internal forces aroused by $T_e(x)$ are equal to those aroused by the true temperature.

Example The thickness of the dam $L = 30$ m, diffusivity $a = 3.0$ m²/month, period $P = 12$ month, the upstream face is the bottom of the reservoir, the amplitude of variation of the temperature of the upstream face $A_u = 0$ and that of the downstream face $A_d = 20°C$. Try to compute the equivalent linear temperature of the dam due to the annual variation of the temperature of the downstream face $T_d = 20 \cos \omega(\tau - 6.50)°C$.

The computed results are shown in Figure 14.4.

14.3 Temperature Loading on Arch Dam for Variable Water Level

14.3.1 Computation of Surface Temperature of Dam for Variable Water Level

The water temperature is dependent on the depth y of the water; from Figure 14.3, it is clear that

$$y = z - z_0 \tag{14.10}$$

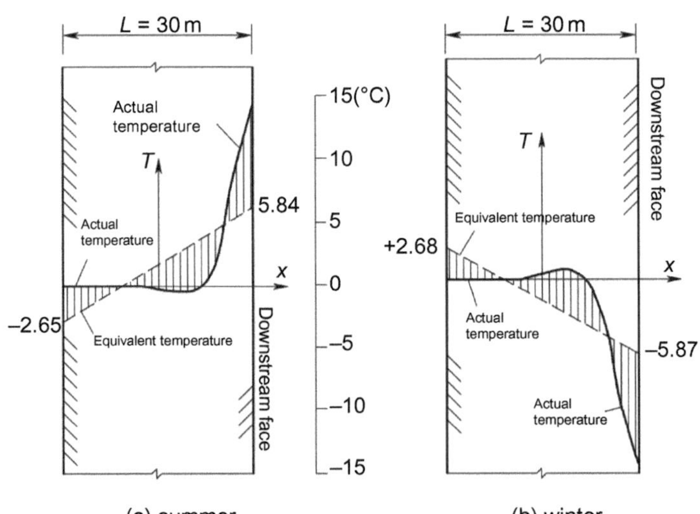

Figure 14.4 Example: The equivalent linear temperature and the actual temperature: (a) summer and (b) winter.

where

y—the depth of water
z—the elevation of water surface
z_0—the elevation of the computed section.

If $z < z_0$, the water surface is below the computed section, the temperature of the upstream face $T_u(\tau)$ is equal to air temperature $T_a(\tau)$ and when $z \geq z_0$, $T_u(\tau)$ is equal to the water temperature T_w; thus the temperature of the upstream face of the dam is given by the following equation:

$$\left.\begin{array}{lll} \text{When} & z - z_0 \geq 0, & T_u(\tau) = T_w(z - z_0, \tau) \\ & z < z_0, & T_u(\tau) = T_a(\tau) \end{array}\right\} \tag{14.11}$$

The temperature of the upstream face of one dam at elevation 570 m and the elevation of water surface are shown in Figure 14.5, in which T_{u1} is the temperature computed by Eq. (14.11) with the water surface elevation varying with time and T_{u2} is the temperature of the upstream face if the elevation of water surface is assumed to be fixed at 600 m. It is clear that T_{u1} is higher than T_{u2}, because from March to July the upstream face is in contact with air with higher temperature.

Assuming that the temperature of the upstream face varies periodically with time, it may be expressed by Fourier series as follows:

$$T_u(\tau) = T_{um} + \sum_{i=1}^{n} B_n \cos\left[\frac{2n\pi}{P}(\tau - \tau_0 - \varepsilon)\right] + \sum_{i=1}^{n} C_n \sin\left[\frac{2n\pi}{P}(\tau - \tau_0 - \varepsilon)\right] \tag{14.12}$$

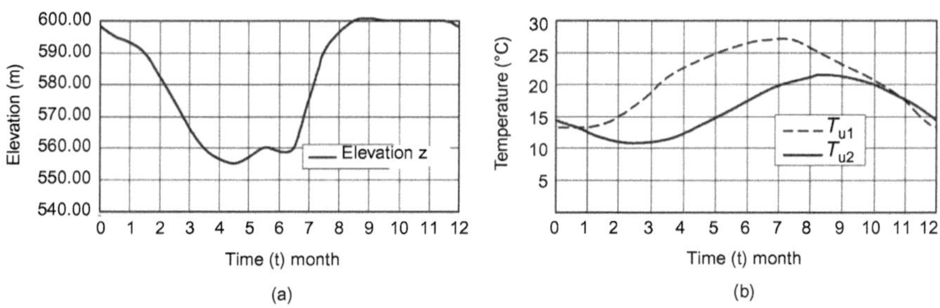

Figure 14.5 The elevation of water surface and the temperature of the upstream face of dam: (a) z—elevation of upstream water surface, (b) T_{u1}, T_{u2}—temperature of upstream face at elevation 570 m.

where

$$\left.\begin{aligned} B_n &= \frac{2}{P}\int T\cos\left[\frac{2n\pi}{P}(\tau-\tau_0-\varepsilon)\right]\mathrm{d}\tau \\ C_n &= \frac{2}{P}\int T\sin\left[\frac{2n\pi}{P}(\tau-\tau_0-\varepsilon)\right]\mathrm{d}\tau \end{aligned}\right\} \tag{14.13}$$

By means of the relation

$$\left.\begin{aligned} a\cos\phi \pm b\sin\phi &= \rho\cos(\phi \mp \theta) \\ \rho = \sqrt{a^2+b^2},\quad \theta &= \tan^{-1}(b/a) \end{aligned}\right\} \tag{14.14}$$

Equation (14.12) may be transformed into the following equation:

$$\left.\begin{aligned} T_{\mathrm{u}}(\tau) &= T_{\mathrm{um}} + \sum_{n=1}^{8} A_{\mathrm{un}}\cos\left[\frac{2n\pi}{P}(\tau-\tau_0-\varepsilon-\theta_{\mathrm{un}})\right] \\ A_{\mathrm{un}} &= \sqrt{B_{\mathrm{un}}^2 + C_{\mathrm{un}}^2},\quad \theta_{\mathrm{un}} = \frac{P}{2n\pi}\tan^{-1}(C_{\mathrm{un}}/B_{\mathrm{un}}) \end{aligned}\right\} \tag{14.15}$$

The temperature of the downstream face may be expressed by

$$T_{\mathrm{d}}(\tau) = T_{\mathrm{dm}} + \sum_{n=1}^{8} A_{\mathrm{d}n}\cos\left[\frac{2n\pi}{P}(\tau-\tau_0-\theta_{\mathrm{d}n})\right] \tag{14.16}$$

14.3.2 Temperature Loading on Arch Dam for Variable Water Level

When the upstream surface temperature T_{u} and the downstream surface temperature T_{d} are expressed by Eqs. (14.15) and (14.16), the temperature loading on arch dam $T_{\mathrm{m}2}$ and $T_{\mathrm{d}2}$ may be computed as follows:

$$T_{\mathrm{m}2} = \sum_{n=1}^{\infty} \frac{\rho_{1n}}{2}\{A_{\mathrm{d}n}\cos[\omega_n(\tau-\tau_0-\theta_{1n})] + A_{\mathrm{u}n}\cos[\omega_n(\tau-\tau_0-\varepsilon-\theta_{1n})]\} \tag{14.17}$$

$$T_{\mathrm{d}2} = \sum_{n=1}^{\infty} \rho_{2n}\{A_{\mathrm{d}n}\cos[\omega_n(\tau-\tau_0-\theta_{2n})] - A_{\mathrm{u}n}\cos[\omega_n(\tau-\tau_0-\varepsilon-\theta_{2n})]\} \tag{14.18}$$

where

$$\left.\begin{aligned}
\rho_{1n} &= \frac{1}{\eta_n}\sqrt{\frac{2(\text{ch}\,\eta_n - \cos\eta_n)}{\text{ch}\,\eta_n + \cos\eta_n}} \\
\rho_{2n} &= \sqrt{a_n^2 + b_n^2} \\
\theta_{1n} &= \frac{1}{\omega_n}\left[\frac{\pi}{4} - \tan^{-1}\left(\frac{\sin\eta_n}{\text{sh}\,\eta_n}\right)\right] \\
\theta_{2n} &= \frac{1}{\omega_n}\tan^{-1}(b_n/a_n) \\
a_n &= \frac{6}{\rho_{1n}\eta_n^2}\sin(\omega_n\theta_{1n}) \\
b_n &= \frac{6}{\eta_n^2}\left[\frac{1}{\rho_{1n}}\cos(\omega_n\theta_{1n}) - 1\right] \\
\eta_n &= \sqrt{\frac{n\pi}{aP}}L, \quad \omega_n = \frac{2n\pi}{P}
\end{aligned}\right\} \tag{14.19}$$

Taking $P = 12$ month, then $\omega_n = n\pi/6$.

T_{m2} and T_{d2} are respectively the mean temperature and the equivalent linear temperature difference of the dam.

Example 1 One arch dam, the maximum dam height $H = 278$ m, the elevation of normal water surface is 600 m, the limited elevation of water surface in flood period is 560 m, the annual mean air temperature is 19.7°C, and the amplitude of annual variation of air temperature is 8.05°C.

Due to sunshine, the increment of the mean temperature of the downstream face is $\Delta T_m = 2.2$°C, the increment of the amplitude of variation of temperature $\Delta A = 0.50$°C, the diffusivity of concrete $a = 2.05$ m^2/month, and the thickness of the dam $L = 33.7$ m. The temperature loadings of elevation 570 m are computed as follows:

1. The water temperature $T_w(y, \tau)$ is computed by the one-dimensional numerical model.
2. The upstream face temperature T_u at elevation 570 m is computed by Eq. (14.11).
3. The upstream face temperature T_u is expressed by Fourier series Eq. (14.12). $T_u(1)$, $T_u(3)$, and $T_u(6)$ given in Table 14.1 are the computed temperatures when one term, three terms, and six terms are taken in Eq. (14.12).
4. Assuming that the closing temperature $T_{m0} = 14.7$°C and $T_{d0} = 0$, the temperature loadings are given by Eqs. (14.17) and (14.18) which are varying with time. The temperature loadings at February and August are the maximum and the minimum are given in Table 14.2.

Table 14.1 Upstream Surface Temperature T_u and Temperatures Given by Fourier Series $T_u(1)$, $T_u(3)$, $T_u(6)$ (°C)

τ (month)		0.5	1.5	2.5	3.5	4.5	5.5	6.5	7.5	8.5	9.5	10.5	11.5
Surface temperature T_u		13.20	13.80	16.40	21.10	23.90	25.80	27.10	27.10	24.50	22.00	19.40	15.00
By series	T_u (1)	13.93	14.41	16.59	19.90	23.43	26.26	27.62	27.14	24.96	21.65	18.11	15.29
	T_u (3)	13.04	13.52	16.86	20.84	23.85	25.95	27.16	26.78	24.79	22.06	18.96	15.50
	T_u (6)	13.18	13.83	16.38	21.13	23.88	25.83	27.07	27.12	24.47	22.02	19.38	15.02

Table 14.2 Example 1: Temperature Loading T_m and T_d ($L = 33.7$ m)

Elevation of Water Surface	Method for Computing Water Temperature	T_{m1}	T_{m0}	T_{m2}		T_m		T_{d1}	T_{d0}	T_{d2}		T_d	
				Feb.	Aug.	Feb.	Aug.			Feb.	Aug.	Feb.	Aug.
Constant	Simple fine, 6 terms in Eq. (14.12)	17.78	14.7	−0.68	0.68	2.40	3.76	8.24	0	−1.99	1.99	6.25	10.23
		21.36	14.7	−0.95	0.92	5..71	7.58	1.17	0	−0.40	0.72	0.77	1.89
Variable	Simple, 1 term in Eq. (14.12)	21.08	14.7	−0.92	0.92	5.47	7.30	1.73	0	−0.45	0.45	1.28	2.18

From Table 14.2, it is clear that: (1) the temperature loadings computed for constant elevation of water surface are far different from those computed for variable elevation of water surface; (2) the temperature loadings computed by Eq. (14.12) with six terms are close to those given by Eq. (14.12) with one term.

Example 2 The fundamental data are the same as Example 1, but $L = 5$ m, $a = 3.0$ m^2/month, and the computed results are given in Table 14.3.

14.4 Temperature Loadings on Arch Dams in Cold Region with Superficial Thermal Insulation Layer [64, 75]

For arch dams in cold region, the influence of temperature loading on the stresses and internal forces is higher than water pressure. In order to reduce the thermal stress, it is suggested to adopt superficial thermal insulation layer for thin arch dams in cold region.

The boundary condition of temperature field of arch dam with superficial thermal insulation layer is shown in Figure 14.6. The temperature of the external surface of the downstream thermal insulation layer is

$$T_D = T_{Dm} + A_D \cos \omega(\tau - \tau_0) \tag{14.20}$$

where

T_{Dm}—annual mean temperature
A_D—the amplitude of variation of temperature
$\omega = 2\pi/P$—the circular frequency
$P = 12$ month—period of variation of temperature
τ_0—the time of the maximum temperature, generally $\tau_0 = 6.5$ month.

The temperature of the external surface of the upstream thermal insulation layer is

$$T_U = T_{Um}(y) + A_U(y)\cos \omega(\tau - \tau_0 - \varepsilon) \tag{14.21}$$

where

$T_{Um}(y)$—the annual mean temperature of the water
$A_U(y)$—the amplitude of annual variation of water temperature
ε—the phase difference between temperature of water and air
y—depth of water.

The cross section with thickness L and depth of water y is analyzed in the following.

14.4.1 T_{m1} and T_{d1} for the Annual Mean Temperature Field $T_1(x)$

T_{m1} and T_{d1} are respectively the average temperature and the equivalent linear temperature difference along the thickness of the annual mean temperature $T_1(x)$ of the

Table 14.3 Example 2: Temperature Loading T_m and T_d ($L = 5$ m)

Elevation of Water Surface	Method for Computing Water Temperature	T_{m1}	T_{m0}	T_{m2}		T_m		T_{d1}	T_{d0}	T_{d2}		T_d	
				Feb.	Aug.	Feb.	Aug.			Feb.	Aug.	Feb.	Aug.
Constant	Simple	17.78	14.7	5.73	5.73	−2.65	8.81	8.24	0	−1.94	1.94	6.30	10.18
	fine, 6 terms in Eq. (14.12)	21.36	14.7	−6.91	6.69	−0.25	13.35	1.17	0	−0.25	1.38	0.92	2.55
Variable	Simple, 1 term in (14.12)	Eq. 21.08	14.7	−6.86	6.86	−0.48	13.24	1.73	0	−0.17	0.17	1.56	1.90

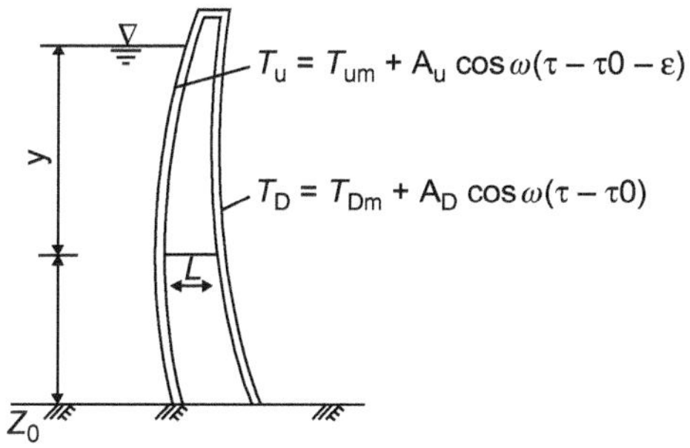

Figure 14.6 Boundary condition for temperature field of arch dam with superficial thermal insulation layer.

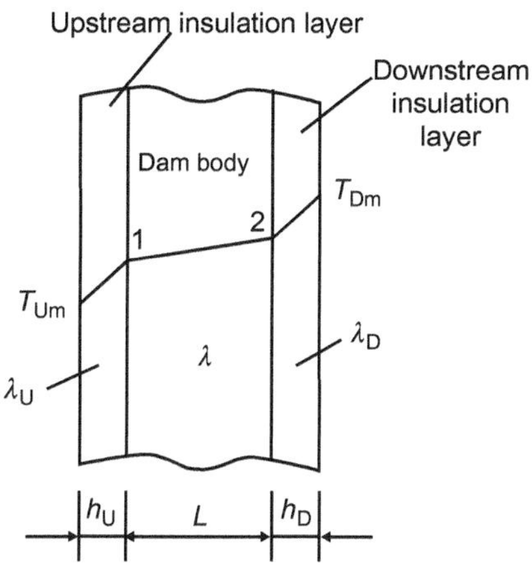

Figure 14.7 The annual mean temperature $T_1(x)$.

dam which is a function of x, as shown in Figure 14.7. The thickness of the dam body is L and the conductivity is λ; the thickness and conductivity of the upstream insulation layer are h_u and λ_u; the thickness and conductivity of the downstream insulation layer are h_D and λ_D. The distribution of the annual mean temperature $T_1(x)$ in the plate consisting of three layers is analyzed in the following. The upstream face annual mean temperature T_{Um} and the downstream face annual mean temperature T_{Dm} are constant, i.e., they do not vary with time, so the annual mean temperature $T_1(x)$ of any point x does not vary with time, thus

$$\frac{\partial T_1}{\partial \tau} = a\frac{\partial^2 T_1}{\partial x^2} = 0$$

Hence, $T_1(x)$ is a linear function of x in each one of the three layers, but on the plane of contact of two layers, the following conditions of balance of heat flow must be satisfied:

On the contact plane 1:

$$\lambda_U \cdot \frac{T_1 - T_{Um}}{h_U} = \lambda \cdot \frac{T_2 - T_1}{L} \tag{14.22}$$

On the contact plane 2:

$$\lambda \cdot \frac{T_2 - T_1}{L} = \lambda_D \cdot \frac{T_{Dm} - T_2}{h_D} \tag{14.23}$$

From the above two equations, we get

$$T_1 = \frac{(1 + \rho_1\rho_2)T_{Um}}{1 + \rho_1 + \rho_1\rho_2} + \frac{\rho_1 T_{Dm}}{1 + \rho_1 + \rho_1\rho_2} \tag{14.24}$$

$$T_2 = \frac{T_{Um}}{1 + \rho_1 + \rho_1\rho_2} + \frac{(\rho_1 + \rho_1\rho_2)T_{Dm}}{1 + \rho_1 + \rho_1\rho_2} \tag{14.25}$$

where

$$\rho_1 = \frac{\lambda_D h_U}{\lambda_U h_D}, \quad \rho_2 = \frac{\lambda_U L}{\lambda h_U} \tag{14.26}$$

T_1 and T_2 are the temperatures at the upstream and downstream face of the dam body respectively and the temperature is a linear function of x, so

$$T_{m1} = \frac{T_1 + T_2}{2}, \quad T_{d1} = T_2 - T_1 \tag{14.27}$$

Example 1 $T_{Um} = 6.0°C$, $T_{Dm} = 4.2°C$, $L = 5.00$ m, $\lambda = 9.0$ kJ/(m h °C), $h_D = 0.05$ m, $h_U = 0.03$ m, $\lambda_U = \lambda_D = 0.1256$ kJ/(m h °C).

From Eq. (14.26),

$$\rho_1 = 0.600, \quad \rho_2 = 2.326$$

From Eqs. (14.24) and (14.25),

$$T_1 = 5.64°C, \quad T_2 = 4.80°C$$

From Eq. (14.27),

$$T_{m1} = 5.22°C, \quad T_{d1} = -0.84°C$$

14.4.2 Exact Solution of T_{m2} and T_{d2} for the Yearly Varying Temperature Field $T_2(x,\tau)$

In order to simplify the solution, the effect of the superficial thermal insulation layer is considered by the equivalent surface conductance as follows:

$$\beta_1 = \frac{1}{1/\beta_0 + h_U/\lambda_U}, \quad \beta_2 = \frac{1}{1/\beta_0 + h_D/\lambda_D}$$

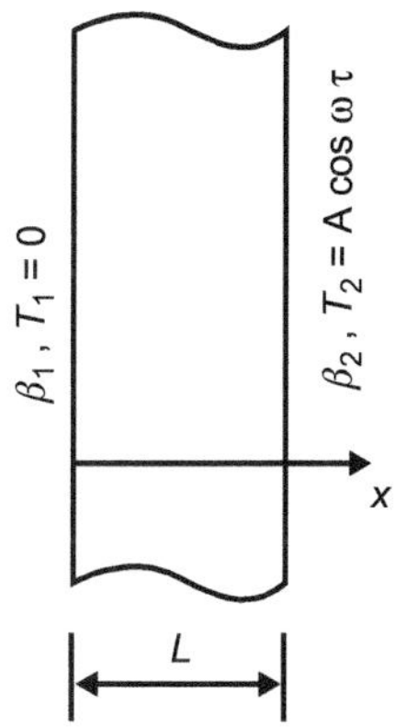

Figure 14.8 Plate.

where

β_0—the surface conductance between the insulation layer and water or air
β_1 and β_2—the equivalent surface conductance of the upstream and downstream face, respectively.

First, consider the plate shown in Figure 14.8, the equation of heat conduction is

$$\frac{\partial T}{\partial \tau} = a\frac{\partial^2 T}{\partial x^2} \tag{14.28}$$

The boundary condition:

$$\text{When} \quad x = 0, \quad \lambda\frac{\partial T}{\partial x} = \beta_1(T - 0) \tag{14.29}$$

$$\text{When} \quad x = L, \quad -\lambda\frac{\partial T}{\partial x} = \beta_2(T - A\cos\omega\tau) \tag{14.30}$$

The problem is solved by the method of complex variable, let

$$T(x, \tau) = \mathrm{Re}[T^0(x, \tau)] \tag{14.31}$$

$$T^0(x, \tau) = [f_1(x) + \mathrm{i}f_2(x)]e^{\mathrm{i}\omega\tau} \tag{14.32}$$

where Re is the real part, $\mathrm{i} = \sqrt{-1}$. By substituting Eq. (14.31) into Eqs. (14.28)–(14.30), $T^0(x, \tau)$ is obtained. Taking the real part, we get

$$T(x, \tau) = A[g_1(qx)\cos\omega\tau + g_2(qx)\sin\omega\tau] \tag{14.33}$$

$$T_{\mathrm{m}} = \frac{1}{L}\int_0^L T\,\mathrm{d}x = \frac{A}{L}(g_3\cos\omega\tau + g_4\sin\omega\tau) \tag{14.34}$$

$$S_0 = \int_0^L Tx\,\mathrm{d}x = A(g_5 \cos \omega\tau + g_6 \sin \omega\tau) \tag{14.35}$$

$$T_\mathrm{d} = \frac{12}{L^2}\left(S_0 - \frac{L^2}{2}T_\mathrm{m}\right) = \frac{12A}{L^2}\left[\left(g_5 - \frac{L}{2}g_3\right)\cos \omega\tau + \left(g_6 - \frac{L}{2}g_4\right)\sin \omega\tau\right] \tag{14.36}$$

where

$$\left.\begin{aligned}
g_1 &= (a_1b_1 + a_2b_2)/(a_1^2 + a_2^2),\quad g_2 = (a_2b_1 - a_1b_2)/(a_1^2 + a_2^2)\\
g_3 &= (a_1a_3 + a_2a_4)/(a_1^2 + a_2^2),\quad g_4 = (a_2a_3 - a_1a_4)/(a_1^2 + a_2^2)\\
g_5 &= (a_1a_6 - a_2a_5)/[2q^2(a_1^2 + a_2^2)],\quad g_6 = (a_1a_5 + a_2a_6)/[2q^2(a_1^2 + a_2^2)]\\
a_1 &= d_1 - d_2s_4 + s_3(d_3 - d_4),\quad a_2 = d_1s_4 + d_2 + s_3(d_3 + d_4)\\
a_3 &= \frac{1}{2q}(2s_1d_1 + d_3 + d_4 - 1),\quad a_4 = \frac{1}{2q}(2s_1d_2 - d_3 + d_4 + 1)\\
a_5 &= s_1(-2\eta d_2 - d_3 + d_4 + 1) + \eta(d_3 - d_4) - d_1\\
a_6 &= s_1(2\eta d_1 - d_3 - d_4 + 1) + \eta(d_3 + d_4) - d_2\\
b_1(\zeta) &= s_1(f_1 - f_2) + f_3,\quad b_2(\zeta) = s_1(f_1 + f_2) + f_4\\
f_1(\zeta) &= \mathrm{ch}\,\zeta \cos \zeta,\quad f_2(\zeta) = \mathrm{sh}\,\zeta \sin \zeta\\
f_3(\zeta) &= \mathrm{sh}\,\zeta \cos \zeta,\quad f_4(\zeta) = \mathrm{ch}\,\zeta \sin \zeta\\
d_1 &= \mathrm{sh}\,\eta \cos \eta,\quad d_2 = \mathrm{ch}\,\eta \sin\eta,\quad d_3 = \mathrm{ch}\,\eta \cos \eta,\quad d_4 = \mathrm{sh}\,\eta \sin \eta\\
s_1 &= \lambda q/\beta_1,\quad s_2 = \lambda q/\beta_2,\quad s_3 = s_1 + s_2,\quad s_4 = 2s_1s_2\\
q &= \sqrt{\pi/aP},\quad \zeta = qx,\quad \eta = qL
\end{aligned}\right\} \tag{14.37}$$

Now, we consider the cross section of an arch dam as shown in Figure 14.9; for the upstream face, the temperature of water or air is $T = A_\mathrm{U} \cos \omega(\tau - \tau_0 - \varepsilon)$, the surface conductance is β_U; for the downstream face, the temperature of the air is

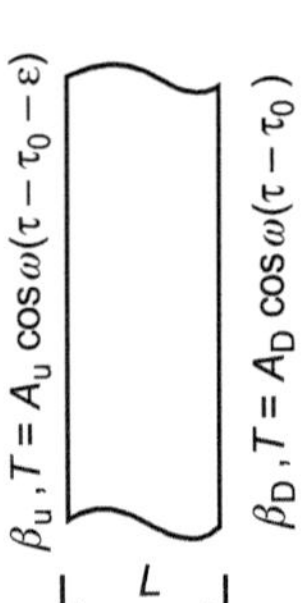

Figure 14.9 A cross section of an arch dam.

$T_D = A_D \cos \omega(\tau - \tau_0)$, the surface conductance is β_D. By means of Eqs. (14.34) and (14.36), we get

$$T_{m2} = k_{mD} A_D \cos \omega(\tau - \tau_0 - \theta_{mD}) + k_{mU} A_U \cos \omega(\tau - \tau_0 - \varepsilon - \theta_{mU}) \quad (14.38)$$

$$T_{d2} = k_{dD} A_D \cos \omega(\tau - \tau_0 - \theta_{dD}) - k_{dU} A_U \cos \omega(\tau - \tau_0 - \varepsilon - \theta_{dU}) \quad (14.39)$$

where

$$k_{mD} = k_{mU} = \frac{1}{L}\sqrt{g_3^2 + g_4^2} \quad (14.40)$$

$$\theta_{mD} = \theta_{mU} = \frac{1}{\omega}\tan^{-1}\left(\frac{g_4}{g_3}\right) \quad (14.41)$$

$$k_{dD} = k_{dU} = \frac{12}{L^2}\sqrt{(g_5 - g_3 L/2)^2 + (g_6 - g_4 L/2)^2} \quad (14.42)$$

$$\theta_{dD} = \theta_{dU} = \frac{1}{\omega}\tan^{-1}\left(\frac{g_6 - g_4 L/2}{g_5 - g_3 L/2}\right) \quad (14.43)$$

Referring to Figures 14.8 and 14.9, take β_1 and β_2 as follows:

$$\left.\begin{array}{ll} \text{For downstream face} & A = A_D,\ \beta_1 = \beta_U,\ \beta_2 = \beta_D \\ \text{Or upstream face} & A = A_U,\ \beta_1 = \beta_D,\ \beta_2 = \beta_U \end{array}\right\} \quad (14.44)$$

As $\beta_1 \neq \beta_2$, so $k_{mU} \neq k_{mD}$, $\theta_{mU} \neq \theta_{mD}$

Example 2 The variation of air temperature of the downstream face is $T_D = 19.3 \cos \omega(\tau - 6.50)$, the variation of the water temperature of the upstream face is $T_U = 6.17 \cos \omega(\tau - 6.50 - 1.64)$, i.e., $T_D = 19.30°C$, $T_U = 6.17°C$, $\varepsilon = 1.64$ month. $\beta_D = 2.437$ kJ/(m^2 h °C), $\beta_U = 4.185$ kJ/(m^2 h °C), $\omega = 2\pi/P = \pi/6$, $a = 3.0$ m^2/month, $q = \sqrt{\pi/aP} = 0.2954$, $\eta = qL = 1.477$, $\lambda = 9.00$ kJ/(m^2 h °C).

From Eqs. (14.37)–(14.44), we get

$$k_{mD} = 0.2254,\quad \theta_{mD} = 1.948 \text{ month},\quad k_{dD} = 0.3878,\quad \theta_{dD} = 0.662 \text{ month}$$

$$k_{mU} = 0.3034,\quad \theta_{mU} = 1.860 \text{ month},\quad k_{dU} = 0.5219,\quad \theta_U = 0.313 \text{ month}$$

$$T_{m2} = 4.35 \cos \omega(\tau - 8.45) + 1.87 \cos \omega(\tau - 10.0)$$

$$T_{d2} = 7.48 \cos \omega(\tau - 7.16) - 3.22 \cos(\tau - 8.45)$$

14.4.3 Approximate Solution of T_{m2} and T_{d2} for the Yearly Varying Temperature Field $T_2(x,\tau)$

As the above exact solutions of T_{m2} and T_{d2} are rather complicated, an approximate solution is given in the following. Consider the cross section of arch dam and the boundary conditions shown in Figure 14.9.

At first, by means of Eq. (3.8), the solution of the quasi-steady temperature field of semi-infinite solid with the third kind of boundary condition, the approximate surface temperatures of the dam body are given as follows:

On upstream face:

$$T_U = k_U A_U \cos\omega(\tau - \tau_0 - \varepsilon - \xi_U) \tag{14.45}$$

On downstream face:

$$T_D = k_D A_D \cos\omega(\tau - \tau_0 - \xi_D) \tag{14.46}$$

where

$$\left.\begin{aligned} k_U &= [1 + 2q\lambda/\beta_U + 2(q\lambda/\beta_U)^2]^{-1/2} \\ \xi_U &= \frac{1}{\omega}\tan^{-1}\left(\frac{1}{1+\beta_U/\lambda q}\right) \\ k_D &= [1 + 2q\lambda/\beta_D + 2(q\lambda/\beta_D)^2]^{-1/2} \\ \xi_D &= \frac{1}{\omega}\tan^{-1}\left(\frac{1}{1+\beta_D/\lambda q}\right) \end{aligned}\right\} \tag{14.47}$$

Then, for a plate with known surface temperature T_U and T_D, from Eqs. (7.30) and (7.31), the mean temperature T_{m2} and the equivalent linear temperature difference T_{d2} across the thickness of plate are given in the following:

$$T_{m2} = \frac{\rho_1}{2}\left[k_D A_D \cos\omega(\tau - \tau_0 + \xi_D - \theta_1) + k_U A_U \cos\omega(\tau - \tau_0 - \varepsilon + \xi_U - \theta_1)\right] \tag{14.48}$$

$$T_{d2} = \rho_2[k_D A_D \cos\omega(\tau - \tau_0 + \xi_D - \theta_2) - k_U A_U \cos\omega(\tau - \tau_0 - \varepsilon + \xi_U - \theta_2)] \tag{14.49}$$

where

$$\left.\begin{aligned}
\rho_1 &= \frac{1}{\eta}\sqrt{\frac{2(\text{ch}\ \eta - \cos\ \eta)}{\text{ch}\ \eta + \cos\ \eta}} \\
\rho_2 &= \sqrt{a_1^2 + b_1^2} \\
\theta_1 &= \frac{1}{\omega}\left[\frac{\pi}{4} - \tan^{-1}\left(\frac{\sin\ \eta}{\text{sh}\ \eta}\right)\right] \\
\theta_2 &= \frac{1}{\omega}\tan^{-1}(b_1/a_1) \\
a_1 &= \frac{6}{\rho_1\eta^2}\sin\ \omega\theta_1 \\
b_1 &= \frac{6}{\eta}\left(\frac{1}{\rho_1}\cos\ \omega\theta_1 - 1\right) \\
\eta &= \sqrt{\frac{\pi}{aP}}L = qL, \quad \omega = \frac{2\pi}{P}
\end{aligned}\right\} \tag{14.50}$$

Example 3 The basic data is the same as example 1, $\beta_U = 4.185$ kJ(m^2 h °C), $\beta_D = 2.437$ kJ(m^2 h °C), $\lambda = 9.00$ kJ(m^2 h °C), $L = 5.00$ m, $a = 3.0$ m^2/month, $q = \sqrt{\pi/aP} = 1.477$, $\lambda/\beta_U = 2.150$ m, $\lambda/\beta_D = 3.693$ m, $\xi_U = 0.9184$ month, $\xi_D = 0.7077$ month, from Eqs. (14.48) and (14.49), we get

$$\begin{aligned}
T_{m2} &= 3.76\cos\ \omega(\tau - 8.06) + 1.62\cos\ \omega(\tau - 9.49) \\
T_{d2} &= 8.14\cos\ \omega(\tau - 7.56) - 3.50\cos\ \omega(\tau - 8.98)
\end{aligned}$$

Experience shows that the precision of the approximate solution is good when $L \geq 5.0$ m and the precision is lower for a very thin dam.

14.5 Measures for Reducing Temperature Loadings of Arch Dam

Generally speaking, the temperature loadings will increase the stresses, especially the tensile stresses in the arch dams. From Eq. (14.4), for reducing temperature loadings, it is necessary to change $T_0(x)$, $T_1(x)$, and $T_2(x)$. There are two measures to reduce the temperature loadings, namely, optimizing the grouting temperature and superficial thermal insulation.

14.5.1 Optimizing Grouting Temperature

1. Control of the mean grouting temperature T_{m0}. It is favorable to control

$$T_{m0} \le T_{m1} \tag{14.51}$$

by piping cooling before the grouting of transverse joints of the dam.

2. Control of the equivalent linear temperature difference T_{d0} of grouting temperature.

The positive equivalent linear temperature difference $T_d > 0$ will induce tensile stress at the upstream face of arch at abutment, so it is necessary to reduce $T_d = T_{d1} + T_{d2} - T_{d0}$. It is more difficult to control T_{d1} and T_{d2} than T_{d0}, it is suggested to control T_{d0} such that

$$T_{d0} \ge 0 \text{ or } T_{d0} = T_{d1} \tag{14.52}$$

Generally the condition $T_{d0} = 0$ is adopted in practice.

14.5.2 Superficial Thermal Insulation [71, 72]

The superficial thermal insulation is the most effective measure for reducing T_{m2} and T_{d2}. The compound insulation plate consisting of foamed polymer layer and protective layer as shown in Figure 12.32 may be adopted.

Some reinforced concrete thin arch dams were constructed in the cold region in Norway. As shown in Figure 14.10, the reinforced concrete insulation walls with thickness 10–15 cm were set on the downstream face of the dam, the distance between the wall and the dam is 80 cm. Experience shows that the air temperature behind the wall may be kept above 0°C when the external air temperature is −30°C.

14.6 Temperature Control of RCC Arch Dams [47, 63]

14.6.1 RCC Arch Dams without Transverse Joint

The steady temperature T_f of the dam is high in the south part of China, it is possible to construct a small arch dam without transverse joint if the whole dam can be

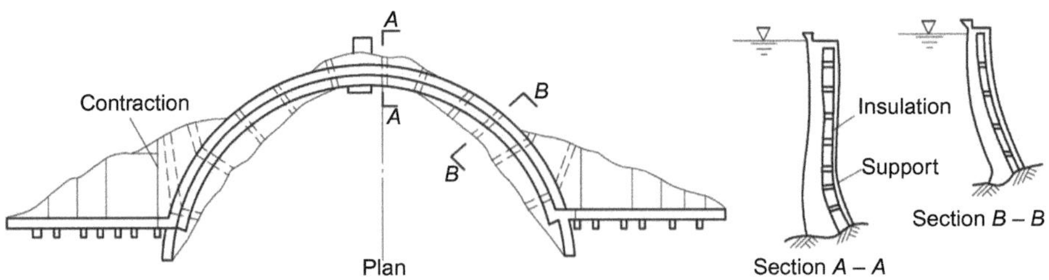

Figure 14.10 Thermal insulation wall of an arch dam in a cold region.

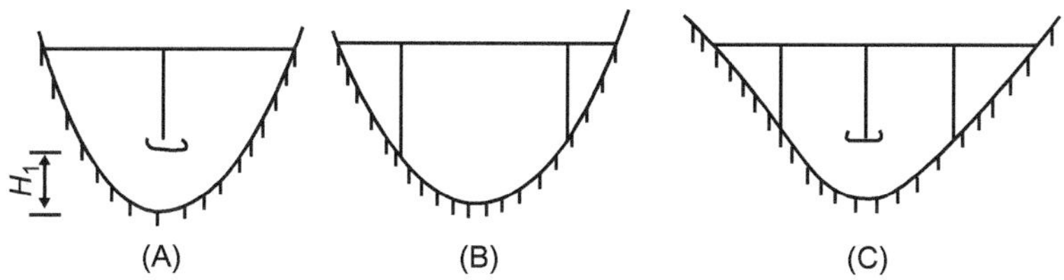

Figure 14.11 Transverse joints of RCC arch dams.

constructed in the months of low air temperature and the tensile stresses of the dam are within the allowable range.

14.6.2 RCC Arch Dam with Transverse Joints

If the whole dam cannot be constructed in the months of low temperature or the damsite is located in a cold region with very low steady temperature T_f, it is necessary to set transverse joints in the dam; otherwise too large tensile stresses will appear in the dam:

1. The principle of design of transverse joints of RCC arch dams
   As shown in Figure 14.11, the lower part of the dam, i.e., $y \leq H_1$, is constructed in the months of low temperature without transverse joint and in the higher part $y \geq H_1$ there are one to two transverse joints. As the temperature rise in RCC is less than that in conventional concrete, the distances between transverse joints in RCC arch dams are bigger than those in conventional concrete arch dams. The concrete in the dam must be cooled to the steady temperature before the grouting of the transverse joints.
2. Two types of transverse joint in RCC arch dams
   There are two types of transverse joint in RCC arch dams: (1) the common transverse joint which cuts the whole section of the dam and (2) the inducing joint which cuts only a part of the section of the dam.
   As shown in Figure 14.12, there are two conventional transverse joints and two inducing joints in Shapai RCC arch dam. The conventional transverse joint cuts the whole section of the dam and the inducing joint cuts only 20% of the section of the dam:
   a. Structure of inducing joint in Shapai RCC arch dam
   The inducing joint is constructed by a pair of gravity type concrete joint forms with length 1 m and height 0.30 m which is equal to the height of placing layer of RCC. In the process of construction of the dam, firstly the gravity type concrete joint forms are fixed on the surface of old concrete, and they are disconnected in two directions: in the horizontal direction, the length of form is 1.0 m and the interval is 0.5 m; in the vertical direction, the height of form is 0.30 m and the interval is 0.60 m (one form is set in every three layers). The height of joint grout region is 6.0 m and three independent grouting pipes are set in each region. Rubber sleeves are put on the grouting pipes for repeated grouting.
   b. Structure of transverse joint of Shapai RCC arch dam
   The transverse joint cuts the whole section of the dam. A series of pairs of gravity type concrete joint forms are set continuously in both the vertical and the horizontal directions.

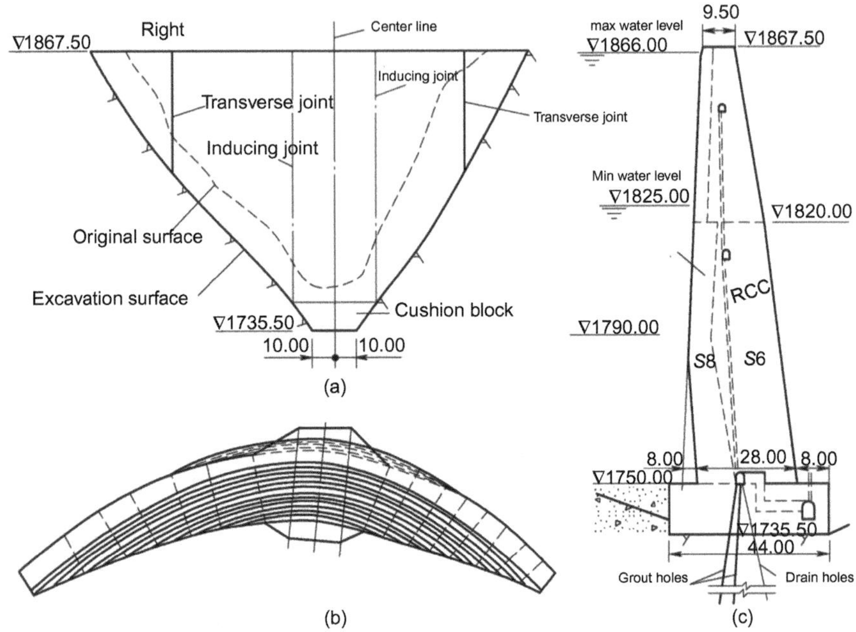

Figure 14.12 Shapai RCC arch dam: (a) downstream elevation, (b) plan, and (c) vertical cross section at the crown.

14.7 Observed Thermal Stresses and Deformations of Arch Dams

In Figure 14.13 are shown the observed deformations and the displacements computed from the practical water elevations and the temperature variations of Isolado arch dam. It is apparent that the temperature variations have remarkable influence on the deformations of arch dams.

The observed temperatures and stresses in the horizontal arch of Kimishiba arch dam are shown in Figure 14.14 from which it is clear that the distribution of temperature and stress are nonlinear and there are initial stresses before the impounding of water. Figure 14.15 shows the practical observed temperatures and stresses in the cantilevers of Kimishiba arch dam. Comparing (b) and (d), although the upstream water levels at both times are close, but due to the change of temperature, the distribution of stress is quite different.

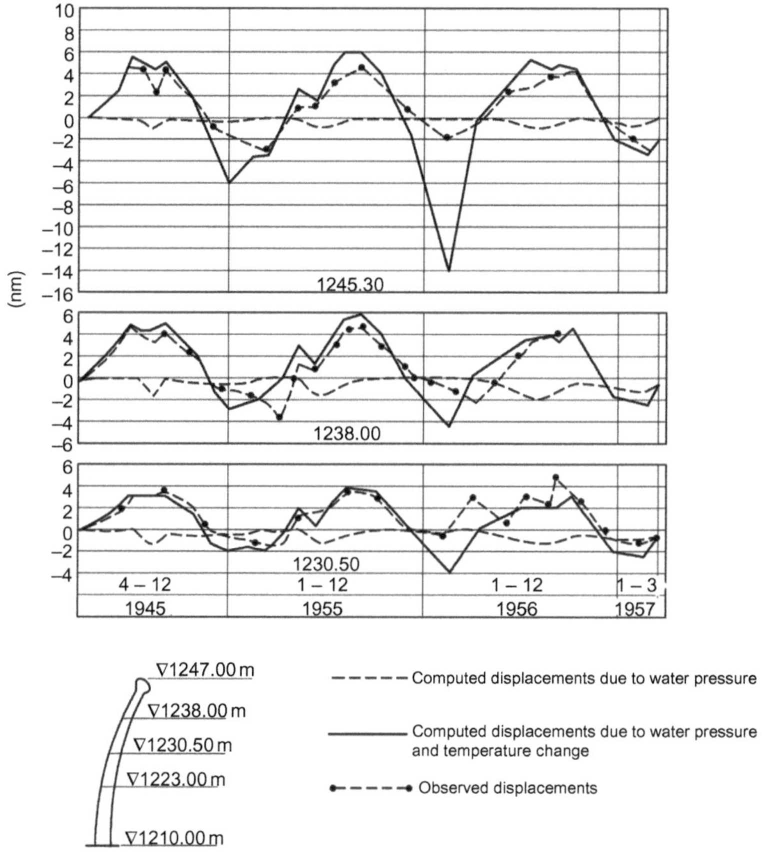

Figure 14.13 Comparison of the observed and the computed displacements of Isolado arch dam.

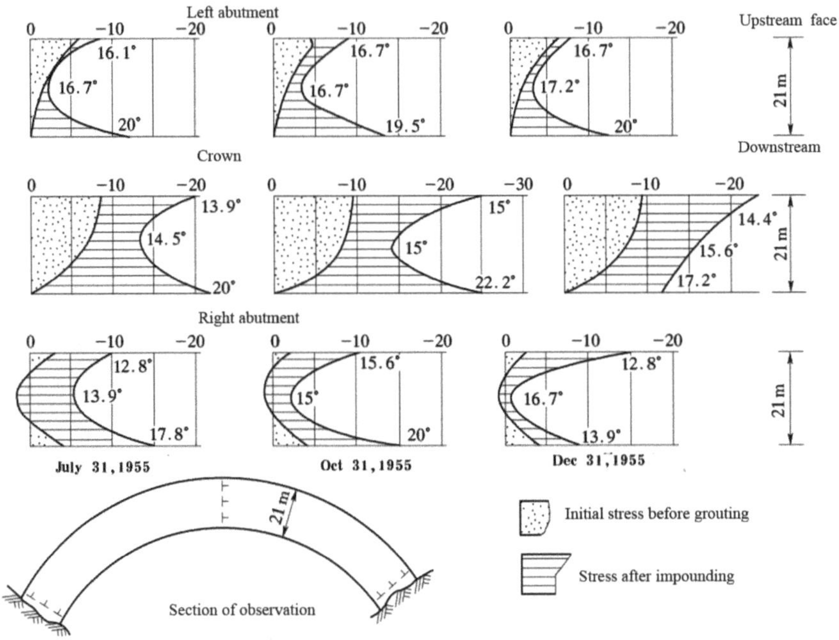

Figure 14.14 Observed temperatures and stresses of Kimishiba arch dam (unit of temperature: °C; unit of stress: 0.10 MPa; tensile stress is positive, compressive stress is negative; the stresses are observed by stress meter, the tensile stress is not accurate).

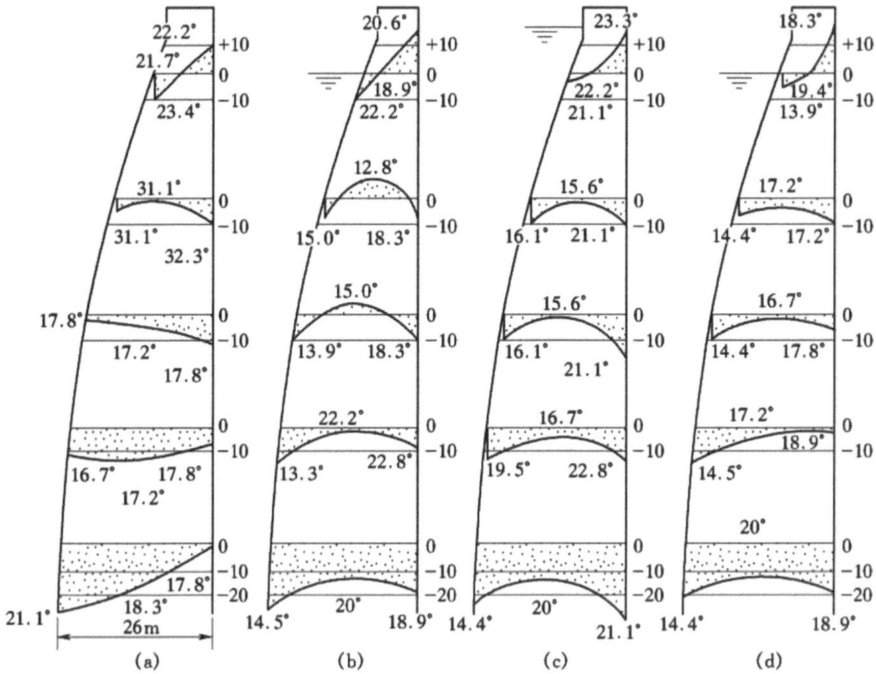

Figure 14.15 Observed temperatures and stresses in the cantilevers of Kimishiba arch dam (unit of temperature: °C; unit of stress: 0.10 MPa; tensile stress is positive, compressive stress is negative): (a) April 26, 1955, (b) July 31, 1955, (c) October 15, 1955, and (d) December 31, 1955.

15 Thermal Stresses in Docks, Locks, and Sluices

Docks, locks, and sluices are generally constructed on soft foundations, the restraints of which are small; rare cracks appear in the bottom plates of these structures, but experiences show that there may be deep cracks in the walls or piers as shown in Figure 15.1. As the bottom plates are constructed earlier, there are temperature differences between the bottom plates and the walls or the piers; due to the restraint of the bottom plate, tensile stresses may appear in the walls and piers which can induce deep cracks.

Strictly speaking, the analysis of thermal stresses in docks, locks, and sluices is a three-dimensional problem of elasticity; as the shape of the structure is rather complex, it is very difficult to get a theoretical solution. The problem had not been studied before 1974. The writer studied the problem in 1974 and got a practical method for stress analysis which had been widely applied in the field of hydraulic and transport engineering in China and will be described in the following [34,84].

15.1 Self-Thermal Stresses in Walls of Docks and Piers of Sluices

The thermal stresses in docks, locks, and sluices may be divided into two parts: the self-stress and the restraint stress. The self-stress is the stress in the wall or pier due to self-restraint and the restraint stress is the stress in the wall or pier due to the restraint of the bottom plate. The actual stress is the sum of the two parts.

The coordinate system is taken as Figure 15.2. Assuming that the temperature varies only in the direction of thickness and is symmetrical about the mid-plane, i.e., $T(z) = T(-z)$, the self-stress of the wall or pier may be computed by the following formula:

$$\sigma_x = \sigma_y = E_1 \alpha (T_m - T) \tag{15.1}$$

$$T_m = \frac{1}{t_1} \int T \, dz \tag{15.2}$$

Thermal Stresses and Temperature Control of Mass Concrete. DOI: http://dx.doi.org/10.1016/B978-0-12-407723-2.00015-4

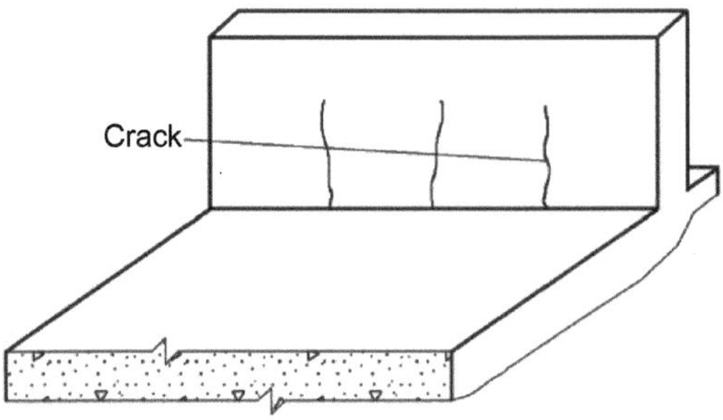

Figure 15.1 Cracks on the wall of a dock.

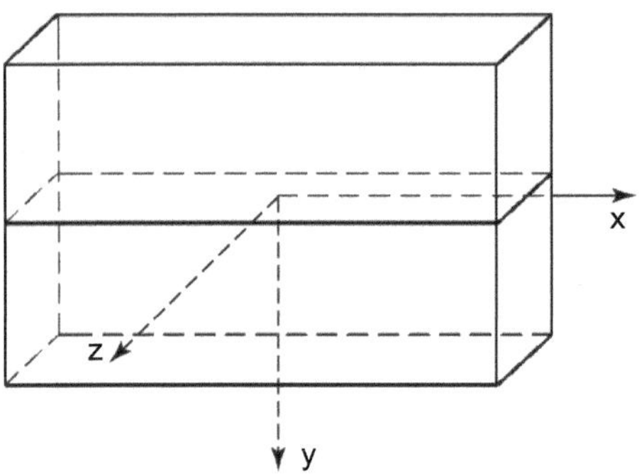

Figure 15.2 Coordinate system.

where T_m is the mean temperature in the direction of thickness, t_1 and E_1 are respectively the thickness and modulus of elasticity of the wall or pier. If the distribution of temperature is uniform, namely $T_m = T$, then the self-stress is equal to zero.

15.2 Restraint Stress in the Wall of Dock

15.2.1 General Theory for the Restraint Stress in the Wall of Dock

A dock is shown in Figure 15.3. Only the thermal stress is considered, and there is no external load and all the external surfaces are free. Referring to Figure 15.4, the bottom plane is a free plane, the normal stress σ_y and shearing stress τ_{yz}, τ_{yx} all of which are equal to zero. On the plane of contact 1-1, there are three stress components: σ_y, τ_{yz}, and τ_{yx}, the magnitudes of which depend on the condition of continuity of deformation of the wall and the bottom plate. The bottom plate is a thin plate, with sizeable rigidity in the plane xz and can produce a large restraint to the displacement in the x direction of the bottom of the wall, thus the shearing stress τ_{yx} is large. The bending rigidity of the bottom plate in the plane xy is very small, so the normal stress σ_y on plane 1-1 is small and can be neglected. Similarly, the bending rigidity of the wall in the plane xz is very small; the shearing stress τ_{yx} on plane 1-1 is small and can be neglected. Thus, there is only one stress component τ_{yx} on plane 1-1. Similarly, only the shearing stress τ_{zx} is in need of consideration on the contact surface 2-2. These assumptions are consistent with the theory of I beam in engineering mechanics. Hence, only the stress components τ_{yx} and τ_{zx}

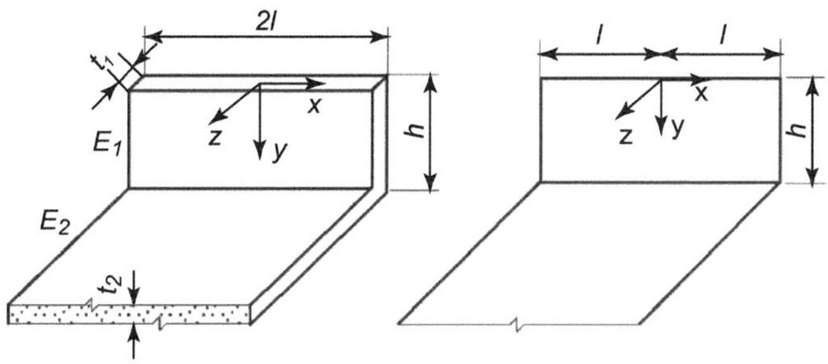

Figure 15.3 A dock and the schematic diagram for stress analysis.

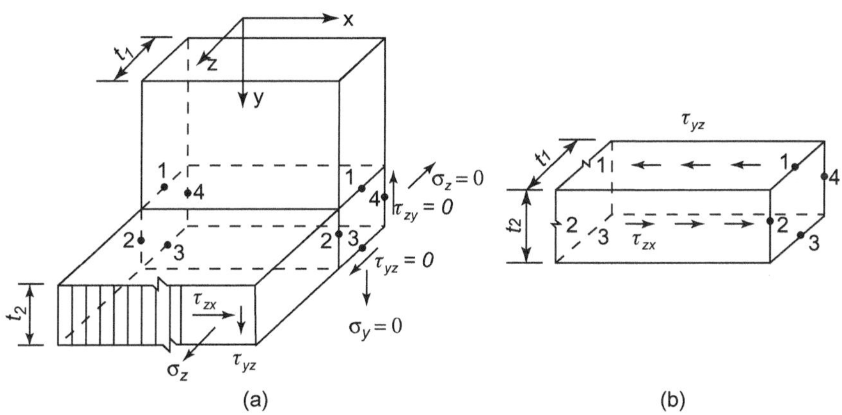

Figure 15.4 The joint of the bottom plate and the wall of the dock.

must be considered in the joint of the wall and the bottom plate as shown in Figure 15.4(b).

For the wall of the dock, the stress function is taken as follows:

$$F = \sum_{i=1}^{\infty} \cos \lambda_i x(A_i \text{ ch } \lambda_i y + B_i \text{ sh } \lambda_i y + C_i y \text{ ch } \lambda_i y + D_i y \text{ ch } \lambda_i y) \tag{15.3}$$

The stress components in the wall are

$$\begin{aligned}
\sigma_x &= \frac{\partial^2 F}{\partial y^2} = \sum_{i=1}^{\infty} \lambda_i \beta_i(y) \cos \lambda_i x \\
\sigma_y &= \frac{\partial^2 F}{\partial x^2} = -\sum_{i=1}^{\infty} \lambda_i^2 \phi_i(y) \cos \lambda_i x \\
\tau_{xy} &= -\frac{\partial^2 F}{\partial x \partial y} = \sum_{i=1}^{\infty} \lambda_i \psi_i(y) \sin \lambda_i x
\end{aligned} \tag{15.4}$$

The displacements of the wall are

$$
\begin{aligned}
u &= \frac{1}{E_1}\sum_{i=1}^{\infty}[\beta_i(y) + \mu\phi_i(y)]\sin \lambda_i x \\
v &= -\frac{1}{E_1}\sum_{i=1}^{\infty}[\eta_i(y) + \mu\psi_i(y)]\cos \lambda_i x
\end{aligned}
\tag{15.5}
$$

where

$$
\begin{aligned}
\beta_i(y) &= A_i\lambda_i \text{ ch } \lambda_i y + B_i\lambda_i \text{ sh } \lambda_i y + C_i(2 \text{ sh } \lambda_i y + \lambda_i y \text{ ch } \lambda_i y) \\
&\quad + D_i(2 \text{ ch } \lambda_i y + \lambda_i y \text{ sh } \lambda_i y) \\
\phi_i(y) &= A_i \text{ ch } \lambda_i y + B_i \text{ sh } \lambda_i y + C_i y \text{ ch } \lambda_i y + D_i y \text{ sh } \lambda_i y \\
\psi_i(y) &= A_i\lambda_i \text{ sh } \lambda_i y + B_i\lambda_i \text{ ch } \lambda_i y + C_i(\text{ch } \lambda_i y + \lambda_i y \text{ sh } \lambda_i y) \\
&\quad + D_i(\text{sh } \lambda_i y + \lambda_i y \text{ ch } \lambda_i y) \\
\eta_i(y) &= A_i\lambda_i \text{ sh } \lambda_i y + B_i\lambda_i \text{ ch } \lambda_i y + C_i(\lambda_i y \text{ sh } \lambda_i y - \text{ch } \lambda_i y) \\
&\quad + D_i(\lambda_i y \text{ ch } \lambda_i y - \text{sh } \lambda_i y)
\end{aligned}
\tag{15.6}
$$

where the coefficients A_i, B_i, C_i, D_i and the characteristic number λ_i are determined by the boundary conditions of the wall. At the top of the wall, i.e., when $y = 0$, we have

$$\sigma_y = 0 \tag{15.7}$$

$$\tau_{xy} = 0 \tag{15.8}$$

at the ends of the wall, i.e., when $x = \pm l$.

$$\sigma_x = 0 \tag{15.9}$$

$$\int_0^h \tau_{xy} \, dy = 0 \tag{15.10}$$

at the bottom of the wall, i.e., when $y = h$.

$$\sigma_y = 0 \tag{15.11}$$

$$t_1\tau_{xy} = t_2\overline{\tau}_{zx} \tag{15.12}$$

where $\overline{\tau}_{zx}$ and t_2 are the shearing stress and the thickness of the bottom plate. From Eq. (15.7),

$$A_i = 0 \tag{15.13}$$

From Eq. (15.8),

$$B_i = -C_i/\lambda_i \tag{15.14}$$

From Eq. (15.9),

$$\lambda_i = (2i-1)\pi/2l \tag{15.15}$$

From Eq. (15.10),

$$C_i = q_i D_i \tag{15.16}$$

$$q_i = \lambda_i h \text{ sh } \lambda_i h/(\text{sh } \lambda_i h - \lambda_i h \text{ ch } \lambda_i h) \tag{15.17}$$

From the condition of balance, it is evident that Eq. (15.10) is equivalent to Eq. (15.11); thus the result derived from Eq. (15.11) is also Eq. (15.16). Substituting Eqs. (15.13)–(15.16) into Eqs. (15.4) and (15.5), we get

$$\begin{aligned}
\sigma_x &= \sum \lambda_i D_i \cos \lambda_i x \cdot [q_i(\text{sh } \lambda_i y + \lambda_i y \text{ ch } \lambda_i y) + 2 \text{ ch } \lambda_i y + \lambda_i y \text{ sh } \lambda_i y] \\
\sigma_y &= -\sum \lambda_i D_i \cos \lambda_i x \cdot [q_i(\lambda_i y \text{ ch } \lambda_i y - \text{sh } \lambda_i y) + \lambda_i y \text{ sh } \lambda_i y] \\
\tau_{xy} &= \sum \lambda_i D_i \sin \lambda_i x \cdot [q_i \lambda_i y \text{ sh } \lambda_i y + \text{sh } \lambda_i y + \lambda_i y \text{ ch } \lambda_i y]
\end{aligned} \tag{15.18}$$

$$u = \frac{1}{E_1} \sum_{i=1}^{\infty} D_i \sin \lambda_i x \cdot [q_i(\text{sh } \lambda_i y + \lambda_i y \text{ ch } \lambda_i y) + 2 \text{ ch } \lambda_i y + \lambda_i y \text{ sh } \lambda_i y] \tag{15.19}$$

where the coefficient D_i is determined by the condition of contact of the wall and the bottom plate.

The width of the bottom plate of a dock is large and the width of a lock or a sluice is rather small. The solution for the bottom plate with large or moderate width is given in the following and the solution for the bottom plate with small width will be given in Section 15.4.

15.2.2 Computation for Wide Bottom Plate

Assuming that the bottom plate is a semi-infinite strip as shown in Figure 15.5, the stress function for which is

$$\overline{F} = \sum \cos \lambda_i x(\overline{C}_i + \overline{D}_i z)e^{-\lambda_i z} \tag{15.20}$$

The stress components are

$$\left.\begin{aligned}
\overline{\sigma}_x &= \sum \lambda_i \cos \lambda_i x(\lambda_i \overline{C}_i - 2\overline{D}_i + \overline{D}_i \lambda_i z)e^{-\lambda_i z} \\
\overline{\sigma}_z &= -\sum \lambda^2{}_i \cos \lambda_i x(\overline{C}_i + \overline{D}_i z)e^{-\lambda_i z} \\
\overline{\tau}_{xz} &= -\sum \lambda_i \sin \lambda_i x(\lambda_i \overline{C}_i - \overline{D}_i + \lambda_i \overline{D}_i z)e^{-\lambda_i z}
\end{aligned}\right\} \tag{15.21}$$

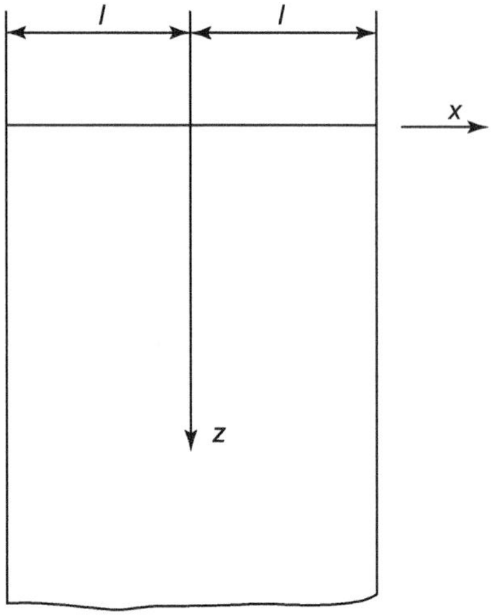

Figure 15.5 Semi-infinite strip.

The displacements are

$$\begin{aligned}\overline{u} &= \frac{1}{E_2}\sum \sin \lambda_i x\left[(1+\mu)\lambda_i(\overline{C}_i + \overline{D}_i z) - 2\overline{D}_i\right]e^{-\lambda_i z}\\ \overline{v} &= \frac{-1}{E_2}\sum \cos \lambda_i x\left[(1+\mu)\lambda_i(\overline{C}_i + \overline{D}_i z) + (1-\mu)\overline{D}_i\right]e^{-\lambda_i z}\end{aligned} \tag{15.22}$$

The boundary conditions of the bottom plate are when $z = 0$,

$$\sigma_z = 0 \tag{15.23}$$

When $x = \pm l$

$$\sigma_x = 0 \tag{15.24}$$

From Eq. (15.23),

$$\overline{C}_i = 0 \tag{15.25}$$

Substituting the above equations into Eqs. (15.31) and (15.22), we get

$$\overline{\tau}_{zx} = \sum \overline{D}_i \lambda_i \sin \lambda_i x(1 - \lambda_i z)e^{-\lambda_i z} \tag{15.26}$$

$$\overline{u} = \frac{1}{E_2}\sum \overline{D}_i \sin \lambda_i x[(1+\mu)\lambda_i z - 2]e^{-\lambda_i z} \tag{15.27}$$

The condition of balance of forces at the joint of the wall and the bottom plate when $z = 0$ and $y = h$ is,

$$t_1 \tau_{xy}|_{y=h} = t_2 \overline{\tau}_{zx}|_{z=0} \tag{15.28}$$

The condition of continuity of deformation is

$$u|_{y=h} + u_{\mathrm{T}} = \overline{u}|_{z=0} \tag{15.29}$$

where $u_{\mathrm{T}} = \alpha T_{\mathrm{m}} x$ is the free thermal displacement of the wall when there is no restraint of the bottom plate, let u_{T} be expanded into Fourier series, we have

$$u_{\mathrm{T}} = \sum \overline{u}_i \sin \lambda_i x \tag{15.30}$$

where

$$\overline{u}_i = \frac{2}{l} \int_0^l \alpha T_{\mathrm{m}}\, x \sin \lambda_i x \,\mathrm{d}x = \frac{4\alpha T_{\mathrm{m}}(-1)^{i-1}}{(2i-1)\pi\lambda_i} \tag{15.31}$$

Substituting Eqs. (15.17) and (15.26) into Eq. (15.28), we get

$$\overline{D}_i = [q_i \lambda_i h \,\mathrm{sh}\, \lambda_i h + \mathrm{sh}\, \lambda_i h + \lambda_i h \,\mathrm{ch}\, \lambda_i h] D_i t_1 / t_2 \tag{15.32}$$

Substituting (15.19), (15.27), (15.32) into (15.29), we have

$$\lambda_i D_i = 4E_1 \alpha T_{\mathrm{m}} (-1)^i / [(2i-1) H_{1i} \pi] \tag{15.33}$$

where

$$H_{1i} = (q_i + \rho_1)(\mathrm{sh}\, \lambda_i h + \lambda_i h \,\mathrm{ch}\, \lambda_i h) + (1 + \rho_1 q_i)\lambda_i h \,\mathrm{sh}\, \lambda_i h + 2 \,\mathrm{ch}\, \lambda_i h \tag{15.34}$$

$$\rho_1 = 2t_1 E_1 / (t_2 E_2) \tag{15.35}$$

When $i \geq 2$, sh $\lambda h \approx$ ch λh, thus

$$q_i \cong \lambda_i h / (1 - \lambda_i h) \tag{15.36}$$

$$H_{1i} \cong \mathrm{sh}\, \lambda_i h \cdot [(q_i + \rho_1)(1 + \lambda_i h) + (1 + \rho_1 q_i)\lambda_i h + 2] \tag{15.37}$$

Substituting $\lambda_i D_i$ in Eq. (15.18), the thermal stress in the wall of the dock is derived. In the computation of thermal stress, the temperature of the bottom is assumed to be zero; thus, T_{m} is the difference between the mean temperatures of the wall and the bottom plate.

Example 1 The height of the wall of the dock is $h = 8.00$ m, the width $2l = 15.00$ m, the thickness $t_1 = 1.60$ m, the modulus of elasticity is E_1, the uniform temperature drop is $-T$, the coefficient of expansion is α, the thickness of the bottom plate is $t_1 = 1.85$ m, the length is semi-infinite, modulus of elasticity is E_2, and the computed results are expressed by dimensionless number $\sigma_x/E\alpha T$ in Figure 15.6, in which the dotted curve represents the stresses computed by the simplified method of Section 15.6.

From Figure 15.6, it is clear that, under the action of uniform temperature drop $-T$ of the wall of the dock, the lower two-thirds of the wall are subjected to tension, the maximum tensile stress is $\sigma_x = 0.62E\alpha T$, the upper one-third of the wall is subjected to compression, and the maximum compressive stress $\sigma_x = -0.30E\alpha T$ appears at the top.

15.2.3 Computation for Bottom Plate with Moderate Width

The bottom plate with moderate width shown in Figure 15.7 may be analyzed as if it is a rectangular plate with length $2l$, width s, and thickness t_2. Taking the stress function as follows:

$$\overline{F} = \sum \cos \lambda_i x \cdot (A_i \text{ ch } \lambda_i z + B_i \text{ sh } \lambda_i z + C_i z \text{ ch } \lambda_i z + D_i z \text{ sh } \lambda_i z) \tag{15.38}$$

The stress components are:

$$\left.\begin{aligned} \overline{\sigma}_x &= \sum \lambda_i \beta_i(z) \cos \lambda_i x \\ \overline{\sigma}_z &= -\sum \lambda^2{}_i \phi_i(z) \cos \lambda_i x \\ \overline{\tau}_{xz} &= \sum \lambda_i \psi_i(z) \text{sh} \lambda_i x \end{aligned}\right\} \tag{15.39}$$

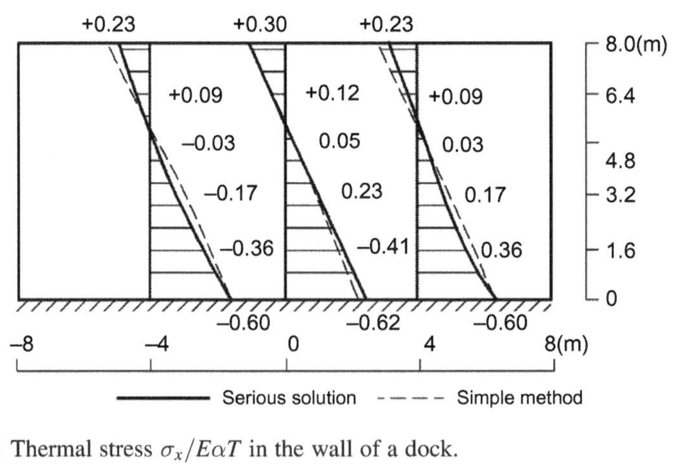

Figure 15.6 Thermal stress $\sigma_x/E\alpha T$ in the wall of a dock.

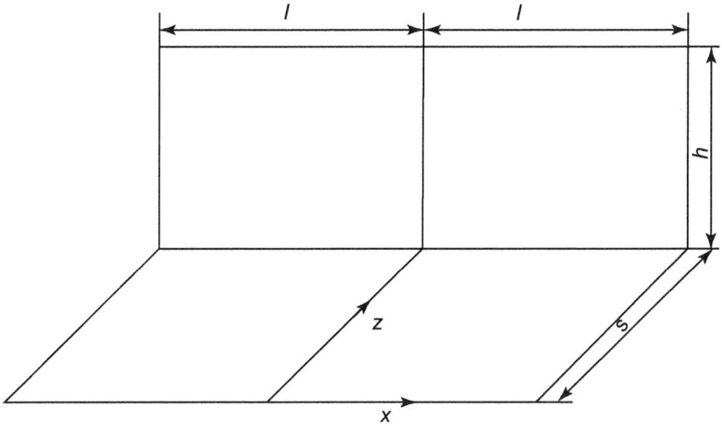

Figure 15.7 Sketch for analyzing a dock with bottom plate of moderate width.

The displacement components are

$$\begin{aligned}\overline{u} &= \frac{1}{E_2}\sum\left[\beta_i(z) + \mu\phi_i(z)\right]\sin \lambda_i x \\ \overline{w} &= -\frac{1}{E_2}\sum\left[\eta_i(z) + \mu\phi_i(z)\right]\cos \lambda_i x\end{aligned} \tag{15.40}$$

where $\beta_i(z)$, $\phi_i(z)$, $\psi_i(z)$, $\eta_i(z)$ are similar to the functions in Eq. (15.6). Substituting y in Eq. (15.6) by z, we get the correspondent functions in Eqs. (15.39) and (15.40).

On the joint of the wall and the bottom plate, the condition of balance and continuity of deformation are

$$t_1\tau_{xy}|_{y=h} = t_2\overline{\tau}_{zx}|_{z=s} \tag{15.41}$$

$$u|_{y=h} + u_T = \overline{u}|_{z=s} \tag{15.42}$$

After the coefficients D_i and $\overline{D}_i$ are determined by the above two equations, the problem is solved.

15.3 Restraint Stress in the Piers of Sluices

A sluice is shown in Figure 15.8. The pier is a rectangular plate; the expressions of the stresses and displacements of which are the same as those of the wall of a dock and can be expressed by Eqs. (15.18) and (15.19). The bottom plate consists of two semi-infinite plates or two rectangular plates jointed at $z = 0$, the stresses and

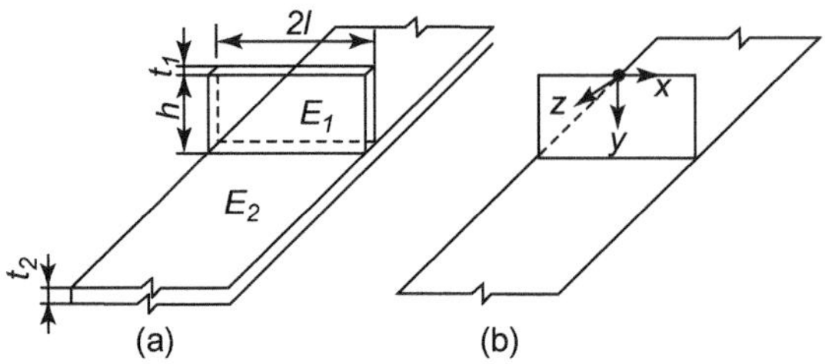

Figure 15.8 A sluice and the schematic diagram for computation.

displacements of which can still be expressed by Eqs. (15.21), (15.22), (15.39), and (15.40), but the boundary conditions are different from the bottom plate of a dock.

From the condition of a joint at $z = 0$, the boundary conditions of a wide bottom plate are as follows:

$$\text{When} \quad z = 0, \quad \overline{w} = 0 \tag{15.43}$$

$$\text{When} \quad x = \pm l, \quad \overline{\sigma}_x = 0 \tag{15.44}$$

From Eqs. (15.43) and (15.22), we get $\lambda_i \overline{C}_i = -(1-\mu)\overline{D}_i/(1+\mu)$. From Eq. (15.44), we get $\lambda_i l = (2i-1)\pi/2$; substituting them into Eqs. (15.21) and (15.22), the shearing stress and displacement of the bottom plate of a sluice are derived:

$$\overline{\tau}_{zx} = -\sum \lambda_i \overline{D}_i \sin \lambda_i x[\lambda_i z - 2/(1+\mu)]\text{e}^{-\lambda_i z}$$
$$\overline{u}_i = [(1+\mu)/E_2]\sum \overline{D}_i \sin \lambda_i x \cdot [\lambda_i z - (3-\mu)/(1+\mu)]\text{e}^{-\lambda_i z}$$

Let $z = 0$ in the above two equations, we get the shearing stress and the displacement in the following:

$$\overline{\tau}_{zx}|_{z=0} = [2/(1+\mu)]\sum \lambda_i \overline{D}_i \sin \lambda_i x \tag{15.45}$$

$$\overline{u}|_{z=0} = -[(3-\mu)/E_2]\sum \overline{D}_i \sin \lambda_i x \tag{15.46}$$

At the joint of the pier and the bottom plate, the condition of balance of forces is

$$t_i \tau_{xy}|_{y=h} = 2t_2 \overline{\tau}_{zx}|_{z=0} \tag{15.47}$$

The condition of continuity of deformation is

$$u|_{y=h} + u_{\text{T}} = \overline{u}|_{z=0} \tag{15.48}$$

Substitution of Eqs. (15.17) and (15.45) into Eq. (15.47) yields

$$\overline{D}_i = (1+\mu)t_1 D_i(q_i \lambda_i h \text{ sh } \lambda_i + \text{sh } \lambda_i h + \lambda_i h \text{ ch } \lambda_i h)/4t_2 \tag{15.49}$$

Substitution of Eqs. (15.19), (15.30), and (15.36) into Eq. (15.48) yields

$$\lambda_i D_i = 4(-1)^i E_1 \alpha T_{\text{m}}/[(2i-1)\pi H_{2i}] \tag{15.50}$$

$$H_{2i} = (q_i + \rho_2)(\text{sh}\ \lambda_i h + \lambda_i h\ \text{ch}\ \lambda_i h) + (1 + \rho_2 q_i)\lambda_i h\ \text{sh}\ \lambda_i h + 2\ \text{ch}\ \lambda_i h \quad (15.51)$$

$$\rho_2 = (3 - \mu)(1 + \mu)t_1 E_1/(4t_2 E_2) \quad (15.52)$$

where q_i is given in (15.17). Substituting $\lambda_i D_i$ given by (15.50) into (15.17), the stresses in the pier will be obtained.

For $i \geq 2$, because sh $\lambda_i h \cong$ ch $\lambda_i h$, Eq. (15.51) may be simplified as follows

$$H_{2i} = \text{sh}\ \lambda_i h[(q_i + \rho_2)(1 + \lambda_i h) + (1 + \rho_2 q_i)\lambda_i h + 2] \quad (15.53)$$

Example 2 The height of the pier of a sluice is $h = 8$ m, width $2l = 15$ m, thickness $t_1 = 1.5$ m, and modulus of elasticity E_1. The thickness of the bottom plate is $t_2 = 1.5$ m, width $2l = 15$ m, modulus of elasticity $E_2 = E_1 = E$, Poisson's ratio $\mu = 1/6$. Try to compute the thermal stresses in the pier due to uniform temperature drop $-T_m$ by the above method. The results of computation are expressed in $\sigma_x/E\alpha T_m$ in Figure 15.9, from which it is clear that these stresses are similar to but greater than those in the wall of a dock.

The above-mentioned method of computation may also be applied to the bottom plate of moderate width, but the condition of balance equation (15.41) should be replaced by the following equation:

$$t_1\tau_{xy/y=h} = 2t_2\overline{\tau}_{zx/z=s} \quad (15.54)$$

15.4 Restraint Stress in the Wall of Dock or the Pier of Sluice on Narrow Bottom Plate

The wall or pier on a narrow bottom plate is shown in Figure 15.10. As the bottom plate is narrow, it is assumed that the stresses in the bottom only change in the x direction and do not change in the z direction.

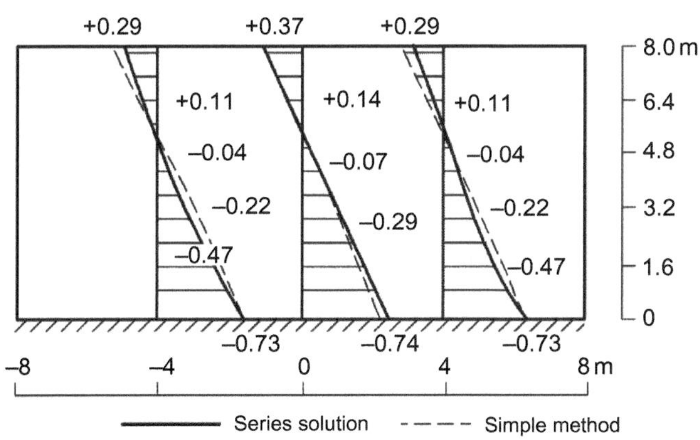

Figure 15.9 The thermal stress $\sigma_x/E\alpha T_m$ in the pier of a sluice.

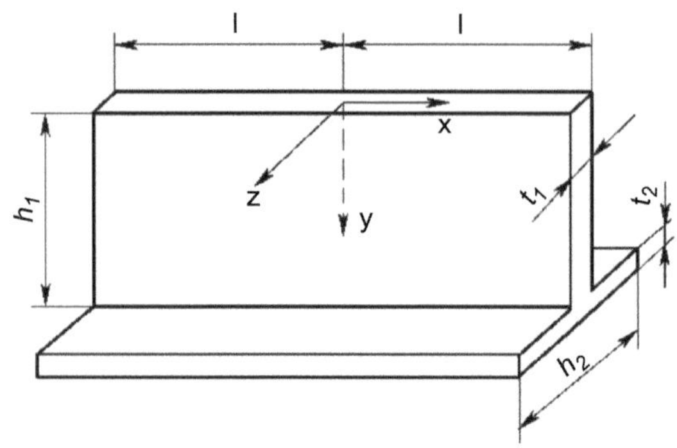

Figure 15.10 The wall or pier on a narrow bottom plate.

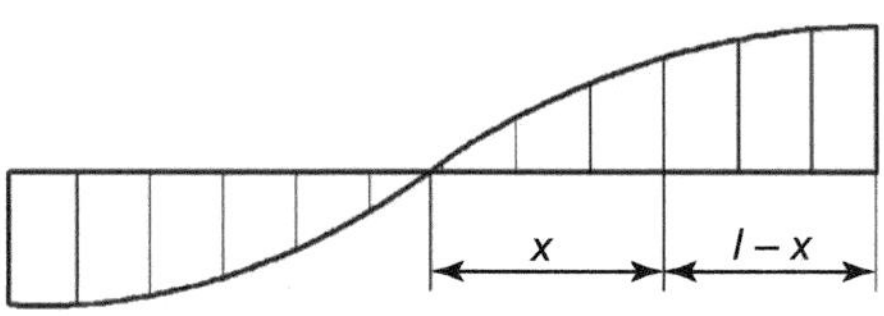

Figure 15.11 The distribution of shearing stress in a narrow bottom plate.

As shown in Figure 15.11, the shearing stress in the bottom plate is an odd function of x and can be expressed by

$$Q = \sum_{i=1}^{\infty} \lambda_i \overline{D_i} \sin \lambda_i x \tag{15.55}$$

After integration along the x direction, the normal force of the bottom plate is

$$N = -\sum_{i=1}^{\infty} \overline{D_i} \cos \lambda_i x \tag{15.56}$$

Thus, the tensile stress in the bottom plate is

$$\overline{\sigma_x} = \frac{N}{h_2 t_2} = -\frac{1}{h_2 t_2} \sum \overline{D_i} \cos \lambda_i x = E_2 \frac{\partial \overline{u}}{\partial x} \tag{15.57}$$

After integration of the above equation, the displacement of the bottom plate is

$$\overline{u} = -\frac{1}{E_2 h_2 t_2} \sum \frac{1}{\lambda_i} \overline{D_i} \sin \lambda_i x \tag{15.58}$$

The stresses and displacements of the wall and the pier are given by Eqs. (15.18) and (15.19). The condition of balance of force at the joint of the wall or pier with the bottom plate is

$$t_1 \tau_{xy}|_{y=h} = Q \tag{15.59}$$

and the condition of continuity of deformation is

$$u|_{y=h} + u_T = \bar{u}|_{z=0} \tag{15.60}$$

Substituting Eqs. (15.17), (15.19), (15.30), (15.55), and (15.58) into the above two equations, we get

$$\lambda_i D_i = 4E_1 \alpha T_m (-1)^i / [(2i-1)\pi H_{3i}] \tag{15.61}$$

$$H_{3i} = (q_i + \rho_3)(\text{sh}\ \lambda_i h + \lambda_i h\ \text{ch}\ \lambda_i h) + (1 + \rho_3 q_i)\lambda_i h\ \text{sh}\ \lambda_i h + 2\ \text{ch}\ \lambda_i h \tag{15.62}$$

$$\rho_3 = t_1 E_1 / (t_2 E_2 \lambda_i h_2) \tag{15.63}$$

Substitution of $\lambda_i D_i$ into Eq. (15.17) will yield the restraint stress in the wall or pier.

Example 3 There is a uniform temperature drop $-T\,°C$ in the wall of a dock, the height of which is $h_1 = 8$ m, the thickness $t_1 = 1.60$ m. The length $2l = 15$ m; the width of the bottom plate is $h_2 = 2.50$ m, the modulus of elasticity $E_2 = E_1 = E$, the Poisson's ratio $\mu = 1/6$, and the temperature in the bottom plate is zero. The stresses computed by the above-mentioned method are shown by solid lines in Figure 15.12.

15.5 Simplified Computing Method

15.5.1 T beam

The wall or pier on a narrow bottom plate is simplified as a T beam shown in Figure 15.13. The origin of coordinates is placed on the centroid of the cross section with weight E, namely

$$\int E(y) b y\, dy = 0$$

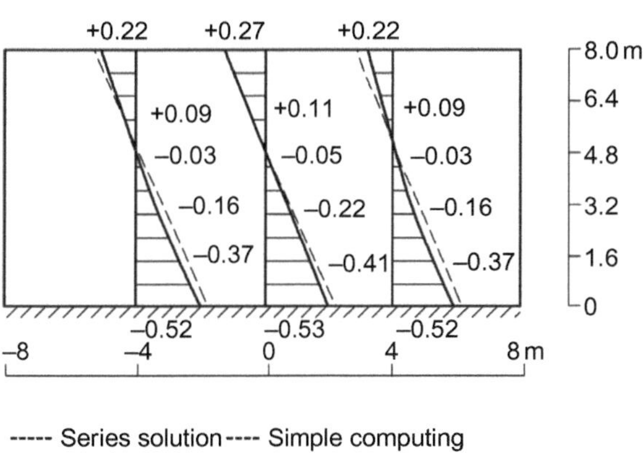

Figure 15.12 The thermal stress $\sigma_x/E\alpha T_m$ in the wall or pier on a narrow bottom plate.

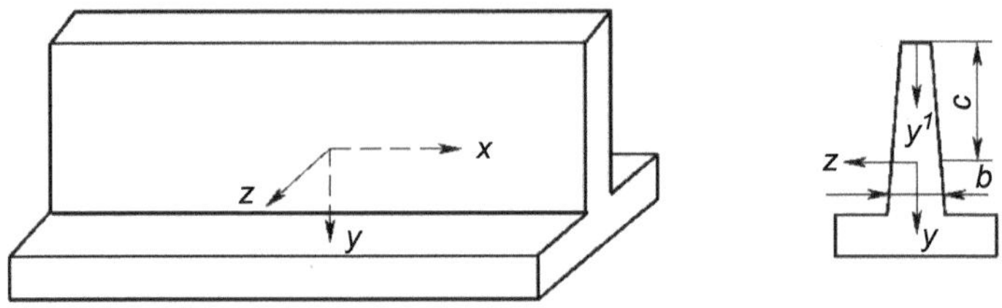

Figure 15.13 The coordinate system for a simplified method.

The distance c between the centroid and the top of the wall is given by

$$c = \int Eby'\,\mathrm{d}y' / \int Eb\,\mathrm{d}y' \tag{15.64}$$

where $y' = y + c$ is the ordinate of the centroid from the top of the wall. E and b are functions of y'. If the cross section is divided into n layers in the direction of height, for the ith layer the modulus of elasticity is E_i, the width is b_i, the height is $\Delta y'_i$, and the ordinate of the centroid is y'_i, then the distance c between the centroid of the whole section and the top of the wall is given by

$$c = \sum_{i=1}^{n} E_i b_i y'_i \Delta y'_i / \sum_{i=1}^{n} E_i b_i \Delta y'_i \tag{15.65}$$

If the temperature $T(y,z)$ is a function of y and z and is symmetrical about the plane $z=0$, namely, $T(y,z) = T(y,-z)$, according to the hypothesis of plane section, the thermal stress may be computed by

$$\sigma_x = E(y)\alpha[T_m + \psi y - T(y,z)] \tag{15.66}$$

$$T_m = \iint E(y)T(y,z)\mathrm{d}y\,\mathrm{d}z / \int E(y)b(y)\mathrm{d}y \tag{15.67}$$

$$\psi = \iint E(y)T(y,z)y\,\mathrm{d}y\,\mathrm{d}z / \int E(y)b(y)y^2\,\mathrm{d}y \tag{15.68}$$

where T_m is the mean temperature with weight $E(y)$, ψ is the equivalent temperature gradient with weight $E(y)$, and the origin of coordinate is the centroid with weight $E(y)$. For arbitrary temperature field, T_m and ψ can be computed from Eqs. (15.67) and (15.68) by numerical integration.

It should be pointed out that Eq. (15.66) can be used to compute the thermal stresses in walls or piers with variable thickness and the self-stress and the restraint stress can be computed together.

Example 4 Example 3 is computed by the above-mentioned simplified method, the results of computation are shown by dotted lines in Figure 15.12. It is evident that the stresses given by the two methods are close to each other.

15.5.2 Simplified Computation of Thermal Stresses in Dock

From Eqs. (15.17) and (15.33), it is clear that the thermal stresses of the wall of a dock are determined by coefficient D_i which is determined by H_{1i}. Comparison between Eqs. (15.34) and (15.62) shows that the difference of H_{1i} and H_{3i} is determined by ρ_1 and ρ_3. If an equivalent width h_2 is given by the condition

$$\rho_1 = \rho_3 \tag{15.69}$$

then the wall of a dock on a semi-infinite bottom plate can be substituted by a T beam, the width of the bottom plate of which is h_2, and the thermal stresses in the T beam can be computed by the simplified equation (15.66) (Figure 15.14).

Substitution of Eqs. (15.35) and (15.63) into Eq. (15.69) yields the equivalent width of bottom plate as follows:

$$h_2 = \frac{1}{2\lambda} = \frac{l}{(2i-1)\pi} \tag{15.70}$$

where h_2 is related to i. If the stress is computed by Eq. (15.61), h_2 is given by Eq. (15.70). Now the stress is computed by T beam formula, we may take $i = 1$, thus

$$h_2 = l/\pi \tag{15.71}$$

Example 5 The basic data are the same as example 1. From Eq. (15.71), we get $h_2 = 8.0/\pi = 2.55$ m, from Eq. (15.66), we get the thermal stresses as shown in Figure 15.6 by dotted lines. It is clear that the stresses given by the simplified method are close to those given by the series method.

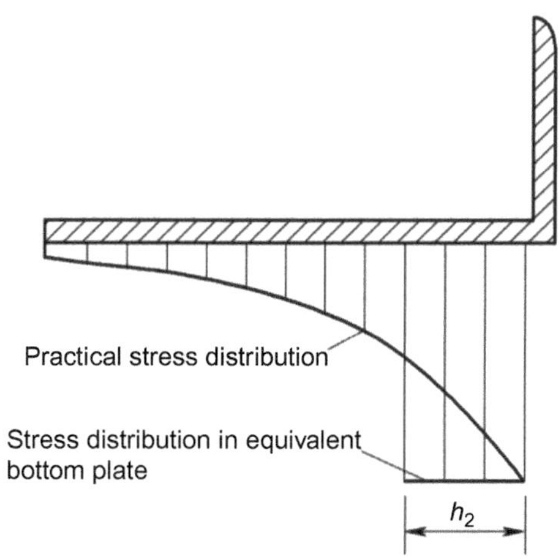

Figure 15.14 The equivalent width of the bottom plate of a dock.

15.5.3 Simplified Method for Thermal Stresses in Sluices

From Section 15.3, it is evident that the differences between the thermal stresses in the pier on a narrow bottom plate and those in the pier on a bottom plate with infinite length depend on the value of ρ_2 and ρ_3. Choose an equivalent width of bottom plate h_2 such that

$$\rho_2 = \rho_3 \tag{15.72}$$

and substitute Eqs. (15.52) and (15.63) in the above equation, we get

$$h_2 = \frac{4}{(3-\mu)(1+\mu)\lambda_i} = \frac{8l}{(3-\mu)(1+\mu)(2i-1)\pi} \tag{15.73}$$

Let $i = 1$ and $\mu = 1/6$, we have

$$h_2 = 0.773l \tag{15.74}$$

Now the actual sluice may be replaced by a T beam the width of the bottom plate of which is h_2 as shown in Figure 15.15 and the thermal stresses may be computed by the simplified formula (15.66).

Example 6 The basic data are the same as example 2. From Eq. (15.74), $h_2 = 0.773 \times 8 = 6.18$ m, the thermal stresses given by Eq. (15.66) are shown by dotted lines in Figure 15.9.

Example 7 As shown in Figure 15.16, there is a uniform temperature drop in the wall of a dock. From Eq. (15.71), $h_2 = 9.00/\pi = 2.87$ m. The total width of the bottom plate is $2.87 + 2.15 + 2.35 = 7.37$ m. According to the actual dimensions in Figure 15.16(b), the thermal stresses in the wall of the dock given by Eq. (15.66) are shown in Figure 15.16(a), and the maximum tensile stress is $\sigma_x = -0.66E\alpha T$.

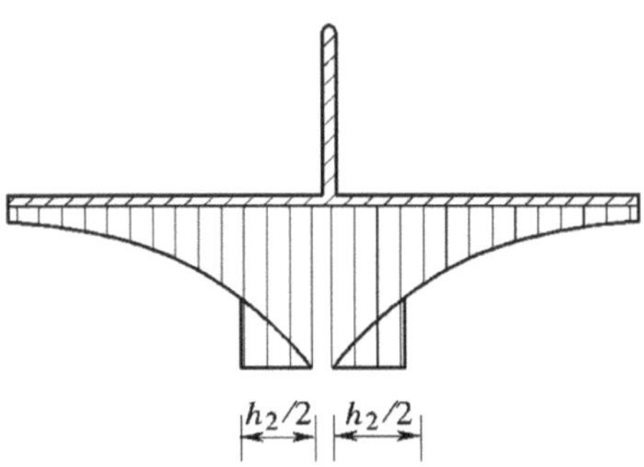

Figure 15.15 The equivalent width of a bottom plate of a sluice.

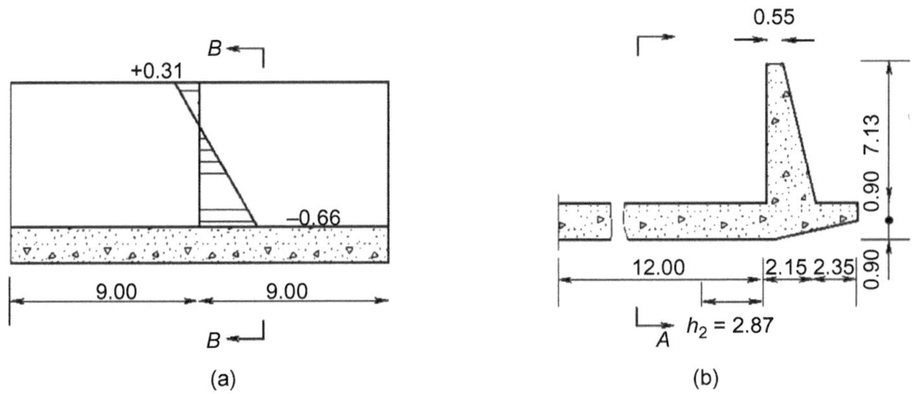

Figure 15.16 The thermal stresses in a dock: (a) longitudinal section A − A and $\sigma_x/E\alpha T$ and (b) cross section B−B.

15.5.4 Simplified Method for E(y, τ) Varying with Age τ

In order to consider the process of construction and the variation of E with age τ of concrete, it is necessary to adopt the increment method. The elastic stress increments in $\Delta\tau_i$ are given by

$$\Delta\sigma_{xi} = E(y, \tau_i)\alpha[\Delta T_{\mathrm{m}}(\tau_i) + \Delta\psi(\tau_i)y - \Delta T(y, z, \tau_i)] \tag{15.75}$$

where $\Delta T(y,z,\tau_i) = T(y,z,\tau_{i+1}) - T(y,z,\tau_i), E(y,\tau_i) = [E(y,\tau_i) + E(y,\tau_{i+1})]/2, \Delta T_{\mathrm{m}}(\tau_i)$ and $\Delta\psi(\tau_i)$ are computed by formulas similar to Eqs. (15.67) and (15.68).

The viscoelastic thermal stress is computed by

$$\sigma_x(t) = \sum \Delta\sigma_x(\tau_i)K(t, \tau_i) \tag{15.76}$$

where $K(t, \tau_i)$ is the relaxation coefficient of concrete.

15.6 Thermal Stresses in a Sluice by FEM

The thermal stresses in a sluice are computed systematically by three-dimensional FEM with the program SAPTIS compiled by Prof. Zhang Guoxin.

15.6.1 Thermal Stress due to Hydration Heat of Cement in Construction Period

The dimensions of the sluice are shown in Figure 15.17. For the pier, length × height × thickness = 10.5 m × 6.0 m × 1.0 m; for the bottom plate,

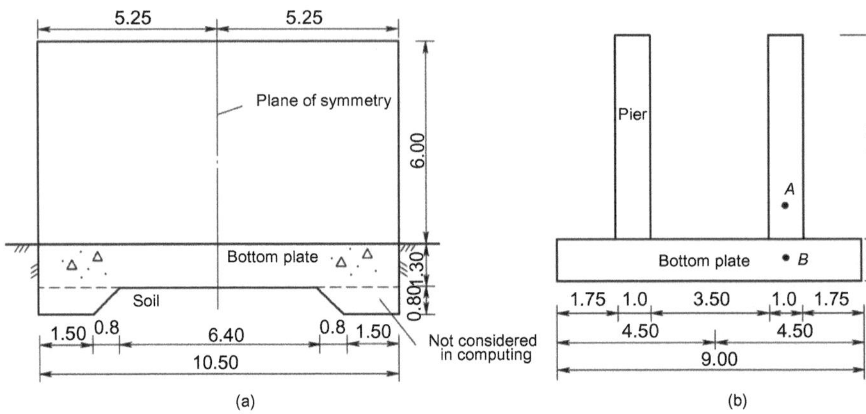

Figure 15.17 The dimensions of a sluice (m): (a) longitudinal section and (b) transverse section.

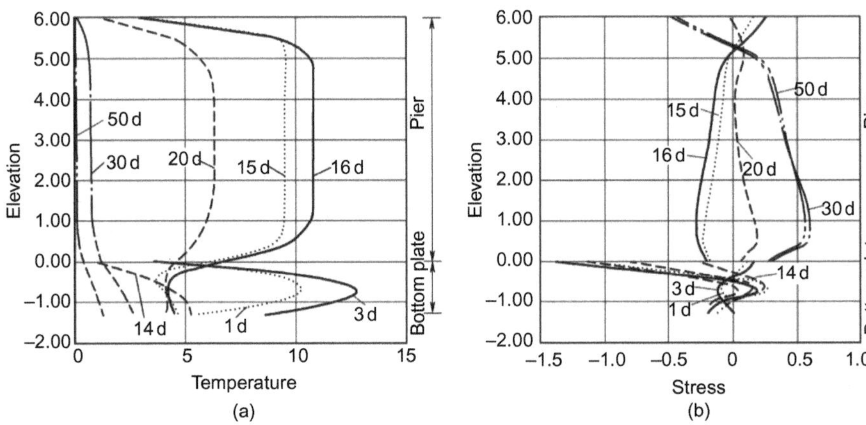

Figure 15.18 Temperatures and stresses in the mid-plane of pier of the sluice in the direction of flow in the period of construction due to hydration heat of cement.

length × width × thickness = 10.5 m × 9.0 m × 1.30 m. Due to symmetry, only 1/4 is taken in the analysis.

For the concrete, the diffusivity $a = 0.10\ \text{m}^2/\text{day}$, the adiabatic temperature rise $\theta(\tau) = 30\tau/(1.70 + \tau)$°C, Young's modulus $E(\tau) = 35{,}000\tau/(3.30 + \tau)$ MPa, Poisson's ratio $\mu = 0.167$, the unit creep is given by Eq. (6.21), the surface conductance of the exposed surface is $\beta = 70\ \text{kJ/(m}^2\ \text{h °C)}$ and that of the surface covered by timber 1 cm thick is $\beta = 38\ \text{kJ/(m}^2\ \text{h °C)}$.

For the soil foundation, there is no hydration heat, no creep, $E_f = 50$ MPa, $\mu = 0.25$.

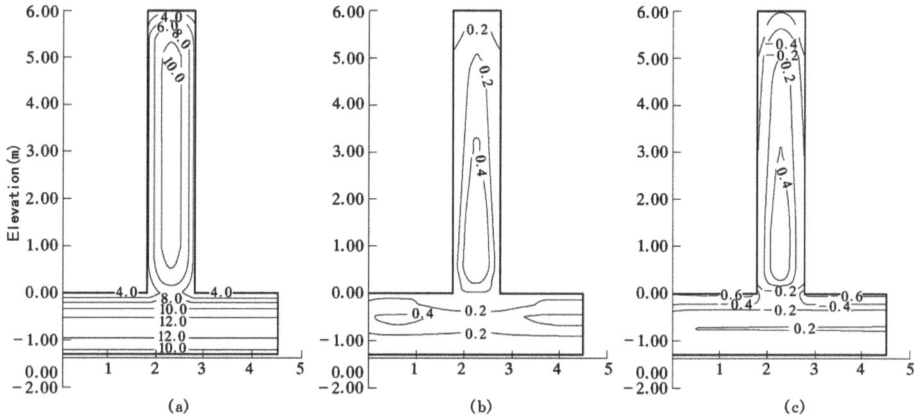

Figure 15.19 Temperatures and stresses in the central transverse section of a sluice in the construction period due to hydration heat of cement: (a) envelopes of temperatures, (b) envelopes of stress σ_x, and (c) final stress σ_x.

Taking air temperature $T_a = 0°C$, initial temperature of concrete $T_0 = 0°C$, and $\theta(\tau) = 30\tau/(1.7 + \tau)°C$. The bottom plate is poured first and the pier is placed 14 days later. The computed results are shown in Figures 15.18 and 15.19. From the comprehensive computed results it is clear that the maximum tensile stress due to hydration heat of cement, annual variation of air temperature and cold wave are respectively 0.59, 0.49, and 1.56 MPa, each one of which is lower than the tensile strength of concrete and will not induce cracking alone but the maximum comprehensive tensile stress is 1.99 MPa which is possible to induce cracking in the sluice.

16 Simulation Analysis, Dynamic Temperature Control, Numerical Monitoring, and Model Test of Thermal Stresses in Massive Concrete Structures

16.1 Full Course Simulation Analysis of Concrete Dams

In the construction process, the concrete dam is divided into many blocks which are further divided into many horizontal layers with a thickness of 1.5–3.0 m. Generally, there are several years from the beginning to the completion of construction of a dam. Due to the variation of air temperature, the hydration heat, and pipe cooling, the variation of temperature field is very complex. It is difficult to compute the thermal stresses in the construction period by the traditional methods of structural mechanics. As the number of the monitoring instruments is small, it is also difficult to give the actual stress field of a concrete dam in the construction process by the instrumental monitoring. Only the finite element method (FEM) can be used to give the simulation computation of concrete dams considering the process of construction and all the factors which influence the temperature and stress fields of the dam. If necessary, it can give the factor of safety by overload computation.

The peculiarities of simulation computation are as follows: (1) adopt the incremental FEM; (2) simulate the whole course of construction, the dam is divided into blocks and layers; (3) consider the following factors—the variation of the ambient and the interior temperatures, the variation of the mechanical and thermal properties with age of concrete, the artificial cooling, etc.; (4) the influence of the opening and grouting of joints; (5) give the factor of safety by overload computation, if necessary.

The scope of simulation computation depends on the purpose of analysis and the type of structure. Generally, only one dam block is computed for a gravity dam. In the analysis of the thermal stresses in the course of construction of arch dams, only one dam block or three adjacent dam blocks are computed. In the analysis of the stress state of an arch dam after impounding of water, it is necessary to conduct simulation computation of the whole dam [26, 28, 35, 37, 52, 57, 78].

The envelopes of the first principal stresses of the 3D whole course simulation computation of the Jinghong gravity dam are shown in Figure 16.1.

Thermal Stresses and Temperature Control of Mass Concrete. DOI: http://dx.doi.org/10.1016/B978-0-12-407723-2.00016-6

16.2 Dynamic Temperature Control and Decision Support System of Concrete Dam

In the past, the allowable temperature differences and maximum temperature of concrete are given in the design report of concrete dams. Technical measures are taken in the stage of construction to satisfy the needs of design. Experience shows that this is not sufficient for the temperature control of high concrete dams.

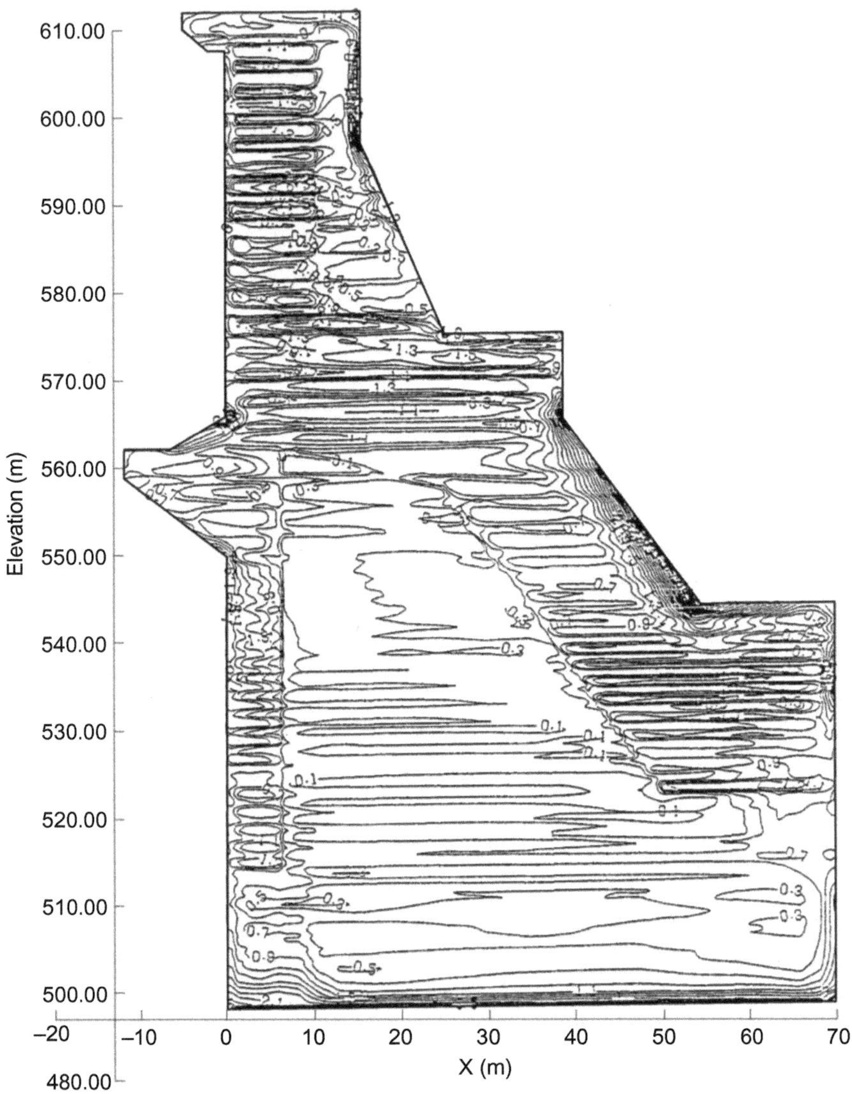

Figure 16.1 The envelopes of the first principal stresses of the 3D simulation computation of Jinghong gravity dam (stress unit: MPa).

In the construction process, there may be some changes of the conditions assumed in the design but generally there is no analysis and monitoring of these changes. This may be the cause of cracking of concrete. A decision-support system of dynamic temperature control of high concrete dams has been developed by Zhu Bofang, Zhang Guoxin, and Xu Ping. The functions of the system are as follows:

1. Whole course simulation computation of the temperature and stress fields of the dam by 3D FEM. The output of the system includes the actual geometrical figure of the shape of the dam blocks, the temperature and stress fields at any time.
2. Back analysis of the thermal properties of concrete and the effect of superficial thermal insulation are conducted in the process of construction to give the practical thermal properties of concrete and the actual effect of the superficial thermal insulation.
3. Forecast of the temperature and stress fields of the dam. In the process of construction, based on the actual temperature and stress fields of the concrete blocks which have been constructed, according to the predetermined schedules of progress and technical measures of temperature control, simulation computation is conducted to predict the temperature and stress fields in the future and check the effect of the technical measures and the schedule of progress.
4. Decision support of the control of temperatures and thermal stresses of the dam. The experiences of the experts in the world about the design, construction and temperature control, the specifications, and examples of design and construction of concrete dams are collected and systematized. Based on the above data and the results of simulation computation of the temperature and stress fields, information will be given to the engineers to help them to modify the measures of temperature control and construction schedule to prevent the cracking of a dam.
5. Operation platform and database for comprehensive management of the system.

This system had been successfully applied in the Zhougongzhai concrete arch dam with a height of 126.5 m. The dam was constructed from December 2003 to April 2006. The first principal stresses on January 15, 2005 are shown in Figure 16.2.

16.3 Numerical Monitoring of Concrete Dams

At present concrete dams are monitored by instruments in the period of construction and operation. Instrumental monitoring is important and can be used to judge

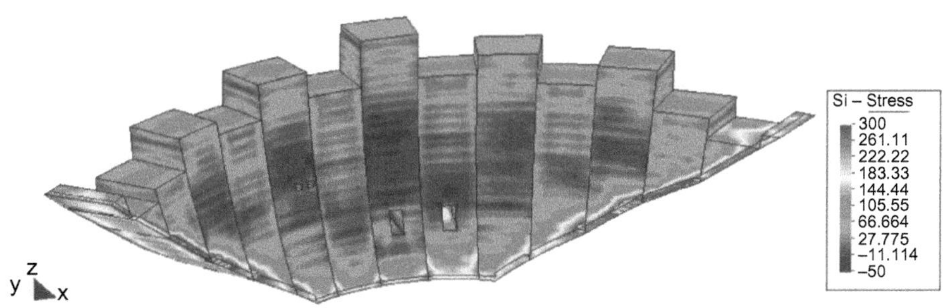

Figure 16.2 The first principal stress of Zhougongzhai concrete arch dam on January15, 2005 (unit: 10^{-2} MPa).

whether the dam is working normally, but there are some drawbacks, so it is suggested to add numerical monitoring to instrumental monitoring of concrete dams.

16.3.1 The Drawbacks of Instrumental Monitoring

1. There are few cross sections of monitoring.
   Due to the restraint of many practical factors, generally there are two to four cross sections embedded with instruments. Practically, there is no instrument in most dam blocks.
2. It is difficult to give the whole picture of temperature and stress fields even in the cross section embedded with instruments.
   A concrete dam is constructed layer by layer. The variation of temperature and stress fields in each layer is very complex. In order to monitor the temperature and stress variations, at least three rows of instruments must be embedded in each layer of concrete. For a dam block with 100 layers, there are 300 rows of instruments to be embedded. Practically, there are only 3–4 rows of instruments in the observed cross section, so it is impossible to give the whole picture of temperature and stress fields and factor of safety even in the monitored section.

16.3.2 Numerical Monitoring

It is suggested to add numerical monitoring in the construction of concrete dams. The content of numerical monitoring includes the following:

1. Whole dam and whole course simulation computation of the temperature field and stress field by 3D FEM.
2. Loading, including temperature, self-weight, water pressure, seepage flow, and initial stress.
3. Properties of materials, including the elastic, inelastic, and creep deformations and the influences of joints.
4. In addition to the room test, the actual properties of materials are determined by back analysis of observed results of the prototype.
5. Forecast the temperature and stress fields of the dam.

16.3.3 The Important Functions of Numerical Monitoring

1. The function of numerical monitoring in the period of construction
   By numerical monitoring, we can understand the temperature and stress fields of the dam in the period of construction and technical measures may be adopted to prevent possible cracks, if there are any problems.
2. The function of numerical monitoring in the period of operation
   It is possible to give appraisals of a dam's work frequently in the period of operation from the results of numerical monitoring.

A concrete gravity-arch dam with maximum height 76.3 m and crest length 419 m was constructed in 1968–1972. There was no efficient temperature control in the construction period of the dam so some cracks appeared in the dam. We gave a whole dam and whole course inelastic finite element simulation analysis to the dam. All the

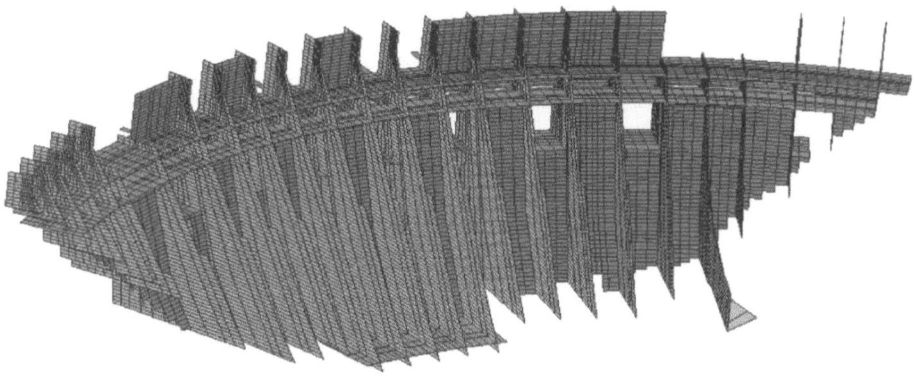

Figure 16.3 The transverse joints and cracks in the computing model of a concrete gravity-arch dam.

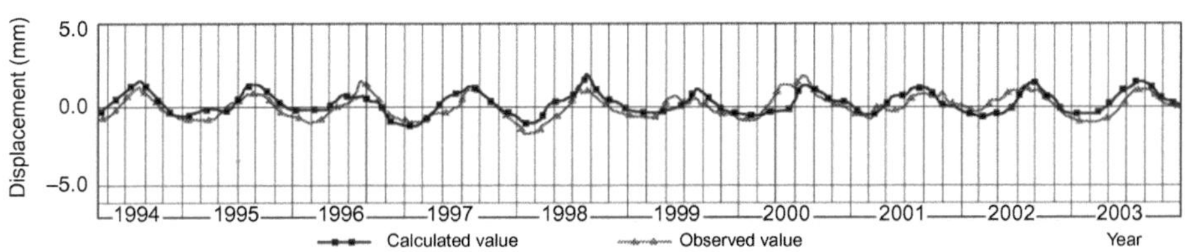

Figure 16.4 Comparison of the observed displacements with the displacements given by simulation computation of FEM of a concrete gravity-arch dam.

faults in the foundation, the joints and principal cracks in the dam, and the construction process are simulated in the computation. The joints and cracks in the computing model are shown in Figure 16.3. The computed displacements and the observed displacements of the dam are shown in Figure 16.4; they are close to each other.

Based on the results of simulation computation, an overload analysis is given by FEM: (1) With consideration of stress history and influence of joints, the safety factor is $k = 1.11$ (in winter) and 1.14 (in summer). (2) Without consideration of stress history and influence of joints, $k = 1.67$. (3) If efficient temperature control was conducted in the period of construction as required by specifications of concrete dam, the safety factor is $k = 3.26$. The failure of the downstream face of the dam under loading is shown in (Figure 16.5).

16.4 Model Test of Temperature and Stress Fields of Massive Concrete Structures

If the adiabatic temperature rise of concrete is $\theta(t) = \theta_0(1 - e^{-st})$, the differential equation of heat conduction and the initial and boundary conditions of the prototype are as follows:

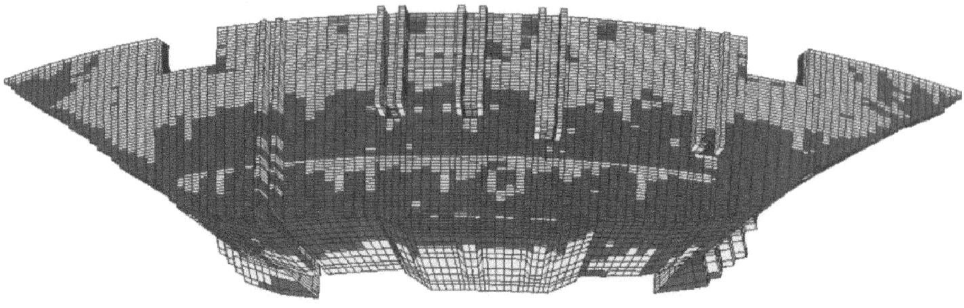

Figure 16.5 Failure of the downstream face of a concrete gravity-arch dam after overloading.

Equation of heat conduction:

$$\frac{\partial T}{\partial t} = a\left(\frac{\partial^2 T}{\partial x^2} + \frac{\partial^2 T}{\partial y^2} + \frac{\partial^2 T}{\partial z^2}\right) + \frac{\theta_0 \partial(1 - e^{-st})}{\partial t} \tag{16.1}$$

Initial condition:

$$T(0, x, y, z) = T_0(x, y, z) \tag{16.2}$$

Boundary condition:

$$\lambda \frac{\partial T}{\partial n} + \beta(T - T_c) = 0 \tag{16.3}$$

If the adiabatic temperature rise of the material of the model is $\theta_m(t_m) = \theta_{0m}(1 - e^{-s_m t_m})$, the differential equation and the conditions of the model are:

Equation of heat conduction:

$$\frac{\partial T_m}{\partial t_m} = a_m\left(\frac{\partial^2 T_m}{\partial x_m^2} + \frac{\partial^2 T_m}{\partial y_m^2} + \frac{\partial^2 T_m}{\partial z_m^2}\right) + \frac{\theta_{0m} \partial(1 - e^{-s_m t_m})}{\partial t_m} \tag{16.1a}$$

Initial condition:

$$T_m(0, x_m, y_m, z_m) = T_0(x_m, y_m, z_m) \tag{16.2a}$$

Boundary condition:

$$\lambda_m \frac{\partial T_m}{\partial n_m} + \beta_m(T_m - T_{cm}) = 0 \tag{16.3a}$$

Let

$$\left.\begin{aligned} C_T = T/T_m, \quad & C_t = t/t_m, \quad C_L = x/x_m, \quad C_a = a/a_m \\ C_\theta = \theta_0/\theta_{0m}, \quad & C_\lambda = \lambda/\lambda_m, \quad C_\beta = \beta/\beta_m, \quad C_s = s/s_m \end{aligned}\right\} \tag{16.4}$$

where the subscript "m" represents the model.

Substitution of Eq. (17.51) into Eqs. (16.1)–(16.3) yields

$$\frac{C_T}{C_t}\frac{\partial T_\mathrm{m}}{\partial t_\mathrm{m}} = \frac{C_a C_T}{C_L^2}\cdot a_\mathrm{m}\left(\frac{\partial^2 T_\mathrm{m}}{\partial x_\mathrm{m}} + \frac{\partial^2 T_\mathrm{m}}{\partial y_\mathrm{m}^2} + \frac{\partial^2 T_\mathrm{m}}{\partial z_\mathrm{m}^2}\right) + \frac{C_\theta}{C_t}\cdot\frac{\theta_{0\mathrm{m}}\partial(1-\mathrm{e}^{-C_s C_t s_\mathrm{m} t_\mathrm{m}})}{\partial t_\mathrm{m}} \tag{16.1b}$$

$$C_T C_L T_\mathrm{m}(0, x_\mathrm{m}, y_\mathrm{m}, z_\mathrm{m}) = C_T C_L T_0(x_\mathrm{m}, y_\mathrm{m}, z_\mathrm{m}) \tag{16.2b}$$

$$\frac{C_\lambda C_T}{C_L}\lambda_m \frac{\partial T_m}{\partial n_m} + C_\beta C_T\cdot\beta_m(T_m - T_{cm}) = 0 \tag{16.3b}$$

Comparing Eqs. (16.1b)–(16.3b) with Eqs. (16.1a)–(16.3a), it is clear that, in order to make the temperature field of the model similar to that of the prototype, the following conditions of similarity must be satisfied:

$$\frac{C_T}{C_t} = \frac{C_a C_T}{C_L^2} = \frac{C_\theta}{C_t},\quad \frac{C_\lambda C_T}{C_L} = C_\beta C_T,\quad C_s C_t = 1 \tag{16.5}$$

$$C_t = C_L^2/C_a,\quad C_\beta = C_\lambda/C_L,\quad C_\theta = C_T,\quad C_s = 1/C_t \tag{16.6}$$

There are eight parameters and four equations in Eq. (16.5), thus four parameters may be given freely and the other four parameters must be computed by Eqs. (16.5) and (16.6). For example, let $C_T = 1$, $C_a = 1.1$, $C_\lambda = 1.1$, $C_L = 20$, from Eq. (16.6), we have:

$$C_t = 20^2/1.1 = 364,\quad C_\beta = 1.1/20 = 0.055,\quad C_\theta = 1,\quad C_s = 1/364$$

In order to satisfy the similarity conditions, for the period of annual variation ($P = 1$ year) in the prototype, the period of variation in the model is $P_\mathrm{m} = 1/364$ year $=$ 1d; for the duration of cold wave $Q = 2-4$d in the prototype, in the model $Q_\mathrm{m} = 0.13 - 0.26$ h; in the prototype $\beta = 80$ kJ/(m^2 h °C), in the model, $\beta = 4.40$ kJ/(m^2 h °C). These conditions are possible to be fulfilled in the model, but it is difficult to make the rate of hydration heat in the model $s_\mathrm{m} = s/364$. Thus, it is possible to make a model test of the temperature field for the period of operation and it is difficult to conduct a model test of the temperature field for the period of construction.

In the construction period, the thermal stresses are related to the modulus of elasticity $E(\tau)$ and unit creep $C(t,\tau)$ which are functions of time t and age τ. It is difficult to make $E(\tau)$ and $C(t,\tau)$ to satisfy the similarity conditions. So it is difficult to do a model test for thermal stress in the period of construction.

The modulus of elasticity of concrete is nearly constant in the period of operation, so it is possible to do a model test for elastic thermal stress of mass concrete in the period of operation. Because the creep of concrete is difficult to satisfy the similarity condition, it is difficult to do a model test for viscoelastic thermal stress even in the period of operation.

In 2010, one engineer had made a 1:7 model test for the thermal stresses in the period of construction, and the material for model is also concrete, so $C_t = 1.0$, $C_a = 1.0$, $C_s = 1.0$, from Eq. (16.5), $C_L = 1.0$. Thus, the model must have the same dimensions as the prototype, and the model with $C_L = 7$ is not similar to the prototype.

17 Pipe Cooling of Mass Concrete

17.1 Introduction [3, 8, 32, 44, 46, 53, 80, 82, 83, 85, 86]

Pipe cooling was first used in the construction of the Hoover Dam in 1933. Coils of steel pipe with outer diameter 25 mm and thickness 1.5−1.8 mm are laid on the surface of lift of concrete before placing the new concrete as shown in Figure 17.1. The vertical spacing of the pipe generally is equal to the height of lift of concrete and the horizontal spacing is 1.5−3.0 m. The concrete is cooled by water flowing in the pipes. In the vertical cross section of concrete, if the pipes are laid in the form of a hexagon as shown in Figure 17.2 with horizontal spacing S_1 and vertical spacing S_2, then $S_2 = S_1 \cos 30°$, $S_1 = 1.1547S_2$ and the area of the hexagon is S_1S_2. From the condition $\pi b^2 = S_1S_2$, we get [3, 8]

$$b = \sqrt{S_1S_2/\pi} \tag{17.1}$$

b is the radius of a circle the area of which is equal to the area of the hexagon. The concrete cooled by each pipe is a cylinder with outer radius b. Due to symmetry, the surface of the cylinder is thermally insulated. The model of computation is a hollow cylinder with outer radius b given by Eq. (17.1) and inner radius c which is equal to the exterior radius of the cooling pipe.

The effect of cooling is best for the hexagonal arrangement of pipes, but in the construction of dams the actual arrangement generally is rectangular, the effect of cooling is somewhat lower. According to the results of computation, the cooling area must be increased 7% to consider this effect. Hence the radius b and diameter D of the cooling cylinder are given by (Figure 17.3).

$$b = \sqrt{1.07S_1S_2/\pi} = 0.5836\sqrt{S_1S_2}, \quad D = 2b = 1.1672\sqrt{S_1S_2} \tag{17.2}$$

where S_1—horizontal spacing of pipe and S_2—vertical spacing of pipe.

Generally the cooling of concrete is conducted in two stages. The early-stage cooling is conducted immediately after placing the concrete to reduce the temperature rise of concrete due to hydration heat of cement. The late-stage cooling is conducted before the grouting of contraction joints in the dam to reduce the temperature of concrete to the final steady temperature. In recent years, one to two intermediate stages of cooling are added for high concrete dams to reduce the

Thermal Stresses and Temperature Control of Mass Concrete. DOI: http://dx.doi.org/10.1016/B978-0-12-407723-2.00017-8

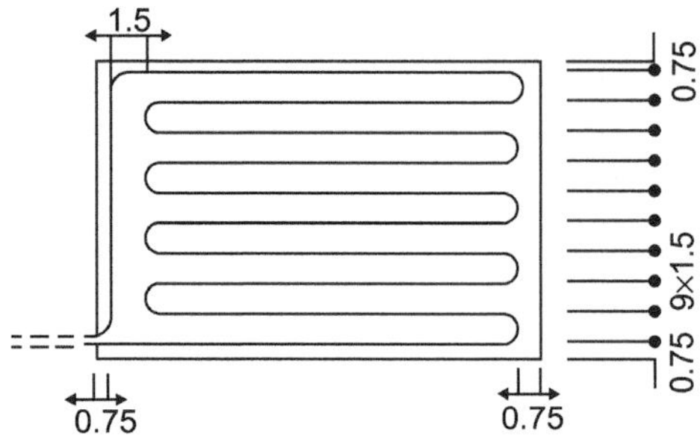

Figure 17.1 Schematic diagram for horizontal arrangement of cooling pipes.

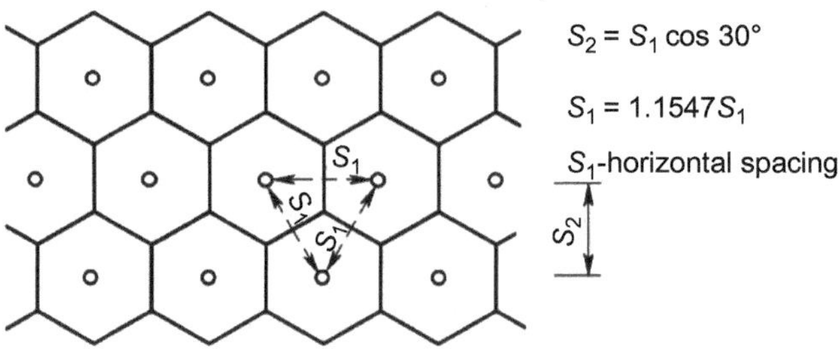

Figure 17.2 Hexagonal arrangement of cooling pipes.

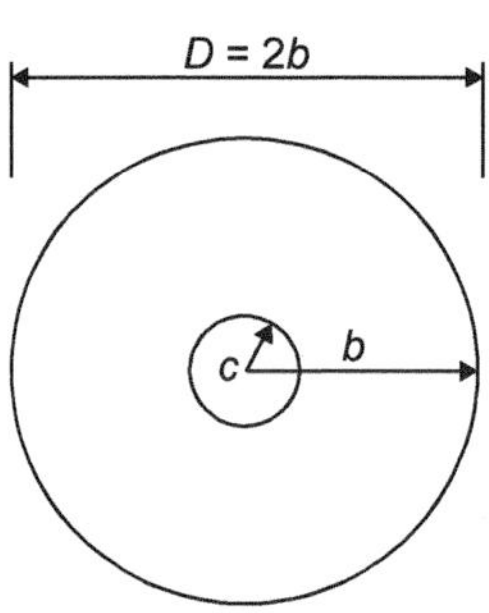

Figure 17.3 Scheme for computation.

thermal stresses, but the method of computation is the same as the late-stage cooling.

17.2 Plane Temperature Field of Pipe Cooling in Late Stage

17.2.1 Plane Temperature Field of Concrete Cooled by Nonmetal Pipe in Late Stage [3, 8, 32]

The computing model is shown in Figure 17.4. The inner surface of the pipe is in contact with water, so its temperature is equal to the water temperature. The temperature of the outer surface of the pipe is equal to the temperature T_c of the inner

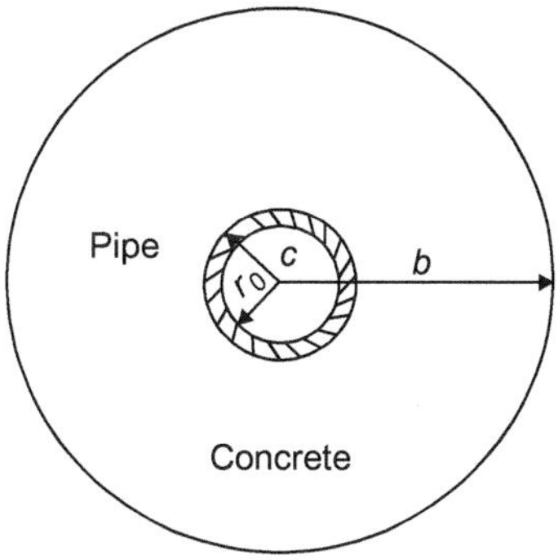

Figure 17.4 Computing model for nonmetal cooling pipe.

surface of concrete. Assuming that the water temperature is zero, the boundary condition of the pipe is: when $r = r_0$, $T = 0$; when $r = c$, $T = T_c$ and the radial flux of heat is as follows:

$$q = -\frac{\lambda_1 T_c}{c \ln(c/r_0)} = -k_5 T_c \tag{17.3}$$

where

$$k_5 = \frac{\lambda_1}{c \ln(c/r_0)} \tag{17.4}$$

in which

λ_1—coefficient of heat conduction of pipe
c—outer radius of pipe
r_0—inner radius of pipe.

The flux of heat q given by Eq. (17.3) must be equal to the heat flux of the inner surface of the concrete, i.e., $q = -\lambda[\partial T/\partial r]_{r=c}$, thus, the boundary conditions of the concrete cylinder are as follows:

$$\left.\begin{array}{lll} \text{when} & \tau = 0,\ c < r < b & T(r,0) = T_0 \\ \text{when} & \tau > 0,\ r = c & -\lambda \dfrac{\partial T}{\partial r} + k_5 T = 0 \\ \text{when} & \tau > 0,\ r = b & \dfrac{\partial T}{\partial r} = 0 \end{array}\right\} \tag{17.5}$$

The equation of heat conduction is

$$\frac{\partial T}{\partial \tau} = a\left(\frac{\partial^2 T}{\partial r^2} + \frac{1}{r}\frac{\partial T}{\partial r}\right) \tag{17.6}$$

where a, λ, T_0 are respectively the diffusivity, conductance, and the initial temperature of concrete.

By Laplace transformation, the solution is as follows [32]:

$$T(r,t) = T_0 \sum_{n=1}^{\infty} \frac{2\mathrm{e}^{-a\alpha_n^2\tau}}{\alpha_n b} \cdot \frac{J_1(\alpha_n b)Y_0(\alpha_n r) - Y_1(\alpha_n b)J_0(\alpha_n r)}{R_1(\alpha_n b)} \frac{n!}{r!(n-r)!} \tag{17.7}$$

$$\begin{aligned} R_1(\alpha_n b) &= -\frac{\lambda}{k_5 b}\alpha_n b \\ &\left\{\frac{c}{b}[J_1(\alpha_n b)Y_0(\alpha_n c) - J_0(\alpha_n c)Y_1(\alpha_n b)] + [J_0(\alpha_n b)Y_1(\alpha_n c) - J_1(\alpha_n c)Y_0(\alpha_n b)]\right\} \\ &+ \frac{c}{b}[J_1(\alpha_n b)Y_1(\alpha_n c) - J_1(\alpha_n c)Y_1(\alpha_n b)] + [J_0(\alpha_n c)Y_0(\alpha_n b) - J_0(\alpha_n b)Y_0(\alpha_n c)] \end{aligned} \tag{17.8}$$

in which $\alpha_n b$ is the root of the following characteristic equation

$$\frac{\lambda}{k_5}\alpha_n[J_1(\alpha_n c)Y_1(\alpha_n b) - J_1(\alpha_n b)Y_1(\alpha_n c)] + [J_1(\alpha_n b)Y_0(\alpha_n c) - J_0(\alpha_n c)Y_1(\alpha_n b)] = 0 \tag{17.9}$$

The mean temperature is

$$T_{\mathrm{m}} = T_0 \sum_{n=1}^{\infty} H_n \mathrm{e}^{-\alpha_n^2 b^2 \alpha\tau/b^2} = T_0 F(\tau) \tag{17.10}$$

where

$$H_n = \frac{4bc}{b^2 - c^2} \cdot \frac{Y_1(\alpha_n b)J_1(\alpha_n c) - J_1(\alpha_n b)Y_1(\alpha_n c)}{\alpha_n^2 b^2 R_1(\alpha_n b)} \tag{17.11}$$

Equation (17.10) converges so quickly that only one term is required in practical computation with error $\leq 1\%$ and $H_1 \cong 1.00$, thus, the mean temperature may be computed by

$$T_{\mathrm{m}} = T_0 \mathrm{e}^{-\alpha_1^2 b^2 \alpha\tau/b^2} \tag{17.12}$$

The characteristic root $\alpha_1 b$ is given in Table 17.1 and Figure 17.5.

Let the temperature of cooling water be T_{w} and the initial temperature of concrete be T_0, the mean temperature of concrete is given by

$$T_{\mathrm{m}} = T_{\mathrm{w}} + (T_0 - T_{\mathrm{w}})\mathrm{e}^{-\alpha_1^2 b^2 \alpha\tau/b^2} \tag{17.13}$$

Table 17.1 The Characteristic Root $\alpha_1 b$ for Cooling of Concrete by Nonmetal Pipe

b/c	$\lambda/(k_5 b)$					
	0	**0.010**	**0.020**	**0.030**	**0.040**	**0.050**
20	0.926	0.888	0.857	0.827	0.800	0.778
50	0.787	0.734	0.690	0.652	0.620	0.592
80	0.738	0.668	0.617	0.576	0.542	0.512

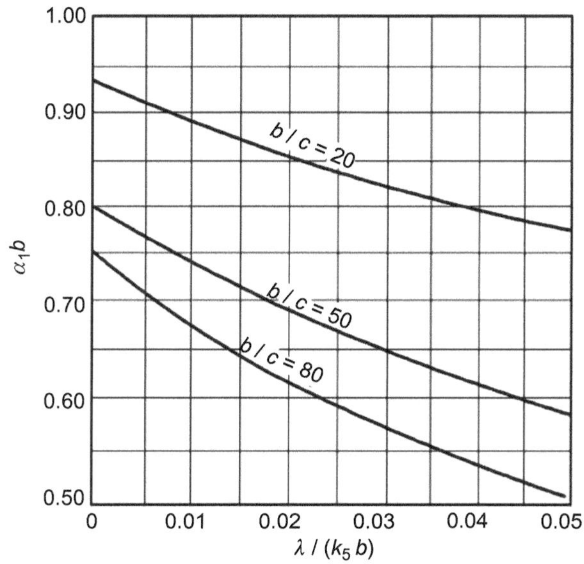

Figure 17.5 The characteristic root $\alpha_1 b$ for cooling of concrete by nonmetal pipe.

Example 1 Plane problem for cooling of concrete by nonmetal pipe, the diffusivity of concrete $a = 0.0040\ \text{m}^2/\text{h}$, the conductivity of concrete $\lambda = 8.37$ kJ/(m h °C), the outer radius of the cooling cylinder $b = 0.845$ m, the inner radius $c = 0.0160$ m, the outer radius of the polythene pipe $c = 1.60$ cm, the inner radius $r_0 = 1.40$ cm, the conductivity of pipe $\lambda_1 = 1.66$ kJ/(m h °C). The initial temperature of concrete $T_0 = 20°\text{C}$, the water temperature $T_w = 0°\text{C}$. Try to compute the variation with time of the mean temperature of concrete.

From Eq. (17.4),

$$k_5 = \frac{\lambda_1}{c \cdot \ln(c/r_0)} = \frac{1.66}{0.016 \cdot \ln(0.016/0.014)} = 777.0$$

$$\frac{\lambda}{k_5 b} = \frac{8.37}{777.0 \times 0.845} = 0.01275$$

$$\frac{b}{c} = \frac{0.845}{0.0160} = 52.81$$

and from Figure 17.5, $\alpha_1 b = 0.712$, substitute it into Eq. (17.13), we get the mean temperature of concrete as follows:

$$T_m = 20e^{-0.002840\tau}$$

in which the unit of time τ is h. If the concrete is cooled by steel pipe with outer radius $c = 1.60$ cm, $k_1 = \infty$, $\lambda/(k_5 b) = 0$, $\alpha_1 b = 0.783$, from Eq. (17.12), the mean temperature of concrete is

$$T'_m = 20e^{-0.003435\tau}$$

the variations of T_m and T'_m with time are shown in Figure 17.6.

17.2.2 Plane Temperature Field of Concrete Cooled by Metal Pipe in Late Stage

The initial temperature of the concrete cylinder is T_0, the outer surface is insulated, as the conductivity of metal pipe is considerable, the temperature of the inner surface of concrete is practically equal to water temperature $T_w = 0°C$.

The equation of heat conduction is

$$\frac{\partial T}{\partial \tau} = a\left(\frac{\partial^2 T}{\partial r^2} + \frac{1}{r}\frac{\partial T}{\partial r}\right) \tag{17.14}$$

The initial and boundary conditions are as follows:

$$\begin{array}{lll} \text{when} & \tau = 0,\ c \le r \le b & T(r,0) = T_0 \\ \text{when} & \tau > 0,\ r = c & T(c,\tau) = 0 \\ \text{when} & \tau > 0,\ r = b & \dfrac{\partial T}{\partial r} = 0 \end{array} \tag{17.15}$$

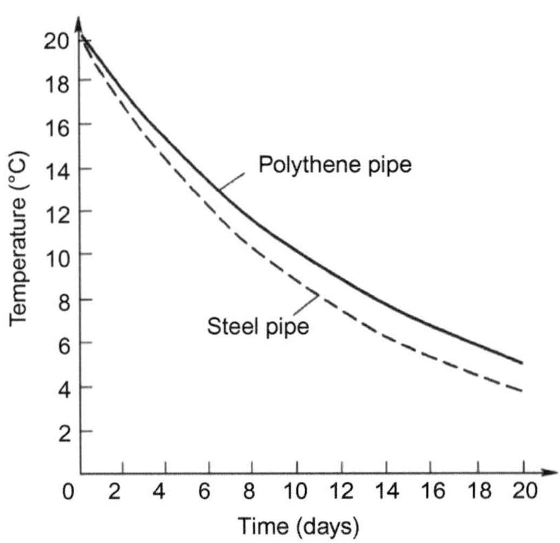

Figure 17.6 Example of plane problem for pipe cooling of concrete without internal source of heat.

The solution of the above problem given by the U.S. Bureau of Reclamation converges slowly. The following solution is given by the author by Laplace transformation:

$$T(r,t) = T_0 \sum_{n=1}^{\infty} \frac{2e^{-\alpha_n^2 b^2 a\tau/b^2}}{\alpha_n b} \cdot \frac{J_1(\alpha_n b)Y_0(\alpha_n r) - Y_1(\alpha_n b)J_0(\alpha_n r)}{R(\alpha_n b)} \tag{17.16}$$

$$\begin{aligned} R(\alpha_n b) = (c/b)[J_1(\alpha_n b)Y_1(\alpha_n c) - J_1(\alpha_n c)Y_1(\alpha_n b)] \\ + [J_0(\alpha_n c)Y_0(\alpha_n b) - J_0(\alpha_n b)Y_0(\alpha_n c)] \end{aligned} \tag{17.17}$$

where

J_0, J_1—the first kind of Bessel functions of zero order and first order
Y_0, Y_1—the second kind of Bessel functions of zero order and first order
$\alpha_n b$—the root of the following characteristic equation.

$$J_1(\alpha_n b)Y_0(\alpha_n c) - J_0(\alpha_n c)Y_1(\alpha_n b) = 0 \tag{17.18}$$

when $b/c = 100$, the first 5 roots of the above equation are $\alpha_n b = 0.716691$, 4.289947, 7.546395, 10.766313, 13.972000. The first root $\alpha_1 b$ may be computed by Eq. (17.40). The distributions of $T(r,\tau)/T_0$ for various time intervals of a hollow cylinder with $b/c = 100$ are shown in Figure 17.7. The mean temperature is

$$T_m = \int_c^b 2\pi r T(r,\tau)dr \Big/ \int_c^b 2\pi r\, dr = T_0 \sum_{n=1}^{\infty} H_n e^{-\alpha_n^2 b^2 a\tau/b^2} = T_0 F(\tau) \tag{17.19}$$

where

$$H_n = \frac{4bc}{b^2 - c^2} \cdot \frac{Y_1(\alpha_n b)J_1(\alpha_n c) - J_1(\alpha_n b)Y_1(\alpha_n c)}{\alpha_n^2 b^2 R(\alpha_n b)} \tag{17.20}$$

The solution Eq. (17.19) converges so quickly that practically only one term is required with error $\leq 1.0\%$ and $H_1 \cong 1.00$, thus taking $H_1 = 1.00$, the mean temperature is given by

$$T_m = T_0 e^{-\alpha_1^2 b^2 a\tau/b^2} \tag{17.21}$$

when $b/c = 100$, $\alpha_1 b = 0.7167$, thus

$$T_m = T_0 e^{-0.7167^2 a\tau/b^2} = T_0 e^{-0.5136 a\tau/b^2} \tag{17.22}$$

The approximate value of T_m given by Eq. (17.22) and the precise T_m are given in Table 17.2. It is clear that the precision of Eqs. (17.21) and (17.22) is very high.

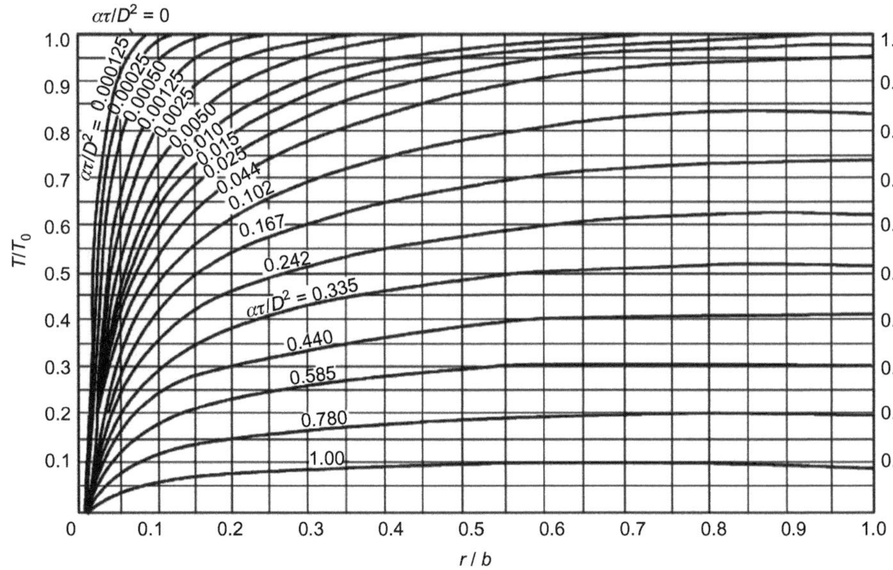

Figure 17.7 The distribution of temperature of a hollow cylinder ($b/c = 100$) at different time.

The metal cooling pipe is a special case of nonmetal pipe. Let the conductivity of pipe $\lambda_1 = \infty$, then $k_5 = \infty$, the $R_1(\alpha_n b)$ in Eq. (17.8) is equal to the $R(\alpha_n b)$ in Eq. (17.17), the characteristic Eq. (17.9) degenerates to Eq. (17.18).

17.3 Spatial Temperature Field of Pipe Cooling in Late Stage

17.3.1 Method of Solution of the Spatial Problem of Pipe Cooling

In the above computation of plane problem, the temperature of water is assumed to be constant. Practically, due to absorption of heat from concrete, the temperature of water will increase gradually and the temperature of water at the outlet will be higher than that at the inlet. This is a spatial problem which is difficult to solve. In the construction of a concrete dam, the length of the cooling pipes is 200–500 m, and the spacing of the pipes is 1.5–3.0 m which is far smaller than the length. The heat is conducted primarily in the plane perpendicular to the axis of the pipes and the temperature gradient of concrete in the direction of axis of the pipes may be neglected. Thus the actual temperature in the concrete can still be analyzed as a plane problem but the influence of the variation of water temperature must be taken into account. The key for the spatial problem is the computation of water temperature which increases gradually along with the length of pipe.

Table 17.2 Mean Temperature T_m of a Cylinder ($b/c = 100$, Initial Temperature T_0)

$a\tau/b^2$		0.0025	0.025	0.100	0.250
T_m/T_0	Exact value	0.975	0.811	0.440	0.128
	Eq. (17.22)	0.979	0.814	0.440	0.128
	Error	0.41%	0.37%	0	0

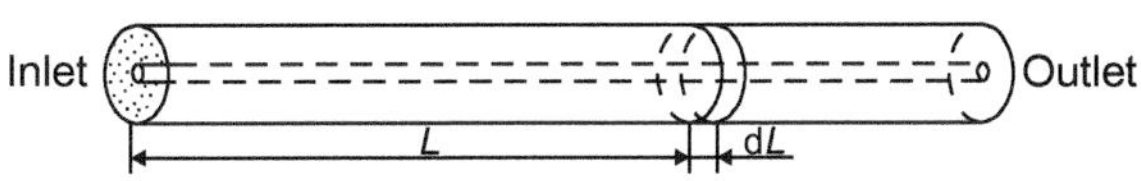

Figure 17.8 The spatial problem of pipe cooling.

As shown in Figure 17.8, let T_0 be the initial temperature of concrete, T_w be the temperature of water at the inlet of the pipe, T_{Lw} be the temperature of water at the section where the length of pipe is L, T_{Lm} be the mean temperature of concrete at the section where the length of pipe is L, and T_m be the mean temperature of concrete in the range of length of pipe $0-L$, defined three variables as follows:

$$X = \frac{T_m - T_w}{T_0 - T_w}, \quad Y = \frac{T_{Lw} - T_w}{T_0 - T_w}, \quad Z = \frac{T_{Lm} - T_w}{T_0 - T_w} \tag{17.23}$$

hence

$$\left.\begin{aligned} T_m &= T_w + X(T_0 - T_w) \\ T_{Lw} &= T_w + Y(T_0 - T_w) \\ T_{Lm} &= T_w + Z(T_0 - T_w) \end{aligned}\right\} \tag{17.24}$$

The variables X, Y, and Z will be computed in the following. As the problem is linear, the flux of heat may be divided into three parts as follows:

1. Q_1—the amount of heat flowed from the concrete into the water;
2. Q_2—the amount of heat flowed from the warmed water into the concrete with zero initial temperature;
3. Q_3—the amount of heat absorbed by the water in the range of length $0-L$.

The amount of heat flows into the water in unit of time from the concrete cylinder of unit length is

$$\frac{\partial Q_1}{\partial L} = 2\pi c\lambda\left(-\frac{\partial T}{\partial r}\right)_{r=c} = \lambda(T_0 - T_w)R(t) \tag{17.25}$$

The amount of heat flows into the water is equal to the loss of heat of concrete, so $\partial Q_1/\partial L$ can also be computed by

$$\frac{\partial Q_1}{\partial L} = \pi(b^2 - c^2)c_1\rho\left(-\frac{dT_m}{dt}\right) = \lambda(T_0 - T_w)R(t) \tag{17.26}$$

Substitution of Eq. (17.7) into (17.25) or substitution Eq. (17.10) into Eq. (17.26) yields

$$R(t) = \frac{4\pi c}{b}\sum \frac{e^{-\alpha_n^2 b^2 at/b^2}}{R_1(\alpha_n b)}[Y_1(\alpha_n b)J_1(\alpha_n c) - J_1(\alpha_n b)Y_1(\alpha_n c)] \tag{17.27}$$

The above series converges so quickly that only one term is required in engineering computation. Substituting Eq. (17.12) into Eq. (17.26), we get

$$R(t) \cong \pi\left(1 - \frac{c^2}{b^2}\right)\alpha_1^2 b^2 e^{-\alpha_1^2 b^2 at/b^2} \tag{17.28}$$

The increase of water temperature from time τ to $\tau + d\tau$ is $\Delta T_{Lw} = (T_0 - T_w)(\partial Y/\partial\tau)d\tau$. The flux of heat from water to concrete due to ΔT_{Lw} at time t is

$$\frac{\partial Q_2}{\partial L} = \lambda(T_0 - T_w)R(t-\tau)\frac{\partial Y}{\partial \tau}d\tau \tag{17.29}$$

At time t, the amount of heat absorbed by the water in the range $0-L$ is

$$Q_3 = c_w \rho_w q_w (T_0 - T_w)Y(t, L) \tag{17.30}$$

From balance of heat,

$$Q_3 = \int_0^L \frac{\partial Q_1}{\partial L}dL - \int_0^L \frac{\partial Q_2}{\partial L}dL \tag{17.31}$$

Substituting Eqs. (17.26), (17.29), and (17.30) into Eq. (17.31) and canceling $T_0 - T_w$, we obtain the fundamental equation for computing the water temperature as follows:

$$\eta q_w Y(t, L) = LR(t) - \int_0^t \int_0^L R(t-\tau)\frac{\partial Y}{\partial \tau}d\tau\, dL \tag{17.32}$$

where

$$\eta = c_w \rho_w / \lambda$$

Equation (17.32) is an integral equation. Now the time t is divided into n time intervals Δt_i, $i = 1 - n$; the length L is divided into m subregions ΔL_j, $j = 1 - m$, we have

$$\int_{t_{i-1}}^{t_i}\int_{L_{j-1}}^{L_j} R(t-\tau)\frac{\partial Y}{\partial \tau}d\tau\, dL \cong R(t - t_{i-0.5})\Delta Y(t_{i-0.5}, L_{j-0.5})\Delta L_j \tag{17.33}$$

Substituting the above equation into Eq. (17.32), we get

$$\eta q_{\mathrm{w}} Y(t, L) = LR(t) - \sum_i \sum_j R(t - t_{i-0.5}) \Delta Y(t_{i-0.5}, L_{j-0.5}) \Delta L_j \tag{17.34}$$

Since $\Delta Y(t_{i-0.5}, L_{j-0.5})$ is unknown, the above equation must be solved by method of iteration. For example, for the first time interval and first subregion ($t = 0 \sim t_1$, $L = 0 \sim L_1$), i.e., ($i = 1$, $j = 1$), from Eq. (17.34), we have

$$\eta q_{\mathrm{w}} Y(t, L) = LR(t) - R(t - t_{i-0.5}) \Delta Y(t_{0.5}, L_{0.5}) \Delta L_1 \tag{17.35}$$

at first, let $\Delta Y(t_{0.5}, L_{0.5}) = 0$, from Eq. (17.35), we get the first approximate value as follows:

$$\eta q_{\mathrm{w}} Y^{(1)}(t, L) = LR(t)$$

hence we have

$$Y^{(1)}(t_1, L_{0.5}) = \frac{1}{\eta q_{\mathrm{w}}} L_{0.5} R(t_1)$$

$$Y^{(1)}(0, L_{0.5}) = \frac{1}{\eta q_{\mathrm{w}}} L_{0.5} R(0)$$

Thus, the first approximate value of $\Delta Y(t_{0.5}, L_{0.5})$ is

$$\Delta Y^{(1)}(t_{0.5}, L_{0.5}) = \frac{L_{0.5}}{\eta q_{\mathrm{w}}} [R(t_1) - R(0)] \tag{17.36}$$

Substitution of Eq. (17.36) into Eq. (17.35) yields the second approximate value. Repeating the above computation, $Y(t, L)$ of higher and higher precision will be obtained.

In Eqs. (17.32) and (17.34), q_{w} is known but it may be a known value varying with time.

The mean temperature of the section of concrete at L is ZT_0, from Eq. (17.19), if the influence of the increase of water temperature is neglected, $ZT_0 = T_0 F(t)$; if the influence of increase of water temperature is taken into account, ZT_0 is given by the following equation:

$$Z(t, \xi) T_0 = T_0 F(t) + \int_0^t T_0 [1 - F(t - \tau)] \frac{\partial Y}{\partial \tau} \mathrm{d}\tau$$

or

$$Z(t, \xi) = F(t) + \sum [1 - F(t - \tau)] \Delta Y(\tau, \xi) \tag{17.37}$$

The mean temperature of concrete in the range of length $0-\underline{L}$ is XT_0, X is given by

$$X(t,\xi) = \frac{1}{\xi}\int_0^{\xi} Z(t,\xi)\mathrm{d}\xi \tag{17.38}$$

where $\xi = \lambda L/(c_w \rho_w q_w)$ and $\mathrm{d}\xi = \lambda \mathrm{d}L/(c_w \rho_w q_w)$.

Thus, Y, Z, X are determined by Eqs. (17.32), (17.37), and (17.38).

17.3.2 Spatial Cooling of Concrete by Metal Pipe in Late Stage

The X, Y, Z for spatial cooling of concrete by metal pipe with $b/c = 100$ at late stage are given by U.S. Bureau of Reclamation as shown in Figures 17.9–17.11 [3].

When $b/c \neq 100$, instead of the diffusivity a, we must use an equivalent diffusivity a' as follows:

$$a' = \left(\frac{\alpha_1 b}{0.7167}\right)^2 a = 1.947(\alpha_1 b)^2 a \tag{17.39}$$

in which 0.7167 is the characteristic root for $b/c = 100$ and $\alpha_1 b$ is the characteristic root for $b/c \neq 100$ given by Figure 17.12 or by the following equation:

$$\alpha_1 b = 0.926 \exp\left[-0.0314\left(\frac{b}{c}-20\right)^{0.48}\right], \quad 20 \le \frac{b}{c} \le 130 \tag{17.40}$$

The equivalent diffusivity of concrete may also be given by the following approximate formula suggested by U.S. Bureau of Reclamation:

$$a' = \frac{a \ln 100}{\ln(b/c)} \tag{17.41}$$

Equation (17.39) is more precise than Eq. (17.41). The computed results of these two formulas are shown in Table 17.3, from which it is clear that the error is not too big when $50 \le b/c \le 130$.

Example 1 For the concrete, the initial temperature $T_0 = 30$°C, the diffusivity $a = 0.0040$ m^2/h, the conductivity $\lambda = 8.37$ kJ/(m h °C). The outer radius of pipe $c = 1.25$ cm, the arrangement of the cooling pipes is rectangular, both the horizontal spacing s_1 and vertical spacing s_2 are equal to 1.50 m. For the cooling water, the temperature at the inlet $T_w = 10$°C, the specific heat $c_w = 4.187$ kJ/(kg °C), the density $\rho_w = 1000$ kg/m^3, the discharge $q_w = 15$ L/min = 0.90 m^3/h. The length of pipe $L = 200$ m. Try to compute the mean temperature T_m of concrete of length L, the

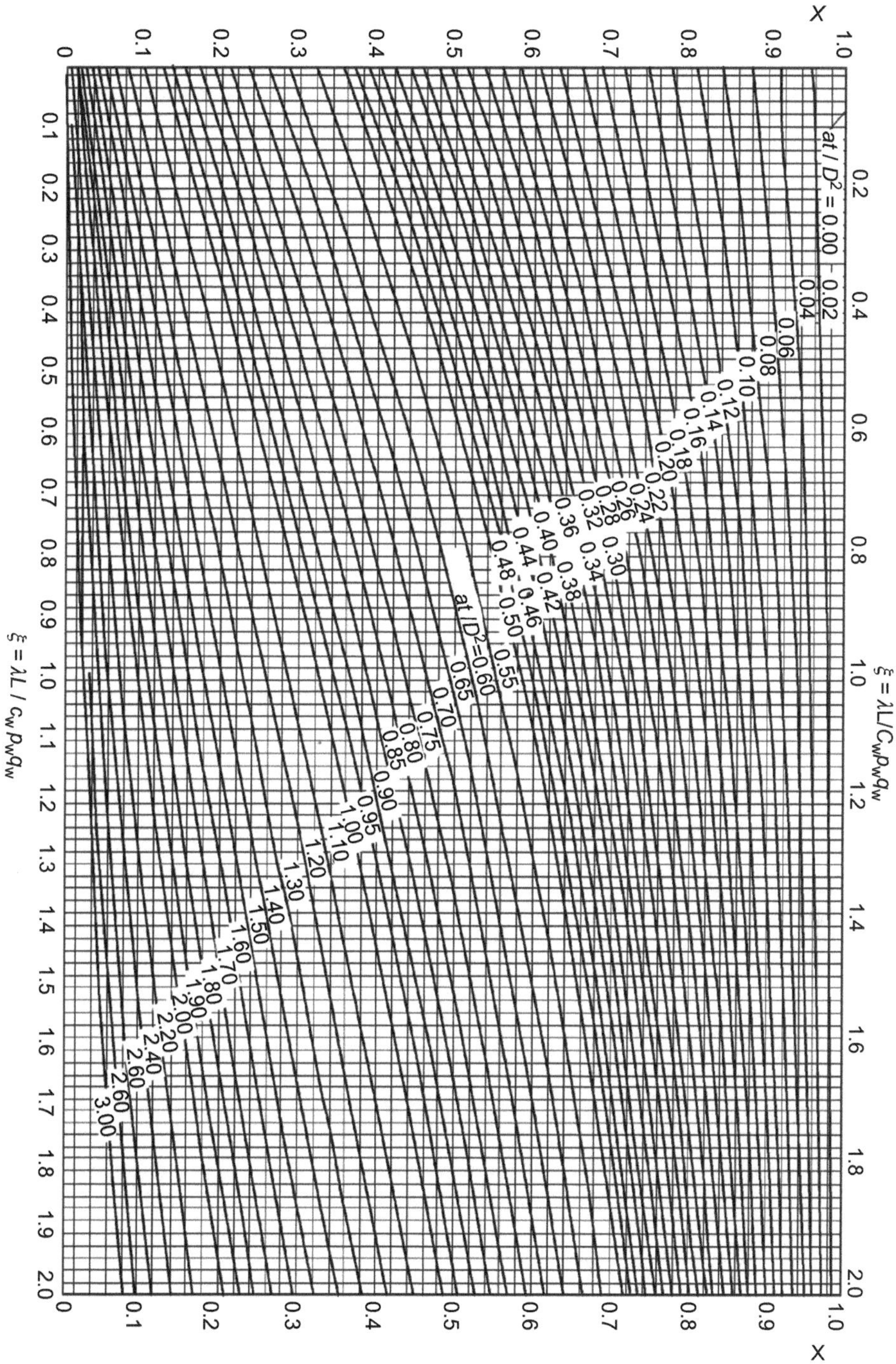

Figure 17.9 *X* for cooling of concrete by metal pipe in late stage (*b*/*c* = 100).

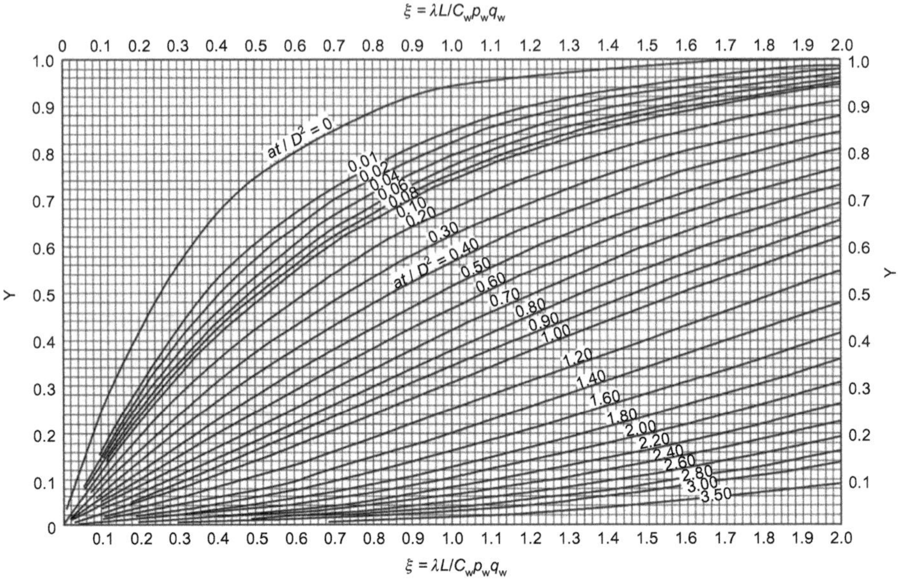

Figure 17.10 *Y* for cooling of concrete by metal pipe in late stage ($b/c = 100$).

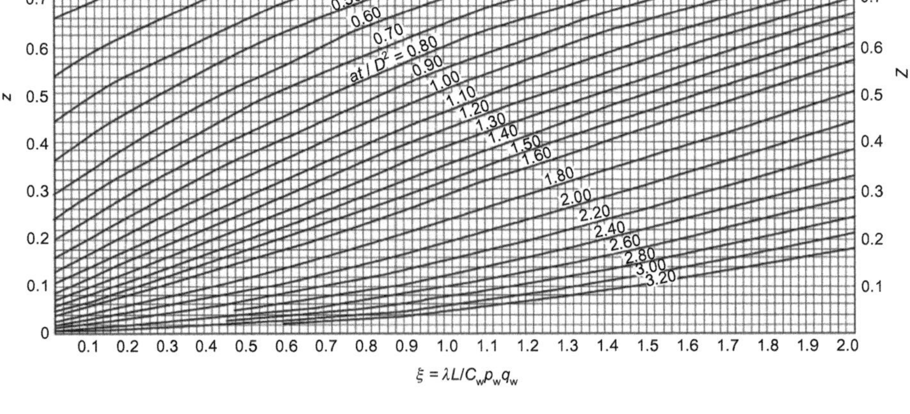

Figure 17.11 *Z* for cooling of concrete by metal pipe in late stage ($b/c = 100$).

water temperature T_{Lw} at the outlet, and the mean temperature of concrete of the section at length L.

From Eq. (17.2), $b=\sqrt{1.07\times 1.5\times 1.5/\pi}=0.875$ m, $b/c=0.875/0.0125=70.0$.

From Eq. (17.40), the characteristic root is

$$a_1 b=0.926\,\exp[-0.0314(70-20)^{0.48}]=0.754$$

From Eq. (17.29), the equivalent diffusivity is

$$a'=1.947\times 0.754^2\times 0.096=0.1063\ \mathrm{m^2/day}$$

$$a't/D^2=0.1063\times 20/(2\times 0.875)^2=0.694$$

$$\xi=\frac{\lambda L}{c_w\rho_w q_w}=\frac{8.37\times 200}{4.187\times 1000\times 0.90}=0.444$$

From Figures 17.9–17.11, we get $X=0.349$, $Y=0.187$, and $Z=0.460$. From Eq. (17.24), we have

$$T_m=T_w+X(T_0-T_w)=10+0.349\times(30-10)=17.0°\mathrm{C}$$

$$T_{Lw}=T_w+Y(T_0-T_w)=10+0.187\times(30-10)=13.74°\mathrm{C}$$

$$T_{Lm}=T_w+Z(T_0-T_w)=10.0+0.460(30-10)=19.2°\mathrm{C}$$

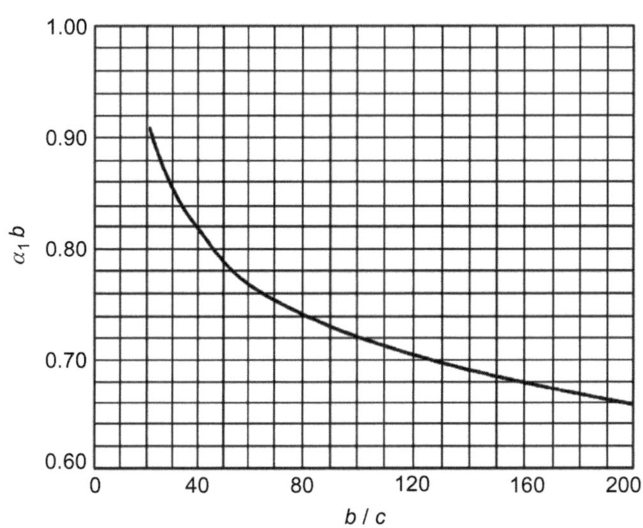

Figure 17.12 The characteristic root $\alpha_1 b$ for cooling of concrete by metal pipe.

Table 17.3 Comparison of the Equivalent Diffusivity

b/c		20	50	80	100	130
$\alpha_1 b$		0.926	0.787	0.738	0.7167	0.692
a'/a	Eq. (17.39)	1.669	1.206	1.060	1.00	0.932
	Eq. (17.41)	1.537	1.177	1.051	1.00	0.946
	Error	−7.9%	−2.4%	−0.85%	0	1.47%

17.3.3 Spatial Cooling of Concrete by Nonmetal Pipe in Late Stage [53]

1. Method I

Comparing Eq. (17.12) for nonmetal pipe with Eq. (17.21) for metal pipe, it is clear that the structure of the two formulas is the same, the difference lies in the value of characteristic root $\alpha_1 b$ which may be found in Figure 17.5. A more precise formula for $\alpha_1 b$ will be derived in the following.

Consider a cooling pipe with outer radius c and inner radius r_0, when $r = c$, $T = T_c$ and when $r = r_0$, $T = 0$, the solution of steady temperature is

$$T(r) = T_c \frac{\ln(r/r_0)}{\ln(c/r_0)} \tag{17.42}$$

The temperature gradient is

$$\frac{\partial T}{\partial r} = \frac{T_c}{r \ln(c/r_0)} \tag{17.43}$$

The original problem is a composite cylinder consisting of two parts: the outer part is *a* concrete cylinder with outer radius b, inner radius c and conductivity λ; the inner part is the cooling pipe with outer radius c, inner radius r_0, and conductivity λ_1. Instead of the original composite cylinder, we will give an equivalent homogeneous concrete cylinder with outer radius b, inner radius r_1, and conductivity λ and the surface temperature at $r = r_1$ is equal to the water temperature. Thus, the new problem can be solved by the conventional method for concrete cooled by metal pipe with outer radius r_1. The key lies in the computing of r_1.

Assuming that the flux of heat of the new problem is equal to that of the original problem when $r = c$

$$\frac{\lambda T_c}{c \ln(c/r_1)} = \frac{\lambda_1 T_c}{c \ln(c/r_0)}$$

so

$$\ln(c/r_1) = \eta \ln(c/r_0) = \ln(c/r_0)^\eta$$

hence

$$r_1 = c\left(\frac{r_0}{c}\right)^\eta \tag{17.44}$$

where $\eta = \lambda/\lambda_1$. As $r_0/c < 1$, thus $r_1 < c$. The original problem of concrete cooled by nonmetal pipe is substituted by a new problem of concrete cooled by metal pipe.

Substituting r_1 in Eq. (17.40), we get the characteristic root $\alpha_1 b$ for the nonmetal pipe as follows:

$$\alpha_1 b = 0.926\ \exp\left\{-0.0314\left[\frac{b}{c}\left(\frac{c}{r_0}\right)^{\eta} - 20\right]^{0.48}\right\} \tag{17.45}$$

From Eq. (17.39), the equivalent diffusivity a' is given as follows:

$$a' = 1.67a \cdot \exp\left\{-0.0628\left[\frac{b}{c}\left(\frac{c}{r_0}\right)^{\eta} - 20\right]^{0.48}\right\} \tag{17.46}$$

2. Method II

Let

$$\frac{\ln(b/r_1)}{\lambda} = \frac{\ln(b/c)}{\lambda} + \frac{\ln(c/r_0)}{\lambda_1}$$

Substituting $\ln(b/c)$ in Eq. (17.39) by $\ln(b/r_1)$, we get the equivalent diffusivity a' of concrete as follows:

$$a' = \frac{a \ln 100}{\ln(b/c) + (\lambda/\lambda_1)\ln(c/r_0)} \tag{17.47}$$

Example 2 Concrete is cooled by polythene pipe. For concrete, $a = 0.0040$ m^2/h, $\lambda = 8.37$ kJ/(m h °C), $b = 0.845$ m.

For the polythene pipe, the outer radius $c = 1.60$ cm, the inner radius $r_0 = 1.40$ cm, the conductivity $\lambda_1 = 1.66$ kJ/(m h °C). Try to compute the equivalent diffusivity of concrete a'.

$$\eta = \lambda/\lambda_1 = 8.37/1.66 = 5.04$$

Method I, from Eq. (17.46)

$$a' = 1.67 \times 0.0040 \times \exp\left\{-0.0628\left[\frac{0.845}{0.016}\left(\frac{0.016}{0.014}\right)^{5.04} - 20\right]^{0.48}\right\} = 0.00395\ \text{m}^2/\text{h}$$

Method II, from Eq. (17.47)

$$a' = \frac{0.0040 \ln 100}{\ln(0.845/0.016) + (8.37/1.66)\ln(0.016/0.014)} = 0.00397\ \text{m}^2/\text{h}$$

For the same polythene pipe with $c = 1.60$ cm, $r_0 = 1.40$ cm, and $\eta = 5.04$, the values of a'/a given by the two methods for different b/c are shown in Table 17.4.

It is clear that the results given by two methods are close to each other when $20 \leq b/c \leq 80$, but when $b/c > 100$, the precision of method B is rather low.

17.4 Temperature Field of Pipe Cooling in Early Stage

The pipe cooling of early stage is conducted immediately after placing the concrete, so it is necessary to take the hydration heat of cement into account. Let T_w be the water temperature, T_0 be the initial temperature of concrete, and $\theta(\tau)$ be the adiabatic temperature rise due to hydration heat. The temperature field may be divided into two parts: the first part is due to the influence of $T_0 - T_w$ which can be computed by the above-mentioned method for cooling in late stage, the second part is due to the influence of the adiabatic temperature rise $\theta(\tau)$ whose computing method will be given in the following [8, 32].

17.4.1 Plane Problem of Pipe Cooling of Early Stage

Consider the pipe cooling of a concrete cylinder, the initial temperature of concrete T_0 and the water temperature T_w are equal to zero, and the adiabatic temperature rise due to hydration heat of cement is

$$\theta(\tau) = \theta_0(1 - e^{-m\tau}) \tag{17.48}$$

$$\frac{\partial \theta}{\partial \tau} = \theta_0 m e^{-m\tau} \tag{17.49}$$

The equation of heat conduction is

$$\frac{\partial T}{\partial \tau} = a\left(\frac{\partial^2 T}{\partial r^2} + \frac{1}{r}\frac{\partial T}{\partial r}\right) + \theta_0 m e^{-m\tau} \tag{17.50}$$

At first consider the concrete cooled by metal pipe, the initial and the boundary conditions are as follows:

$$\left.\begin{array}{lll} \text{when} & \tau = 0,\ c \leq r \leq b & T(r,0) = 0 \\ \text{when} & \tau > 0,\ r = c & T(c,\tau) = 0 \\ \text{when} & \tau > 0,\ r = b & \dfrac{\partial T}{\partial r} = 0 \end{array}\right\} \tag{17.51}$$

The solution given by Laplace transformation [32] is as follows:

$$\begin{aligned} T &= \theta_0 e^{-(b\sqrt{m/a})^2 at/b^2}\left[\frac{Y_1(b\sqrt{m/a})J_0(r\sqrt{m/a}) - J_1(b\sqrt{m/a})Y_0(r\sqrt{m/a})}{Y_1(b\sqrt{m/a})J_0(c\sqrt{m/a}) - J_1(b\sqrt{m/a})Y_0(c\sqrt{m/a})} - 1\right] \\ &\quad + 2\theta_0 \sum_{n=1}^{\infty} \frac{e^{-\alpha_n^2 b^2 at/b^2}}{\left[1 - \alpha_n^2 b^2/(b\sqrt{m/a})^2\right]\alpha_n b} \cdot \frac{Y_0(\alpha_n r)J_1(\alpha_n b) - Y_1(\alpha_n b)J_0(\alpha_n r)}{R(\alpha_n b)} \end{aligned} \tag{17.52}$$

Table 17.4 Ratio a'/a of Equivalent Diffusivity Given by Two Methods

b/c	20	50	80	100	130
Method A, Eq. (17.46)	1.288	1.004	0.858	0.788	0.705
Method B, Eq. (17.47)	1.255	1.008	0.910	0.873	0.842
Error (%)	2.5	0.5	6.1	10.8	19.5

in which $R(\alpha_n b)$ is given by Eq. (17.17) and $\alpha_n b$ is the root of Eq. (17.18).

The mean temperature is

$$T_m = \theta_0 e^{-(b\sqrt{m/a})^2\cdot at/b^2}\left[\frac{2bc}{(b^2-c^2)b\sqrt{m/a}}\times\frac{J_1(b\sqrt{m/a})Y_1(c\sqrt{m/a})-J_1(c\sqrt{m/a})Y_1(b\sqrt{m/a})}{J_0(c\sqrt{m/a})Y_1(b\sqrt{m/a})-J_1(b\sqrt{m/a})Y_0(c\sqrt{m/a})}-1\right]$$
$$+\frac{4\theta_0 bc}{b^2-c^2}\sum_{n=1}^{\infty}\frac{e^{-(\alpha_n b)^2\cdot at/b^2}}{[1-\alpha_n^2 b^2/(b\sqrt{m/a})^2]\alpha_n^2 b^2}\cdot\frac{J_1(\alpha_n c)Y_1(\alpha_n b)-J_1(\alpha_n b)Y_1(\alpha_n c)}{R(\alpha_n b)} \quad (17.53)$$

The above series converges so quickly that only one term is required in practical computation.

Equations (17.52) and (17.53) represent an exact solution. A convenient approximate solution will be given in the following. In the time interval $d\tau$, the adiabatic temperature rise is $d\theta$ which may be considered as an initial temperature difference, due to pipe cooling, from Eq. (17.21), the mean temperature at time t will be

$$dT_m = e^{-\alpha_1^2 a(t-\tau)}d\theta \quad (17.54)$$

Integrating from 0 to t, we get an approximate formula for the mean temperature of first-stage cooling as follows:

$$T_m = \int_0^t e^{-\alpha_1^2 a(t-\tau)}\cdot\frac{\partial\theta}{\partial\tau}d\tau = \frac{m\theta_0}{m-a\alpha_1^2}(e^{-a\alpha_1^2 t}-e^{-mt}) \quad (17.55)$$

If $\alpha_1 b$ is the roof of the characteristic equation Eq. (17.9) for cooling by non-metal pipe, the above equation may be used to compute the temperature of concrete cooled by nonmetal pipe.

Example Plane problem of first-stage pipe cooling. Comparison between metal pipe and nonmetal pipe. For concrete, diffusivity $a = 0.0040\ m^2/h$, conductivity $\lambda = 8.37$ kJ/(m h °C), outer radius $b = 0.845$ m, inner radius $c = 0.0160$ m. For polythene pipe, outer radius $c = 1.60$ cm, inner radius $r_0 = 1.40$ cm, conductivity

$\lambda_1 = 1.66$ kJ/(m h °C). Initial temperature of concrete $T_0 = 0$, water temperature $T_w = 0$, the adiabatic temperature rise of concrete is

$$\theta(\tau) = 25(1 - e^{-0.35\tau})$$

namely, $\theta_0 = 25°C$, $m = 0.35$ (l/day). For polythene pipe, $\alpha_1 b = 0.712$, $\alpha_1 = \alpha_1 b/b = 0.712/0.845 = 0.8426$, substitute it into Eq. (17.55), the mean temperature rise due to hydration heat cooled by polythene pipe is

$$\begin{aligned} T_m &= \frac{0.350 \times 25}{0.350 - 0.096 \times 0.8426^2}(e^{-0.096 \times 0.8426^2 t} - e^{-0.350t}) \\ &= 31.04(e^{-0.06815t} - e^{-0.350t}) \end{aligned}$$

If steel pipe is used, $\alpha_1 b = 0.783$, $\alpha_1 = 0.9266$, the mean temperature rise due to hydration heat is

$$\begin{aligned} T'_m &= \frac{0.350 \times 25}{0.350 - 0.096 \times 0.9266^2}(e^{-0.096 \times 0.9266^2 t} - e^{-0.350t}) \\ &= 32.70(e^{-0.8242t} - e^{-0.350t}) \end{aligned}$$

The computed results are shown in Figure 17.13.

17.4.2 Spatial Problem of Pipe Cooling of Late Stage

The water temperature at pipe length L is

$$T_{wL} = T_w + U(t, L) \tag{17.56}$$

The heat absorbed by the water in unit time is

$$Q_3 = c_w \rho_w q_w U(t, L) \tag{17.57}$$

The heat flows into the water from concrete of unit length in unit time is

$$\frac{\partial Q_1}{\partial L} = \lambda(T_0 - T_w)R(t) + \lambda \int_0^t R(t - \tau)\frac{\partial \theta}{\partial \tau}\mathrm{d}\tau \tag{17.58}$$

From time τ to $\tau + \mathrm{d}\tau$, the water temperature increases $(\partial U/\partial \tau)\mathrm{d}\tau$, and the amount of heat flows from water to concrete at time t in unit time and unit length is

$$\frac{\partial Q_2}{\partial L} = \lambda R(t - \tau)\frac{\partial U}{\partial \tau}\mathrm{d}\tau \tag{17.59}$$

From Eq. (17.31) of balance of heat, we get

$$c_{\mathrm{w}}\rho_{\mathrm{w}}q_{\mathrm{w}}U(t,L) = \lambda(T_0 - T_{\mathrm{w}})LR(t) + \lambda\int_0^t\int_0^L R(t-\tau)\frac{\partial\theta}{\partial\tau}\mathrm{d}\tau\,\mathrm{d}L - \lambda\int_0^t\int_0^L R(t-\tau)\frac{\partial\theta}{\partial\tau}\mathrm{d}\tau\,\mathrm{d}L \tag{17.60}$$

Let $\eta = c_{\mathrm{w}}\rho_{\mathrm{w}}/\lambda$, we have

$$\eta q_{\mathrm{w}}U(t,L) = LR(t) + \sum_i\sum_j R(t-t_{i-0.5})\Delta\theta(t_{i-0.5},L_{j-0.5})\Delta L_j - \sum_i\sum_j R(t-t_{i-0.5})\Delta U(t_{i-0.5},L_{j-0.5})\Delta L_j \tag{17.61}$$

Solving the above equation, the water temperature will be obtained, then we can compute the temperature of concrete. The mean temperature of concrete is given by

$$T_{\mathrm{m}} = X'\theta_0 \tag{17.62}$$

For $b/c = 100$, $b\sqrt{m/a} = 1.5$, and $b\sqrt{m/a} = 2.0$, coefficient X' is shown in Figure 17.14.

In practical engineering, besides pipe cooling, there is loss of heat from the horizontal surface of concrete lift. This is a complicated problem which will be analyzed by the equivalent equation of heat conduction in Section 17.6.

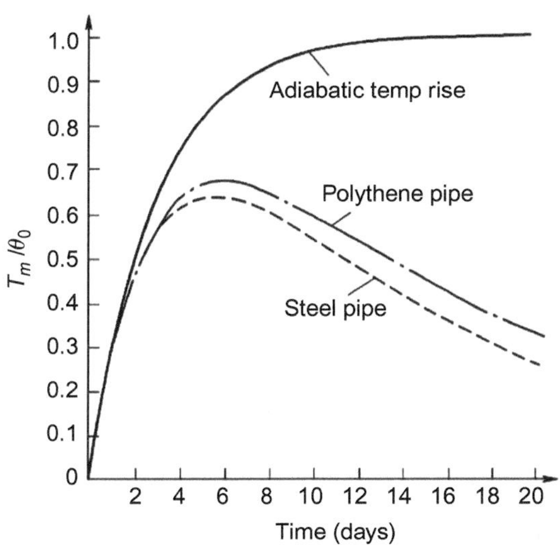

Figure 17.13 Examples of plane problem of pipe cooling of concrete with internal source of heat.

17.5 Practical Formulas for Pipe Cooling of Mass Concrete

In recent years, the design of pipe cooling of mass concrete has become finer. Besides the initial temperature of concrete and the water temperature, it is necessary to consider more factors, such as the hydration heat of cement and the spacing of pipes, thus, there is a great need for practical formulas for computation which will be given in the following [8, 53].

17.5.1 Mean Temperature of Concrete Cylinder with Length L

Consider the pipe cooling of a concrete cylinder with diameter D, length L, initial temperature T_0, adiabatic temperature rise $\theta(\tau)$, and water temperature T_w at the inlet. The mean temperature of the cylinder to length L is

$$T_1(t) = T_{\mathrm{w}} + (T_0 - T_{\mathrm{w}})\phi(t) + \sum \Delta\theta(\tau_i)\phi(t - \tau_i - 0.5\Delta\tau_i) \tag{17.63}$$

Two formulas are given for the pipe cooling function $\phi(t)$. The first formula is

$$\left.\begin{aligned} \phi_1(t) &= \exp(-k_1 z^s) \\ k_1 &= 2.08 - 1.174\xi + 0.256\xi^2 \\ s &= 0.971 + 0.1485\xi - 0.0445\xi^2 \end{aligned}\right\} \tag{17.64}$$

The second formula is

$$\left.\begin{aligned} \phi_2(t) &= \mathrm{e}^{-k_2 z} \\ k_2 &= 2.09 - 1.35\xi + 0.320\xi^2 \end{aligned}\right\} \tag{17.65}$$

where

$$\xi = \frac{\lambda L}{c_{\mathrm{w}}\rho_{\mathrm{w}}q_{\mathrm{w}}}, \quad z = \frac{gat}{D^2} \tag{17.66}$$

in which g is a coefficient to consider the influence of b/c and the material of pipe. There are two formulas for g:

$$g = a'/a = 1.67\exp\left\{-0.0628\left[\frac{b}{c}\left(\frac{c}{r_0}\right)^{\eta} - 20\right]^{0.48}\right\} \tag{17.67}$$

or

$$g = a'/a = \frac{\ln 100}{\ln(b/c) + (\lambda/\lambda_1)\ln(c/r_0)} \tag{17.68}$$

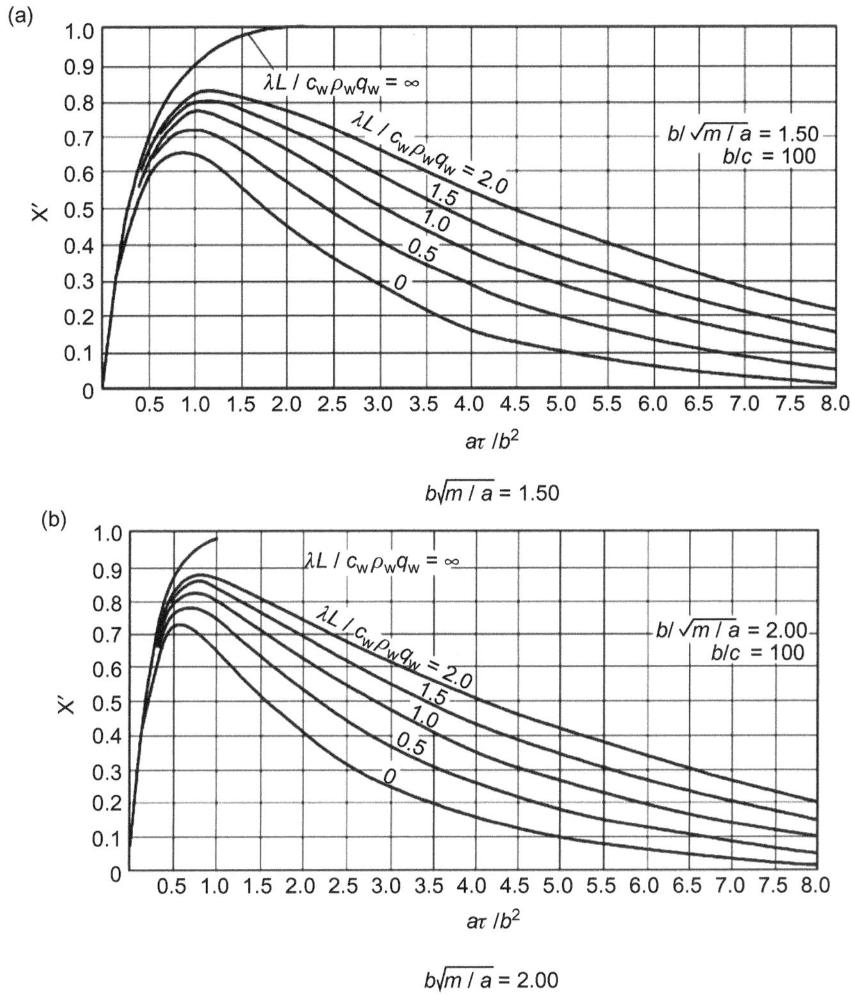

Figure 17.14 Mean temperature of concrete in early-stage pipe cooling $T_m = X'\theta_0$.

Let $b/c = 100$, $\lambda_1 = \infty$, $\eta = \lambda/\lambda_1 = 0$; from the above equations, $g = 1.00$; from Figure 17.9, we get ϕ_3; from Eqs. (17.64) and (17.65), we get ϕ_1 and ϕ_2. The comparison of ϕ_1, ϕ_2, and ϕ_3 is given in Table 17.5. It is evident that the precision of ϕ_1 is high from beginning to end, and the precision of ϕ_2 is high when $z \leq 2.0$ but low when $z \geq 2.0$. When the spacings of pipe are 1.5 m × 1.5 m, $z > 2.0$ is equivalent to $t > 60$ days, thus, generally speaking the precision of ϕ_2 is good.

Substituting $z = gat/D^2$ into Eqs. (17.64) and (17.65), we have

$$\phi_1(t) = e^{-p_1 t^s} \quad (17.69)$$

$$\phi_2(t) = e^{-p_2 t} \tag{17.70}$$

where

$$p_1 = k_1(ga/D^2)^s, \quad p_2 = k_2 ga/D^2 = 0.734k_2 ga/(S_1 S_2) \tag{17.71}$$

in which S_1 and S_2 are the spacings of pipes.

Example 1 Pipe cooling in late stage, polythene pipe, spacing 1.5 m × 1.5 m, outer radius of pipe $c = 0.016$ m, inner radius $r_0 = 0.014$ m, conductivity $\lambda_1 = 1.66$ kJ/(m h °C), length $L = 300$ m. The diffusivity of concrete $a = 0.10\ \mathrm{m^2/day}$, conductivity $\lambda = 8.37$ kJ/(m h °C). The specific heat of water $c_w = 4.187$ kJ/(kg °C), density $\rho_w = 1000\ \mathrm{kg/m^3}$, discharge $q_w = 1.00\ \mathrm{m^3/h}$, time $t = 30$ days.

$$b = 0.5836\sqrt{s_1 s_2} = 0.875\ \mathrm{m}, \quad D = 2b = 1.75\ \mathrm{m}$$

$$g = 1.67\exp\left\{-0.628\left[\frac{0.875}{0.016}\left(\frac{0.016}{0.014}\right)^{8.37/1.66} - 20\right]^{0.48}\right\} = 0.977$$

$$z = \frac{0.977 \times 0.10 \times 30}{1.75^2} = 0.957$$

$$\xi = \frac{8.37 \times 300}{4.187 \times 1000 \times 1.00} = 0.600$$

$$k_1 = 2.08 - 1.174 \times 0.600 + 0.256 \times 0.60^2 = 1.468$$

$$s = 0.971 + 0.1485 \times 0.60 - 0.0445 \times 0.60^2 = 1.044$$

$$\phi_1(t) = \exp[-1.468 \times 0.957^{1.044}] = 0.25$$

$$k_2 = 2.09 - 1.35 \times 0.600 + 0.320 \times 0.600^2 = 1.395$$

$$\phi_2(t) = \mathrm{e}^{-1.395 \times 0.957} = 0.26$$

17.5.2 Mean Temperature of the Cross Section of Concrete Cylinder

For a concrete cylinder with insulated surface, diameter D, length L, initial temperature T_0, adiabatic temperature rise $\theta(\tau) = \theta_0 f(\tau)$, water temperature T_w, the mean temperature of the cross section at length L is given by

$$T_{1L}(t) = T_w + (T_0 - T_w)\phi_L(t) + \theta_0 \psi_L(t) \tag{17.72}$$

$$\phi_L(t) = \mathrm{e}^{-p_3 t} \tag{17.73}$$

$$\psi_L(t) = \sum \Delta f(\tau) \cdot \mathrm{e}^{-p_3(t-\tau-0.5\Delta\tau)} \tag{17.74}$$

$$p_3 = k_3 ga/D^2 = 0.734k_3 ga/(s_1 s_2) \tag{17.75}$$

Table 17.5 Comparison Between the Cooling Functions

$z = at/D^2$	$\xi = 0$			$\xi = 1.0$			$\xi = 2.0$		
	ϕ_1	ϕ_2	ϕ_3	ϕ_1	ϕ_2	ϕ_3	ϕ_1	ϕ_2	ϕ_3
0.10	0.801	0.811	0.808	0.906	0.899	0.897	0.940	0.935	0.940
0.20	0.647	0.658	0.655	0.814	0.809	0.808	0.877	0.875	0.875
0.30	0.524	0.534	0.532	0.727	0.728	0.727	0.816	0.818	0.822
0.50	0.346	0.352	0.352	0.576	0.589	0.585	0.701	0.715	0.715
0.75	0.207	0.209	0.212	0.426	0.452	0.438	0.576	0.605	0.590
1.00	0.125	0.124	0.125	0.313	0.346	0.327	0.470	0.512	0.485
2.00	0.017	0.015	0.018	0.087	0.120	0.095	0.200	0.262	0.203
3.00	0.002	0.002	0.000	0.023	0.042	0.025	0.082	0.130	0.075

Note: ϕ_1 and ϕ_2 are computed results, ϕ_3 is obtained from Figure 17.4.

$$k_3 = 2.05 - 2.14\xi + 0.65\xi^2 \tag{17.76}$$

in which

$$\xi = \frac{\lambda L}{c_w \rho_w q_w}$$

17.5.3 Time of Cooling

Let T_0 be the initial temperature of concrete, T_w be the water temperature at the inlet of pipe, from Eqs. (17.63), (17.69), and (17.70), we have

$$\phi(t) = \frac{T(t) - T_w}{T_0 - T_w} = e^{-p_1 t^s} = e^{-p_2 t} \tag{17.77}$$

Hence, we get two formulas for computing the time t for the mean temperature of a cylinder reducing from T_0 to $T(t)$ as follows:

$$t = -\frac{1}{p_2} \ln\left[\frac{T(t) - T_w}{T_0 - T_w}\right] \tag{17.78}$$

or

$$t = \left\{-\frac{1}{p_1} \ln\left[\frac{T(t) - T_w}{T_0 - T_w}\right]\right\}^{1/s} \tag{17.79}$$

Example 2 The basic data are the same as example 1, the initial temperature $T_0 = 30°C$, try to compute the time t required from T_0 to $T(t) = 20°C$.

Let $T_w = 10°C$, $k_1 = 1.468$, $s = 1.044$, $g = 0.977$, $D = 1.75$ m, $k_2 = 1.395$,

$$p_1 = k_1(ga/D^2)^s = 1.468 \times (0.977 \times 0.10/1.75^2)^{1.044} = 0.04025$$

$$p_2 = k_2 ga/D^2 = 1.395 \times 0.977 \times 0.10/1.75^2 = 0.0445$$

From Eq. (17.78),

$$t = -\frac{1}{0.0445}\ln\left[\frac{20-10}{30-10}\right] = 15.6 \text{ days}$$

From Eq. (17.79),

$$t = \left\{-\frac{1}{0.04025}\ln\left[\frac{20-10}{30-10}\right]\right\}^{1/1.044} = 15.3 \text{ days}$$

17.5.4 Formula for Water Temperature

A formula is given in the following for computing the temperature of water as follows:

$$T_{wL} = T_w + (T_0 - T_w)w(\xi, z) + \sum \Delta\theta(\tau_i)w(\xi, z') \tag{17.80}$$

$$w(\xi, z) = [1 - (1-g)e^{-\xi}](1 - e^{-2.70\xi})\exp[-2.40z^{0.50}e^{-\xi}] \tag{17.81}$$

$$w(\xi, z') = [1 - (1-g)e^{-\xi}](1 - e^{-2.70\xi})\exp[-2.40(z')^{0.50}e^{-\xi}] \tag{17.82}$$

where

$$z = gat/D^2, \quad z' = ga(t - \tau_i - 0.5\Delta\tau_i)/D^2 \tag{17.83}$$

Example 3 The basic data are the same as example 1, $g = 0.977$, try to analyze the relation between the coefficient $w(\xi, z)$, the discharge q_w and time t.

$$\xi = \frac{8.37 \times 300}{4.187 \times 1000 q_w} = 0.600/q_w \quad (q_w \text{in m}^3/h, \text{ to match the unit of } \lambda)$$

$$z = \frac{0.977 \times 0.10t}{1.75^2} = 0.0319t \quad (t \text{ in } d, \text{ to match the unit of } a)$$

From Eq. (17.76), the coefficient of water temperature is

$$w(q_w, t) = (1 - 0.023e^{-0.60/q_w}) \cdot (1 - e^{-1.62/q_w})\exp[1 - 0.43\sqrt{t}e^{-0.60/q_w}] \tag{17.84}$$

The computed results are shown in Figure 17.15.

Example 4 Pipe cooling of the early stage, the basic data are the same as example 1, the adiabatic temperature rise $\theta(\tau) = 25\tau/(2.0+\tau)$°C, the initial temperature $T_0 = 12$°C, the water temperature at inlet $T_w = 9$°C, try to compute the water temperature T_{Lw} at the outlet where $L = 300$ m. There are two schemes: (i) Scheme I, $\tau = 20$ days, $q_w = 2.0$ m^3/h; (ii) Scheme II, $\tau = 0$–8 days, $q_w = 2.0$ m^3/h; $\tau = 8$–2 days, $q_w = 0.50$ m^3/h or 0.30 m^3/h. From Eq. (17.80),

$$T_{wL} = 9.0 + (12.0 - 9.0)w(q_w, t - 0) + \sum \Delta\theta(\tau_i)w(q_w, t - \tau_{i-0.5}) \tag{17.85}$$

$$w(q_{w^2}t - \tau) = (1 - 0.023\,e^{-0.60/q_w})(1 - e^{-1.62/q_w})\exp[1 - 0.43\sqrt{t-\tau} \cdot e^{-0.60/q_w}] \tag{17.86}$$

where

$$\Delta\theta(\tau_i) = \theta(\tau_i) - \theta(\tau_{i-1})$$

The computed results are shown in Figure 17.16. The premise of Eqs. (17.81) and (17.82) is that q_w is constant thus the precision will be lower when they are used for q_w varying with time. It would be better to use Eqs. (17.34) and (17.61).

17.6 Equivalent Equation of Heat Conduction Considering Effect of Pipe Cooling [46,53]

In the computation of the temperature field in the construction of concrete dams, it is necessary to take into account the effect of pipe cooling as well as the effect of natural surface cooling. The best way is to use the equivalent equation of heat conduction considering the effect of pipe cooling proposed by the author.

17.6.1 Temperature Variation of Concrete with Insulated Surface and Cooling Pipe

Consider a concrete cylinder with insulated surface, diameter D, length L, initial temperature T_0, water temperature at inlet T_w, adiabatic temperature rise $\theta(\tau) = \theta_0 f(\tau)$, under the simultaneous action of the adiabatic temperature rise and the pipe cooling, the mean temperature of the cylinder is

$$T_1(t) = T_w + (T_0 - T_w)\varphi(t) + \int_0^t \varphi(t-\tau)\frac{\partial\theta}{\partial\tau}d\tau \tag{17.87}$$

From Eqs. (17.69) and (17.70), it is evident that there are two expressions of $\phi(t)$ as follows:

$$\phi_1(t) = \mathrm{e}^{-p_1 t^s}, \quad \phi_2(t) = \mathrm{e}^{-p_2 t}$$

Both of them can be used in Eq. (17.87), but $\phi_2(t)$ is more convenient in integration and differentiation and the precision of computation is also good; thus $\phi_2(t)$ will be used in the following and omit the subscript. Substitution of $\phi(t) = \mathrm{e}^{-pt}$ in Eq. (17.87) yields

$$T_1(t) = T_w + (T_0 - T_w)\mathrm{e}^{-pt} + \int_0^t \mathrm{e}^{-p(t-\tau)} \frac{\partial \theta}{\partial \tau} \mathrm{d}\tau \tag{17.88}$$

where

$$\phi(t) = \mathrm{e}^{-pt} \tag{17.89}$$

$$p = p_2 = k_2 g a / D^2 \tag{17.90}$$

k_2 and g are given in Eqs. (17.65), (17.67), and (17.68).

Now consider the third term in the right part of Eq. (17.88), namely

$$T_3(t) = \int_0^t \mathrm{e}^{-p(t-\tau)} \frac{\partial \theta}{\partial \tau} \mathrm{d}\tau = \theta_0 \psi(t) \tag{17.91}$$

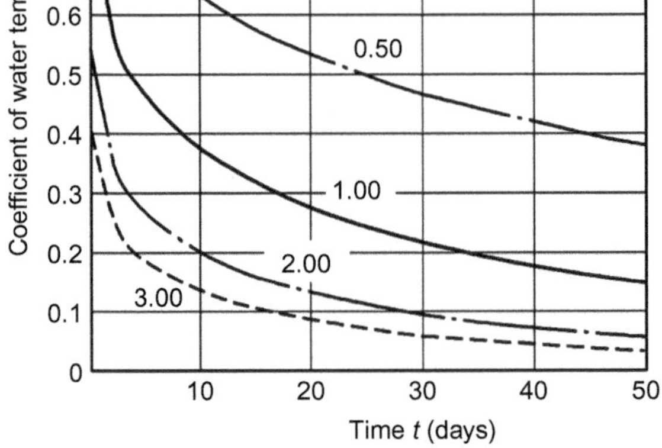

Figure 17.15 Example 3, coefficient of water temperature $w(q_w, t)$ in cooling of late stage (spacing of pipe 1.5 m × 1.5 m, $L = 300$ m).

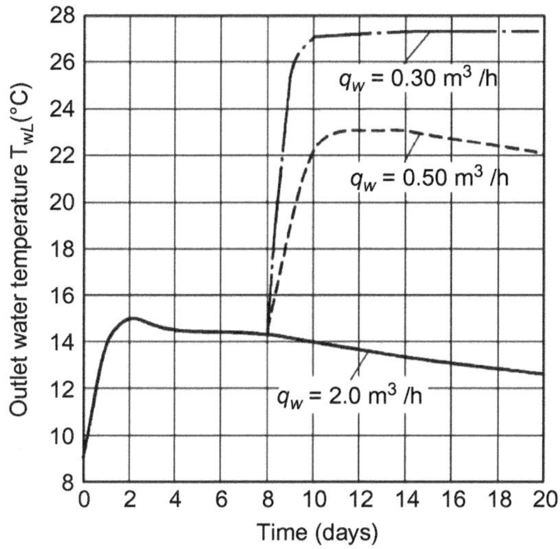

Figure 17.16 Example 4, the water temperature at the outlet in the early-stage cooling of concrete.

which is computed in the following three cases:

1. when

$$\theta(\tau) = \theta_0(1 - \mathrm{e}^{-m\tau}) \tag{17.92}$$

$$\psi(t) = \frac{m}{m-p}(\mathrm{e}^{-pt} - \mathrm{e}^{-mt}) \tag{17.93}$$

2. when

$$\theta(\tau) = \theta_0\tau/(n+\tau) \tag{17.94}$$

$$\psi(t) = np\,\mathrm{e}^{-p(n+t)}\left\{\frac{\mathrm{e}^{pn}}{np} - \frac{\mathrm{e}^{p(n+t)}}{p(n+1)} - E_i(pn) + E_i[p(n+t)]\right\} \tag{17.95}$$

$$E_i(px) = \int \frac{\mathrm{e}^{px}}{x}\mathrm{d}x$$

$E_i(px)$ is an integral of the exponential which may be taken from a mathematical handbook or computed by the following formula:

$$\int_c^d \frac{\mathrm{e}^{px}}{x}\mathrm{dx} = \sum \mathrm{e}^{p(x+0.5\Delta x)}\ln\left(\frac{x+\Delta x}{x}\right) \tag{17.96}$$

3. when

$$\theta(\tau) = \theta_0 f(\tau) \tag{17.97}$$

where $f(\tau)$ is an arbitrary function of time, the function $\psi(t)$ may be given by numerical integration as follows:

$$\psi(t) = \sum e^{-p(t-\tau-0.50\Delta\tau)}\Delta f(\tau) \tag{17.98}$$

Generally speaking, the mean temperature of concrete is given by

$$T_1(t) = T_w + (T_0 - T_w)\phi(t) + \theta_0\psi(t) \tag{17.99}$$

where T_0—initial temperature of concrete, T_w—water temperature at the inlet of pipe, $\phi(t)$ is given by Eq. (17.89), $\psi(t)$ is given by Eqs. (17.93), (17.95), and (17.98).

Example For concrete, diffusivity $a = 0.10\ \text{m}^2/\text{day}$, conductivity $\lambda = 8.37$ kJ/(m h °C), $p = k_2ga/D^2$, $\theta(\tau) = \theta_0(1 - e^{-0.40\tau})$; for polythene pipe, $c = 1.60$ cm, $r_0 = 1.40$ cm, conductivity $\lambda = 1.66$ kJ/(m h °C), $\xi = \lambda L/(c_w\rho_w q_w) = 0.50$. The computed $\phi(t)$ and $\psi(t)$ are shown in Figures 17.17–17.19.

17.6.2 Equivalent Equation of Heat Conduction Considering the Effect of Pipe Cooling

By differentiating Eq. (17.99), we get the rate of increase of temperature of a concrete cylinder with insulated surface, adiabatic temperature rise $\theta(\tau) = \theta_0 f(\tau)$, and pipe cooling in the following:

$$\frac{\partial T_1}{\partial t} = (T_0 - T_w)\frac{\partial \phi}{\partial t} + \theta_0\frac{\partial \psi}{\partial t} \tag{17.100}$$

Substituting $\partial\theta/\partial\tau$ in Eq. (2.1) by $\partial T_1/\partial t$ in Eq. (17.100), we get the equivalent equation of heat conduction considering the effect of pipe cooling as follows:

$$\frac{\partial T}{\partial t} = a\left(\frac{\partial^2 T}{\partial x^2} + \frac{\partial^2 T}{\partial y^2} + \frac{\partial^2 T}{\partial z^2}\right) + (T_0 - T_w)\frac{\partial \phi}{\partial t} + \theta_0\frac{\partial \psi}{\partial t} \tag{17.101}$$

There are three parts in the right part of the above equation, the first part represents the influence of the flux of heat across the surface of concrete, the second and the third terms represent the influences of the initial temperature of concrete, pipe cooling, and the hydration heat of cement.

Since the radius of cooling pipe is only 1.0–1.6 cm, if the finite elements are directly used to compute the influence of cooling pipe, the dimensions of the elements near the pipe must be of the order 1.0–1.6 cm, the total number of elements will be so enormous that the computation is very difficult. Thus, the equivalent equation of heat conduction Eq. (17.101) is widely applied in the analysis of temperature field and thermal stresses in mass concrete.

17.7 Theoretical Solution of the Elastocreeping Stresses Due to Pipe Cooling and Self-Restraint

The thermal stresses due to pipe cooling are the sum of the self-stresses and the restraint stresses. The former will be described in the following.

17.7.1 The Elastic Thermal Stress Due to Self-Restraint

Consider a circular cylinder free of external restraint, the cross section is shown in Figure 17.4 and the temperature field is given by Eq. (17.7). If the modulus of elasticity is constant, the elastic thermal stress is given by the following formula:

$$\left.\begin{aligned} \sigma_r &= \frac{E\alpha}{1-\mu}\left(\frac{r^2-c^2}{b^2-c^2}\int_c^b Tr\,\mathrm{d}r - \int_c^r Tr\,\mathrm{d}r\right) \\ \sigma_\theta &= \frac{E\alpha}{(1-\mu)r^2}\left(\frac{r^2+c^2}{b^2-c^2}\int_c^b Tr\,\mathrm{d}r + \int_c^r Tr\,\mathrm{d}r - Tr^2\right)\sigma_z \\ \sigma_z &= \frac{E\alpha}{1-\mu}(T_\mathrm{m}-T) \end{aligned}\right\} \tag{17.102}$$

where σ_r, σ_θ, σ_z are respectively the radial, tangential, and axial stress. From Eq. (17.7), we get the integral

$$\int_c^r Tr\,\mathrm{d}r = T_0 \sum \frac{2b\mathrm{e}^{-a\alpha_n^2\tau}}{\alpha_n^2 b^2 R_1(\alpha_n b)}\cdot$$
$$\{r[J_1(\alpha_n b)Y_1(\alpha_n r) - Y_1(\alpha_n b)J_1(\alpha_n r)] + c[Y_1(\alpha_n b)J_1(\alpha_n c) - J_1(\alpha_n b)Y_1(\alpha_n c)]\} \tag{17.103}$$

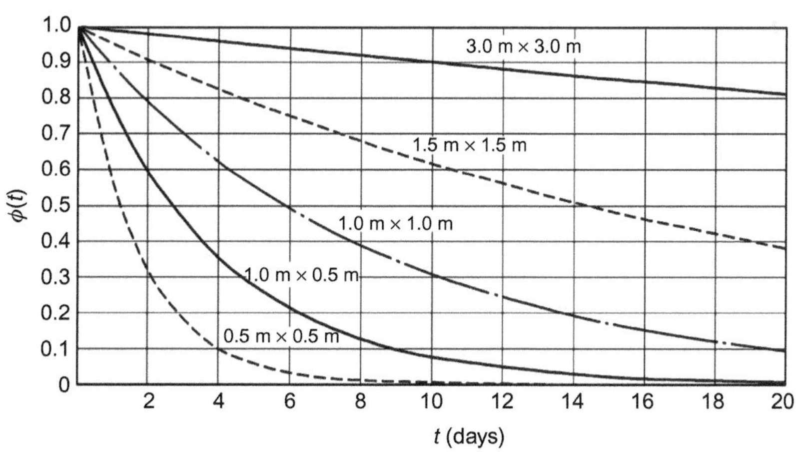

Figure 17.17 $\phi(t) = \mathrm{e}^{-pt}$ ($a = 0.10\ \mathrm{m^2/day}$, $p = k_2 ga/D^2$).

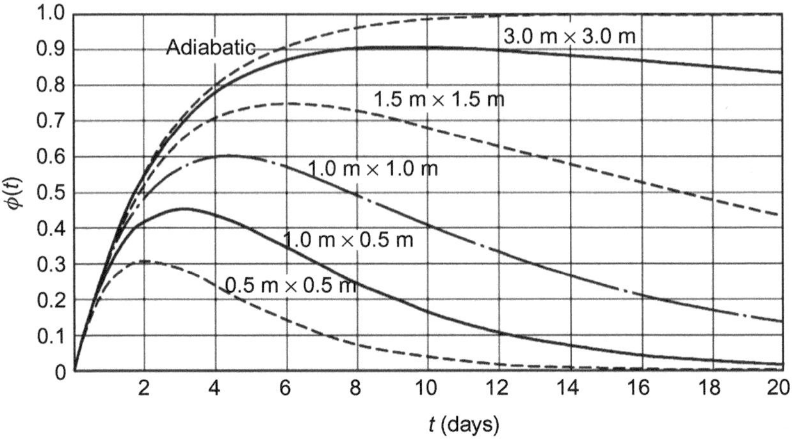

Figure 17.18 $\psi(t)$ for $\theta(\tau) = \theta_0(1 - e^{-0.40\tau})$.

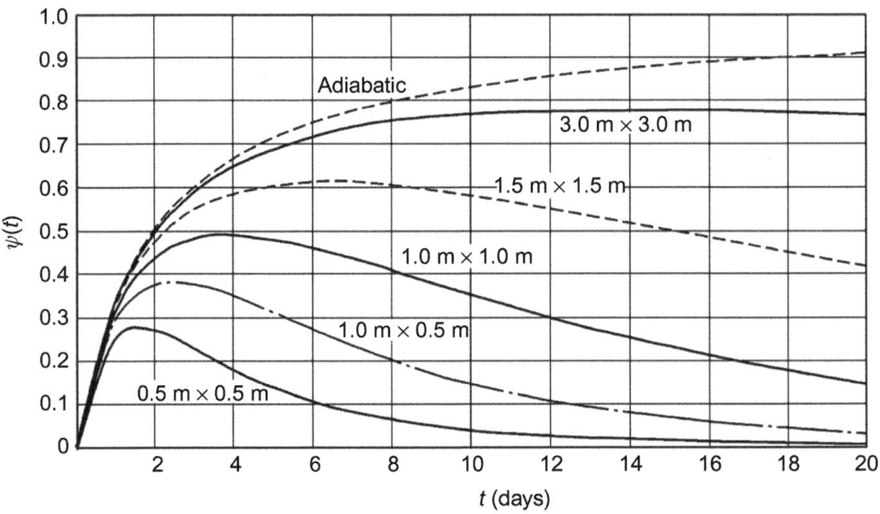

Figure 17.19 $\psi(t)$ for $\theta(\tau) = \theta_0\tau(2.0 + \tau)$.

The stresses at any point can be computed by Eqs. (17.102) and (17.103), generally the maximum stress appears at inner surface $r = c$ where

$$\sigma = \sigma_\theta = \sigma_z = \frac{E\alpha}{1-\mu}(T_m - T_c), \quad \sigma_r = 0 \tag{17.104}$$

in which T_m is the mean temperature given by Eq. (17.12) and T_c is the temperature at $r = c$ given by the following formula:

$$T_c = T_0 \sum m_n e^{-\alpha_1^2 b^2 a\tau / b^2} \tag{17.105}$$

$$m_n = \frac{2[J_1(\alpha_n b)Y_0(\alpha_n c) - Y_1(\alpha_n b)J_0(\alpha_n c)]}{\alpha_n b R_1(\alpha_n b)} \tag{17.106}$$

Equation (17.105) converges quickly and only one term is required, let T_0 be the initial temperature of concrete and T_w be the water temperature, from Eq. (17.104), we get the thermal stress at the inner surface of concrete as the following:

$$\sigma(t) = -\frac{E\alpha(T_w - T_0)(1 - m_1)}{1 - \mu} e^{-p_2(t-\tau)} \tag{17.107}$$

$$p_2 = \alpha_1^2 b^2 a / b^2 \tag{17.108}$$

For polythene pipe with inner radius 14 mm and outer radius 16 mm, we may take $m_1 = 0.20$. For steel pipe, $m_1 = 0$. For three-dimensional problem, Eq. (17.71) must be used with $p_2 = k_2 g a / D^2$.

If the initial temperature of concrete is T_0 and the water temperature varies step by step, when $t = \tau_1, \tau_2, \tau_3, \ldots, T_w = T_{w1}, T_{w2}, T_{w3}, \ldots$, the elastic thermal stresses at the inner surface are as follows

$$\sigma(t) = -\sum_i \frac{E\alpha(1 - m_1)\Delta T_i}{1 - \mu} e^{-p_2(t-\tau_i)} \tag{17.109}$$

where

$$\Delta T_1 = T_{w1} - T_0, \quad \Delta T_2 = T_{w2} - T_{w1}, \quad \Delta T_3 = T_{w3} - T_{w2}$$

17.7.2 The Elastocreeping Thermal Stress Due to Self-Restraint

Since the temperature varies with time, the elastic stress is variable even if the water temperature is constant. For the first step water temperature T_{w1}, the stress increment at Δt_j on the inner surface of the concrete is

$$\Delta\sigma_{1j} = \sigma(t_j + \Delta t_j) - \sigma(t_j) = -\frac{E(t_j)(1 - m_1)\alpha\Delta T_1}{1 - \mu}(e_j^{-p_2\Delta t} - 1)e^{-p_2(t_j-\tau_1)} \tag{17.110}$$

Under the action of the stepwise water temperature $T_{w1}, T_{w2}, T_{w3}, \ldots$, the elasto-creeping thermal stress is

$$\sigma^*(t) = -\sum_i \frac{(1 - m_1)\alpha\Delta T_i}{1 - \mu} \sum_j E(t_j)(e_j^{-p2\Delta t} - 1)e^{-p_2(t_j-\tau_i)} K(t, t_j) \tag{17.111}$$

where $K(t, t_j)$ is the stress relaxation coefficient. The computed results show that the depth of tensile stress around the pipe is about 35 cm (1.5 m × 1.5 m pipe spacing) to 75 cm (3.0 m × 3.0 m pipe spacing) which is near to the depth of tensile stress induced by cold wave.

17.7.3 A Practical Formula for the Elastocreeping Thermal Stress Due to Self-Restraint

For late-stage pipe cooling, the tensile stress at the boundary of a hole may be computed by

$$\sigma^*(t) = -\frac{E(\tau)\alpha(T_w - T_0)(1 - m_1)k_4}{1 - \mu} e^{-p_2 s(t-\tau)} \tag{17.112}$$

where T_0—initial temperature, T_w—water temperature, p_2 is given by Eq. (17.108)k_4 and s are coefficients to consider the influence of creep. For polythene pipe with outer radius 16 mm and inner radius 14 mm, $k_4 = 0.81$, $m_1 = 0.20$, $s = 1.30$. For steel pipe, $k_4 = 1.00$, $m_1 = 0$, $s = 1.30$.

Example 1 Spacing of pipe 1.5 m × 1.5 m, polythene pipe with inner radius 14 mm and outer radius 16 mm, the initial temperature of concrete $T_0 = 30°C$, the diffusivity $a = 0.10\ m^2/day$; when $\tau = 90-110$ days, $T_w = 19°C$; when $\tau = 110-150$ days, $T_w = 9°C$, the elastocreeping thermal stresses at the surface given by Eq. (17.112) are shown in Figure 17.20.

17.7.4 Reducing Thermal Stress by Multistage Cooling with Small Temperature Differences—Theoretical Solution

In the past generally one-step cooling with $T_0 - T_w = 20 - 25°C$ was used in the late-stage cooling which induced large tensile stress. In order to reduce the tensile stress, it is suggested to use multistage cooling with small temperature differences.

Example 2 The initial temperature of concrete $T_0 = 30°C$, the final steady temperature of dam $T_f = 10°C$. The spacing of cooling pipe 1.5 m × 1.5 m, diffusivity $a = 0.10\ m^2/day$, polythene pipe with inner and outer radius 14 and 16 mm, respectively. There are four schemes. Scheme 1, one-step cooling, cooling time $\tau_1 = 90-150$ days, $T_{w1} = 9°C$. Scheme 2, two-step cooling, $\tau_1 = 90-110$ days, $T_{w1} = 19°C$; $\tau_2 = 110-150$ days, $T_{w2} = 9°C$. Scheme 3, three-step cooling, $\tau_1 = 90-110$ days, $T_{w1} = 23°C$; $\tau_2 = 110-130$ days, $T_{w2} = 16°C$; $\tau_3 = 130-150$ days, $T_{w3} = 9°C$. Scheme 4, six step cooling $\tau_1 = 30-50$ days, $T_{w1} = 26°C$; $\tau_2 = 50-70$ days, $T_{w2} = 22.5°C$; $\tau_3 = 70-90$ days, $T_{w3} = 19.0°C$; $\tau_4 = 90-110$ days, $T_{w4} = 15.5°C$; $\tau_5 = 110-130$ days, $T_{w5} = 12°C$; $\tau_6 = 130-150$ days, $T_{w6} = 9°C$.

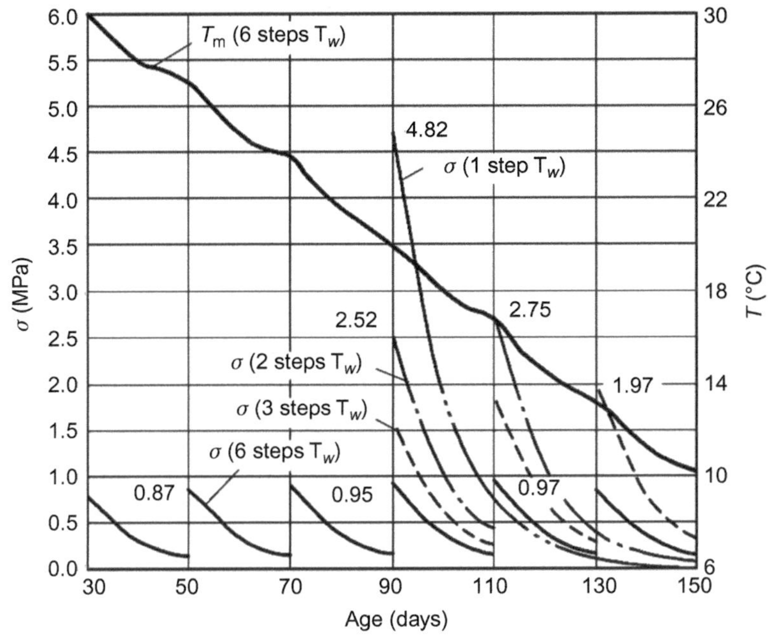

Figure 17.20 Example 2, the elastocreeping thermal stresses at the inner surface of concrete in late-stage cooling (tensile stress is positive), theoretical solution.

Table 17.6 Example 2, the Maximum Elastocreeping Stress at Inner Surface of Concrete (MPa)

Steps of Water Temperature	1	2	3	6
Stress (MPa)	4.82	2.75	1.97	0.97

All the schemes are computed by Eq. (17.112), the thermal stresses at the inner surface of concrete of all the schemes and the mean temperature of concrete of Scheme 4 (six step cooling) are shown in Figure 17.20. The maximum elastocreeping stresses at the inner surface of concrete for all the schemes are shown in Table 17.6. It is clear that the maximum tensile stress is reduced from 4.82 to 0.97 MPa if the one step cooling is replaced by six step cooling.

17.7.5 The Elastocreeping Self-Stress Due to Pipe Cooling and Hydration Heat of Cement

Assuming that the adiabatic temperature rise is $\theta(\tau)$, both the initial temperature of concrete and the water temperature are zero, the elastocreeping stress at the inner surface of concrete is

$$\sigma(t) = -\sum_i \frac{(1-m_1)\alpha\Delta\theta(t_i)}{1-\mu}\sum_j E(T_j)(\mathrm{e}^{-p_1\Delta t_j}-1)\mathrm{e}^{-p_1(t_j-\tau_i)}K(t,t_j) \quad (17.113)$$

Under the action of hydration heat and pipe cooling, the mean temperature of concrete first increases and afterward decreases and the surface stress first is tensile and afterward becomes compressive. Let $t \to \infty$ in Eq. (17.113), the final compressive stress at the inner surface of concrete may be estimated as follows:

$$\sigma(\infty) = \frac{(1-m_1)E(\tau_1)\alpha\eta\theta_0 K_p}{1-\mu} \tag{17.114}$$

where $\eta = T_r/\theta_0$ is the coefficient of temperature rise due to hydration heat given in Figure 17.20, τ_1 is time of temperature drop, approximately $\tau_1 \cong 10$ days, K_p is the stress relaxation coefficient. In the early age, the temperature will increase and induce some compressive stress which will offset part of the tensile stress produced in the late age; k_r is a coefficient to consider this factor.

For example, pipe spacing $1.5\ \text{m} \times 1.5\ \text{m}$, $\eta = 0.75$, $k_r = 0.60$, $K_p = 0.40$, $E(10) = 20,000$ MPa, from Eq. (17.114), the compressive stress at the surface of a hole at late age is -0.86 MPa.

Of course, the residual compressive stress at the inner surface of concrete induced by the hydration heat at early age will offset one part of the tensile stress due to pipe cooling at late age.

17.8 Numerical Analysis of Elastocreeping Self-Thermal Stress of Pipe Cooling

In practical engineering, the cooling pipes sometimes are put on the surface of old concrete or rock foundation, the concrete cylinder consists of new and old concrete or concrete and rock. It is difficult to get a theoretical solution in this case. We shall show how to compute the elastocreeping thermal stress of pipe cooling by the three-dimensional finite element method (3D FEM) [44].

17.8.1 Computing Model

As shown in Figure 17.21, the computing model is a rectangular prism with height s_1, width s_2, and length $L = 28$ m and a cooling pipe at the center. s_1 and s_2 are

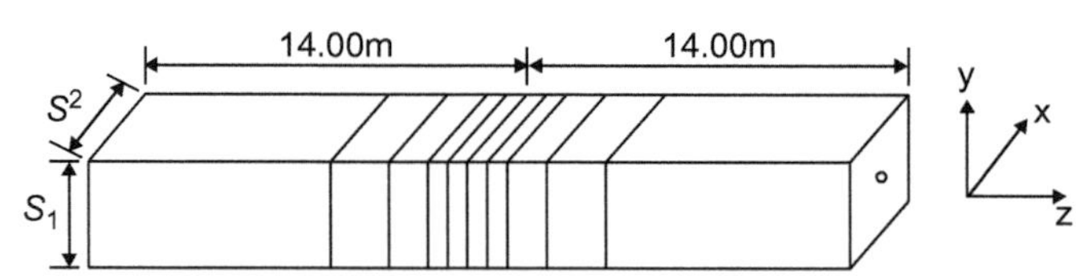

Figure 17.21 Computing model of new concrete.

respectively the vertical and horizontal spacing of the pipe. The cross sections of the computing model are shown in Figure 17.22. Model A consists of one kind of concrete, Model B consists of new and old concrete with time interval 20 days, and Model C consists of concrete and rock.

For concrete, the diffusivity $a = 0.10\ \text{m}^2/\text{day}$, the conductivity $\lambda = 8.37$ kJ/(m h °C), the coefficient of expansion $\alpha = 1 \times 10^{-5}/°\text{C}$, the Poisson's ratio $\mu = 0.167$, the modulus of elasticity $E(\tau) = 35,000[1 - \exp(-0.40\tau^{0.34})]$ MPa, the unit creep is given by Eq. (6.51), the external surface of concrete is insulated. For the rock, $E = 35{,}000$ MPa, $\mu = 0.25$, $C(t, \tau) = 0$. For the polythene cooling pipe, $c = 16$ mm, $r_0 = 14$ mm, $\lambda = 1.66$ kJ/(m h °C). Due to symmetry, only 1/4 of the prism is taken to be computed by 3D FEM, the minimum dimension of the element near the pipe is 12 mm.

17.8.2 Elastocreeping Stresses in 60 days Early Pipe Cooling

Consider the pipe cooling in 60 days early age of concrete, $T_0 = 0$, $T_w = 0$, $\theta(\tau) = 25\tau/(2.1 + \tau)°\text{C}$, spacing of pipe 1.5 m × 1.5 m. The elastocreeping stresses at the surface of hole are shown in Figure 17.23. They are tensile stresses in the early age and become compressive stresses in the late age.

17.8.3 Elastocreeping Stresses in 20 days Early Pipe Cooling

The concrete is cooled in the first 20 days and the cooling is stopped after $\tau = 20$ days. The stresses are shown in Figure 17.24.

17.8.4 Elastocreeping Stresses in Late Pipe Cooling

The initial temperature of concrete $T_0 = 30°\text{C}$, the final steady temperature $T_f = 10°\text{C}$, water temperature $T_w = 9°\text{C}$. The concrete is cooled by pipe from $\tau = 90$ days. The tangential stress σ_x at point a is shown in Figure 17.25. The distribution of σ_x in section ab of Model A is given in Figure 17.26. From Figure 17.26(a), it is evident that the stress is close to zero when $x > 0.25$ m. Hence an important conclusion is derived: the temperature field and stress field given by

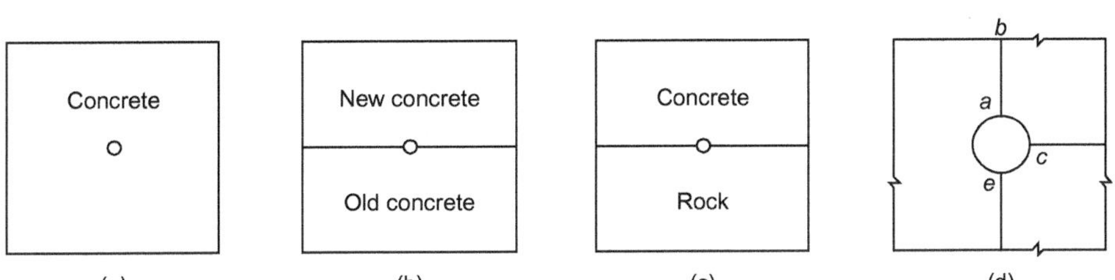

Figure 17.22 Cross sections of computing model.

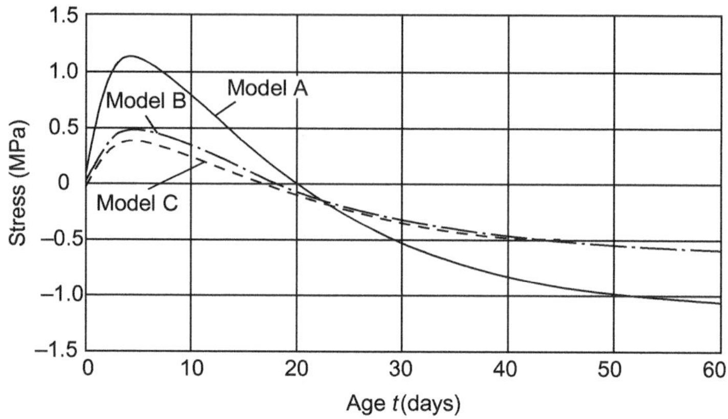

Figure 17.23 Sixty days early pipe cooling, elastocreeping tangential stress σ_x at point *a* (pipe spacing 1.5 m × 1.5 m, $\theta_0 = 25°C$).

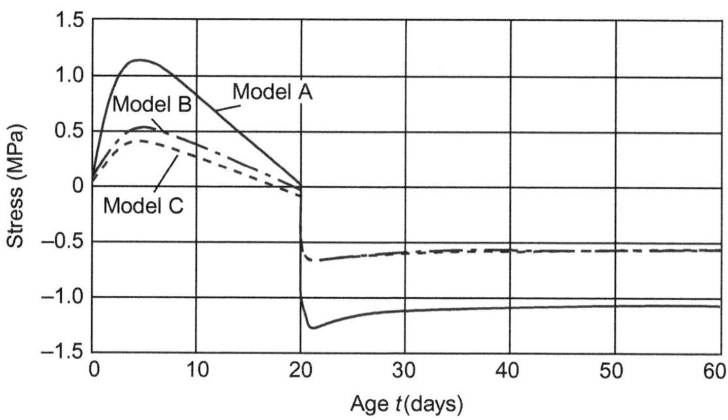

Figure 17.24 Twenty days early pipe cooling, elastocreeping, tangential stress σ_x at inner surface (pipe spacing 1.5 m × 1.5 m, $\theta_0 = 25°C$).

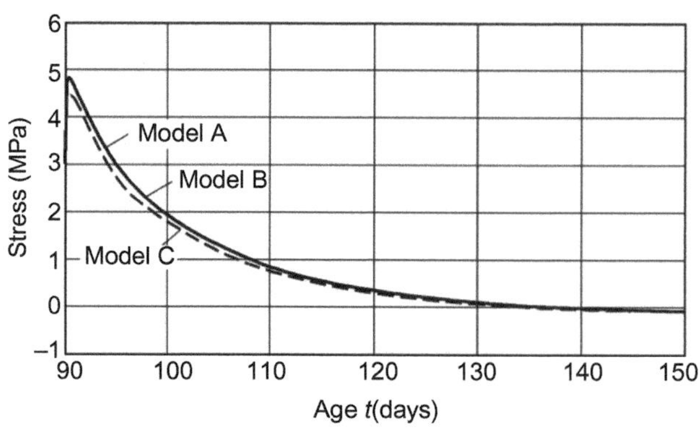

Figure 17.25 Elastocreeping stress σ_x at point *a* in late pipe cooling ($T_0 = 30°C$, $T_w = 9°C$, pipe spacing 1.5 m × 1.5 m).

the equivalent equation of heat conduction considering the effect of pipe cooling are close to the practical temperatures and stresses except the small region near the pipe (Figure 17.26).

The maximum elastocreeping stress appears at the instant of beginning of flow of water, the values of which are shown in Table 17.7.

17.8.5 New Method of Cooling—Multistep Early and Slow Cooling with Small Temperature Differences—Numerical Analysis [82, 83]

Experience shows that the best way to prevent cracking of mass concrete is to use the author's method—multistep early and slow cooling with small temperature differences to replace the traditional method—quick and late cooling with big temperature differences. It has been proved in Section 17.7 by theoretical method that the new method can reduce the tensile stress a great deal and in the following it will be proved again by the numerical method.

Let the pipe spacing be 1.5 m × 1.5 m, computing Model A, initial temperature $T_0 = 30°C$, the final steady temperature $T_f = 10°C$, three cases are computed as follows:

1. 1 step water temperature, $\tau = 90 - 160$ days, $T_w = 9°C$;
2. 2 step water temperature, $\tau = 90 - 120$ days, $T_w = 19°C$; $\tau = 120 - 160$ days, $T_w = 9°C$;
3. 4 step water temperature, $\tau = 30 - 60$ days, $T_w = 24°C$; $\tau = 60 - 90$ days, $T_w = 19°C$; $\tau = 90 - 120$ days, $T_w = 14°C$; $\tau = 120 - 160$ days, $T_w = 9°C$.

The computed results are shown in Figure 17.27. It is evident that the new method can reduce the tensile stress a great deal. This conclusion is the same as that given by theoretical solution in Section 17.7.

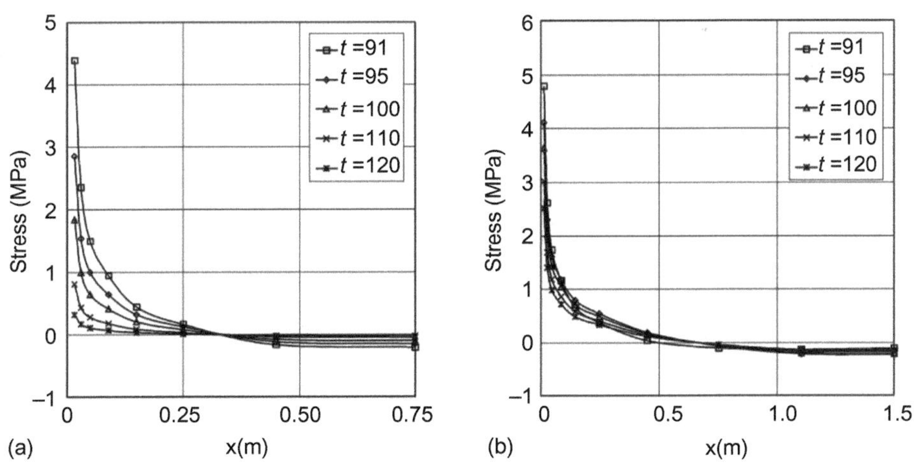

Figure 17.26 Distribution of tangential stress σ_x in cross section *ab* of Model A in late pipe cooling ($T_0 = 30°C$, $T_w = 9°C$, cooling begins at $t = 90$ days). (a) Pipe spacing 1.5 m × 1.5 m and (b) pipe spacing 3.0 m × 3.0 m.

Table 17.7 The Maximum Stress at the Inner Surface in Late Cooling (MPa) ($T_0 = 30°C$, $T_w = 9°C$)

Pipe Spacing (m × m)	Maximum Stress			Computing Model			Depth of Tensile Stress (m)
		Point	Stress	A	B	C	
1.5 m × 1.5 m	Axial stress	a	σ_z	5.00	5.00	4.96	0.43
		c	σ_z	5.00	5.06	6.00	
		e	σ_z	5.00	5.09	7.12	
	Tangential stress	a	σ_x	4.81	4.81	4.52	0.33
		c	σ_y	4.81	4.86	5.76	
		e	σ_x	4.81	4.89	7.05	
3.0 m × 3.0 m	Axial stress	a	σ_z	5.25	5.25	5.23	0.87
		c	σ_z	5.25	5.33	6.40	
		e	σ_z	5.25	5.36	7.73	
	Tangential stress	a	σ_x	5.05	5.06	4.74	0.70
		c	σ_y	5.05	5.11	6.12	
		e	σ_x	5.05	5.14	7.58	

17.9 The FEM for Computing Temperatures and Stresses in Pipe Cooled Concrete

17.9.1 Pipe Cooling Temperature Field Solved Directly by FEM

1. Fundamental equation

The differential equation of heat conduction of concrete cooled by pipe is

$$\frac{\partial T}{\partial t} = a\left(\frac{\partial^2 T}{\partial x^2} + \frac{\partial^2 T}{\partial y^2} + \frac{\partial^2 T}{\partial z^2}\right) + \frac{\partial \theta}{\partial t} \tag{17.115}$$

Initial condition: when

$$t = 0, \quad T(x, y, z, o) = T_0(x, y, z) \tag{17.116}$$

Boundary condition: on surface of concrete,

$$-\lambda \frac{\partial T}{\partial n} = \beta(T - T_a) \tag{17.117}$$

On boundary of pipe,

$$-\lambda \frac{\partial T}{\partial n} = k_5(T - T_w) \tag{17.118}$$

where

n—normal
β—surface conductance of concrete

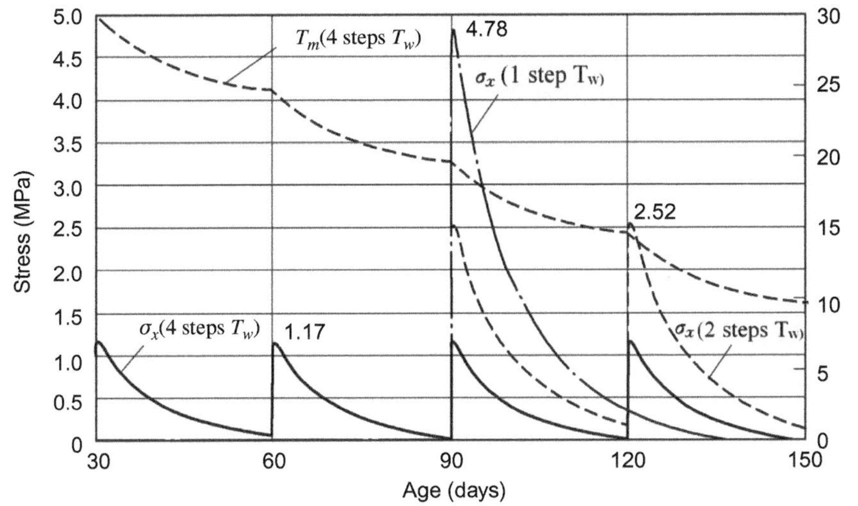

Figure 17.27 Model A, the elastocreeping stress σ_x at the inner surface of concrete and the mean temperature T_m in the late cooling, numerical solution ($T_0 = 30°C$, $T_w = 9°C$, pipe spacing 1.5 m × 1.5 m).

T_a—air temperature
k_5—coefficient of heat conductance of pipe given by Eq. (17.4)
T_w—water temperature.
Discretizing Eq. (17.115) by FEM, we get a system of linear equations as follows:

$$\left([H] + \frac{1}{s\Delta\tau_n}[R]\right)\{T_{n+1}\} + \left(\frac{1-s}{s}[H] - \frac{1}{s\Delta\tau_n}[R]\right)\{T_n\} + \frac{1-s}{s}\{F_n\} + \{F_{n+1}\} = 0 \tag{17.119}$$

The matrices $[H]$ and $[R]$ are given in Eq. (9.19) and Eq. (9.24), and the elements of $\{F_n\}$ and $\{F_{n+1}\}$ are given in the following:

$$F_i = \sum_e \left(-f_i \frac{\partial\theta}{\partial\tau} - p_i^e T_a - \overline{p_i^e} T_w\right) \tag{17.120}$$

$$p_i^e = \frac{\beta}{\lambda}\iint_{\Delta c} N_i \, ds \tag{17.121}$$

$$\overline{p_i^e} = \frac{k_5}{\lambda}\iint_{\Delta c} N_i \, ds \tag{17.122}$$

where T_a—air temperature and T_w—temperature of cooling water.

2. Temperature of water in the cooling pipe

We only know the water temperature T_w at the inlet of pipe, due to absorption of heat from concrete, T_w will increase along the length of pipe, the variation of T_w depends on the temperature field of concrete, thus T_w is practically unknown. This is the difficulty for solving the pipe cooling problem directly by FEM. This problem was solved by the author and Cai Jiangbo in 1989 [41] as follows.

The temperature of water T_w at the inlet of the pipe is known, the variation of T_w along the pipe will be computed in the following.

Cutting a series of cross sections as shown in Figure 17.28, due to absorption of heat from the concrete between two cross sections, the increment of water temperature is

$$\Delta T_{wi} = \frac{\lambda \Delta L}{2c_w\rho_w q_w}\left[\left(\int_{B_0}\frac{\partial T}{\partial r}\mathrm{d}s\right)_i + \left(\int_{B_0}\frac{\partial T}{\partial r}\mathrm{d}s\right)_{i+1}\right] \tag{17.123}$$

where B_0 is the contact surface between concrete and water, the water temperature along the pipe is

$$T_{wi} = T_{w1} + \sum_{j=1}^{i-1}\Delta T_{wj} \quad i = 2, 3, 4, \ldots \tag{17.124}$$

in which T_{wi} is the water temperature at the inlet of the pipe. As T_{wj} for $j \geq 2$ is unknown, the above equation must be computed by the iteration method as follows:

The first iteration: assuming that all the water temperatures T_{wi} are equal to T_{w1} at the inlet of the pipe, solve the temperature field by Eq. (17.119) and compute the first approximate water temperature $T_{wi}^{(1)}$ by Eqs. (17.123) and (17.124).

The second iteration: let the water temperature along the cooling pipe be equal to $T_{wi}^{(1)}$, solve Eq. (17.119) again and compute the second approximate water temperature $T_{wi}^{(2)}$ by Eq. (17.124).

This method converges so quickly that only two to three iterations are required.

17.9.2 Equivalent FEM for Computing the Temperatures and Stresses in Mass Concrete Block with Cooling Pipe

Consider a massive concrete block with cooling pipe as shown in Figure 17.29, along the planes of symmetry between pipes cut a prism with one cooling pipe as shown in Figure 17.29(b).

First step, assuming that the external surface of the cooling block is insulated, due to symmetry, the planes of symmetry between cooling pipes are also insulated,

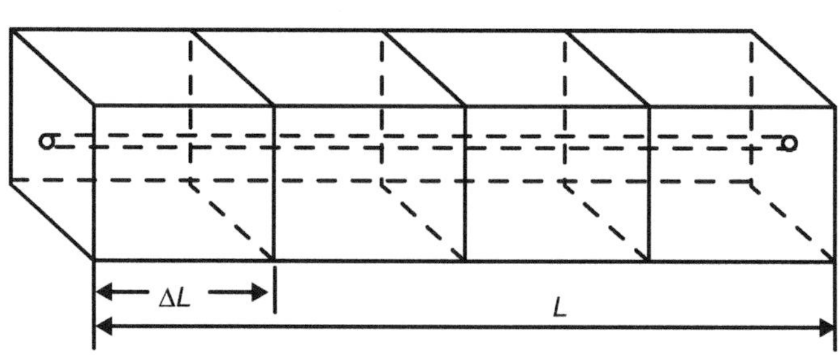

Figure 17.28 The concrete prism cooled by a water pipe.

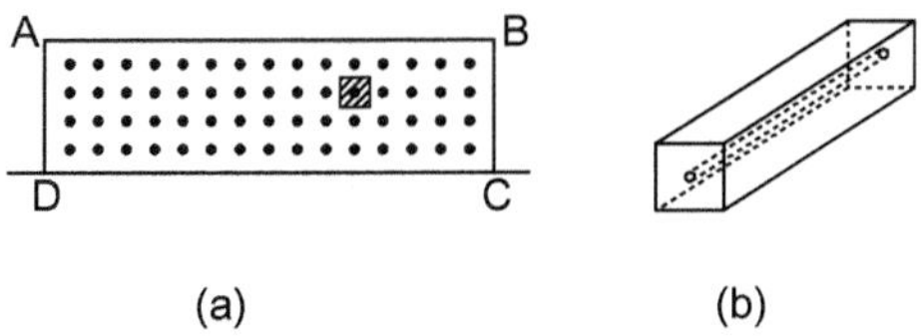

Figure 17.29 Concrete dam block and cooling prism. (a) Original model, cross section of a dam block and (b) submodel, cooling prism with insulated surface.

thus, the external surface of each cooling prism is insulated. In this case, the temperature field of the block is equal to the temperature $T_1(x,y,z,t)$ of the cooling prism with insulated surface, in which z is the direction in the length of pipe. The mean temperature in the xy plane of the cooling prism is $T_{1\mathrm{m}}(z,t)$ which varies along the length of pipe and may be expressed as $T_{1\mathrm{m}}(L,t)$ and can be computed as follows:

$$T_{1\mathrm{m}}(L,t) = (T_0 - T_\mathrm{m})\Phi_L(L,t) + \theta_0\psi_L(L,t) \tag{17.125}$$

The stresses in the cooling prism are $\sigma_1(x,y,z,t)$.

By differentiating Eq. (17.125), we have

$$\frac{\partial T_{1\mathrm{m}}(L,t)}{\partial t} = (T_0 - T_\mathrm{m})\frac{\partial \Phi_L(L,t)}{\partial t} + \theta_0\frac{\partial \psi_L(L,t)}{\partial t} \tag{17.126}$$

Second step, the external surfaces of the dam block return to the natural condition, the temperature field of the block is determined by the second equivalent equation of heat conduction with pipe cooling as follows:

$$\frac{\partial T}{\partial \tau} = a\left(\frac{\partial^2 T}{\partial x^2} + \frac{\partial^2 T}{\partial y^2} + \frac{\partial^2 T}{\partial z^2}\right) + (T_0 - T_\mathrm{w})\frac{\partial \Phi_L(L,t)}{\partial \tau} + \theta_0\frac{\partial \psi_L(L,t)}{\partial t} \tag{17.127}$$

This equation may be solved by FEM with common mesh, the influence of cooling pipe is considered by the function Φ_L and ψ_L, thus, the computation is simplified remarkably. The temperatures and stresses in this case are $T_2(x,y,z,t)$ and $\sigma_2(x,y,z,t)$, respectively.

The actual temperatures and stresses in the concrete block are given by

$$\left.\begin{aligned} T(x,y,z,t) &= T_1(x,y,z,t) + T_2(x,y,z,t) \\ \sigma(x,y,z,t) &= \sigma_1(x,y,z,t) + \sigma_2(x,y,z,t) \end{aligned}\right\} \tag{17.128}$$

The temperatures and stresses given by Eq. (17.128) are precise values. $T_1(x,y,z,t)$ and $\sigma_1(x,y,z,t)$ are computed by a submodel as shown in Figure 17.29 (b) or by theoretical Eq. (17.96), and $T_2(x,y,z,t)$ and $\sigma_2(x,y,z,t)$ of the original model are computed by FEM with common mesh, so the computation is very simple.

For further simplification, integrating $\partial\phi_L/\partial t$ and $\partial\psi_L/\partial t$ along the length of pipe, the mean value $\partial\phi/\partial t$ and $\partial\psi/\partial t$ of length L is obtained. Substituting them in

Eq. (17.127), we get the first kind of equivalent equation of heat conduction considering the effect of pipe cooling as follows:

$$\frac{\partial T}{\partial \tau} = a\left(\frac{\partial^2 T}{\partial x^2} + \frac{\partial^2 T}{\partial y^2} + \frac{\partial^2 T}{\partial z^2}\right) + (T_0 - T_w)\frac{\partial \Phi}{\partial t} + \theta_0 \frac{\partial \psi}{\partial t} \tag{17.129}$$

which is the same as Eq. (17.126). The error induced by the simplification is small. For example, for a cooling pipe with length $L = 300$ m, the temperature difference between the water at the inlet and outlet is about 4–6°C, the temperature gradient is about 0.013–0.020°C/m. The pipe is arranged in snakelike form. The length of one coil is about $2 \times 20 = 40$ m, the maximum difference of water temperature between two adjacent pipes is about 0.5–0.8°C which has some influence on the temperature field of the original model but the influence is small. The direction of flow of cooling water is changed every 12 or 24 h in practical engineering, and the influence of difference of water temperature between adjacent pipe is reduced further.

17.9.3 Comparison Between the Direct Method and the Equivalent Method for Pipe Cooling

When the temperature field with cooling pipes is solved by direct method, there are three difficulties: (i) As the radius of cooling pipe is 1–2 cm, so the number of elements of a dam block is very large. (ii) Due to the variation of the temperature of water along the length of pipe, a series of sections with spacing L as shown in Figure 17.28 must be computed. (iii) The spacing between cooling pipes is 0.5–3.0 m; there are many cooling pipes in a dam block.

As a result, it is very difficult to compute the temperature field of mass concrete with cooling pipes by the direct method.

When the temperature field with cooling pipes is solved by the equivalent Eq. (17.129), the influence of cooling pipe is considered by the function ϕ and ψ, the above-mentioned three difficulties no longer exist, the problem may be solved by common mesh of elements.

A concrete cylinder is computed by the equivalent Eq. (17.129) with FEM, the results of computation are very close to the theoretical results of Section 17.3. Hence, the precision of the equivalent method is good.

Now, the equivalent Eq. (17.129) is extensively applied in dam engineering.

17.10 Three Principles for Pipe Cooling [79–84]

The variation of the temperature and stress of concrete in the period before grouting of contraction joints are shown in Figure 17.30. The first purpose of pipe cooling is to reduce the temperature of concrete to the final steady temperature T_f

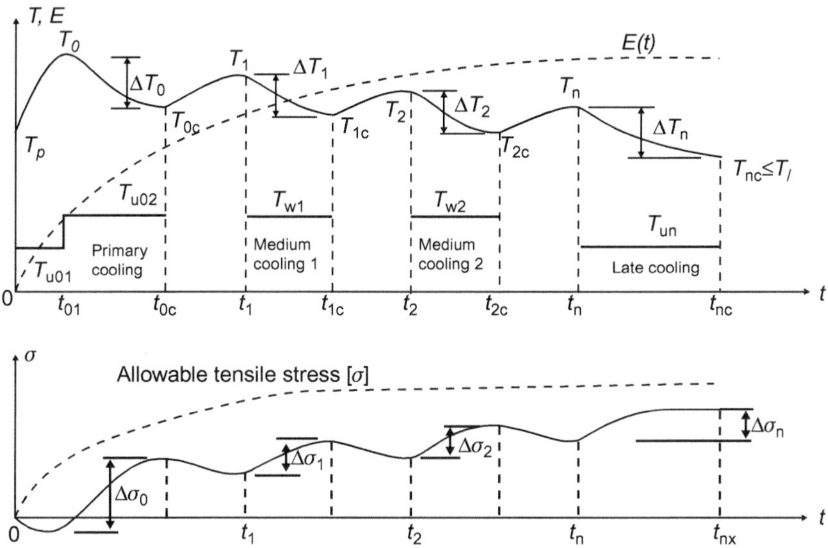

Figure 17.30 Variation of temperature and stress of concrete in the period before grouting of contraction joints.

before the grouting of contraction joints. The second purpose of pipe cooling is to reduce the tensile stress by control of the maximum temperature $T_0 = T_p + T_r$, where T_p is the placing temperature of concrete and T_r is the temperature rise due to hydration heat of cement.

Only the above-mentioned two factors are taken into account in the traditional design of temperature control of concrete dams. Due to the increase of dimensions of the dam block, large tensile stress may appear in the cooling process; in recent years, the mode of pipe cooling has changed to reduce the tensile stress.

In order to prevent cracking, the following three principles of pipe cooling are suggested by the author:

Principle I: Early cooling immediately after placing the concrete to reduce the maximum temperature of concrete. In the lowest part of concrete on rock foundation, the spacing of polythene pipe may be reduced to 1.0 m × 0.5 m, which is sufficient to control the maximum temperature. Part of the cooling pipe may be closed 2–3 days after the maximum temperature has appeared to avoid too rapid temperature drop in the late age.

Principle II: Improving the condition of restraint. In the medium and late stage of cooling, it is suggested that $\Delta H \geq 0.40L$, where ΔH is the height of cooling region and L is the length of concrete block. The temperature gradient in the vertical direction must be controlled.

Principle III: Disperse the temperature difference. It is suggested to adopt the new type of cooling: early cooling and slow cooling with small temperature difference. One to two medium coolings are added between the early cooling and the late cooling; in each step of cooling the difference between the initial temperature of concrete and the water temperature $T_0 - T_w \leq 10°C$.

As shown in Figure 17.30, the task of pipe cooling is to reduce the temperature of concrete from the maximum temperature $T_0 = T_p + T_r$ to the final steady temperature T_f. The difference of temperature is $\Delta T = T_0 - T_f$. In the past, generally there was one or two-step cooling, the late-stage cooling or the early-stage cooling plus the late-stage cooling. The temperature differences are so immense that cracks may appear. In order to prevent cracking of concrete, it is suggested to adopt a new type of cooling, i.e., early and slow cooling with small temperature differences in three to four stages which possess the following benefits: (i) make full use of the effect of creep of concrete; (ii) make use of the low modulus of elasticity of concrete in early age; (iii) the tensile stress of the inner surface of concrete due to sudden change of water temperature is reduced due to dispersing of temperature difference.

17.11 Research on the Pattern of Early Pipe Cooling

The modulus of concrete is small in the early age, so in order to make use of this property, the maximum temperature of concrete should be reduced as low as possible in the early age. The principles for the early pipe cooling of concrete are as follows:

1. The tensile stress in the process of early cooling should not exceed the allowable tensile stress.
2. At the end of early cooling, the temperature T_{oc} of concrete should be as low as possible so the temperature drop in the late age is small.
3. The control of temperature is simple.

The time interval between concrete lifts and the spacing of pipe are the two important factors which influence the thermal stresses.

As shown in Figure 17.31, a concrete block with width 60 m, height 60 m is computed, the thickness of the lower 8 lifts is 1.5 m, the thickness of the upper 16 lifts is 3.0 m, $\theta(\tau) = 25\tau/(1.8 + \tau)°C$, $E(\tau) = 35{,}000\tau/(5.0 + \tau)$ MPa, the unit creep is given in Eq. (6.51), $\mu = 0.167$. For the rock foundation, $E = 35{,}000$ MPa, $\mu = 0.25$. The initial temperature of concrete and rock is 12°C, the air temperature is 12°C. The allowable tensile stress of concrete is $[\sigma] = 1.50\tau/(9.0 + \tau)$ MPa, τ is the age of concrete of each lift, t is the time from the placing of concrete of the first lift, polythene pipe with length 200 m, and discharge of cooling water 1.20 m^3/h. There are four stages of cooling. The time of cooling of the first, the second, and the third stages is 20 days, the last stage of cooling is stopped when the temperature of concrete reduces to the steady temperature 12°C.

Fourteen schemes are computed and the results are given in Table 17.8 and Figures 17.32–17.34. It is evident that Schemes 3 and 9 are good schemes.

17.12 Research on the Pattern of the Medium and the Late Cooling

The pattern of the medium and the late cooling have important influence on the thermal stress. The factors of influences include: (i) the height of cooling region, (ii) the temperature gradient in the vertical direction, (iii) the spacing of pipes, and (iv) the number of stages of cooling. The influence of the height of cooling region has been discussed in Section 11.3, the other factors will be discussed in the following [82, 83].

17.12.1 The Influence of Temperature Gradient on the Thermal Stress

The thermal stresses will reduce if the temperature varies step by step in the vertical direction as shown in Figure 17.35. The stress coefficients are as follows:

$$\eta = \frac{\sigma(1-\mu)}{E\alpha\Delta T}$$

For four kinds of temperature difference ($c/L = 0.10$, $b/L = 0.50$) computed by FEM are shown in Figure 17.36.

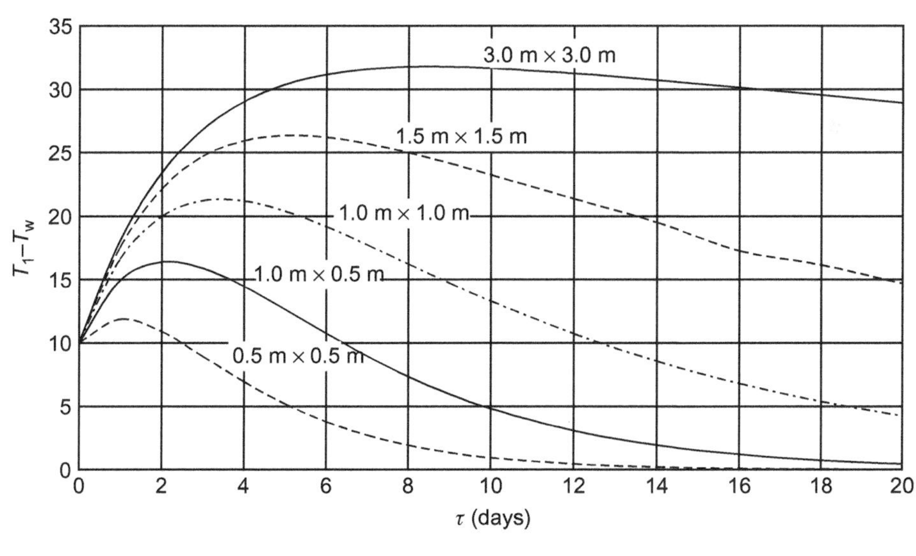

Figure 17.31 Example, a dam block.

Table 17.8 Computed Results for 14 Schemes

Pipe Spacing (m × m)	Time Interval (days)	Scheme	Water Temperature in Early Cooling (°C)				Water Temperature in Medium Cooling (°C)		Water Temperature in Late Cooling (°C)	Early Cooling			Medium and Late Cooling		Good Scheme
			Age (days) $0-t_{01}$	Water Temperature (°C) T_{w01}	Age (days) $t_{01}-t_{02}$	Water Temperature (°C) T_{w02}	First Step	Second Step		Max Temperature in Course of Cooling (°C)	Temperature at the End of Early Cooling (°C)	Maximum Tensile Stress (MPa)	Maximum Tensile Stress in Course of Cooling (MPa)	Maximum Tensile Stress at End of Cooling (MPa)	
1.5 × 1.5	5	1	0–5	9	5–20	9	16	16	9	25.76	18.89	0.78	1.26	0.87	√
		2		9		16	16	16	9	25.79	22.61	0.38	1.55	1.14	
	10	3	0–5	9	5–20	9	16	16	9	25.61	18.1	0.74	1.13	0.77	√
		4		9		20	16	16	9	25.72	23.53	0.25	1.54	1.16	
		5		9		16	16	16	9	25.68	21.61	0.39	1.39	1.02	
		6		9		Stop	16	16	9	27.79	25.45	0.25	1.65	1.16	
	20	7	0–5	9	5–20	9	16	16	9	25.55	16.12	1.17	1.05	0.70	
		8		9		20	16	16	9	25.63	20.56	0.65	1.35	0.96	
		9	0–12	9	12–30	16	16	16	9	25.66	17.82	0.92	1.10	0.78	√
1.5 × 0.75	5	10	0–7	9	7–20	9	16	16	9	23.43	13.16	1.01	0.88	0.51	
		11		9		16	16	16	9	23.43	18.45	0.45	1.29	0.84	
	10	12	0–7	9	7–20	9	16	16	9	23.38	13.26	1.03	0.71	0.48	
		13		9		16	16	16	9	23.45	18.3	0.43	1.12	0.81	
		14		9	7–30	16	16	16	9	23.45	17.31	0.54	1.04	0.75	√
	20	15	0–7	9	7–20	9	16	16	9	23.35	12.51	1.35	0.67	0.49	
		16		9		16	16	16	9	23.39	17.01	0.71	0.95	0.74	√

17.12.2 The Influence of Pipe Spacing on the Thermal Stress

The rate of cooling and hence the thermal stress are influenced by the pipe spacing. This is an important factor which is neglected in the past. A concrete block with $H = L = 60$ m on rock foundation is computed by FEM, the height of cooling region is $b = 0.30L$, the concrete is cooled from $T_0 = 30°C$ to $T_f = 10°C$, the stresses on the central section after cooling are shown in Figure 17.37. Corresponding to four kinds of pipe spacing 1.0 m × 0.50 m, 1.0 m × 1.0 m, 1.5 m × 1.5 m, and 3.0 m × 3.0 m, the maximum stresses are respectively $\sigma_x = 3.49$, 3.26, 3.05, and 2.03 MPa. The smaller the pipe spacing, the quicker the cooling and the larger the tensile stress.

17.12.3 The Influence of the Number of Stages of Pipe Cooling

The thermal stresses are influenced by the number of stages of pipe cooling. The more the stages of cooling, the smaller the stresses. The comparison between the two-stage cooling and the three-stage cooling is given by an example in the following.

Consider a dam block with length $L = 60$ m, height $H = 60$ m, thickness of concrete lift 3.0 m, time interval between lifts 7 days, $\theta(\tau) = 25\tau/(1.7 + \tau)°C$, placing temperature $T_p = 25°C$, air temperature $T_a = 25°C$, initial temperature of rock 25°C, pipe spacing: the lower part $y = 0–6$ m, 1.0 m × 0.5 m; $y = 6–60$ m, 1.5 m × 1.5 m. There are three schemes: (i) Scheme 1, two-stage cooling, for the first stage, $\tau = 0 - 20$ days, $T_w = 20°C$; the second stage of cooling begins at $\tau = 240$ days, $T_w = 10°C$, cooling stopped when the temperature of concrete drops to 12°C. (ii) Scheme 2, three-stage cooling, first stage, $\tau = 0 - 20$ days, $T_w = 20°C$; second stage, $\tau = 150 - 200$ days, $T_w = 15°C$; third stage, $t \geq 240$ days, $T_w = 10°C$, cooling stopped when $T_f = 12°C$. (iii) Scheme 3, three-stage cooling + temperature adjustment, the first stage, $\tau = 0 - 25$ days, $T_w = 20°C$; the second stage, $\tau = 150 - 200$ days, $T_w = 16°C$; the third stage including 2 steps, $\tau = 240 - 270$d, $T_w = 13°C$, $\tau > 270$ days, $T_w = 11°C$, cooling stopped when temperature of concrete reduced to $T_f = 12°C$.

The envelope of the tensile stresses is shown in Figure 17.38. The maximum stresses are given in Table 17.9. For the two-stage scheme, the maximum stress is 2.73 MPa which exceeds the allowable stress 2.10 MPa. For the three-stage scheme, the maximum stress is 1.90 MPa which is lower than the allowable stress.

17.13 Strengthen Cooling by Close Polythene Pipe

17.13.1 Effect of Cooling by Close Pipe [80, 82]

Instead of steel pipe, the polythene pipes are widely used in the cooling of mass concrete in recent years. The placing and bending of polythene pipe are easy, thus, the spacing of pipes may be very small which can strengthen the cooling of concrete remarkably.

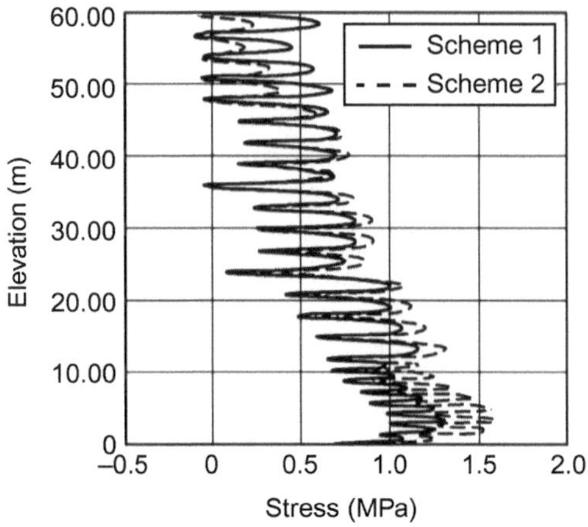

Figure 17.32 Envelop of σ_x on the mid plane of dam block, Scheme 1 with constant water temperature and Scheme 2 with variable water temperature.

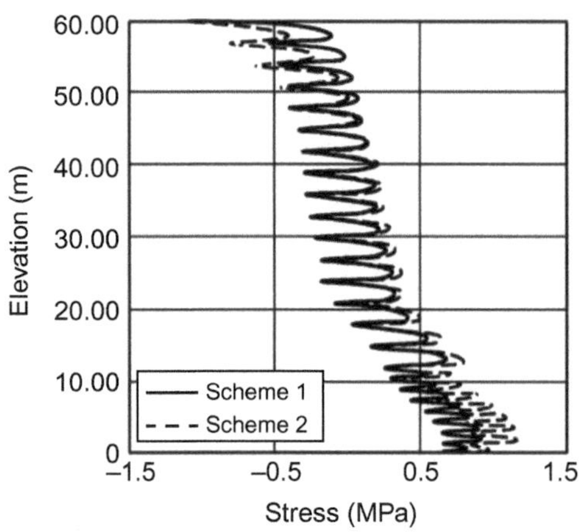

Figure 17.33 The final stress σ_x on the mid plane of dam block, Scheme 1 (constant water temperature) and Scheme 2 (variable water temperature).

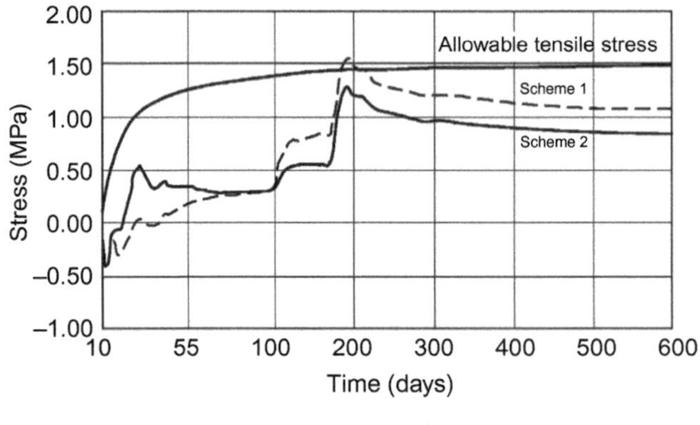

Figure 17.34 The variation of σ_x of the midpoint of the center line of third lift of the dam block.

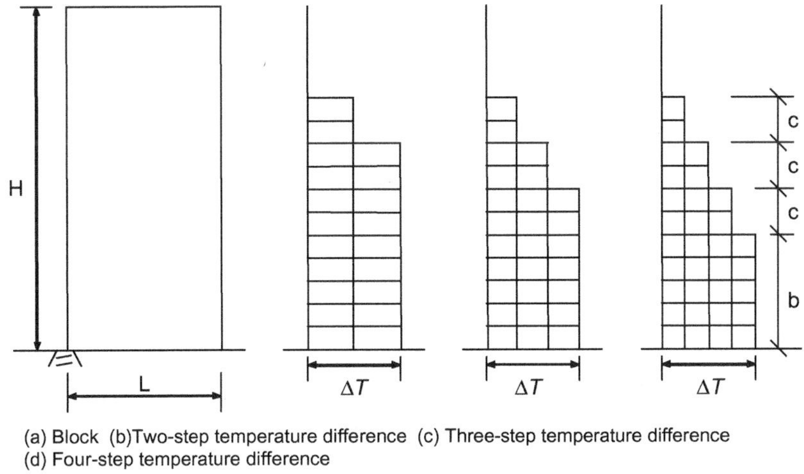

Figure 17.35 Stepwise temperature difference in late cooling.

Under the influence of pipe cooling, initial temperature difference and hydration heat, the temperature T_1 of concrete cylinder with insulated surface is given by Eq. (17.88). $T_1 - T_w$ for pipe cooled concrete cylinder with exponential and hyperbolic hydration heat are shown in Figures 17.39 and 17.40. It is clear that the influence of pipe spacing is remarkable.

For exponential type of hydration heat, from Eq. (17.88) and $dT_1/dt = 0$, the time t_1 before T_1 reaches its maximum value is

$$t_1 = \frac{1}{m-p}\ln\left\{\frac{\theta_0 m^2}{p(m-p)[T_0 - T_w + \theta_0 m/(m-p)]}\right\} \tag{17.130}$$

Substitution of t_1 into (17.88) will yield the maximum T_1. Let $\theta_0 = 25°C$, $m = 0.40$ (1/d), the difference between the maximum temperature T_1 and T_w is shown in Table 17.10.

From Eq. (17.88), $T_1 = T_w + (T_1 - T_w)$ and $T_w = T_0 - (T_0 - T_w)$, let $T_0 = 12°C$, from Table 17.10, we get the maximum temperature T_1 of concrete for different spacing of pipes as shown in Table 17.11. The maximum temperature $T_1 = 12 + 25 = 37°C$ when there is no pipe cooling. From Table 17.11, the decrease of the maximum temperature is 6.3–8.6°C for pipe spacing 1.5 m × 1.5 m, 10–13.7°C for pipe spacing 1.0 m × 1.0 m, and 13–19°C for pipe spacing 1.0 m × 0.5 m. The effect of pipe cooling is apparent. The maximum temperature and the temperature difference of concrete block on rock foundation can be reduced a great deal by pipe cooling.

17.13.2 Influence of Cooling of Pipe with Small Spacing on the Thermal Stress

The temperature difference is reduced by the use of cooling pipe with small spacing, thus, the thermal stresses are also reduced. An example is given in the

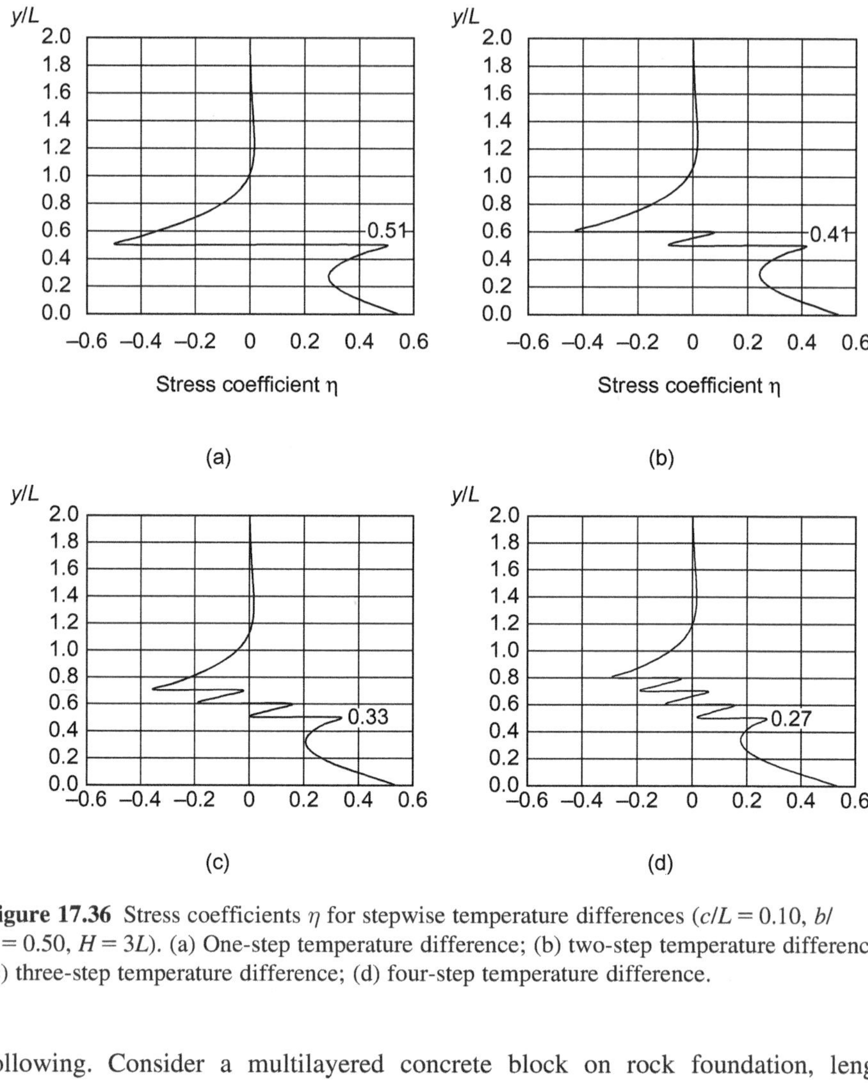

Figure 17.36 Stress coefficients η for stepwise temperature differences ($c/L = 0.10$, $b/L = 0.50$, $H = 3L$). (a) One-step temperature difference; (b) two-step temperature difference; (c) three-step temperature difference; (d) four-step temperature difference.

following. Consider a multilayered concrete block on rock foundation, length $L = 60$ m, height $H = 60$ m, thickness of lift 3.0 m, time interval between lifts 7 days, diffusivity $a = 0.10\ \text{m}^2/\text{day}$, surface conductance $\beta = 70\ \text{kJ/(m}^2\ \text{h °C)}$, $\mu = 0.167$, $E(\tau) = 35{,}000[1 - \exp(-0.40\tau^{0.34})]$ MPa, unit creep is given by Eq. (6.51), $\alpha = 1 \times 10^{-5}(1/°\text{C})$, the allowable tensile stress $[\sigma_t] = 2.10\tau/(7.0 + \tau)$ MPa, $\theta(\tau) = 25\tau/(1.7 + \tau)$°C. There is no creep in the foundation, $E_f = 35{,}000$ MPa, $\mu = 0.25$. Initial temperature of concrete and rock $T_0 = 25$°C, air temperature $T_a = 25$°C.

Pipe spacing: height $y = 0–6$ m, $1.0\ \text{m} \times 0.5$ m; $y = 6–60$ m. $1.5\ \text{m} \times 1.5$ m. There are three stages of cooling, first stage, $\tau = 0–20$ days, $T_w = 20$°C; second stage, $\tau = 150–200$ days, $T_w = 15$°C; third stage, $\tau \geq 240$ days, $T_w = 11.5$°C, cooling is stopped when the temperature of concrete decreases to $T_f = 12$°C.

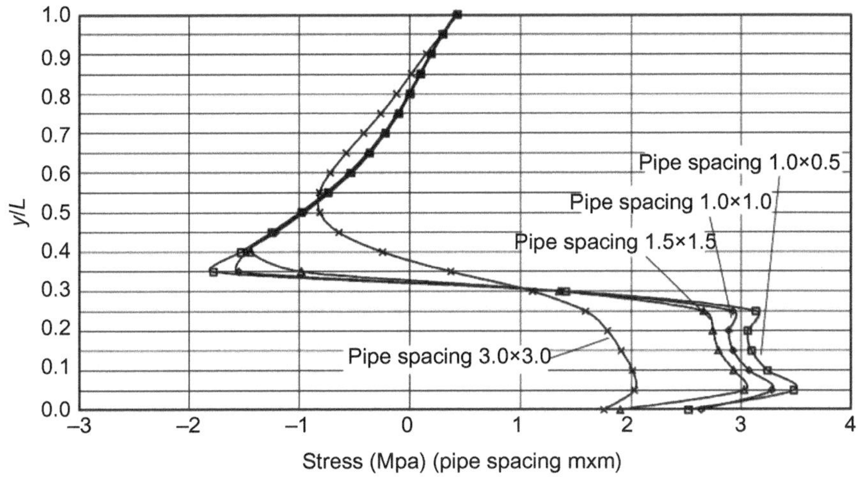

Figure 17.37 Thermal stresses on the central section after pipe cooling ($H = L = 60$ m, $b/L = 0.30$).

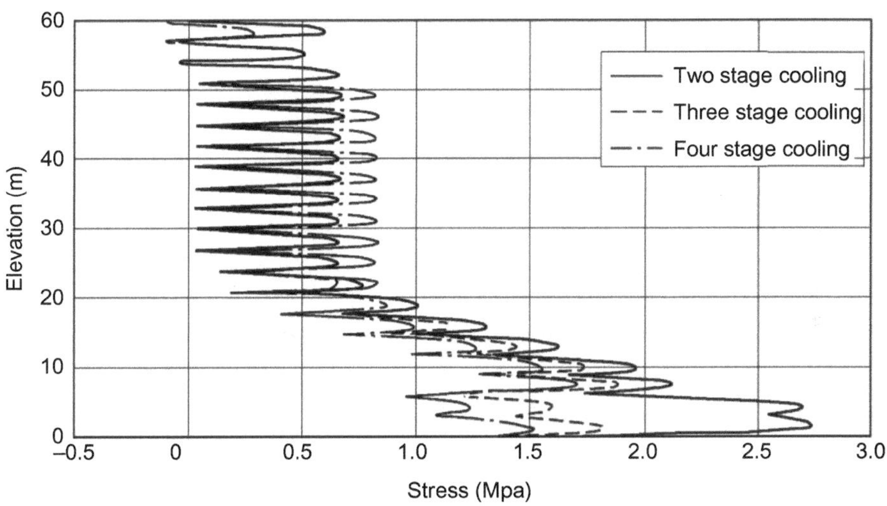

Figure 17.38 The envelopes of stresses for different schemes of pipe cooling.

Computed by 3D FEM, the results are shown in Figures 17.41 and 17.42, from which the following important conclusions are derived:

1. The length of block is 60 m, the thickness of lift is 3.0 m, the velocity of rising of concrete is high, there is no precooling of concrete, the initial temperature of concrete and air temperature is 25°C. Although the above-mentioned conditions are not favorable, the

Table 17.9 The Maximum Tensile Stress for Different Schemes in Late Cooling (MPa)

Pattern of Cooling	Two-Stage Cooling	Three-Stage Cooling	Three-Stage Cooling with Temperature Adjustment
Maximum tensile stress	2.73	1.90	1.72

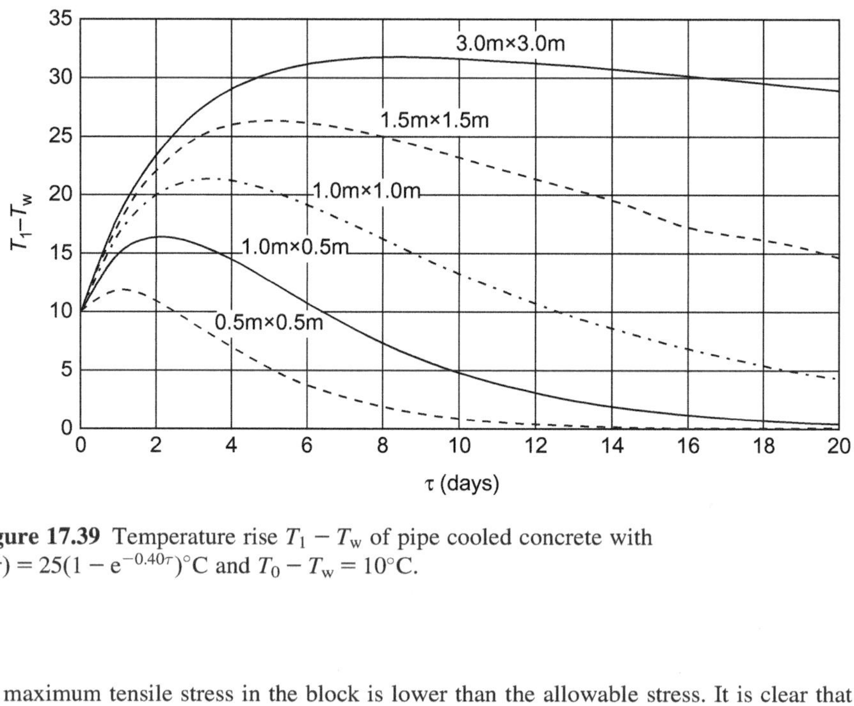

Figure 17.39 Temperature rise $T_1 - T_w$ of pipe cooled concrete with $\theta(\tau) = 25(1 - e^{-0.40\tau})$°C and $T_0 - T_w = 10$°C.

maximum tensile stress in the block is lower than the allowable stress. It is clear that the artificial cooling by close polythene pipe in the range $y = 0$–6 m above the rock surface can effectively control the thermal stress with minimal expense.

2. Artificial cooling by close polythene pipe in the range $h = 0$–$0.10L$ can effectively control the tensile stress due to foundation temperature difference or upper–lower temperature difference in the concrete block with minimal expense.
3. By means of cooling of close polythene pipe, it is possible to save the expensive precooling of concrete in some cases.

17.13.3 The Principle for Control of Pipe Spacing and Temperature Difference $T_0 - T_w$

There are two methods to reduce the maximum temperature of concrete in the early age: closing the pipe spacing and reducing the water temperature, but if the water temperature is too low or $T_0 - T_w$ is too big, it is possible to induce large tensile

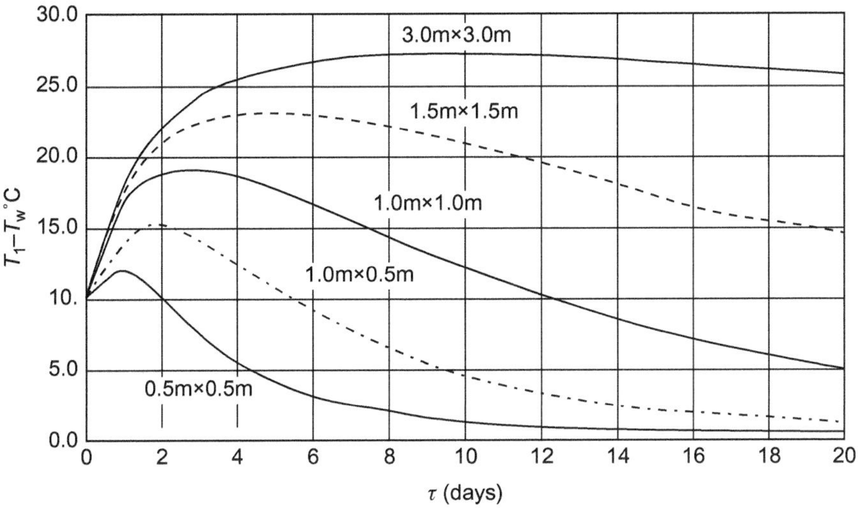

Figure 17.40 Temperature rise $T_1 - T_w$ of pipe cooled concrete with $\theta(\tau) = 25\tau/(2+\tau)$°C and $T_0 - T_w = 10$°C.

stress in the concrete. The principle for the early cooling is that the pipe spacing is small and the temperature difference $T_0 - T_w$ is suitable which must be determined by computation. For the computed example in the section, $T_0 = T_a = 25$°C, for $y = 0$–6 m, the pipe spacing is 1.0 m × 0.5 m, the temperature difference $T_0 - T_w = 5$°C is suitable. In the medium and late pipe cooling, if the pipe spacing is small, it is necessary to strictly control the temperature difference $T_0 - T_w$. The water temperature T_w should be reduced step by step so as to avoid too large $T_0 - T_w$ which will induce large tensile stress. In the medium or late pipe cooling, or 2 days after the appearance of the maximum temperature in the early pipe cooling, one part of the pipe coils may be closed to reduce the rate of cooling.

17.14 Advantages and Disadvantages of Pipe Cooling

Although pipe cooling has been relevant for some time, the appropriate research was insufficient 10 years ago. People knew its advantages but did not know some of its disadvantages. As a result, serious cracks had appeared in some concrete blocks after pipe cooling [85].

Example Concrete block on rock foundation, length $L = 60$ m, height $H = 60$ m, the thickness of the lower 8 lifts 1.5 m, the thickness of the upper 16 lifts 3.0 m, time interval between lifts 10 days, diffusivity $a = 0.10$ m^2/day, surface conductance $\beta = 70$ kJ/(m^2 h °C), modulus of elasticity $E(\tau) = 35{,}000\tau/(5.0+\tau)$ MPa, unit creep is given by Eq. (6.32), $\mu = 0.167$, $\theta(\tau) = 25\tau/(1.80+\tau)$°C, initial temperature of concrete and rock 12°C, air temperature 12°C. For rock foundation,

Table 17.10 The Maximum $T_1 - T_w$ (°C) for Different Pipe Spacing ($\theta_0 = 25$°C)

Pipe spacing (m × m)		0.5 × 0.5	1.0 × 0.5	1.0 × 1.0	1.5 × 1.5	3.0 × 3.0
T_0-T_w (°C)	0°C	7.66	11.28	14.99	18.72	22.67
	5°C	9.44	13.68	18.07	22.51	27.22
	10°C	11.86	16.39	21.32	26.37	31.78

Table 17.11 The Maximum Temperature T_1(°C) for Different Pipe Spacing ($T_0 = 12$°C, $\theta_0 = 25$°C)

Pipe spacing (m × m)		0.5 × 0.5	1.0 × 0.5	1.0 × 1.0	1.5 × 1.5	3.0 × 3.0
T_w (°C)	12°C	19.66	23.28	27.00	30.72	34.67
	7°C	16.44	20.68	25.07	29.51	34.22
	2°C	13.86	18.39	23.32	28.37	33.78

$E_f = 35{,}000$ MPa, $\mu_f = 0.25$. Four schemes are computed by FEM program SAPTIS compiled by Professor Zhang Guoxin:

1. Scheme 1, natural cooling, no cooling pipe,
2. Scheme 2, early pipe cooling,
3. Scheme 3, early pipe cooling + late pipe cooling.
   Pipe spacing 1.5 m × 1.5 m, early cooling, $\tau = 0$–20 days, $T_w = 9$°C; late cooling, $\tau \geq 90$ days, $T_w = 9$°C, pipe cooling is stopped when the temperature of concrete decreases to 12°C.
4. Scheme 4, early cooling + two medium cooling + late cooling.

The computed results are shown in Table 17.12. The following points may be pointed out: (i) Comparing Scheme 2 with Scheme 1, due to early pipe cooling, the maximum temperature is reduced from 30.7°C to 25.6°C, the maximum tensile stress is reduced from 1.43 MPa to 0.80 MPa, the effect of early pipe cooling is apparent. (ii) Compare Scheme 3 with Scheme 2, the maximum temperature is the same, but the maximum stress 1.26 MPa in Scheme 3 is bigger than 0.80 MPa in Scheme 2. This is due to the quicker temperature drop in late cooling in Scheme 3; there is no sufficient time for creep to develop which is a disadvantage of pipe cooling. (iii) There are four stages of pipe cooling in Scheme 4, the maximum tensile stress is small and close to that in Scheme 2. (iv) There is no late pipe cooling in Schemes 1 and 2, which cannot be applied to arch dams, because the temperature of concrete cannot decrease to the final steady temperature T_f before grouting of contraction joints. Hence, the advantages and disadvantages of pipe cooling may be summarized in the following:

The advantages of pipe cooling are as follows:

1. The maximum temperature of concrete can be controlled.
2. The temperature of the dam can be reduced to the final steady temperature before grouting of joints in the dam.

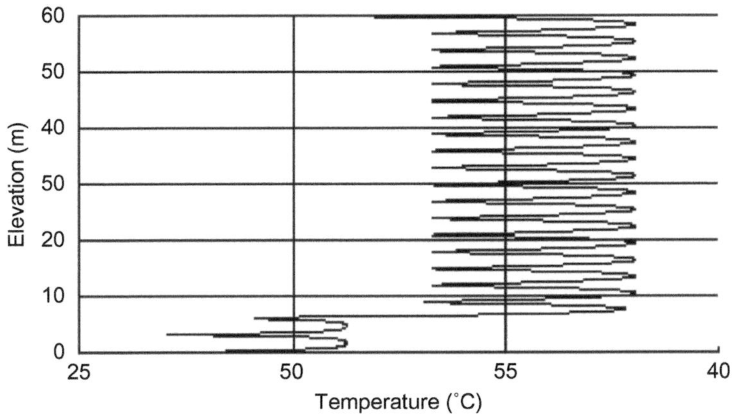

Figure 17.41 The envelope of temperatures on the central section of the concrete block.

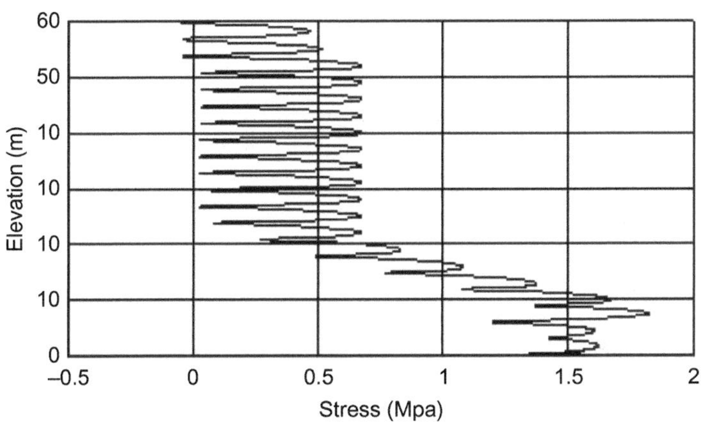

Figure 17.42 The envelope of thermal stress on the central section of the concrete block (allowable tensile stress 2.1 MPa).

The disadvantages of pipe cooling are as follows:

1. The temperatures of concrete block decrease rapidly, there is no sufficient time for the creep of concrete to display, thus for the same temperature drop, the tensile stress will be bigger in pipe cooling.
2. Local tensile stress will appear near the inner surface of concrete due to nonlinear distribution of temperature during pipe cooling.

The best way is to adopt multistage cooling with small temperature difference as the Scheme 4 in Table 17.12.

Table 17.12 Comparison of Natural Cooling with Three Schemes of Pipe Cooling

Scheme	Pipe Cooling	Maximum Temperature (°C)	Maximum Stress in Course of Cooling (MPa)	Final Maximum Stress (MPa)
1	No pipe cooling	30.7	1.43	1.38
2	Early pipe cooling	25.6	0.80	0.72
3	Early cooling + late cooling	25.6	1.26	0.76
4	Early cooling + two medium cooling + late cooling	25.6	0.99	0.76

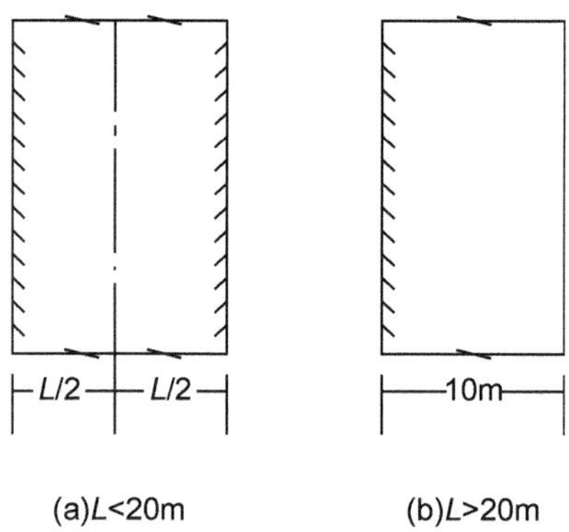

Figure 17.43 Schematic diagram for computation.

17.15 Superficial Thermal Insulation of Mass Concrete During Pipe Cooling in Hot Seasons

Sometimes pipe cooling must be conducted in hot seasons. As the air temperature is high, it is difficult to cool the concrete in the region 5–7 m near the surface of dam block to the final steady temperature. In order to overcome this drawback, it is necessary to adopt superficial thermal insulation [86].

Generally the depth of influence of superficial temperature does not exceed 10 m, if the thickness of dam $L < 20$ m, the computing model may be an infinite plate with thickness L as shown in Figure 17.43(a), the two lateral surfaces of which are in contact with air. If the dam is thicker than 20 m, the computing model may be an infinite plate with $L = 10$ m, one side being insulated and the other side being in contact with air, as shown in Figure 17.43(b). It is convenient to compute the variation of temperature by the finite difference method.

Example Thickness of dam block is 30 m. The computing model is an infinite plate with thickness $L = 10$ m with the left side in contact with air and the right

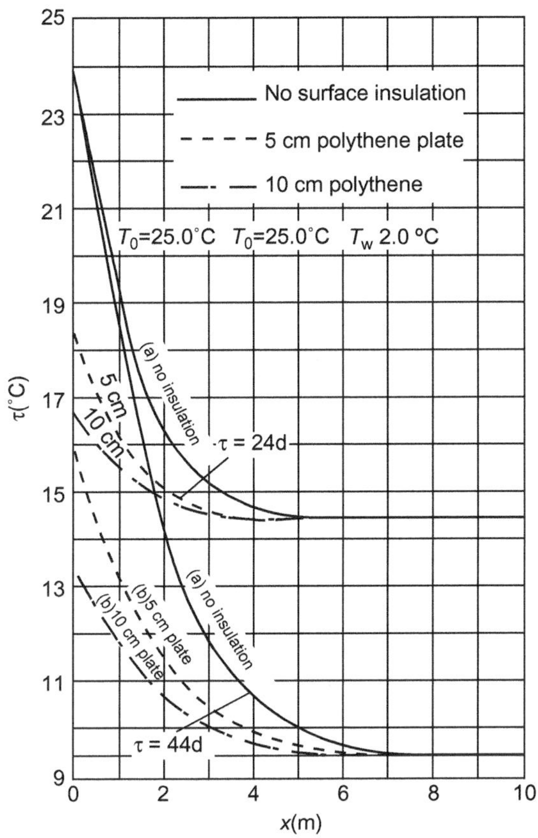

Figure 17.44 The effect of superficial thermal insulation on the temperature field of concrete cooled by pipe in hot season.

side insulated. $a = 0.10\ \mathrm{m^2/day}$, $\lambda = 200$ kJ/(m d °C), $\beta = 1000$ kJ/($\mathrm{m^2}$ d °C), $T_0 = 25°\mathrm{C}$, $T_a = 25°\mathrm{C}$, $T_w = 2°\mathrm{C}$, $\lambda L / c_w \rho_w q_w = 200 \times 200/(4.187 \times 1000 \times 21.6) = 0.442$, outer radius of the steel pipe is 1.25 cm. Three cases are computed for the left side: (i) in contact with air, $\beta = 1000$ kJ/($\mathrm{m^2}$ d °C); (ii) insulated by polythene plate with thickness 5 cm and $\lambda_1 = 3.00$ kJ/(m d °C); (iii) insulated by polythene plate with thickness 10 cm. The computed results are shown in Figure 17.44. It is evident that the effect of superficial thermal insulation is apparent.

18 Precooling and Surface Cooling of Mass Concrete

18.1 Introduction

The U.S. Bureau of Reclamation has developed a temperature control method for concrete dams based on cooling pipes and dividing the dam into concrete blocks which are grouted after complete cooling of concrete. This method had been successfully used to build the Hoover Dam in the 1930s. However, this method also has some disadvantages: formwork, cooling pipes, and joint grouting required more man power, and also increased the costs. After World War II, the U.S. Army Corps of Engineers developed a method of constructing dams with precooling aggregate and pouring without longitudinal joint. In the initial phase, because the precooling technology was not mature enough, the concrete placing temperature could only be reduced to 17–18°C. Subsequently, with the gradual improvement of the technology, the placing temperature can be reduced to 7°C, even in the hot summer months [8,23–25].

Now, pipe cooling and precooling are generally independently or simultaneously used in the construction of concrete dams.

Mixing with cooled water or ice is the easiest way to precool the concrete. In 1938, the U.S. Tennessee Valley Authority used mixing cold water to reduce the placing temperature of concrete from 25.6°C to 22.2°C in the construction of the Hiwassee dam in summer. And in 1940, the U.S. Bureau of Reclamation used the method of mixing water with ice to reduce the placing temperature of concrete to 21.1°C in the Friant dam in summer. However, because of the small proportion of water in the heat capacity of concrete, it is not enough to reduce the concrete placing temperature only by the method of mixing with cooled water and ice, and only the method of precooling of aggregate can significantly reduce the temperature of concrete.

A numerical example is given in Table 18.1, which lists the drop of placing temperature ΔT by precooling a variety of raw materials when the precooling temperature is 1°C. Table 18.1 shows that the effect of precooling the stones is the best, followed by precooling the sand and water, and then the effect of precooling the cement. A significant cooling effect could be achieved by using the ice instead of part of mixing water, since the ice will absorb 355 kJ/kg latent heat when melting, which can be seen from Eq. (4.4) and Table 4.1. For example, the temperature of the concrete can be about 6°C lower, by using 50 kg of ice instead of water, so the

Thermal Stresses and Temperature Control of Mass Concrete. DOI: http://dx.doi.org/10.1016/B978-0-12-407723-2.00018-X

Table 18.1 The Cooling Effect of Precooling the Various Raw Materials

Raw Material	The Weight (W) in One Cubic Meter of Concrete (kg)	Specific Heat [kJ/(kg°C)]	The Heat Dissipated by Precooling Raw Materials When Precooling Temperature is 1°C (kJ)	The Temperature Drop of Concrete (°C)
Stone	1600	0.84	1344	0.57
Sand	570	0.84	479	0.20
Water	100	4.19	419	0.18
Cement	150	0.84	126	0.05
Concrete	2420	1.005	2368	1.00

precooling of stones and sand, mixing water with ice are often used as the main precooling measures.

In order that the cooling loss should be reduced to the minimum in the transportation and pouring process, after the concrete leaves the mixer, the concrete should preferably be delivered in the hanging pot. If delivered by car, the sun visor should be installed in summer. When delivered by conveyor belt, the temperature of the concrete rises quickly because of its large exposed area, especially when the temperature is high, so the belt should better be placed in a closed and air-conditioned gallery. Or strictly, control the length of conveyor belt transport and reduce the delivery time of the conveyor belt. Section 18.4.3 shows that the surface temperature of the concrete block rises quickly, so the speed of pouring concrete should be accelerated to shorten the pouring time. For example, in Itaipu Dam, the old concrete was covered with new concrete within half an hour, so as to control the temperature rise of the surface in less than 1–2°C in summer. As shown in Figure 4.3, the stepped pouring method can also be used, when necessary; or cover the concrete with the heat preservation quilt immediately after vibrating, which is unfolded until the new concrete is paved. A method of spraying water can also reduce the surface temperature of the concrete block appropriately.

The calculation methods of mixing temperature and placing temperature of concrete have been given in Chapter 4, and will not be repeated here; in this chapter, we mainly describe the precooling of concrete.

18.2 Getting Aggregates from Underground Gallery

Sieved and washed aggregates should each be stockpiled; in summer days aggregate piling height should be increased and the gallery should be put under the ground and this can reduce aggregate temperature, because the earth's temperature is lower than air temperature in summer. When Danjiangkou dam was under construction, two net material yards had been observed for their temperature, and the results show that: when aggregates are stockpiled for 3 days, the change of internal temperature has been small; for 5 days, the temperature is stable. As is shown in Figure 18.1, the air temperature change of 1 day affects the surface temperature of the material within 1.0 m, while, in 4 m's deep, material's temperature is equal to the underground gallery's temperature which is 23°C. The surface temperature of the aggregate can reach 40–50°C under strong sunlight, and it is close to the air temperature without sunshine. The old gallery of the Tangjiagou material yard in Danjiangkou was built above ground, and the summer temperature of the gallery was 23–24°C; later on, it was rebuilt underground, then in summer, the temperature in gallery is 17–18.5°C, and the temperature of the taken-out material is 19.5–21°C.

To spray water in the aggregate yard is unnecessary. The temperature of the river is lower than the air temperature by 2–3°C in summer, but a few degrees higher than the ground's temperature, so water spraying is not useful.

In short, in China, aggregate should be stockpiled for 6–8 m in summer, and the stockpiling time is 5–7 days. The gallery should be built underground, and the door of the gallery should be open in turns, so that the low temperature aggregate in the low level of the stockpiling can be taken out.

18.3 Mixing with Cooled Water and Ice

The mixing water can be cooled by refrigeration industry equipment, so as to reduce the mixing temperature of concrete. Though the process is relatively simple, the cooling effect is limited, and this can be seen from Eq. (4.4). For the sake of improving the effect of precooling concrete, ice should be used instead of water as far as possible in the mixing.

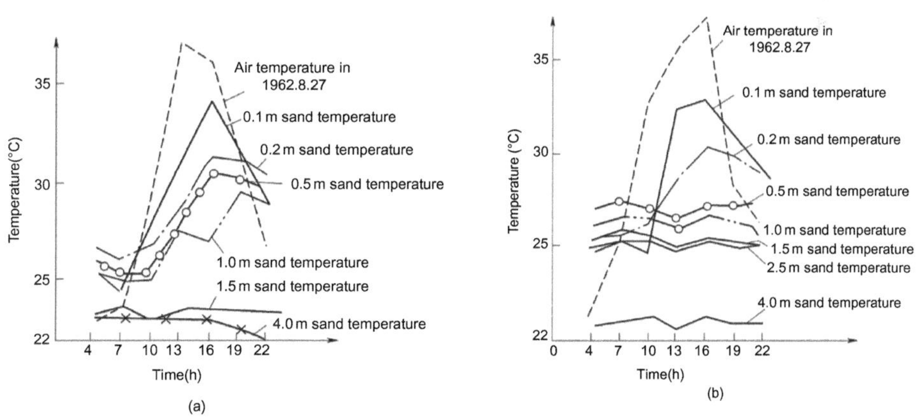

Figure 18.1 Temperature of aggregate pile: (a) temperature of 4–8 cm stone and (b) temperature of sand.

There are two ways of mixing the concrete with ice: one way is that big blocks of ice are first made, which are broken into 2–4 cm by an ice crusher and added to the concrete mixer. We can also store ice in winter in cold areas. According to the experiment, the cooling effect of ice with particle diameter less than 1 cm is lower than that of ice with particle diameter within 2–4 cm by around 20%. If we increase the concrete mixing time by half a minute, the ice with particle diameter within 2–4 cm would melt completely.

The other method is making tube ice or flake ice directly at the worksite, which is then added into the mixer, eliminating the process of crushing the ice. The device has the features of less site area, and could be directly placed in the concrete mixing house. At present, people incline toward the flake ice, about 3 cm long and 2 mm thick, which is easy to melt, so attention should be paid to the insulation of ice storage.

18.4 Precooling of Aggregate

The methods for precooling of aggregate include water cooling, air cooling, and mixed type.

18.4.1 Precooling of Aggregate by Water Cooling

Water cooling lowers the temperature of aggregate by direct contact of cold water and the surface of the aggregate; it can be used in the following two ways:

1. Water immersion precooling

   Lower the temperature of aggregate by immersing the aggregate into cold water, in the specific cooling tank.

   The Detroit dam built from 1950 to 1953, adopted the water immersion precooling. It is required that the placing temperature of concrete should not be higher than 10°C and not lower than 4.5°C. Considering the fact that the temperature of concrete may rise by 2.8°C in transportation, the temperature of concrete leaving the mixer must be no higher than 7.2°C. Therefore, all the raw materials (including stones, sand, cement, and mixing water) need to be precooled, and a 100% water cooling system is used. The temperature of coarse aggregate dropped to 3.3°C, in the cooling tower filled with cold water of 1.7°C. Sand and cement was cooled with a spiral heat exchanger, which is often used in the food industry, and the sand was cooled to 10°C, the cement to 15.6°C. The water was cooled to 1.7°C, and the capacity of the refrigeration plant was 880×10^4 kJ/h. There were five cooling towers, each of 91.8 m^3, four of them for four kinds of particle size of aggregate, respectively, and one for reserve.

   Immersion cooling is at present rarely used, due to the complexity of the equipment and operation process.

2. Spray cooling

   When stones of different particle size are moving slowly on the conveyor belt, which was installed in the cooling room, spray the cold water from above to cool the stones. Sand is usually cooled by air cooling, because of its poor dehydration. For example, the

Rihand dam required that the cooling temperature leaving the mixer was 13°C. The stones were cooled from 33.5°C to 5°C by spraying cold water in the cooling room, and the sand was cooled from 33.5°C to 14°C by air cooling on the conveyer belt, and the cooling water was cooled from 29°C to 2°C. There were four conveyer belts; each for four kinds of stones with different particle size, respectively; the belt was 83 m long, of which 61 m was the horizontal section, and the remaining 22 m was 5:1 slope section. The horizontal section was installed above with spray water pipes, and the slope section was for dehydration. The belt speed was 6–10 m/min. In order to accelerate the cooling, the cooling room was also air-conditioned.

18.4.2 Precooling of Aggregate by Air Cooling

The stones can be cooled in the aggregate storage bin of the concrete mixing house. For example, in the Harlan state dam, the requirement that the placing temperature of concrete not higher than 18°C was achieved by the method of the air cooling and mixing with cooled water. Precooling of aggregate was carried out in the aggregate storage bin of the concrete mixing house. The cooled air, delivered by the refrigeration plant, finds its way to the bottom of the aggregate storage bin through four air supply tubes. The cooled air flows upward through the voids of the aggregate and returns back to the refrigeration plant by the air return pipe on the top of the mixing house, completing a closed loop.

It is not easy to cool the sand in the aggregate storage bin of the concrete mixing house because of its small voids and poor ventilation. The sand used in Harlan state dam is not cooled. In order to cool the sand, the sand on the belt conveyer is blown with cold air in the cooling room, and the sand is continuously turned over with the sand-cast machine.

Several cooling methods must be used to get the best cooling effect. For example: water cooling + air cooling + ice cooling, namely spraying the stones with cold water, blowing the sand with cooled air on the conveyor belt; ventilating the mixer bin by cooled air, and using the ice instead of part of mixing water.

18.4.3 Precooling of Aggregate by Mixed Type of Water Spraying and Air Cooling

Since the water freezes at 0°C, the temperature of cold water must be higher than 0°C, which is often used as 2–4°C in practice; but the temperature of cooled air can be below 0°C, such as −13°C to −17°C. If the temperature of concrete leaving the mixer is required to be 6–7°C, we should better use the air cooling at the last stage.

In hot summer, the placing temperature of concrete could be reduced to 14°C by the water-cooled stones, the air-cooled sand, and the mixing with cold water. To reduce the temperature of the concrete further, it is better to use the mixed cooling mode, namely, (1) spraying stones with cooled water and blowing sand with cooled air on the conveyor belt; (2) cooling the stones further by blowing cooled air in the mixing house; and (3) mixing with ice. This mixed cooling method can make the temperature of concrete leaving the mixer drop to 5–6°C. The Deveau dam, the Itaipu dam, and Chinese Gezhouba dam all adopted the mixed type of water spraying and air cooling for precooling of aggregate.

For example, the process of the mixed type of water spraying and air cooling for precooling of aggregate is as follow: firstly, the aggregate moves slowly on the conveyor belt in the cooling gallery, and the length of the conveyor belt is 150–200 m, with the slope within 1–2%, which is then increased to 5–7% within 20–30 m from the head of the belt, so as to be beneficial to the drainage of the aggregate. The cold water of 2°C, sprayed from the pipe above the conveyor belt, would cool the aggregate. The steel tank under the conveyor belt catches the poured cold water, which flows into the sedimentation tank at the end of the conveyor belt and is then delivered to the cooling pond after clear water is replenished. After being cooled by water spraying on the conveyor belt, the aggregate is removed to the dewatering screen for dewatering, and then sent to the mixing house bin by the conveyor belt. For further cooling, add cooled air of −15°C to the mixing house bin. Finally, add the flake ice to the concrete mixer.

18.4.4 Precooling of Aggregate by Secondary Air Cooling

A new process of secondary air cooling aggregate was proposed for the first time in the Three Gorges stage I Project. After the success in the Three Gorges stage I Project, the process was applied to stages II and III of the Three Gorges on a large scale.

The secondary air cooling can be summarized as: the first air cooling + the second air cooling + ice cooling:

1. The first air cooling, using the cold air of 0 to −5°C, turns the temperature of the aggregate from the natural (such as 28–30°C) down to 8–10°C, in the aggregate regulating bin on the ground.
2. The second air cooling, using the cold air of −13°C to −17°C, turns the temperature of the aggregate down to 0 to −6°C, in the mixing aggregate storage bin.
3. The ice cooling, with the ice instead of part of mixing water, could reduce the mixing outlet temperature of concrete down to 7°C.

18.5 Cooling by Spraying Fog or Flowing Water over Top of the Concrete Block

18.5.1 Spraying Fog over Top of the Concrete Block

When pouring concrete in summer, it is better to form a foggy insulating layer over the top of the concrete block by spraying fog. This could reduce the direct sunlight so as to lower the surface temperature of the concrete block.

There are arc sprayer, straight tube sprayer, T-shaped sprayer, and axial flow fan sprayer. The former two can be used in large concrete block, and the latter two are suitable for small concrete blocks. The performances of these four kinds of sprayers are shown in Table 18.2.

In order to improve the effect of spraying, a fog-spray device was successfully developed and used in the Three Gorges stage II Project. Pressure swirl atomizer atomizes the water into fine droplets, which were then blown onto the surface of the concrete uniformly and formed a fog layer. On one side, the fog droplets would evaporate by absorbing heat; on the other side, the fog layer could reduce the direct sunlight, thereby reducing temperature of the pouring surface. Its working process is as follows.

The pressure water gets into the combined atomizing nozzle through the control valve, the pressure gauge and the filter, and then formed into micro droplets, of which the diameter ranges from 40 to 100 μm. The control valve can adjust the water pressure within 0.3–0.6 MPa, in order to control the atomizing. Oblique flow high-pressure blower generates high-pressure conveying flow, which sends the droplets into the distance. The swinging system, consisting of a low-speed motor, worm gear reducer, eccentric wheel, swing link, and a bracket, drives the fan to reciprocating swing of 3–4 times/minute within 0–90°. The elevation angle bracket installed under the fan, according to the environmental condition, can upgrade the fan within 0–20°.

The fog-spray device comprises atomizing system, conveying flow system, and swinging system and the base, etc. The base is equipped with a solid caster and can be pushed on the surface and change the spray direction. The base bench was

Table 18.2 Comparison of Performance of Four Kinds of Sprayers

Spraying Equipment	Arc Sprayer	Tube Sprayer	T-Shaped Sprayer	Axial Flow Fan Sprayer
Direct use of energy	Pressure water, wind	Pressure water, wind	Pressure water, wind	Water, electricity
Performances of spray	The fog is dense and the effect of fogging is good, the fogging range is large	The fogging range is large, while the effect of fogging is common	The fogging range is small, and the effect of fogging is common. More sprayers should be placed in one block	The fogging range is small, and the effect of fogging is common. More sprayers should be placed in one block
Covering range (m)	5–20	10–20	4–7	6–15
Cooling effect (°C)	3–11	3–6	3–5	3–6

welded by angle steel, with steel plate covered to the external, and can be removed for inspection and maintenance of the swing system in the base box.

The effect of the fog-spray device used in the Three Gorges Project is as follows. (1) The temperature of the concrete is lower than the environment, after the spraying. The temperature can be cooled by 6–10°C within 9–12 m, and 2–3°C with 18.0 m from the nozzle. (2) The cooling effect of the fog-spray device without elevation angle, the elevation angle is 0°, is better than that with some elevation angle. (3) The spraying effect has something to do with the wind direction and wind speed, and when spraying along wind, the cooling effect is best, and also the area of coverage is large. (4) In order to understand the spraying effect on concrete, two 100 mL measuring cylinders are respectively placed 6 and 9 m from the nozzle, and 1 h later the cylinder is checked to see if there were any water drops. This would not affect the quality of the concrete. (5) The experiments show that, although the spraying effect and environmental temperature have certain relations, the atomizing effect on cooling extent of the surface temperature is basically the same. (6) The spraying effect is related to the head pressure, the larger the water pressure, the better the atomizing cooling effect, but the atomizing effect changes a little with continued increase of the water pressure. The test results show that the atomizing effect is good, and the water consumption is small when the water pressure is about 0.6 MPa.

18.5.2 Cooling by Flowing Water over Top of the Concrete Block

In the construction of Toktogul dam in the former Soviet Union, the flowing water is adopted fully to cool the concrete. The flow of water starts immediately after the final set of concrete and the clearance of laitance of the surface, no more than 12 h after pouring the concrete. The water comes from small holes drilled in the pipe and forms a flow layer with thickness 2–8 mm on the surface of the concrete and the flow velocity is less than 0.8 m/s. It is not allowed to interrupt the flowing of water when the temperature is higher than 20°C. The temperature of the flowing water in the pipe is no more than 18°C and that over the top of the concrete is no more than 19°C. In the hot months of July and August, the flowing water over the top of the concrete is 13.5 L/s in each of the 1000 m^2 areas; in June and September, the flow can be reduced to 25%. The average flow for each 1000 m^2 of 8 h in every day is 8 L/s in April, May, and October. The measured water temperature would increase 1°C–3°C in practice. To insure the effect of flowing water, the concrete should be poured in thin layers. The layer thickness of the Toktogul dam is 0.5–1.0 m, and due to the flowing water, the temperature heat rises only 3–5°C due to hydration heat in the hottest month every year. The cooling water can be derived from the drilling holes or the drainage sump.

The calculation of the cooling effect of flowing water over the surface is simple: calculate the concrete lift according to the first kind of boundary condition and let the surface temperature equal to the water temperature.

19 Construction of Dam by MgO Concrete

Generally, the conventional concrete produces $(-40 \text{ to } -60) \times 10^4$ shrinkage deformation, but the MgO microexpansive concrete developed by Cao Zesheng and colleagues in China can produce about 80×10^{-6} expansive deformation which can compensate some tensile stress due to temperature drop. This kind of concrete was applied to Baishan arch dam in 1975 in which the cement containing MgO was adopted, namely MgO is included in the mineral components of the cement. Only a few cement plants can produce this cement because of the limitation of raw materials. After 1985, Cao Zesheng and his colleagues began to research concrete dams in which MgO is mixed in the concrete mixer, and it was applied in the five monoliths of Shuikou Hydropower Station in Fujian province and the whole dam of Qingxi Hydropower Station in Guangdong province. Later it came into use in several medium- and small-sized arch dams such as Shalaohe arch dam, Sanjianghe arch dam, and Yujianxi arch dam in Guizhou province [6,29,31,56,58,59,62,68].

19.1 MgO Concrete

By calcining the magnesite in the rotary kiln under a temperature of 1050 ± 50°C and grinding it, we can get lightly roasted MgO which is mixed with the cementing material in a certain proportion; the concrete added with MgO is then developed.

The hydration of periclase (MgO) crystals generates magnesium hydroxide, with accompanying volume expansion. The process is so slow under usual temperatures that the expansive deformation appears rather late. Temperature also has a significant influence on the rate of expansive deformation of MgO concrete. Li Chengmu adds 4% MgO into the Portland dam cement mixed with 30% fly ash so as to carry out the autogenous volume deformation test under different temperature conditions and the results are shown in Figure 19.1, from which we can see the autogenous volume deformation of MgO under different ages and the influence of ambient temperature.

Apart from the great influence of mixing amount of MgO and temperature on the autogenous volume deformation of MgO concrete, cement type, mixing amount of fly ash, calcining temperature of MgO, and grinding grain size also exert a certain impact.

Thermal Stresses and Temperature Control of Mass Concrete. DOI: http://dx.doi.org/10.1016/B978-0-12-407723-2.00019-1

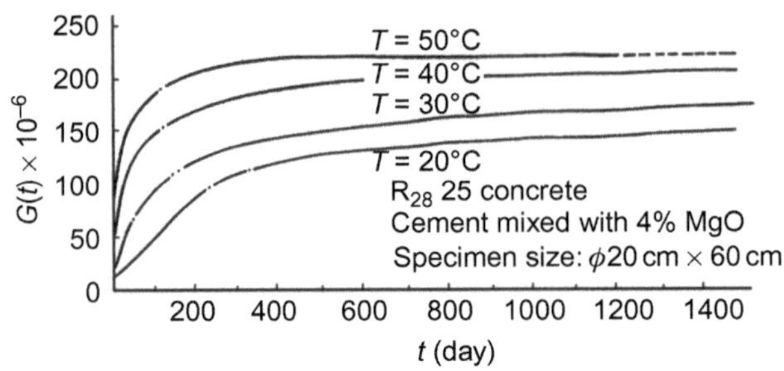

Figure 19.1 The autogenous volume deformation $G(t)$ of the dam concrete, the cement of which is mixed with 4% MgO.

According to the information of Li Chengmu, laboratory experiments show that the self-expansion of the autogenous volume deformation of the concrete added with MgO is mostly at around $(80-120) \times 10^{-6}$ after 1 year in various projects in China and only three projects are over 125×10^{-6} because of excess cement content.

The cement containing MgO produced by a qualified cement plant is of good quality, but because the calcining temperature of cement is generally 1400°C which is too high for MgO, the autogenous expansive deformation is small, and sometimes deformation may be shrinking. When MgO is produced with low calcining temperature 1000–1100°C, the autogenous expansive deformation is large, and the concrete added with MgO is agile for use. But most MgO concrete is produced by small cement plant with shaft kilns and relatively crude technology, so it is hard to guarantee the quality of the product. This type of concrete is mainly used for medium- and small-sized projects. There are two types of MgO concrete, the first type is produced by cement containing MgO in which MgO is contained in the raw materials of cement and calcined with temperature about 1400°C, the second type is produced by putting the lightly roasted MgO in the concrete mixer. The lightly roasted MgO is produced by calcining the magnesite with temperature 1050 ± 50°C. Due to the difference in calcining temperature, the second type of MgO concrete can produce more expansive autogenous deformation.

19.2 Six Peculiarities of MgO Concrete Dams

After systematic research, the author summarizes six peculiarities of MgO concrete; it is very important to master them for the proper application of MgO concrete in the construction of dams [57–62].

19.2.1 Difference Between Indoor and Outdoor Expansive Deformation

The difference between indoor and outdoor expansive deformation refers to the difference between the MgO expansion obtained in the laboratory and the actual MgO expansion of the dam outdoors. People failed to pay attention to this difference in the past–actually the difference is quite large.

It is obvious that the amount of MgO (kg/m^3) is an important factor which affects the autogenous volume deformation of the concrete. The author has researched the relation between the autogenous volume deformation ε (10^{-6}) of 365-day concrete and the content M (kg/m^3) of MgO and obtained the following approximate formula:

$$\varepsilon = 13.59(M - 0.500) \times 10^{-6} \tag{19.1}$$

The indoor specimens must expel large aggregate (e.g. for ψ20x60 cm specimen in which the strain gauge is buried, it is necessary to expel large aggregate over 40 mm) through wet screening and have higher content of MgO per unit volume, so the autogenous volume deformation in actual engineering will be less than the indoor experimental value. When burying an observation instrument on site, the concrete around the non-stressmeter also goes through wet screening, so the autogenous volume deformation measured on the spot is also too large, but this important factor was ignored in the past.

From Eq. (19.1), under the same temperature and material conditions, the content of MgO is M_1 and M_2, the autogenous volume deformation G_1 and G_2 generally meets the following relation:

$$\frac{G_1}{G_2} = \frac{M_1 - g}{M_2 - g} \tag{19.2}$$

Obviously, the value of g is mainly dependent on the cement type, it also has a certain relationship with the mixing amount of fly ash, admixture, and so on. Below is an estimating method, given concrete age τ (τ generally is 1 or 3 years), when $M = M_1$, $G = G_1$; $M = M_2$, $G = G_2$; from the interpolation relationship we get

$$g = \frac{1}{2}\left[(M_1 + M_2) - \frac{(G_1 + G_2)(M_1 - M_2)}{G_1 - G_2}\right] \tag{19.3}$$

For example, the age of concrete is 1 year, when $M_1 = 9.90\ kg/m^3$, $G_1 = 80 \times 10^{-6}$; when $M_2 = 0$ (without MgO), $G_2 = -30 \times 10^{-6}$ (shrinkage). From Eq. (19.3), we get

$$g = \frac{1}{2}\left[9.90 + 0 - \frac{(80 - 30) \times 10^{-6} \times (9.90 - 0)}{(80 + 30) \times 10^{-6}}\right] = 2.70\ kg/m^3$$

Now taking $g = 2.50\ kg/m^3$, let us calculate the influence of wet screening on the autogenous volume deformation of concrete from Eq. (19.2).

Example 1 In 4-graded concrete, the content of MgO accounts for 5.5%, the materials contained in one cubic meter of the concrete are as follows: water 90 kg, cement 119 kg, fly ash 51 kg, water reducing agent 1.19 kg, MgO 9.4 kg, sand 510 kg, gravel: 5–20 mm 359 kg, 20–40 mm 359 kg, 40–150 mm 1077 kg. One cubic meter of the concrete weighs 2575.6 kg and concrete specimens which get rid of 40–150 mm aggregates weigh 1498.6 kg. Suppose the bulk density of stones is 2700 kg/m^3, the volume of 40–150 mm aggregates is 0.399 m^3, the volume of the concrete after wet screening is 0.601 m^3, the MgO content per cubic meter of the concrete after wet screening is $9.4 \times 1.00/0.601 = 15.64$ kg.

Let us take $g = 2.50$ kg/m^3 in Eq. (19.2), the autogenous volume deformation of the specimen to that of the dam body concrete is

$$\frac{G_1}{G_2} = \frac{15.64 - 2.50}{9.40 - 2.50} = 1.904$$

The autogenous volume deformation of the specimen is 1.904 times of natural gradation concrete.

It is clear that, after wet screening, the autogenous volume deformation of the concrete around non-stressmeter and specimen is rather bigger than that of natural gradation concrete in the dam. This difference can be called the difference between indoor and outdoor expansive deformation.

19.2.2 Time Difference

Time difference means that the dam cooling rate is different from the expansive deformation rate of MgO concrete, so the expansive deformation of MgO cannot compensate the tensile stress to a sufficient degree.

In order to compensate dam temperature stress sufficiently, the development of the expansive deformation of MgO concrete and the falling of the temperature must keep pace with each other, but the natural cooling rate of the dam is in contrast to the square of the thickness of the dam, while the inflation of MgO concrete has nothing to do with the thickness of the dam, thus, generally the inflation of MgO and dam cooling do not happen at the same time. This is the time difference which was first proposed by the author. People did not pay attention to this problem in the past, while, in fact, it is an important problem.

Suppose the initial temperature is evenly distributed and the boundary temperature keeps constant, the average temperature change process of the concrete dam body with different thicknesses is shown in Figure 19.2; $1-G(t)/G_0$ is also shown in the figure, G_0 is the final autogenous volume deformation of MgO concrete.

As seen in the figure, when the thickness of the dam L is 10–20 m, the cooling of the dam body is generally synchronous with the expansive deformation caused by MgO; for a thinner dam body, such as when L is 5 m, the autogenous volume expansion has not fully developed when the cooling of the dam body ends; for a thicker dam body, when L is 50 m, the dam body cools so slowly that the autogenous volume expansion is pretty large while the cooling of the dam body is not enough, so the dam has to bear considerable compressive deformation. This is the time difference of dam cooling and autogenous volume deformation. For arch dams, time difference is an important problem.

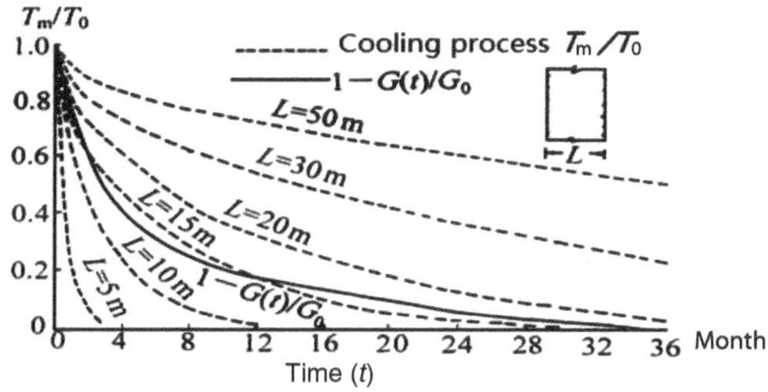

Figure 19.2 The cooling process of dam body of different thickness.

19.2.3 Regional Difference

Without a cooling pipe, the temperature difference in the dam is

$$\Delta T = T_p + T_r - T_f \tag{19.4}$$

where

T_p—the placing temperature
T_r—the temperature rise of hydration heat
T_f—the steady temperature of dam body.

According to common sense, the placing temperature in the north should be lower than that in the south, but in the cold northern areas, a large amount of concrete is poured in the summer instead of winter, thus the placing temperature of the north dam actually is not low.

The average water and air temperature in the north is lower than that in the south, thus, the steady temperature of the dam body in the north is lower than that in the south. The steady temperatures of the dam body are as follows: northeast China 5−10°C, north China, northwest China 8−14°C, central China 14−18°C, south China/southwest China 17−20°C.

Due to the above two reasons, the maximum temperature difference of concrete dams built in north China is always bigger than that in south China if built without strict temperature control measures.

The autogenous volume deformation of MgO concrete depends on temperature. Under the same MgO content, due to the high concrete temperature in south China, the expansive deformation is large; while in north China, low concrete temperature causes low expansive deformation.

In conclusion, it is easy to achieve the temperature control simplified by the autogenous volume deformation of MgO concrete in south China, while it is hard to achieve in north China. When MgO concrete is applied in the construction of the dam, we should pay attention to this "regional difference," namely (i) temperature difference is large in the north while small in the south; (ii) under the same MgO concrete, the autogenous volume expansion is small in the north while large in the south.

19.2.4 Dam Type Difference

Arch dam and gravity dam have the following differences in the dam construction technology with MgO concrete.

Different shape: The dam axis of a gravity dam is a straight line so that the expansive deformation only brings compressive stress and we need not worry about overexpansion. While the dam axis of an arch dam is a curve, either temperature rise or drop will cause tensile stress, the arch dam has to control not only temperature drop but also too large expansive deformation.

Different scope of constraints: The transverse joints of a gravity dam usually are not grouted and the concrete volume deformation only brings stress at restraint areas of foundation or old concrete. Once out of restraint areas, average volume deformation will not bring stress. While an arch dam is different, the whole dam is restrained by the bedrock of the banks from the top to the bottom of the dam after transverse joint grouting so that average volume deformation will cause stress.

Different thickness: A gravity dam is relatively thick while an arch dam is rather thin. The thickness of the upper arch dam body is always only 3–5 m so that the temperature may drop from the maximum to a very low value in a very short time before the expansive deformation of MgO concrete could develop.

The arch dam has to control the temperature during not only the construction period but also the operating period, which includes temperature rise and drop.

19.2.5 Two Kinds of Temperature Difference

There are two kinds of temperature difference, namely the temperature difference of concrete above the foundation and the temperature difference between the surface and the interior of concrete, which are quite different. As the expansive deformation of MgO concrete is uniform, which only produces compressive stress under restraint of rock foundation and cannot compensate the thermal stress caused by cold wave and the interior–exterior temperature difference which must be solved by surface heat insulation; the autogenous volume deformation of MgO concrete can only be used to reduce the thermal stress caused by foundation restraint.

19.2.6 Dilatation Source Difference

The dilatation source difference means the difference between MgO expansion and temperature rise expansion.

We must stress that, though the autogenous volume expansion of MgO concrete and the deformation caused by concrete temperature rise are similar in macroscopic view, they have essential difference in microscopic structure. When the concrete produces deformation owing to temperature changes, the deformations of cement paste and aggregate are generally synchronous, but the autogenous expansion of MgO concrete is different; MgO produces expansion while cement paste and aggregate themselves will not expand. Therefore once the mixing amount of MgO surpasses a specific value, cement paste itself and the interface between it and aggregate may be damaged, which affects the basic properties of concrete, such as strength, extensibility, impermeability, and durability, so, it is definitely necessary to enforce an appropriate limitation on the MgO maximum content of the concrete. According to the national cement standard, MgO contained in the concrete cannot exceed 5.5%. For the concrete added with MgO, it is required for hydraulic concrete that the MgO content cannot exceed 5% of the amount of cementing material.

19.3 The Calculation Model of the Expansive Deformation of MgO Concrete

19.3.1 The Calculation Model of the Expansive Deformation for Test Indoors

The expansive deformation of MgO concrete from the laboratory test can be shown as follows:

$$F(\tau, t) = F_0[1 - \exp(-aT^b\tau^c)] \tag{19.5}$$

where F_0 is the final expansive deformation of the specimen, we generally get it from test curves; a, b, c are three parameters. So, Eq. (19.5) is a three parameters expression, take twice the natural logarithm on both sides of the formula, we can get

$$\ln a + b \ln T + c \ln \tau = \ln[-\ln(1 - F/F_0)] \tag{19.6}$$

Let $\ln \tau$ be the abscissa, $\ln[-\ln(1 - F/F_0)]$ be the ordinate, take $\tau = \tau_1$ and $\tau = \tau_2$, draw two parallel lines through the test points, then work out b by the straight slope, a and c can be found by the intercept of the two straight lines on the ordinate; finally we can take some proper correction for a, b, c, in order to make the calculated value agree better with the experimental value.

19.3.2 The Calculation of the Expansive Deformation of MgO Concrete of Dam Body Outdoors

After taking the difference between indoor and outdoor into consideration, we can calculate the autogenous volume deformation of dam body MgO concrete as follows:

$$G(\tau, T) = k(M)F(\tau, T) \tag{19.7}$$

where

$G(\tau, T)$—the autogenous volume deformation of dam body concrete
$k(M)$—the correction coefficient of MgO content
$F(\tau, T)$—the autogenous volume deformation of concrete specimen measured indoors.

From Eq. (19.2), $k(M)$ can be calculated according to the following formula:

$$k(M) = \frac{M_d - g}{M_s - g} \tag{19.8}$$

where M_d is the MgO content of natural gradation concrete of the dam body, kg/m^3; M_s is the MgO content of the concrete specimen indoors (after wet screening), kg/m^3; g is an experimental constant, calculated from Eq. (19.3), we generally take $g = 2.5\ kg/m^3$.

19.3.3 The Incremental Calculation of the Autogenous Volume Deformation

The increment of the autogenous volume deformation needed in the simulation analysis can be given as follows:

$$\Delta G(\tau, T) = k(M)\frac{dF}{d\tau}\Delta\tau \tag{19.9}$$

$$\frac{dF}{d\tau} = F_0 acT^b \tau^{c-1}\left[1 - F(\tau, T)/F_0\right] \tag{19.10}$$

The coefficients a, b, c in Eq. (19.5) are found under different temperature and age. They can still be used here approximately.

19.4 The Application of MgO Concrete in Gravity Dams

19.4.1 Conventional Concrete Gravity Dams

In conventional concrete gravity dams MgO concrete may be used in the foundation restraint zone, which can simplify the temperature control, but the cracks caused by the interior−exterior temperature difference and cold wave must be solved by the surface thermal insulation.

The maximum thermal stress of a concrete block on rock foundation can be calculated as follows:

$$\sigma = \frac{KRE\alpha(T_p - T_f)}{1-\mu} + \frac{C_r KAE\alpha T_r}{1-\mu} - \frac{KRC_g G(\tau)}{1-\mu} \le \frac{E\varepsilon_t}{k} \tag{19.11}$$

where

K—stress relaxation coefficient
E—modulus of elasticity
μ—Poisson's ratio
C_r—the coefficient to consider early warming influence, the value is about 0.85, but the problem is so complicated, we generally take $C_r = 1.0$ in practical projects; $C_g = 1.0$
$G(\tau)$—the autogenous volume expansion
ε_t—the ultimate extensibility of concrete
k—safety factor
R—the foundation restraint coefficient
A—the foundation influence coefficient of stress due to the temperature rise of hydration heat.

Let us multiply both sides of Eq. (19.11) by $(1-\mu)/(KRE\alpha)$, we will get

$$T_p - T_f + \frac{AT_r}{R} \leq \frac{\varepsilon_t(1-\mu)}{kK\alpha} + \frac{G(\tau)}{\alpha} \tag{19.12}$$

The allowable temperature difference varies with the length of concrete block in the design criterion of concrete gravity dams, which is the influence of A/R on the left side of the above equation. From Eq. (19.12), we know, for micro-expansive concrete, the foundation allowable temperature difference could relax $G(\tau)/\alpha$°C.

If we make use of the allowable temperature difference ΔT regulated in the criterion of gravity dams, the allowable placing temperature T_p of concrete can be calculated by the following equation:

$$T_p = T_f - T_r + \Delta T + G(\tau)/\alpha \tag{19.13}$$

Let the steady temperature T_f of the lower part of gravity dams in northeast China, north China, central China, south China be 6°C, 10°C, 15°C, 19°C, respectively and use the allowable temperature difference regulated in the design specifications of concrete gravity dams, then we get the allowable placing temperature from Eq. (19.13) as shown in Table 19.1. The lowest and highest monthly average temperature in a year are generally as follows: northeast China: −17–21°C; north China: −8–25°C; central China: 6–28°C; south China: 10–28°C.

From Table 19.1, we can get the following points.

If the autogenous volume deformation of the concrete is 100×10^{-6}, the gravity dams without longitudinal joint can be constructed all year round in south China and southwest China, while they can only be constructed in the summer with pre-cooling and water pipe cooling in other areas.

If the autogenous volume deformation of the concrete is 100×10^{-6}, dam blocks under 25 m can be constructed all over the country except for northeast China throughout the year. We can pour 15 m dam blocks in northeast China throughout the year.

Table 19.1 The Allowable Placing Temperature T_p of the Restrained Zones of MgO Concrete Gravity Dam

Placing Type	With Longitudinal Joints	Length of Dam Block (m)	Allowable Temperature Difference ΔT (°C)	Hydration Temperature Rise T_r (°C)	Allowable Placing Temperature T_p(°C)					Allowable Placing Temperature T_p (°C)				
					G/α	Northeast China	North China	Central China	South China	G/α	Northeast China	North China	Central China	South China
Conventional	Yes	15	26	15	10	27	31	36	40	20	37	41	46	50
		25	21	15	10	22	26	31	35	20	32	36	41	45
	No	100	15	15	10	16	20	25	29	20	26	30	35	40
RCC	No	60	13	13	5	11	15	20	24	10	16	20	25	29
		100	11	13	5	9	13	18	22	10	14	18	23	27

Note: The steady temperature T_f used in the calculation: northeast China 6°C, north China 10°C, central China 15°C, south China 19°C.

In other words, as long as the autogenous volume deformation of the concrete is 100×10^{-6}, we can pour concrete dams monolithically without precooling aggregate and water pipe cooling in south China and southwest China, while precooling aggregate and water pipe cooling are still needed in other areas. The above analysis is for hard rock; we can appropriately relax the requirement if the bedrock deformation modulus is lower.

For concrete blocks with length $L \leq 25$ m, it is necessary only to control the foundation temperature difference and generally the upper and lower layer temperature difference will play no role. For the gravity dams without longitudinal joint, due to the large length of dam blocks, except the foundation temperature difference, it is also necessary to control the upper and lower temperature difference properly because it also can cause considerable tensile stress.

There is less cement content and high mixing amount of fly ash in RCC dams, but high mixing amounts of fly ash have some inhibitory effect on the autogenous volume deformation of the concrete, so we generally do not use MgO concrete in the construction of RCC dams. Only a few gravity dams use MgO in the conventional concrete cushion plate above the rock foundation and the upstream anti-seepage body of the RCC dams.

Qingxi hydropower station is located at Dapu county of Guangdong province and its installed capacity is 14,400 kW. The barrage is a normal concrete gravity dam whose maximum height is 51.5 m with a longitudinal joint. It was built from 1990 to 1993, the local annual average temperature is 21.0°C and the monthly mean temperature is 28.3°C in July and 11.8°C in January. The ordinary Portland cement which contains 1.5% MgO and 30% fly ash is used in the dam body and the content of MgO accounting for 5% is added in the foundation restraint zone (under 0.4L). We use 2 cm polystyrene foam board on the upstream face of the dam and plastic bubble cushion on the lateral side and horizontal surface for surface insulation. The 180 day autogenous volume deformation measured by the non-stressmeters in the dam is $27.9–98.6 \times 10^{-6}$ and no deep cracks were found after completion.

19.5 The Application of MgO Concrete in Arch Dams

19.5.1 Arch Dams with Contraction Joints

In arch dams with contraction joints, it is necessary to cool the dam by cooling pipes before grouting of joints. The regulation of the temperature of the dam body mainly depends on the cooling pipes. MgO concrete can be used in the lower part of each monolith to compensate tensile stress caused by temperature difference above the foundation.

Two questions need to be considered, the first one is the age of concrete for the second stage of pipe cooling before grouting. Since we make use of the expansive deformation of MgO to reduce foundation thermal stress, the process is rather slow. From the curve $1 - G(\tau)/G_0$ shown in Figure 19.2, the second pipe cooling should be carried 1 year later due to the slow development of the expansive deformation of MgO; of course, it may be quicker in the south area with higher temperatures.

The second one is the problem of time difference, the thinner thickness of the arch dam, the faster the natural cooling of the dam. Figure 19.2 shows that a dam 5 m thick will be cooled in only 2 months, but it is too early for the development of the expansive deformation of MgO concrete.

In summary, for higher and thicker arch dams with contraction joints, it is better to use MgO concrete; for thin arch dams, detailed simulation analysis should be carried out before the application of MgO concrete.

19.5.2 Arch Dams without Contraction Joints, Time Difference

In the operation period, for any cross section of the dam, the temperature at point x may be expressed approximately by:

$$T(x,t) = T_1(x) + T_2(x)\cos\frac{2\pi}{P}(t - t_0) \tag{19.14}$$

where

x—the distance of point x to the surface
t—time
t_0—the time when the maximum temperature appears
P—1 year
$T_1(x)$—the mean temperature of point x
$T_2(x)$—the amplitude of temperature variation of point x.

The Control of Temperature Load

For conventional arch dams with contraction joints, temperature load T_m (T_d is similar and is omitted here) is calculated by the following formula:

$$T_m = T_{m0} - T_{m1} - T_{m2} \tag{19.15}$$

where

T_{m0}—the average temperature of the dam section when the joints are grouted, it is generally taken as the average value of quasi-steady temperature field along the thickness
T_{m1}—the average value of annual mean temperature $T_1(x)$ along the thickness during the operating period
T_{m2}—the average value of the amplitude of annual variation temperature $T_2(x)$ along the thickness during the operating period.

For MgO concrete arch dams without contraction joints, temperature load T_m is calculated by the following formula:

$$T_m = T_{mc} - G/\alpha - T_{m1} - T_{m2} \tag{19.16}$$

where G is the autogenous volume expansion of the concrete, T_{mc} is the mean value of the maximum temperature of the concrete, $T_{mc} = T_p + C_rT_r$, where C_r is the coefficient with the consideration of early warming influence; we can take

$C_r = 1.0$ in the preliminary calculation, that is $T_{mc} = T_p + T_r$, T_p is the placing temperature shown in Eq. (4.6).

For practical engineering design, the calculation of temperature load should be based on Eq. (19.15), the arch dam stress calculated, and the value of G/α decided by trial calculation. In order to know the situation throughout the country on, let us do some approximate analysis below. Practical experience shows that allowable tensile stress is the key factor to control the section of dam body in either a high arch dam or low arch dam, so, in order to maintain the section of dam body roughly the same, we have to keep roughly the same temperature load; from Eqs. (19.15) and (19.16), we know that, in order to keep the same temperature load, the autogenous volume dilatation needed is as follows:

$$G/\alpha = T_{mc} - T_{m0} = T_p + C_r T_r - T_{m0} \tag{19.17}$$

We can calculate the autogenous volume dilatation needed for the cancelation of transverse joints and pipe cooling from Eq. (19.17). T_{mc} relates to the air temperature, sunlight, cement content, concrete lift thickness, and interval time of the placing concrete, etc. T_{m0} generally takes the average value of the quasi-steady temperature field along the thickness; it is not hard to calculate according to the local specific conditions.

Example 1 Suppose we take some simple temperature control measures in the construction, such as mixing with ice, so $T_p = T_a$, in order to prevent freezing in cold areas, take $T_p = 5.0°C$ in the winter, assume that $C_r = 1.0$, $T_r = 19.0°C$, we can calculate G/α required in the construction of arch dams with MgO concrete in different regions of China from Eq. (19.17), as given in Table 19.2.

From Table 19.2, we can get the following points: (i) due to the lower temperature, the G/α required in the winter is less than that in the summer; (ii) the average temperature T_{m0} of the dam is lower at the bottom of the dam body during the operating period, so the G/α required in the bottom of the dam body is higher than that in the top; (iii) T_{m0} in the south is higher than that in the north, so the G/α required in the south is less than that in the north; (iv) if the G/α of concrete is 15°C, in areas such as south China, central China, and north China, we can pour the concrete from December to March, at the same time the G/α required in northeast China is 17.5°C; (v) if we carry out the construction all year round, the G/α required in south China and other areas are 28°C and 35.0°C; it is equal to two or three times the current autogenous volume deformation of MgO concrete. Also note that, under the same mixing amount of MgO, the G/α in the north is less than that in the south due to the lower temperature in the north.

The Importance of Time Difference

For arch dams without transverse joints, the problem of time difference in Section 19.2.2 is rather important. Time difference is caused by the different rate

Table 19.2 The Autogenous Volume Expansion Required in the Construction of Arch Dams with MgO Concrete

Area	Dam Body Parts	T_{m0} (°C)	*G*/α Required in Different Month (°C)											
			1	2	3	4	5	6	7	8	9	10	11	12
South China	Bottom	19.0	10.6	12.4	17.1	21.2	24.9	26.4	28.0	27.7	25.9	22.0	17.9	12.2
	Top	23.5	6.1	7.9	12.6	16.7	20.4	21.9	23.5	23.2	21.4	17.5	13.4	7.7
Central China	Bottom	14.0	11.0	12.4	17.1	25.9	26.7	31.0	33.2	33.0	28.4	23.1	17.3	12.0
	Top	19.0	6.0	7.4	12.1	16.9	21.7	26.0	28.7	28.0	23.4	18.1	12.3	7.0
North China	Bottom	10.0	14.0	14.0	15.1	21.2	24.7	29.3	31.4	30.1	25.5	18.6	14.0	14.0
	Top	12.0	12.0	12.0	13.1	19.2	22.7	27.3	29.4	28.1	23.2	16.6	12.0	12.0
Northeast China	Bottom	6.5	17.5	17.5	17.5	18.8	25.6	30.3	33.7	32.7	25.6	18.3	17.5	17.5
	Top	9.5	14.5	14.5	14.5	15.8	22.6	27.0	30.7	29.7	22.6	15.3	14.5	14.5

of natural cooling of dam body and the expansive deformation of MgO concrete. For thinner arch dams, the temperature cools down quickly after pouring, and the expansive deformation of MgO concrete is too small at the early age, thus generating relatively great tensile stress. For thicker arch dams, the cooling rate of the dam body is slow, but the inner temperature of the dam body is still rather high when the expansive deformation of MgO concrete is already large, thus leading to excess compressive deformation of the dam body. Excess compressive deformation may cause adverse tensile stress because the dam axis of an arch dam is a curve.

Above all, we should be careful in the construction of MgO concrete arch dams without transverse joints.

19.5.3 Example of Application of MgO Concrete, Sanjianghe MgO Concrete Arch Dam

Sanjianghe arch dam, located at north suburb of Guiyang city, is a double-curvature arch dam. The maximum height of the dam is 71.5 m, arc length of dam crest is 115.5 m, the thickness of dam at crest and base are 4.0 m and 10.44 m, respectively. It was designed by Water Resources and Hydropower Design Institute of Guizhou province and its temperature control analysis is undertaken by China Institute of Water Resources and Hydropower Research.

The local monthly mean temperature is 4.4°C in January and 23.5°C in July. In order to exploit the advantage of low temperature in winter, all the concrete was poured from the middle of November 2002 to April 2003. The allowable tensile stress is given by $\sigma = E\varepsilon_p/K$, where $K = 1.65$. Suppose no transverse joints, all the dam pours monolithically, a three-dimensional finite element simulation method is used to calculate, the required mixing amount of MgO is 3–4% (dam height: 0–30 m) and 8–10% (dam height above 30 m). The 8–10% mixing amount has greatly exceeded the code requirement, so we decide to give up the project without transverse joints and set up two induced joints in the new design project, as Figure 19.3 shows. Calculations show that stress meets the requirements, the

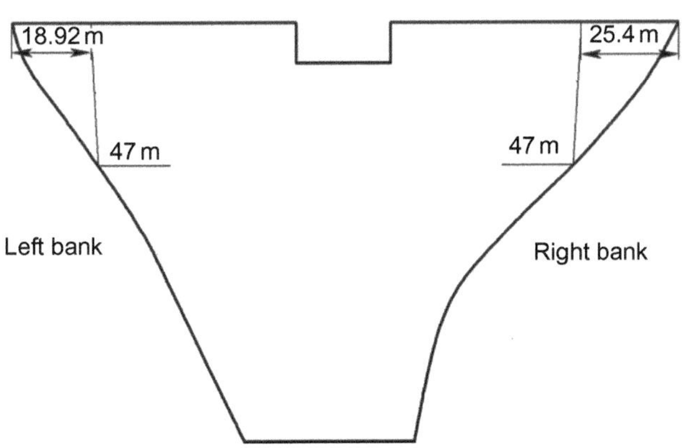

Figure 19.3 Induced joints of Sanjianghe MgO arch dam.

temperature control measures adopted in the construction process are as follows: (i) the mixing amount of MgO is 4.5% in all the dam; (ii) set up two induced joints; (iii) all the dam is poured in the low temperature season (planned placing time is from November 2002 to April 2003, actual placing time is from December 2002 to May 2003); (iv) spray and sprinkling water maintenance; (v) plastic foam board for surface insulation.

This dam was constructed according to Figure 19.3 and was completed in April 2003, i.e. 10 years ago. We only found a hairline surface crack about 1.5 m long in the dam body in 2010 which could be avoided if the surface protection is better; two induced joints have opened. The biggest opening of the induced joint given by calculation is 5.8 mm while the measured value is 5.6 mm. This dam is one of the arch dams ever built which has the fewest cracks. The temperature control of Sanjianghe arch dam is successful.

20 Construction of Mass Concrete in Winter

In civil engineering, the main problem of construction of reinforced concrete structures in winter is to prevent the early-age concrete from freezing; but things are different when constructing mass concrete in winter. There are problems in controlling the temperature difference for preventing cracks besides preventing the early-age concrete from freezing, and there is usually contradiction between the freeze protection and crack prevention. This contradiction must be properly solved, considering the requirement of both freeze protection and crack prevention, which is the main feature of the construction of mass concrete in winter.

Construction of concrete in winter, especially in severe cold regions, generally requires a higher placing temperature to prevent the early-age concrete from freezing, no matter what the construction method is. In the other side, the cold weather makes the stable temperature of dam lower, which would increase the temperature difference above foundation and the temperature difference between surface and interior. Even in winter, the temperature difference would exceed the allowable value if the placing temperature is not controlled, which would fail to meet the requirement of crack prevention. So the contradiction between freeze protection and crack prevention of winter construction is focused on the selection of placing temperature. Practical experience shows that it would cause cracks if you simply choose a higher placing temperature to prevent the early-age concrete from freezing.

In this chapter, the correct design principle of mass concrete construction in winter will be given first, then the correct way to solve the contradiction between freeze protection and crack prevention will be pointed out [40, 50, 64, 71, 72].

20.1 Problems and Design Principles of Construction of Mass Concrete in Winter

20.1.1 Problems of Construction of Mass Concrete in Winter

The first problem to be considered is whether or not to continue construction in winter when constructing concrete dams in cold regions. It is better to stop in winter when constructing the RCCD in a wide valley because it is hard to insulate when placing concrete due to the large placing area. It is feasible to place mass

Thermal Stresses and Temperature Control of Mass Concrete. DOI: http://dx.doi.org/10.1016/B978-0-12-407723-2.00020-8

concrete in winter as constructing concrete arch dam or gravity dam in narrow valley when effective measures, such as automatically rising movable tent, may be taken because the placing area is smaller. If it is decided to place mass concrete in winter, it is necessary to research carefully the technical measures for freeze protection and crack prevention.

20.1.2 Design Principles of Construction of Mass Concrete in Winter

It would come into the period of winter construction when the mean daily air temperature is below 5°C, or the minimum temperature is below −3°C.

Both freeze protection and crack prevention should be taken into account during massive concrete construction in winter, so the following three principles suggested by the author in 1976 should be followed [7]:

1. The optimum placing temperature of concrete is 5−12°C. Therefore, the temperature of fresh concrete at the batch plant could be decided according to the local climate conditions and construction method, as well as considering the heat loss during the transportation and pouring. No frozen block should be contained in sand and other raw materials.
2. The fresh concrete could not suffer freeze injury until it reaches the 50% design strength, or it would lose strength because of the damage of the internal structure.
3. The temperature difference between surface and interior and the maximum temperature of concrete, which influence the temperature difference above foundation and the temperature difference between upper layer and lower layer, could not exceed the specified value.

Among the three principles mentioned above, the first two are about freeze protection and the third one is about crack prevention. The proper thermal insulation method should be chosen according to these principles.

When building a concrete dam in a cold region, there should be a specific study about whether or not to continue construction in winter. Generally speaking, for the conventional concrete dam, winter construction is feasible, but for the RCCD, the winter layoff might be appropriate.

20.2 Technical Measures of Construction of Mass Concrete in Winter

To satisfy the principles of winter construction mentioned above, some technical measures must be taken.

1. Selection of the temperature of fresh concrete at the batch plant and placing temperature

   The placing temperature is the concrete temperature after vibrating and before covered by the second layer of concrete mixture. To prevent the early-age concrete from freezing, the placing temperature of mass concrete is required to be no less than 5°C. The placing temperature of construction of reinforced concrete with small sections is sometimes as high as 30−40°C. For mass concrete, even though the surface temperature is low after placing, the internal temperature will rise sharply because of the heat of hydration of the large body; there is a requirement of freeze protection besides crack prevention. To reduce the temperature difference above foundations and the temperature difference between surface and interior, the placing temperature should be lower, generally no more than 10°C. Therefore, the placing temperature of construction of mass concrete is generally 5–12°C. If the temperature is too low, the surface concrete might suffer freeze damage before it reaches the 50% design strength, and insulation measures should be taken instead of simply raising the placing temperature for freeze protection.

According to local climate conditions and insulation measures, the temperature of fresh concrete at the batch plant could be decided considering the placing temperature and the heat loss during transportation and placing.

2. Preheating of foundation and cold wall

Before placing concrete for the foundation, embedded iron and cold wall such as old concrete and precast concrete formwork, which contact with the new concrete, steam should be used to clear the ice, snow, and frost, and raise the surface temperature. If the internal temperature of bedrock and cold wall is low, preheating would be needed, otherwise the temperature of new concrete near the contact surface would soon drop under 0°C. The temperature, depth, and duration of preheating are decided by temperature calculation. The principle of the calculation is to avoid freezing the new concrete near the contact surface before it reaches the 50% design strength; generally, it should make sure that the temperature of bedrock with depth within 10 cm is above 5°C.

3. Heating the raw material

When air temperature is not lower than −1°C, generally heating the mixing water could satisfy the requirement of temperature of concrete at the exit of the mixer. The water temperature should not exceed 60°C to avoid the false set of cement. When air temperature is lower than −1°C, the water, fine and coarse aggregates should be heated to melt the ice and snow. Overheating and excessive drying should be avoided when heating the aggregate, and the maximum temperature should not exceed 75°C.

Boiler, electric heat, or steam could be used to heat water; using the serpentine pipe to heat the sand, and steam is the most convenient method to heat the stone.

4. Insulation during transportation

The heat loss during transportation is related to the means of transport. The heat loss is generally small by using large transport tanks. If using a dump truck, the exhaust gas can be used to heat the chassis and the truck body should be covered by insulation layer; if using a belt conveyer, it is better to build a completely closed shelter or the heat loss will be large. In addition, the number of transshipments should be minimized.

5. Reduce heat loss during placing

Concrete is placed in layers and each layer has a thickness of 20–50 cm. The heat loss is large during placing due to the thin layer and large radiating surface. Measures to reduce the heat loss are as follows: (i) accelerates the placing speed and shortens the placing time; (ii) EPS insulating layer should be used to insulate the heat; (iii) building automatic rising tent and creating artificial climate in it so that it is absolutely insulated from outside cold air when placing concrete. For many practical dams, the placing of concrete will stop when the air temperature is below −5°C. When placing concrete in lower temperatures, it is better to construct in warm shed. But it is still possible to place concrete around −10°C if there is no wind.

6. Insulation and curing

Strict insulation and curing measures should be taken after placing concrete so that the concrete strength will fully develop. The temperature should not drop below 0°C until the concrete reaches the 50% design strength. At the same time, large temperature differences between surface and interior should be prevented. At present, the foam board seems to be the best insulation method. For details, see Section 11.7.

7. Timely form stripping

The time for form stripping depends on the concrete strength as well as the requirement of crack prevention.

In the United States and Canada, it is required that the surface temperature drop cannot exceed 11°C (20°F) in a day after form stripping. In the former Soviet Union, the temperature difference between surface and interior cannot exceed 20°C. In Krasnoyarsk dam: for C15 concrete with extensibility 0.7×10^{-4}, the temperature difference between surface and interior should not exceed 17°C; for C20 concrete with extensibility $0.9-1.1 \times 10^{-4}$, the temperature difference between surface and interior should not exceed 23°C. In addition, the temperature difference between the concrete surface and the air should not exceed 15°C when stripping the form.

For some dams in Japan, the concrete surface would still be insulated by plastic insulation layer after form stripping. When the former Soviet Union constructed dams in Siberia, the form would not be removed during the whole winter, even though there were still cracks sometimes. The Krasnoyarsk dam provided that the whole exposed surface should be protected by insulation form with surface conductance 2.93 kJ/(m^2 h °C) since September 1 every year; the concrete in the constraint zone should be protected by insulation form with surface conductance 1.67 kJ/(m^2 h °C).

20.3 Calculation of Thermal Insulation of Mass Concrete Construction in Winter

Known conditions: Outside air temperature T_a, concrete placing temperature T_p, adiabatic temperature rise $\theta(\tau)$, the concrete surface temperature T_s is required no less than T_b before $\tau = \tau_R$. Calculate the needed surface insulation capacity, namely the surface conductance β.

This question should be solved by trial method and the following is an effective one.

1. One-dimensional calculation

Since the heat conduction problem is linear, the original problem can be decomposed as follows:

a. Air temperature $T_a = 0$, concrete initial temperature $T_0 = 0$, adiabatic temperature rise $\theta\ (\tau)$, the surface temperature T_{1r} in the condition of $\tau = \tau_R$ can be calculated by the finite difference method.

b. Air temperature $T_a = 0$, concrete initial temperature $T_0 = 1.0$ (unit initial temperature difference), adiabatic temperature rise $\theta(\tau) = 0$, surface conductance β, the surface temperature c in the condition of $\tau = \tau_R$ can be calculated by the finite difference method.

c. Air temperature T_a, concrete placing temperature T_p, adiabatic temperature rise $\theta(\tau)$, surface conductance β, the surface temperature at $\tau = \tau_R$ can be calculated as follows:

$$T_s = T_a + c(T_p - T_a) + T_{1r} \tag{20.1}$$

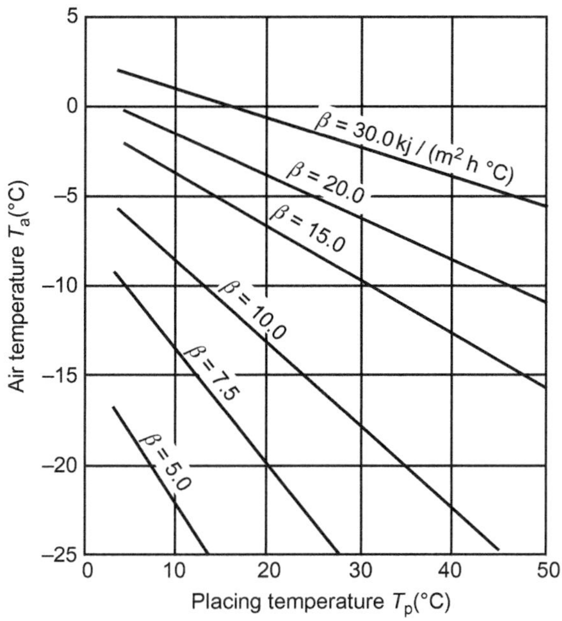

Figure 20.1 Thermal insulation in the winter construction.

Let T_s be equal to the wanted surface temperature T_b, we get:

$$cT_p + (1 - c)T_a = T_b - T_{1r} \tag{20.2}$$

This is a linear equation with T_p and T_a as variables, in which c, T_b, T_{1r} are constants. T_{1r} and c depend on β, thermal properties and geometry of the structure. For a given structure, there will be a line for each value of β. Sequentially let β be equal to $\beta_1, \beta_2, \ldots, \beta_n$, there will be n lines corresponding n values of β, with T_p as abscissa and T_a as coordinate, as shown in Figure 20.1. Accordingly, for any given placing temperature T_p and air temperature T_a, the required surface conductance β can be found in this figure.

Example Wall-type concrete block with thickness of 11.0 m. During winter construction, the surface temperature is required to be no less than $T_b = 5$°C when the concrete age $\tau = 14$ days. The adiabatic temperature rise $\theta(\tau) = 24\ [1 - \exp(-0.32\tau)]$, the diffusivity $a = 0.080$ m^2/day, the conductivity $\lambda = 0.83$k/J (m h °C). Sequentially let the surface conductance be $\beta = 5.0$, 7.5, 10.0, 15.0, 20.0, 30.0 kJ/(m^2 h °C) and calculate the surface temperature T_{1r} caused by adiabatic temperature rise and surface temperature c caused by unit initial temperature difference respectively. Then, we get Figure 20.1. The required β can be found by T_p and T_a.

2. Two- and three-dimensional calculation

For the edge and corner of placing block, there are problems of two-dimensional and three-dimensional heat emissions. The scheme of one-dimensional calculation and the finite difference methods of two-dimensional or three-dimensional or the finite element method could be used to calculate the temperature of edge or corner T_{1r} and c. Then the thermal insulation figure similar to Figure 20.1 could be obtained by substituting T_{1r} and c into Eq. (20.2).

21 Temperature Control of Concrete Dam in Cold Region

In a cold region, there are very low winter temperatures, low mean annual temperatures, and large amplitude of annual temperature variations, and the construction of a dam is usually suspended in winter because of the low temperature, which will cause large temperature differences above foundations, large temperature differences between upper and lower parts, and large temperature differences between surface and interior, which will bring great difficulties to the temperature control of a concrete dam. Furthermore, there are two problems ignored by people when building the RCCD in a cold region. The first one concerns the thin concrete block on the bedrock passing through the winter. Since the construction of concrete dam in a cold region is usually suspended in winter and the last concrete was placed in autumn on both sides of the river that were located in the strong constraint zone of foundation, in the following winter, the low temperature would bring ultracold to the thin concrete block and cause cracks. The second one concerns the problem of the ultracold of bedrock. The surface of rock is in contact with air of low temperature in the winter, the temperature of the bedrock under the first placed concrete in spring is much lower than the steady temperature of the dam, which would increase the temperature difference of concrete above the foundation.

This chapter will show how to solve the temperature control problems when building a concrete dam in a cold region.

21.1 Climate Features of the Cold Region

When building concrete dam in a cold region, the following temperature features are to be considered.

1. The mean annual temperature and the steady temperature of dam are low.

   The annual air temperature and water temperature of northern cold regions are much lower than those of southern regions, so the steady temperature of a dam in the north is lower than that in the south. The steady temperatures of dams in China are Northeast China 5–10°C, North and Northwest China 8–14°C, Center China 14–18°C, South China and Southwest China 17–20°C.
2. The amplitude of annual variation of temperature is large.

Thermal Stresses and Temperature Control of Mass Concrete. DOI: http://dx.doi.org/10.1016/B978-0-12-407723-2.00021-X

The amplitudes of annual variation of air temperature are about: Northeast China 19–20°C, North and Northwest China 14–17°C, Center China 11–12°C, South and Southwest China 6–10°C.

3. The construction is suspended in winter and concrete is placed in summer.

The winter temperature is very low in cold regions, so the construction is usually suspended in winter unless special temperature control measures are taken, and a large amount of concrete, including the concrete in the strong restraint zone above the foundation, is placed in summer.

21.2 Difficulties of Temperature Control of Concrete Dam in Cold Region

1. The temperature difference above the foundation is large

The steady temperature of a dam is low and a large amount of concrete is placed in summer, which will induce a large temperature difference above the foundation.

2. The temperature difference between upper and lower parts of concrete block is large

Since the construction is suspended in winter, the temperature of old concrete will be very low after a long time cooling during the winter and when the new concrete is placed in the following year when weather gets warm, together with the temperature rise caused by the heat of hydration; all these cause large temperature differences between upper and lower parts of concrete blocks.

3. The temperature difference between surface and interior is large

The concrete placed in summer has high interior temperature but the air temperature is very low in winter, which will form large temperature differences between surface and interior, and cause large tensile stress in both the vertical direction and horizontal direction in upstream and downstream surfaces, thus inducing vertical and horizontal cracks.

The stresses due to the temperature difference between surface and interior and the temperature difference between upper and lower parts of concrete blocks mentioned above will be superimposed to some extent, which would cause large vertical tensile stresses on the upstream and downstream surfaces of the dam in the region near the horizontal construction joints suspended in winter, but the tensile strength of construction joints is low, so large amounts of horizontal cracks would appear.

4. Thin concrete blocks above bedrock passing through the winter

The roller compacted concrete is parallelly placed layer by layer between both sides of the valley, thus the concrete placed on both sides before winter is in the strong restraint zone above the foundation, and the concrete block is thin, together with a half-year low temperature of winter, the temperature of concrete drops much lower than the steady temperature of the dam, which results in the situation of thin concrete blocks above rock foundation passing through the winter. Generally, cracks will appear in the concrete block.

5. Ultracold of bedrock

Generally, the starting point of calculation of the temperature difference above the foundation is the steady temperature of the dam, but the temperature of the bedrock at the beginning of spring in a cold region is lower than the steady temperature of the dam due to cooling in the winter.

The first three problems mentioned above, the temperature difference above foundation, the temperature difference between upper and lower layer, and the temperature difference between surface and interior, are more prominent in cold regions though they also exist in common regions; the last two problems mentioned above, the thin concrete blocks above bedrock passing through the winter and the ultracold of bedrock, are special problems in a cold region, and so far, it seems nobody has ever studied them in literature of temperature control of a concrete dam.

21.3 Temperature Control of Concrete Dam in Cold Region

The following section explains how to control temperature of a concrete dam in a cold region by taking an RCCD in Fengman District, Northeast China, as an example. The maximum height of this dam is 94.5 m, and the adiabatic temperature rise is $\theta(\tau) = 23.3\tau/(5.95 + \tau)$.

1. **Control of the temperature difference above foundation and the maximum temperature of concrete**

   Figure 17.40 shows that the maximum temperature $T_p + T_r$ can be controlled to 14−16°C by dense cooling pipes and the steady temperature of a dam in a cold region is about 5−8°C, so the temperature difference can be effectively controlled by densifying the cooling pipes.
2. **Insulation of concrete surface in winter**

   Concrete is placed in hot seasons and the placing process is suspended during winter in cold regions. These will form a large temperature difference between upper and lower layers and cause both horizontal tensile stress in new concrete and vertical tensile stress in the upstream and downstream surfaces near the placing layer. In the winter of the second and third year, when the air temperature is low, the thermal stress between upper and lower layers and the thermal stress between surface and interior may be superimposed and cause horizontal cracks.

   Simulation computation should be carried out for important projects. But a simplified practical algorithm is also needed in the actual engineering design and construction. Here is a practical algorithm given by author. Analysis as half-infinite body $x \geq 0$:

   Heat conduction equation:

$$\frac{\partial T}{\partial t} = a\frac{\partial^2 T}{\partial x^2} \tag{21.1}$$

   Initial condition:

$$\text{when } t = 0, \quad T(x, 0) = T_0 \tag{21.2}$$

   Boundary condition:

$$\text{when } x = 0, \quad \lambda\frac{\partial T}{\partial x} + \beta(T - T_a) = 0 \tag{21.3}$$

   Temperature T_a varies as follows:

$$T_a(t) = T_{a0} - A \sin\left[\frac{\pi t}{2Q}\right] \tag{21.4}$$

The following approximate solution was obtained by the author:

$$T(x,t) = T_0 + (T_{a0} - T_0)f(x,t) - f_2 A\, e^{-qx} \sin\left[\frac{\pi t}{2Q} - qx - M\right] \tag{21.5}$$

in which

$$f_2 = [1 + 1.85q\lambda/\beta + 1.12(\lambda q/\beta)^2]^{-1/2} \tag{21.6}$$

$$M = \tan^{-1}\left[\frac{1}{1 + \beta/(\lambda q)}\right] \tag{21.7}$$

$$q = \sqrt{\pi/(4aQ)} \tag{21.8}$$

$f(x,t)$ can be found in Figure 5.2. Q is the cooling duration, which is generally taken as $Q = 90$ days.

Example The concrete is placed before the middle of October and the initial temperature $T_0 = T_p + T_r = 20°C$, then the air temperature is

$$T_a = 6.8 - 20.2 \sin\left[\frac{\pi t}{2Q}\right]$$

The cooling duration $Q = 3.0$ months, calculate the temperature distribution along the depth in early April of the next year, when $t = 5.5$ months. Calculate two schemes: Scheme A, surface exposed to air, $\beta_0 = 80$ kJ/(m^2 h °C); Scheme B, surface insulated by 8 cm thick XPS plastic board with $\lambda = 0.108$ kJ/(m h °C). The computed results are shown in Figure 21.1, which shows that, without insulation, there will be a large temperature difference between upper and lower layers when placing new upper concrete in early April of the next year because the temperature of lower concrete is low; if insulated with 8 cm XPS board, the temperature of concrete can increase 10–14°C and the temperature difference between upper and lower layers will largely decrease.

3. Control of the ultracooling of bedrock surface

Generally, when calculating the temperature difference above the foundation, the starting point is the steady temperature of dam T_f. In a cold region, when placing concrete in spring, after cooling during the whole winter, the temperature of bedrock is usually lower than the steady temperature of the dam, which would cause the problem of ultracold of the bedrock.

Let the air temperature be expressed by sine function and the annual temperature variation in the bedrock be calculated as it is a semi-infinite body; Figure 21.2 shows the calculation results of Fengman district. After cooling during the whole winter, the temperature of the upper part with depth less than 4.5 m is under 0°C when placing concrete in early April the next year; even though the temperature of the surface of bedrock increases to 4.2°C when placing concrete in middle April, the temperature of the bedrock with depth of 1.0–4.7 m is

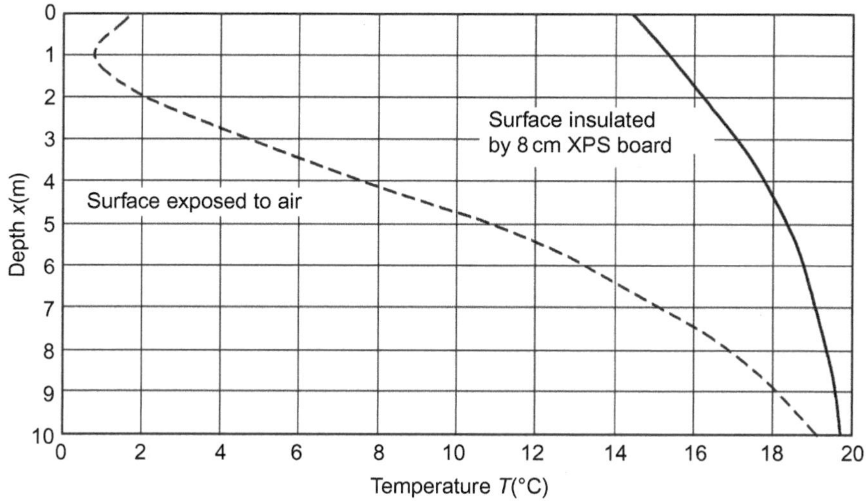

Figure 21.1 Example: the temperature distribution of concrete on 1 April of the next year after passing the winter and the initial temperature 20°C.

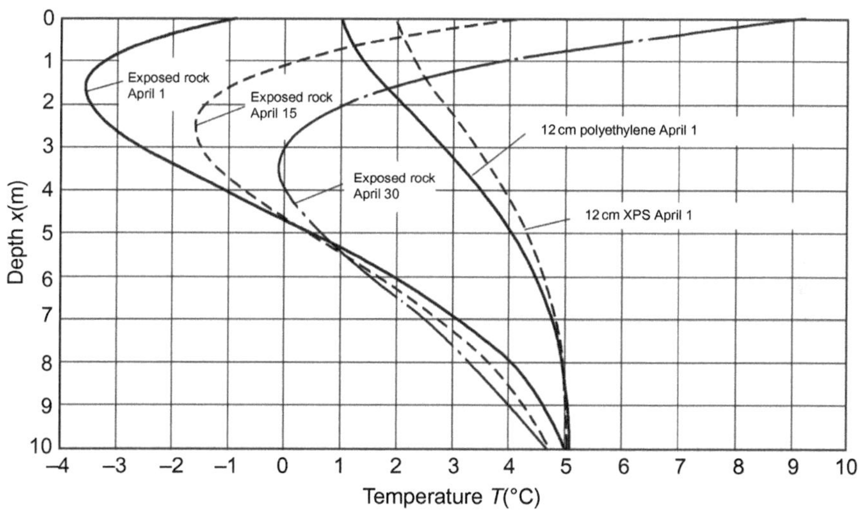

Figure 21.2 The temperature of the surface of bedrock after passing the winter.

still under 0°C, which is much lower than the steady temperature of the dam $T_f = 6$°C. When it comes to the operation period, the temperature of the upper foundation will increase from the negative temperature to the steady temperature of the dam. The rise of the temperature of the foundation will cause tensile stress in the concrete in the restraint zone, which is harmful to crack prevention of the dam. This is the problem of ultracooling of bedrock.

In common regions, the steady temperature of the dam is lower than the mean annual air temperature, which is not a big problem; but in cold regions, the steady temperature of the dam is higher than the mean annual air temperature, meanwhile, the temperature is very low in winter and the low temperature lasts longer, so the problem is more prominent.

The solution to the ultracold of the foundation is thermal insulation of the surface of bedrock before placing concrete. As Figure 21.2 shows, if the surface of rock is insulated by polyethylene layer or XPS foam board with a thickness of 12 cm before winter (the earlier the better), the temperature of bedrock will increase significantly, which is as much as 3–6 °C higher than the bedrock with exposed surface.

4. Thin concrete blocks above bedrock passing the winter

For the normal concrete dam, placing concrete in the strong restraint zone above the foundation should be avoided before winter; for the RCCD, generally parallel placing between the two sides, both sides locate on the bedrock, thus, the concrete placed before winter may be thin blocks above the bedrock, which will cause the problem of the thin concrete blocks above bedrock passing the winter. The RCCD is placed without longitudinal joints; the length of blocks parallel to the flow direction is also very long. Specifically, there are two situations, the first is that there is only one conventional concrete cushion above the bedrock, the second is that there are two layers, a conventional concrete cushion and a layer of roller compacted concrete.

Temperature control measures:

a. Surface insulation

Strict surface insulation measures should be taken 3–5 days after the appearance of the maximum temperature of concrete to control the minimum temperature of concrete in winter.

b. Control the maximum temperature of concrete

Control the maximum temperature of concrete by comprehensive measures to decrease the total temperature difference of concrete and reduce the tensile stress.

The practical measures should be decided by simulation computation, and the following is an example.

Example Figures 21.3 and 21.4 show the thermal stresses of thin concrete blocks above bedrock passing through the winter. The normal concrete cushion has a thickness of 1.50 m and a length of 60 m; bedrock has a depth of 100 m and a length of 260 m. Since the structure is symmetric, take one half of it and calculate as a plain strain problem by the finite element method. For concrete, the adiabatic temperature rise $\theta(\tau) = 25.0\tau/(2.0 + \tau)$°C, the elastic modulus $E(\tau) = 35,000[1 - \exp(-0.40\tau^{0.34})]$ MPa, the unit creep $C(t,\tau) = 6.60(1 + 9.20\ e^{-0.45})\ [1 - e^{-0.30(t-\tau)}] + 14.9\ (1 + 1.70\tau^{-0.45})$

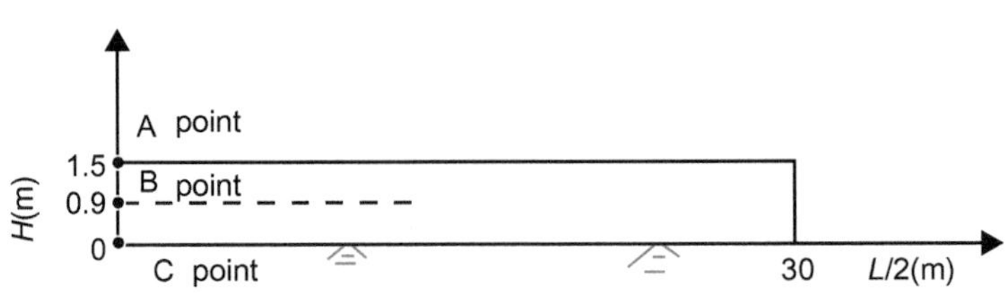

Figure 21.3 Calculation model of the thin block above the bedrock.

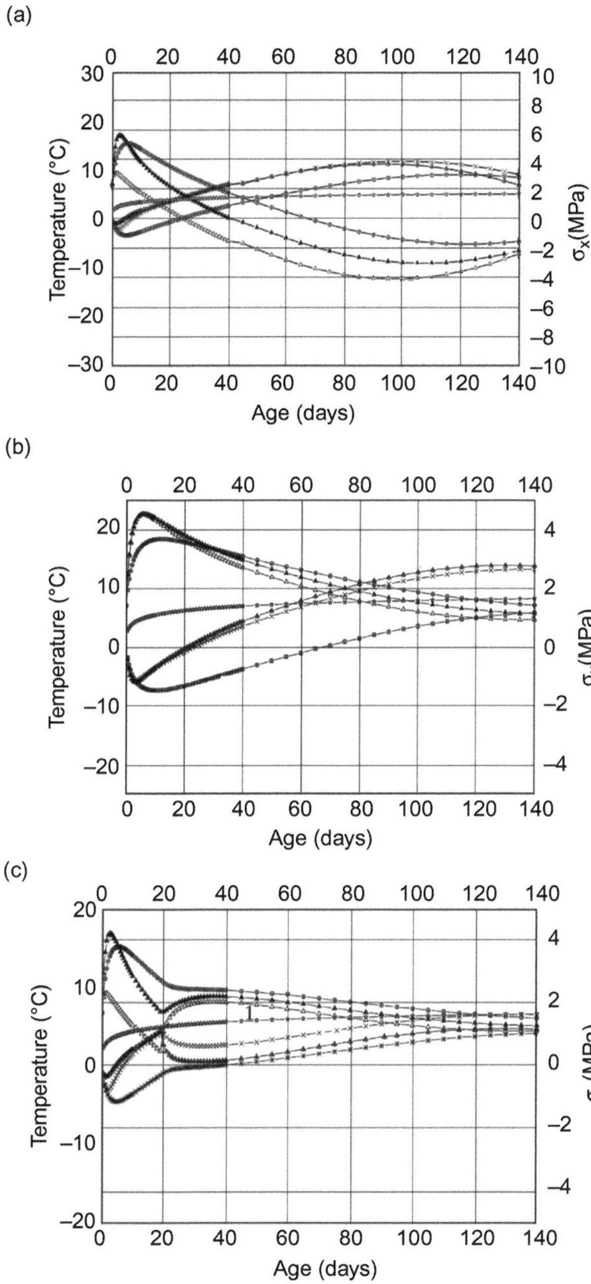

Figure 21.4 The thermal stress of the thin blocks above bedrock passing the winter. (a) Condition A: surface exposed; (b) Condition B: surface insulated over the whole process (15 cm PE); (c) Condition C: surface insulated after 10 days (15 cm PE).

$[1 - e^{-0.0050(t-\tau)}]$ (10^{-6}/MPa), $a = 0.10\ m^2$/day, $\lambda = 10.0$ kJ/(m h °C), $\mu = 0.167$, the initial temperature $T_0 = 7.0$°C, air temperature $T_a = 4.9 - 20.2 \sin(\pi t/180)$°C, t is calculated in days, $\alpha = 1 \times 10^{-5}$/°C. The allowable tensile stress is

$[\sigma_t] = R_t/1.80 = 2.0[1 - \exp(-0.39\tau^{0.30})]$ MPa. For bedrock, the elastic modulus is $E_f = 18,000$ MPa, $\mu = 0.25$, $\alpha = 1 \times 10^{-5}/°C$, and the initial temperature of bedrock is:

$$T(x, 0) = 4.90 + 19.47e^{-0.293x} \sin(0.293x + 0.0353)$$

Calculate three conditions: (i) Condition A: the surface of concrete has no insulation, $\beta = 80$ kJ/(m^2 h °C); (ii) Condition B: the surface of concrete is insulated by polyethylene insulating layer with a thickness of 15 cm, $\lambda_s = 0.148$ kJ/(m h °C), $\beta = 1.45$ kJ/(m^2 h °C); and (iii) Condition C: the surface is exposed for cooling at first and insulated after 10 days, $\beta = 1.45$ kJ/(m^2 h °C).

Figure 21.4 shows the results: Condition A, the surface is exposed; the whole section suffers tensile stress, which is higher than the allowable stress. Condition B, the surface is insulated during the whole process; the tensile stress decreases a lot in winter but the tensile stresses in point A and point B are still higher than the allowable stress. Condition C, the surface is exposed in the first 10 days for cooling to make the maximum interior temperature decrease about 5°C, and 10 days later, the surface is insulated by polyethylene board. Compared to Condition B, the temperature decreases and the tensile stresses in winter are lower than the allowable value in the whole section.

5. Control of the temperature difference between surface and interior

In cold region, a large amount of concrete is placed in summer, which will form large temperature differences between surface and interior and cause cracks in the winter. This is a complex problem, which is better to be solved by simulation computation analysis. The most efficient measure to prevent cracking is superficial thermal insulation. These problems are systematically researched in this chapter and a series of methods are suggested to overcome them.

22 Allowable Temperature Difference, Cooling Capacity, Inspection and Treatment of Cracks, and Administration of Temperature Control

For prevention of cracks in mass concrete, despite the temperature control, other technical measures are needed, including strengthening the crack resistance of concrete, quality control of concrete construction, improvement of the constraint condition of concrete structures, etc., which means that comprehensive technical measures are necessary to prevent crack in mass concrete [8, 12, 13, 23, 42, 52, 56, 69, 72 ,81, 99].

This chapter will illustrate the method to determine the allowable temperature difference of concrete, comprehensive technical measures, allowable temperature differences used by different countries for preventing cracks, and the computing method of cooling capacity in massive concrete construction. Moreover, this chapter will show some practical examples of the temperature control in concrete dams.

22.1 Computational Formula for Concrete Crack Resistance

Figure 22.1 shows two kinds of simplified model for computing concrete crack resistance.

1. Parallel model

As shown in Figure 22.1(a), the strain of each element is $\varepsilon_1 = \varepsilon_2 = \varepsilon_3 = \varepsilon$. When $\varepsilon = \varepsilon_p$, concrete failure happened, the allowable tensile stress $[\sigma]$ may be calculated by the following formula:

$$[\sigma] \leq \frac{E\varepsilon_p}{K_1} \tag{22.1}$$

where $[\sigma]$ is the allowable tensile stress; E is the elastic modulus; ε_p is the extensibility; K_1 is the safety factor; $E\varepsilon_p$ is the virtual tensile strength, its value is slightly larger than the true tensile strength, as shown in Figure 22.2.

Thermal Stresses and Temperature Control of Mass Concrete. DOI: http://dx.doi.org/10.1016/B978-0-12-407723-2.00022-1

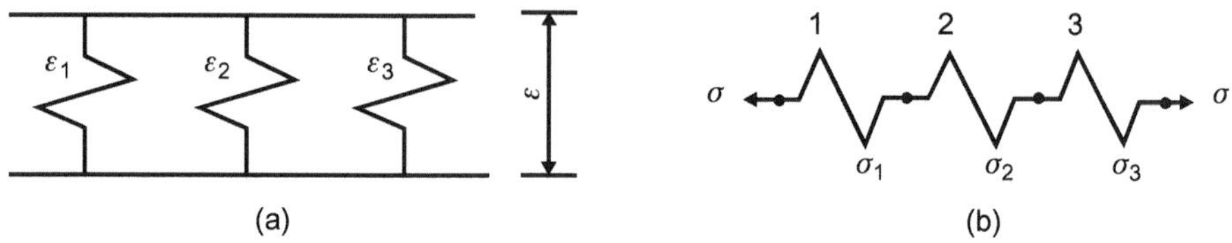

Figure 22.1 Calculation model of crack resistance of concrete: (a) parallel model and (b) serial model.

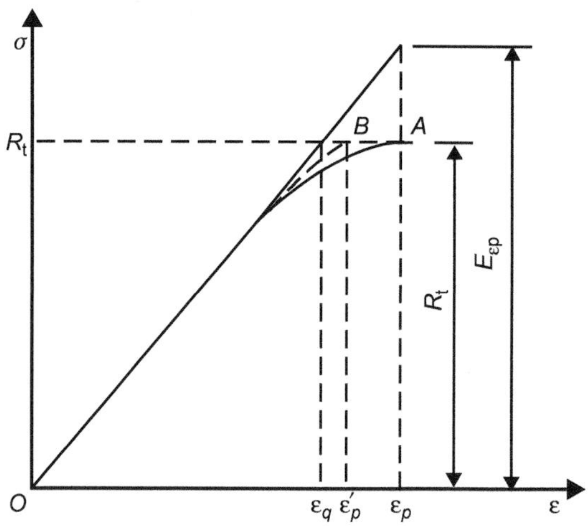

Figure 22.2 Extensibility of concrete brittle fracture.

2. Serial model

As is shown in Figure 22.1(b), stress of each element is $\sigma_1 = \sigma_2 = \sigma_3 = \ldots = \sigma$. When $\sigma = R_t$, concrete failure happened, the allowable tensile stress $[\sigma]$ should be calculated by the following formula:

$$[\sigma] \le \frac{R_t}{K_2} \tag{22.2}$$

where R_t is the concrete axial tensile strength and K_2 is the safety factor.

In reality, the concrete structure is complicated, the structure and stress of the interior of concrete is nonuniform. Before applying of load, micro cracks exist in the concrete. Concrete damage results from the continual development of micro cracks. Of course, the axial tensile strength R_t and the ultimate extensibility obtained from tests have considered the influence of micro cracks in the concrete.

3. Tear of horizontal construction joint

In practical projects, there are lots of horizontal cracks, which are generally torn along the horizontal construction joint. The tensile strength of a horizontal construction joint is smaller than $E\varepsilon_p$, the allowable vertical tensile stress on the joint should be calculated by the following formula:

$$[\sigma_y] = \frac{rR_t}{K_2} \tag{22.3}$$

in which $[\sigma_y]$ is the allowable vertical tensile stress; R_t is the axial tensile strength; and r is the reduction factor. As experience shows $r = 0.5-0.7$.

22.2 Laboratory Test of Crack Resistance of Concrete

A special expensive test machine is needed to obtain the tensile load–strain curve of concrete. At present, in a practical project, a normal testing machine is used to conduct the ultimate tensile test of concrete. The $\sigma-\varepsilon$ relation is represented by curve OA shown in Figure 22.2. Since the measuring instrument is frequently damaged at the time when test sample breaks abruptly, in order to avoid instrument damage, generally instrument is removed when stress reaches 90% of the tensile strength. The rest of the $\sigma-\varepsilon$ curve can only be extended manually, which will introduce error.

From formulae (22.1) and (22.2), we know

$$\frac{R_t}{E\varepsilon_p} = s = \frac{K_2}{K_1} \tag{22.4}$$

Concrete test data of practical projects shows that the ratio s is mostly between 0.75 and 0.95 and its average is about $s = 0.820$. In Figure 22.2, $\varepsilon_q = R_t/E$ is the virtual extensibility, which represents the tensile deformation when the stress–strain relation line extends to tensile strength. $R_q = E\varepsilon_p$ is virtual tensile strength, which represents the stress when the stress–strain relation line extends to the extensibility. Apparently, $\varepsilon_q = s\varepsilon_p$.

22.3 The Difference of Tensile Properties Between Prototype Concrete and Laboratory Testing Sample

22.3.1 Coefficient b_1 for Size and Screening Effect

Let b_1 be the ratio of strength or extensibility of a large sample to that of a small sample.

Normally, testing samples of 10 cm × 10 cm cross section are used for the test of axial tensile strength and extensibility, the maximum grain size of aggregate is 3 cm while the maximum grain size of dam aggregate is 8–15 cm.

From the test results of Yang Chengqiu [8], for the three-graded concrete, b_1 for tensile strength is 0.73 and b_1 for extensibility is 0.70; for the four-graded concrete, b_1 for tensile strength is 0.62 and b_1 for the extensibility is 0.64. For the elastic modulus, b_1 is 1.15 and for the creep, b_1 is 0.80 (three-graded concrete) to 0.70 (four-graded concrete).

From the test result of Li Jinyu [8], b_1 for the axial tensile strength is 0.60, b_1 for the extensibility ratio is 0.57, and the elastic modulus is 1.05. Poisson's ratio of the large sample is 0.23, which is a little higher than the traditional value 0.167.

22.3.2 Time Effect Coefficient b_2

Loading rate has a great impact on concrete strength. In a laboratory static test, a testing sample is destroyed in 1–2 min. Under the effect of earthquake and impulse loading, loading rate is fast and concrete strength is high; on the contrary, in practical dam engineering, rate of loading for water pressure, self-weight, and temperature are slow, the loading time is long, so the concrete strength is lower than the laboratory testing result. For the same concrete testing sample, if the rate of loading is normal and the load is P, then when the sample is applied with $0.9P$, it will be destroyed at about 1 h; when applied with $0.77P$, it will take 30 years to be destroyed. The influence of the time rate of loading to strength can be approximately represented as

$$b_2 = 0.67 + 0.33\exp(-0.06t^{0.160}), \quad t \geq 0.01d \tag{22.5}$$

where t is the time of loading (d), b_2 is the ratio of the strength at loading time t to the strength under standard test rate.

The variation of the thermal stress σ with the time t is shown in Figure 22.3. Δt is the time for thermal stress drop from the maximum σ_{max} to 0.8 σ_{max}. Δt for different thermal stress is approximately as follows: for daily variation, $\Delta t = 3.5$ h; for cold wave, $\Delta t = 0.5Q$ (Q is the time of duration for cold wave); for annual variation, $\Delta t = 1.75$ month; for dam cooling before joint grouting $\Delta t = 50$ d, for the natural cooling of a conventional concrete gravity dam and RCC gravity dam, $\Delta t \geq 5$ years.

From the formula (22.1), the time effect coefficient b_2 for different thermal stresses is approximately: for daily variation, $b_2 = 0.88$; for a cold wave lasting for 3 days, $b_2 = 0.80$; for annual variation and dam cooling before joint grouting, $b_2 = 0.78$; for the natural cooling of a normal concrete gravity dam and RCC (roller compacted concrete) gravity dam, $b_2 = 0.70$.

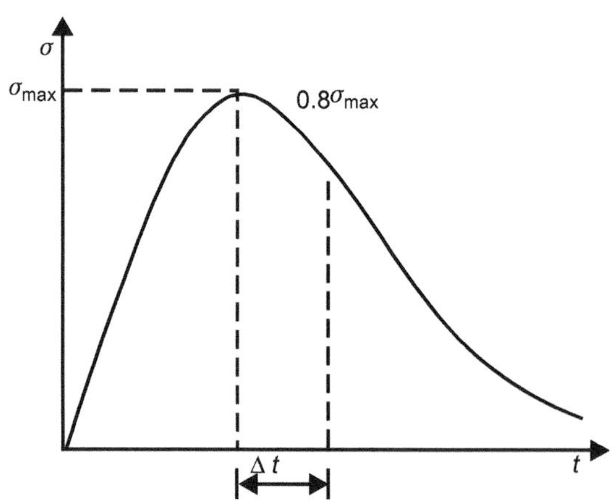

Figure 22.3 Variation of thermal stress.

22.4 Reasonable Value for the Safety Factor of Crack Resistance

22.4.1 Theoretical Safety Factor of Crack Resistance

The tensile strength of prototype concrete is represented as shown below:

$$\overline{R}_t = R_t b_1 b_2 \tag{22.6}$$

where $\overline{R}_t$ is the tensile strength of prototype concrete; R_t is the axial tensile strength of test sample according to the SL352-2006 "Hydraulic Concrete Test Code"; b_1 is the coefficient of size and screening effect and b_2 is the coefficient of effect of hold time of load.

Tensile stress of concrete should not be larger than the tensile strength of the prototype, namely:

$$\sigma \leq b_1 b_2 R_t = R_t / K_{20} \tag{22.7}$$

$$K_{20} = 1/b_1 b_2 \tag{22.8}$$

K_{20} is the theoretical safety factor of tensile stress. Table 22.1 shows an example, from which we know that the theoretical safety factor of tensile stress is 1.56–2.30; the value is relatively large.

22.4.2 Practical Safety Factor of Concrete Crack Resistance

According to the specifications for a concrete gravity dam and arch dam [12–16], the safety factor for compressive stress is $K = 4.0$. On one side, the safety factor for compressive stress is very important, because if concrete is crushed, the dam would break;

Table 22.1 Theoretical Safety Factor of Crack Resistance ($K_{20} = 1/b_1 b_2$)

Maximum grain size of aggregate (mm)			80	150
Coefficient b_1 for size and screening effect			0.73	0.62
Coefficient b_2 for time effect	Daily variation	0.88	1.56	1.83
	Cold wave (3 d)	0.80	1.71	2.01
	Annual variation and cooling before joint grouting	0.78	1.76	2.07
	Natural cooling and continuous placement	0.70	1.96	2.30
	Ignore the time effect	1.00	1.37	1.61

on the other side, compressive strength with a safety factor 4.0 is not difficult for a practical project to achieve. It is easy to accomplish if a proper water–cement ratio is chosen.

If the safety factor for concrete tensile stress is 4.0, cracks will not emerge in practical projects. But the tensile strength of concrete is only about 8% of its compressive strength, if the safety factor is 4.0, the allowable temperature difference would be too small and it is hard to achieve in practice. On the other hand, cracks may have a great impact on the safety and durability of a dam, but if proper treatment is applied, generally dam break would not happen. So, in current design and construction specifications, the safety factor of tensile strength is much lower than 4.0.

Safety factor of crack resistance K_1 in "Design Specifications of Concrete Gravity Dam" SDJ21-78 is only 1.3–1.8, which is rather small. As mentioned above, because of the effect of the testing sample size and wet screening, concrete extensibility in a practical project is only 0.60–0.70 of the laboratory test value. Considering the time effect, if $K_1 = 1.3-1.8$, practical safety factor K_1 is only 0.6–0.8. The small safety factor is the primary reason for large quantity of cracks in mass concrete.

"Design Specifications of Concrete Gravity Dam" SDJ21-78 was edited in the 1970s. At that time, the concrete temperature control of China is low, for example, the precooling aggregate was rarely used. Straw bags are mainly used for surface insulation and their effect is poor. Because of some practical conditions, the safety factor is low. Nowadays, the temperature control level in China is highly improved, the technique of precooling aggregate is mature, and the plastic industry has rapidly developed. Foamed plastic is widely used in surface insulation, the performance is good and the cost is low. Thus, there are suitable conditions to improve the safety factor of concrete crack resistance.

Theoretical safety factor of crack resistance $K_{20} = 1/b_1 b_2$ is calculated according to the concrete mechanical property, but more influential factors should be taken into consideration to determine the practical safety factor of crack resistance. Here is a calculation method designed by the author.

Assume the designed concrete tensile stress to be

$$\sigma_{dt} = \sigma_t a_1 a_2 a_3 a_4 a_5 \tag{22.9}$$

where σ_{dt} is the designed tensile stress; σ_t is the calculated tensile stress; α_1 is the coefficient for importance of the structure, for I, II, III level structure, α_1 should be 1.1, 1.0, 0.9, respectively; α_2 is the factor of importance for the position of tensile stress, for interior and surface of the concrete in the foundation constraint zone and upstream surface, the factor is 1.0 for the lateral face, the factor is 0.9, for downstream surface, the factor is 0.8. α_3 is the overload coefficient. Considering the fact that the drop of air temperature and cold wave may be beyond the computation values and the time of adiabatic temperature rise test, 28 d is short, we take $\alpha_3 = 1.05-1.15$. α_4 is the age factor. Normally in the test of elastic modulus and creep of concrete, the biggest age of loading is only 180 d, experience shows that for late age $\tau > 1$ year, based on these test values, the estimated later elastic modulus is smaller and creep deformation is larger than the actual values. By considering these factors, for early stage, $\tau \leq 1$ year, $\alpha_4 = 1.0$, for later stage, $\tau \geq 3$ year, $\alpha_4 = 1.1-1.2$. α_5 is the correction factor. Considering large amount of engineering experiences and the feasibility of engineering implementation, it is suggested that $\alpha_5 = 0.70-1.00$.

The tensile strength is

$$f_t = b_1 b_2 b_3 R_t \tag{22.10}$$

where f_t is the available tensile strength; b_1 is the coefficient of the size of testing sample and wet screening; b_2 is the coefficient of time duration of loading; b_3 is the coefficient of age of concrete, normally the age of strength test is not greater than 180 d, and the later strength estimated by the curve based on these test results is low. In the natural cooling of gravity dam without longitudinal joint, the major part of the drop of the interior temperature happened 5 years later, lots of test results of drilling core in dam body show that the strength will still slowly increase after 20 years, the later increase of the strength of concrete mixing with fly ash may be more. When $\tau \leq 1$ year, $b_3 = 1.0$; when $\tau \geq 3$ year, $b_3 = 1.2-1.3$.

From $\sigma_{dt} \leq f_t$, the safety factor for crack resistance is

$$K_2 = \frac{a_1 a_2 a_3 a_4 a_5}{b_1 b_2 b_3} \tag{22.11}$$

Calculation example, assume overloading coefficient $a_3 = 1.05$; correction factor, $a_5 = 0.80$; as for I, II, III level structure, $a_1 = 1.1, 1.0, 0.9$; a_2, b_1, b_2, and b_3/b_4 are shown in Table 22.2, the safety factor K_2 is calculated and shown in Table 22.2.

22.4.3 Safety Factors for Crack Resistance in Preliminary Design

In the preliminary design, the safety factors of crack resistance suggested by the author are

$$\begin{aligned} &\text{by formula (22.1):}\ K_1 = 1.6 - 2.2 \\ &\text{by formula (22.2):}\ K_2 = 1.4 - 1.9 \end{aligned} \tag{22.12}$$

Compared with the $K_1 = 1.3-1.8$ adopted in the old design specifications of concrete dams in China, the safety factors mentioned above have increased a lot. According to the experience, cracks are largely reduced if the safety factors mentioned above are chosen and proper temperature control measures are achievable in practical projects. Under current conditions, it has taken account both of the necessity and possibility. According to the author's suggestion to enlarge the safety factor of crack resistance, the newly edited "Design Specifications of Concrete Gravity Dam" of the Ministry of Water Conservancy of China sets $K_1 = 1.5-2.0$. From the developing view, from now on, the tensile stress should be controlled by formula (22.2) and safety factor K_2 should be given by formula (22.11). The advantages of

Table 22.2 Calculation Example for Practical Safety Factor of Tensile Strength K_2

Maximum Size of Aggregate, Size, and Wet Screening Factor b_1					**150 mm ($b_1 = 0.62$)**			**80 mm ($b_1 = 0.73$)**		
Stress Classification and Position		a_2	b_2	b_3/a_4	**I**	**II**	**III**	**I**	**II**	**III**
Stress in zone of foundation restraint	With longitudinal joint	1.0	0.78	1.0	1.91	1.74	1.56	1.62	1.47	1.32
	No longitudinal joint	1.0	0.70	1.1	1.93	1.76	1.58	1.64	1.49	1.34
Stress of upstream surface and lateral surface of foundation restraint zone	Daily variation	1.0	0.88	1.0	1.69	1.54	1.39	1.44	1.31	1.18
	Cold wave (3 d)	1.0	0.80	1.0	1.86	1.69	1.52	1.58	1.44	1.29
	Annual variation	1.0	0.78	1.0	1.91	1.74	1.57	1.62	1.48	1.33
Lateral and top surface outside the restraint zone	Daily variation	0.9	0.88	1.0	1.52	1.39	1.25	1.29	1.18	1.06
	Cold wave (3 d)	0.9	0.80	1.0	1.68	1.52	1.37	1.43	1.29	1.16
	Annual variation	0.9	0.78	1.0	1.72	1.56	1.41	1.46	1.33	1.20
Stress of downstream surface	Daily variation	0.8	0.88	1.0	1.36	1.23	1.11	1.16	0.94	0.94
	Cold wave (3 d)	0.8	0.80	1.0	1.49	1.36	1.22	1.27	1.15	1.04
	Annual variation	0.8	0.78	1.0	1.53	1.39	1.25	1.30	1.18	1.06

using formula (22.2) for crack computation are: (i) the tensile strength test is simple, (ii) the results of tensile strength test are stable, and (iii) the extensibility tests are complicated and they are rarely conducted in the construction process of the dam. The influence of the actual tensile strength in practical construction and the design value can be calculated by the formula (22.2), and temperature control measure can be adjusted when necessary.

22.5 Calculation of Allowable Temperature Difference and Ability of Superficial Thermal Insulation of Mass Concrete

22.5.1 General Formula for Allowable Temperature Difference and Superficial Thermal Insulation

Allowable temperature difference and superficial thermal insulation can be calculated by the formulae (22.1)–(22.3).

During the construction and operation period of a concrete dam, tensile stress is varying due to the changes of loads and temperature. In the three formulae mentioned above, σ is the maximum tensile stress and σ_y is the maximum vertical tensile stress.

Nowadays, the method of finite element simulation calculation is mature. The stresses caused by temperature differences above the foundation, temperature differences between the upper and lower parts of concrete blocks, surface-interior temperature differences, different loads, and different insulation states can be calculated according to the local weather conditions, material properties, and process of construction. By formulae (22.1)–(22.3), it is not difficult to determine the allowable temperature difference and insulation ability.

22.5.2 Approximate Calculation of Allowable Temperature Difference and Insulation Ability

1. Allowable temperature difference above foundation

   The maximum concrete temperature is $T_p + T_r$, where T_p is the placing temperature and T_r is the temperature rise caused by heat of hydration. The minimum temperature is T_f, generally T_f is the steady temperature of dam. If there are holes in the dam, T_f is the water temperature or air temperature in the holes in winter. Besides, there is autogenous deformation of concrete. The allowable temperature difference $T_p + T_r - T_f$ can be calculated by the formula:

$$\frac{K_p R E \alpha}{1-\mu}\left[T_p - T_f + \frac{C_1 A}{R} T_r + C_2 \frac{G}{\alpha}\right] \leq \frac{E\varepsilon_p}{K_1} \text{ or } \frac{R_t}{K_2} \tag{22.13}$$

where

K_p—stress relaxation coefficient caused by creep

R—foundation restraint coefficient

A—foundation influence coefficient as shown in Figure 12.5, it can also be computed by the stress influence line or Eq. (12.23)

C_1—reduction coefficient considering the influence of compressive stress due to temperature rise at the early age of concrete, the value is about 0.70−0.85

C_2—the coefficient considering the process that the autogenous volume deformation varies with the age, the value is approximately $C_2 = 1.00$

G—the autogenous deformation of concrete.

2. The allowable temperature difference between the upper and lower parts of concrete block

The temperature difference between the upper and lower parts of the block can be classified into two cases.

i. Temperature difference between the upper and lower parts of concrete block caused by the annual variation of the placing temperature of concrete.

If there is no special temperature control measure like precooling of aggregate, the placing temperature of concrete will vary with the air temperature, so the maximum concrete temperature will also vary with air temperature. The temperature is high in summer and low in winter. For example, the process of construction for the block shown in Figure 22.4 takes 4 years, there are four peak temperatures along the height.

In order to investigate the relation between the length of concrete block and the stresses caused by the temperature difference between the upper and lower parts of the block, a rectangular concrete block is computed by FEM. The heights of the block and the rock surface are 185 m and 27 m, respectively, the lengths of the block are

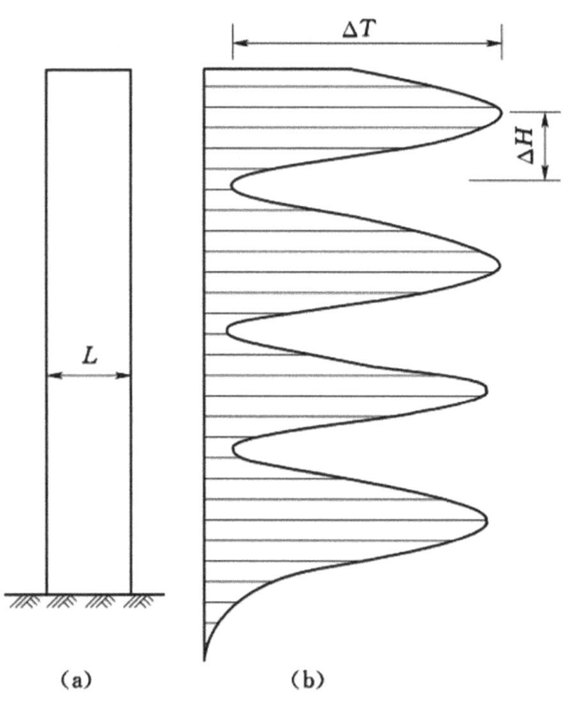

Figure 22.4 Temperature difference between upper and lower parts caused by annual variation of placing temperature: (a) concrete block l and (b) temperature difference.

$L = 20$, 40, 80, 120 m. According to the practical construction schedule, the temperature field and thermal stresses are computed. The results are shown in Figure 12.33. It is evident that the thermal stresses are dependent on the length L of concrete block. The bigger the length of block, the larger the thermal stress. For the block with $L = 20$ m, there are large thermal stresses only in the restraint zone above the foundation, there are very small thermal stresses in the rest of the block. For a concrete gravity dam without longitudinal joints, the length of dam block in the water direction is long, large thermal stresses may be caused by the temperature difference between the upper and lower parts of the block.

Let ΔT be defined as the annual temperature difference between peak temperature in summer and lowest temperature in winter, ΔH be defined as the height difference of the block in summer and winter. The width of the block is L. From Figure 12.33, we can know:

a. when $L \leq 1.5\Delta H$, the thermal stress is small, and it is not mainly caused by temperature difference between the upper and lower parts of the block;

b. when $L \geq 3\Delta H$, thermal stress is large, and it may be largely caused by temperature difference between the upper and lower parts of the block.

ΔH is the height rise of half a year. Assuming that the height rise velocity of the block is 6 m/month, so $\Delta H = 6 \times 6 = 36$ m. When $L \leq 54\ \text{m} = 1.5\Delta H$, temperature difference between the upper and lower parts of the block may not be the main cause of thermal stress, whereas when $L \geq 108$ m, temperature difference between the upper and lower parts of the block may be the important cause of thermal stress.

ii. The temperature difference between the upper and lower parts caused by placing new concrete on old concrete.

If the old concrete was placed a long time ago, the heat of hydration of cement is completely dissipated and the temperature is low. There will be a large temperature difference between the old concrete and new concrete. Since the elastic modulus of old concrete may exceed the elastic modulus of bed rock, the tensile stress produced by temperature difference between the upper and lower parts of concrete may exceed the tensile stress caused by the bedrock restraint. As a result, cracks may emerge and it should be of high concern. The thermal stress can be calculated by the thermal stress influence line.

3. The allowable interior–exterior temperature difference and surface insulation.

There are mainly two causes of the interior–exterior temperature difference: (i) The variation of ambient temperature, including the annual air or water temperature variation and cold wave. The thermal stress caused by these factors can be calculated using the method mentioned in Chapter 11. (ii) Heat hydration of cement. As for the fixed slab, free slab, and foundation beam, the thermal stress can be calculated using the method in Chapters 7 and 8. For the complex cases, the finite element method can be used for calculation.

For concrete dams with longitudinal joints, artificial pipe cooling is needed for joint grouting. After pipe cooling, the interior of the dam is cooled to steady temperature, the exterior–interior temperature difference is not large. The gravity dams without longitudinal joints including the RCC dams do not need pipe cooling since there is no longitudinal joint. The internal temperature drops slowly, so there will be a large exterior–interior temperature difference in winter. Superficial thermal insulation should be emphasized.

22.6 The Allowable Temperature Difference Adopted by Practical Concrete Dam Design Specifications

22.6.1 Regulations of Allowable Temperature Difference in Chinese Concrete Dam Design Specifications

Prescription for allowable temperature difference in Chinese "Design Specifications of Concrete Gravity Dam" and "Design Specifications of Concrete Arch Dam" [12–16].

1. Allowable temperature difference above the foundation

   The temperature difference above the foundation is normally defined as the temperature difference between the peak temperature of concrete and stable temperature of the dam in the restraint zone of the foundation. When the extensibility of concrete with age of 28 d is not smaller than 0.85×10^{-4}, for the concrete block with good construction quality, similar modulus of foundation and concrete, and short interval between concrete lifts, the value of allowable temperature difference of concrete above the foundation is given in Table 22.3.

   Under certain conditions, the temperature of block in construction and operation period (such as deep hole, dam block with wide slot) may be lower than the stable temperature. The influences of these cases should be considered in the design.

   Allowable temperature difference of the following cases should be studied especially:

   a. height–width ratio of block is smaller than 0.5;
   b. long time stopped block in foundation restraint range; and
   c. modulus of bedrock is very high.

   The allowable temperature difference of concrete filled in pond, concrete plug, and concrete placed on steep slope of rock foundation should be more rigorous than Table 22.3. In the cold area, the allowable temperature difference above foundation must be determined and discussed separately.

2. Temperature difference between the upper and lower parts of concrete block

   Temperature difference between the upper and lower parts of the block is defined as: within the range of $L/4$ of the upper and lower parts of the old concrete (age exceeds 28 d), the temperature difference between the maximum temperature of upper part and the average temperature of lower part when new concrete is placed. When the height h of new concrete placed with short interval between lifts is larger than $0.5L$, the temperature difference between upper and lower parts is about 15–20°C. If the surface of the pouring block is always exposed, a small value should be accepted. Standard for temperature difference between the upper and lower parts of mass concrete constructed in a cold region should be studied especially.

3. Abrupt temperature drop of concrete surface

   There are two kinds of superficial concrete cracks, the first kind of crack appeared in the early stage of the pouring process and the abrupt temperature drop of concrete surface at this time is the main cause of cracking. The second kind of crack generally appeared in the period

Table 22.3 Allowable Temperature Difference ΔT (°C) of Concrete Block Above Foundation

Length of the Block L		≤16 m	17–20 m	21–30 m	31–40 m	>40 m
Height above foundation surface h	0–0.2L	26–25	24–22	22–19	19–16	16–14
	0.2–0.4L	28–27	26–25	25–22	22–19	19–17

of cold wave that occurred in the first to third winter after placing of concrete. The cracks are caused by both the cold wave and the low air temperature in winter.

If daily average air temperature continues to decrease over 6°C in 2–4 days, cracks may occur on the exposed surface of concrete with age <28 d (in warm areas, concrete before 5 d age is not easy to crack), so the structure should be implemented with surface protection measures.

4. Interior–exterior temperature difference of longtime-exposed concrete block

   Because of factors such as annual temperature variation, large interior–exterior temperature difference is formed, and cracks may occur at the longtime-exposed concrete later. So for a massive concrete structure, the time and materials for surface protection should be decided according to local temperature conditions. Even in the region free of foundation constraint, it is necessary to prevent cracking due to interior–exterior temperature difference by longtime superficial thermal insulation. The entrances and exits of galleries and holes in the dam must be sealed in winter.

5. Height difference between adjacent blocks

   During the construction, blocks should rise evenly to prevent large height difference and the intermission between pouring process should not be too long. The height difference between adjacent blocks should not exceed 10–12 m to reduce exposing time and avoid the pressing of keys of adjacent blocks which may influence the grouting quality of longitudinal joints.

6. Placing temperature of concrete

   Placing temperature of concrete is the temperature of concrete at 5–10 cm depth after vibrating and before pouring the upper layer of concrete. The placing temperature of concrete should satisfy the provision of allowable temperature difference and maximum temperature in the dam body, and it should not exceed 25–30°C during the construction in summer. For winter construction, the placing temperature should be determined with the principle of no freeze of concrete. If there is no precooling, the concrete above the foundation should be poured in low temperature seasons to reduce the foundation temperature difference.

7. Grouting temperature in dam body

   Generally the steady temperature of the dam body is used for joint grouting temperature. It can be increased a little bit in cold areas after careful research. The suitable time for grouting should be the low temperature seasons.

22.6.2 The Requirement of Temperature Control in "Design Guideline of Roller Compacted Concrete Dam" of China

The history of the RCC dam is short. Due to lack of practical experience, "Design Guideline of Roller Compacted Concrete Dam" (abbreviate to "Guideline" in the following paragraph) suggests calculating concrete thermal stress according to the properties of concrete, construction condition, geological and weather condition, so as to determine the temperature difference above foundation, the maximum temperature and interior–exterior temperature difference. Based on the assumption that (i) the extensibility of RCC is $\varepsilon_{p1} = 0.70 \times 10^{-4}$ and that of the conventional concrete is $\varepsilon_{p2} = 0.80 \times 10^{-4}$, (ii) the allowable temperature differences ΔT_1 of RCC above rock foundation and the allowable temperature difference ΔT_2 of the conventional concrete above rock foundation satisfy the following equation

$$\Delta T_1 / \Delta T_2 = \varepsilon_{p1} / \varepsilon_{p2} = 0.70/0.80 \qquad \text{(a)}$$

Table 22.4 Allowable Temperature Difference Above Foundation of RCC Dam ΔT (°C)

Maximum Length of Concrete Block *L*		**Under 30 m**	**30–70 m**	**Above 70 m**
Height above foundation	0–0.2*L*	18–15.5	14.5–12	12–10
	0.2–0.4*L*	19–17	16.5–14.5	14.5–12

Table 22.5 Allowable Temperature Difference of Concrete Above Foundation of U.S. Bureau of Reclamation (°C)

Length of Concrete Block *L* (m)	**Height $h = 0–0.2L$**	**0.2–0.05*L***	**>0.5*L***
55–73	16.7	19.5	22.2
37–55	19.5	22.2	25.0
27–37	22.2	25.0	Unlimited
18–27	25.0	Unlimited	Unlimited
<18	27.8	Unlimited	Unlimited

From ΔT_1 given by "Design Specifications of Concrete Gravity Dam" of China, the allowable temperature difference ΔT_1 of RCC above rock foundation is given by formula (a) and shown in Table 22.4. It is recommended as shown in Table 22.4.

Allowable temperature difference above foundation of the following cases should be studied especially:

1. Thin structure with height to width ratio *H*/*L* smaller than 0.5.
2. Long rest layer in foundation restraint range.
3. Elastic modulus of rock foundation and concrete is largely different.
4. Concrete filled in pond, concrete plug and concrete on steep slope of foundation.

"Design Specification of Roller Compacted Concrete Dam" SL 314-2004 only prescribes to determine the allowable temperature difference by using temperature control design, but the actual number is not given.

22.6.3 Temperature Control Regulation of Concrete Dam by U.S. Bureau of Reclamation and U.S. Army Corps of Engineers

The allowable temperature difference given by the U.S. Bureau of Reclamation is shown in Table 22.5.

The U.S. Army Corps of Engineers classifies the temperature control of a concrete dam into basic control, level A control, and level B control.

Basic control includes: (i) use cement of moderate heat; (ii) adopt the minimum cement quantity which can satisfy the construction requirement; (iii) thickness of placing lift is 1.5 m; and (iv) when temperature suddenly drops over 14°C, make protection for concrete surface. Low gravity dams lower than 15 m should satisfy the basic requirement.

Level A control, despite the basic control, includes: (i) prohibit pouring concrete in hot daytime; (ii) requirement to place four thin layers with thickness 0.75 m on the surface of rock foundation or old concrete over 15 d, intermission should not be <3 d, intermission of 1.5 m placing lift should not be <5 d; (iii) height difference between adjacent blocks not bigger than 4.6 m; and (iv) from every September to next April, when the top and side surface of the pouring block is open to air more than 30 d, protection is needed to prevent dramatic temperature changes. High gravity dams with height about 15–46 m should adopt part or all measures of A level control.

Level B control, despite the basic control and A level control, includes: (i) limit the concrete placing temperature to 10°C; (ii) choose low heat cement on a hot day; and (iii) use cooling pipes to conduct first stage cooling at local zone in basic constraint range, in order to reduce the temperature rise caused by heat of hydration. A gravity dam over 46 m needs to adopt part or all the above measures. After restricting the placing temperature of concrete, some of the aforementioned rules need to be changed. Such as, concrete can be placed in daytime, it is not suitable to prescribe least intermission for thin layer pouring in hot days, because the concrete absorbs rather than dissipates heat to open air on a hot day. It is not suitable to pour thin layers in this condition, and the maximum intermission should be limited.

The provision of U.S. Army Corps of Engineers does not define the actual value of allowable temperature difference but limits the placing temperature. It actually limits the temperature difference in a certain way. They do not define the relationship between the allowable temperature difference and the length of pouring block, but actually consider the influence of the length of pouring blocks. Since the requirements are related to the dam height, it means that they make regulation for the length of the concrete block.

22.6.4 Temperature Control Requirements of Concrete Dam of Russia

There is no universal regulation for allowable temperature difference of concrete dams in Russia, it depends on the practical condition of the project and is determined after calculation. On this point, it is different from China and the United States. According to the data collected from some practical dams, the allowable temperature differences they used are shown in Table 22.6. From this table, we can see the allowable temperature difference is related to the grade of concrete. It is reasonable. In the design specifications of Chinese gravity dam and arch dam, the allowable temperature differences are not related to the grade of concrete, but are implied in the relationship between concrete extensibility and its grade.

22.7 Practical Examples for Temperature Control of Concrete Dams

22.7.1 Laxiwa Arch Dam

Laxiwa arch dam locates at the cold northwest area of China. The maximum height of the arch dam is 250 m, the bottom width of crown section is 49.0 m, and the maximum

Table 22.6 Allowable Temperature Difference of Russian Concrete Dam

Concrete Grade	Length of Concrete Block L (m)	Allowable Maximum Concrete Temperature T_{max} (°C)	Allowable Temperature Difference Above Foundation ΔT (°C)	Allowable Temperature Difference between Lateral Surface and Center $T_Я - T_{Б,Г}$ (°C)	Allowable Temperature Difference Between Horizontal Top Surface and Center $T_Я - T_{Г,Г}$ (°C)
150	10	38	27–29	25	12
	15	38	22–24	24	11
	20	38	18.5–20.5	23	11
	25	38	16–18	22	10
	30	38	14–16	21	10
200	10	40	29–31	26	14
	15	40	23–25	25	13
	20	40	19.5–21.5	24	13
	25	40	17–19	23	12
	30	40	15–17	22	12
250	10	42	30–32	27	16
	15	42	24.5–26.5	26	15
	20	42	21–22.5	25	15
	25	42	18–20	24	14
	30	42	16–17.5	23	14
300	10	44	32–34	28	18
	15	44	26–28	27	17
	20	44	22–24	26	17
	25	44	19–21	25	16
	30	44	17–19	24	16

width of arch abutment is 55 m. The average local annual air temperature is 7.3°C, the average temperature for July is 18.3°C, for November is 0.1°C, for December is −5.0°C, for January is −6.4°C, for February is −2.4°C, for March is 3.7°C. The concrete is characterized by low water–cement ratio, low water consumption, mixing with fly ash, mixing with water reducing agent, and mixing with air entraining agent. The construction process continues in the whole year. In summer, air-cooled aggregate, cooling water, mixing with ice are used to control the temperature of the concrete at the exit of mixer below 7°C. The vehicle for concrete horizontal transportation is equipped with awning, sponges are set around for thermal insulation and rigorously control the use of cable crane, in order to control the temperature rise below 1°C, control the concrete placing temperature lower than 12°C. The thickness of concrete lift is 1.5 m in the strong restrained region and 3.0 m in the weak restrained region. In the month from May to September, placing of concrete should be avoided in period of high temperature. The time from the concrete leaving the exit of mixer to it being covered by new concrete on the surface of block is controlled to 150 min (June to August) and 180 min (May and September). Fog spray operation should be adopted when pouring concrete in high temperature period and the new concrete is covered with quilt (with 3-cm-thick polyethylene foam plastics roll inside canvas).

Pipe cooling is divided into three stages. Water should be immediately filled into cooling pipe when it is covered by concrete. From May to September, 4–6°C cooling water is used, in other seasons, natural river water is used. Cooling time in first stage is 15–20 d, and the temperature drop rate is controlled below 1°C/d. Natural river water is used in second stage, and its main purpose is to reduce the exterior–interior temperature difference. 4–6°C cooling water is used in final stage to reduce the dam temperature to steady level.

The time interval between two lifts of concrete is 6–8 d from April to October and 5–7 d from November to March, it should not exceed 14 d. The height difference between adjacent dam blocks is not >12 m, the height difference between the highest and lowest parts of the dam should not exceed 30 m.

The strengthened heat storage method is used in winter construction. The exiting temperature of concrete is 12–15°C. In the neighborhood of 3–4 m around formwork, temporary warming shed was set inside which the air temperature is increased by fan heater. The temperature around formwork could be increased to 8°C. For the central place of the pouring area, electric blanket is used. The temperature of concrete surface could be increased to 11°C. The foamed polystyrene boards 5 cm thick were used on the upstream and downstream surfaces for permanent thermal insulation. Two layers of foamed polystyrene soft plate with the thickness of 2 cm are used for surface insulation of the horizontal top surface of concrete block in the construction period.

22.7.2 Toktogulskaya Gravity Dam

The maximum height of Toktogulskaya gravity dam is 215 m, the maximum bottom width is 170 m, the concrete volume is 324×10^4 m^3. It was built from 1969 to 1977. The local annual average air temperature is 8.4°C, the average temperature for January is −14.4°C, and the average temperature for July is 24.4°C. The dam is divided into columnar blocks. In the original design, the spacing of transverse joint and longitudinal joint is 16 m and 15 m. During the construction, the "Toktogulskaya construction method" was applied. The spacing of transverse joint was changed to 32 m. On upstream side, a cutting joint is set. The spacing of longitudinal joint is increased to 30–60 m.

1. Auto rising tent

   In 1950s and 1960s, Russian built lots of concrete dam, such as Brazk, Crasnoyarsk in the Siberia area. Because of the cold weather, the crack problem is serious. When building the Toktogulskaya gravity dam, the so-called Токтогульская construction method is used to solve the crack problem of concrete dam. The core idea of this method is to use auto rising tent to create an artificial climate in order to achieve heat insulation in winter and sun shading in summer. Concrete is placed inside the tent throughout the whole construction process. Figure 22.5 shows the auto rising tent used by this dam. The inside height of the tent is 7 m, which is determined by the height of dump truck used for concrete placing inside the tent. When tent rises, it takes 10–15 min for every 0.5 m height rise; electric heater or steam heater is used in the tent in winter. When the temperature outside tent is −25 to −30°C, the average temperature inside tent is +3 to +5°C, the minimum temperature is −3.5°C, the maximum is 12°C. Because the tent can achieve sun shading in summer, the temperature inside the tent is 1.5–2.0°C lower than the outside temperature, the maximum temperature difference may be up to 4–5°C. The air humidity inside the tent is 20–35% higher than the outside.
2. Surface water cooling on thin layer

   The second important thing for the Toktogulskaya construction method is thin layer placing and surface water cooling. After concrete is placed, cover it with polyethylene film immediately to prevent drying and excessive cooling. After concrete is setting, eliminate cement cream and start surface water cooling immediately. The operation of surface water

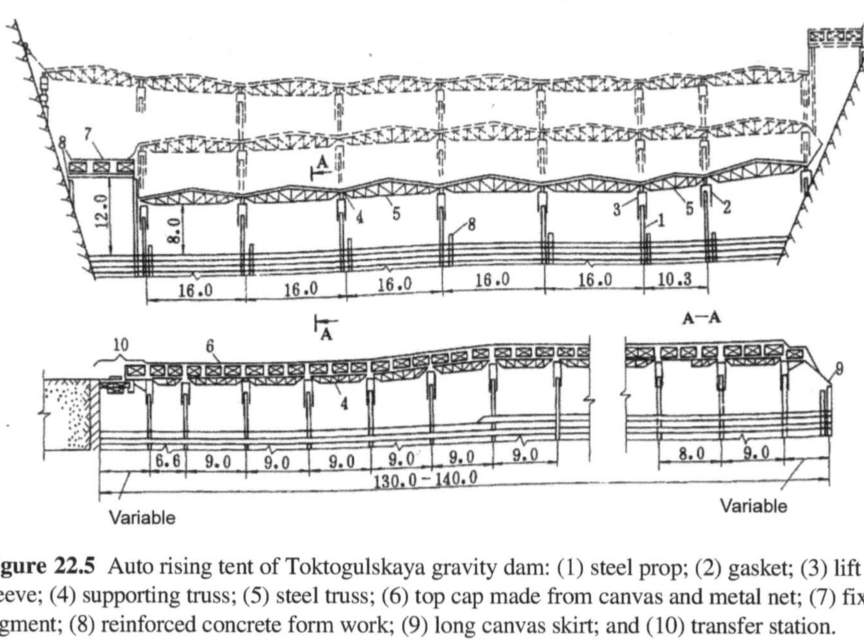

Figure 22.5 Auto rising tent of Toktogulskaya gravity dam: (1) steel prop; (2) gasket; (3) lift sleeve; (4) supporting truss; (5) steel truss; (6) top cap made from canvas and metal net; (7) fixed segment; (8) reinforced concrete form work; (9) long canvas skirt; and (10) transfer station.

cooling should not be later than 12 h after concrete pouring, and continue it until the concrete is covered with newly placed concrete. The surface water comes from the pored pipe, and the leaking water forms the flowing water layer on the concrete surface; when the air temperature is higher than 20°C, water should not be suspended. When the water velocity is not faster than 0.8 m/s, the water layer should be 2–8 mm thick; the flowing water temperature is not higher than 19°C; the water in the pipe should not exceed 18°C, and it comes from the sump of the well of drilling and drainage system. In the hottest month from July to August, the average water flow on every 1000 m^2 surface of concrete is 13.5 L/s. It will be 25% reduced in June and September. In April, May, and October, water flow lasts for 8 h each day and the average discharge of water on 1000 m^2 is 8 L/s. Practical observation shows that temperature of the flowing water generally rises for 1–3°C.

In order to improve the performance of the surface water cooling, the pouring layer should be as thin as possible. In the study of this method, the relationship between the size of pouring block, the strength of concrete and the concrete allowable temperature difference are calculated, the result is shown in Table 22.7. The thickness of pouring layer is decided to be 0.5 m, and enlarge the longitudinal joint spacing to 30 m. When the concrete tensile strength is 1.8–2.0 MPa, the allowable maximum concrete temperature is 26°C. The practical construction result is: in summer from June to August, the temperature in the tent is 22.9–24.0°C, the concrete temperature at the exit of mixer is 21.2–21.9°C, the placing temperature of concrete is 21.7–22.0°C, the maximum temperature inside concrete is 25.3–25.7°C, and temperature rise caused by heat of hydration is 3.8–4.1°C. The grouting temperature of this dam is 6–8°C, so the maximum foundation temperature difference is 17–19.5°C, 16–18°C in general. If the surface water flow is 18°C in the tent, the difference between the concrete maximum temperature and the surface water flow is 7.3–7.7°C, this value is not large. The maximum interior–exterior temperature difference appears on the surface of upstream and downstream of the dam.

3. Reinforced concrete form work

During the construction process, this dam used large amount of reinforced concrete form work, even in longitudinal joint. The form work is equipped with grouting pipeline, grout cell, etc.

Later, the thickness of pouring layer is increased to 1.0 m, and the height difference between adjacent blocks is limited to one layer (0.5–1.0 m). Cooling pipes are set inside the dam, and the spacing between pipes is 1.5 m. These pipes are used for first stage and second stage cooling, and the cooling water is the natural river water. Due to the use of the natural river water for cooling, the cooling capacity of the dam is not large, only 586 kJ/h (1,400,000 kcal/h).

Since thin layer pouring was applied, there are lots of horizontal construction joints. During the construction process, lots of experiments are conducted to research the impact of construction joints too the strength. The in-house test results show that the strength of the joint is 70–75% of the strength of the concrete; for the test of the concrete core, strength of joint is 40–50% of the strength of the whole concrete; shearing test shows that the adhesive strength is 1.0–1.2 MPa, friction coefficient *f* is 1.65–1.70.

4. Concrete pouring

Concrete is transported by dump truck from mixing plant to the bucket of the dumping trestle bridge outside the tent. It is further dumped by chute to the dump trucks inside the warm shed. Concrete pouring is conducted by dump trucks. Placing, vibrating, and cream cleaning are all mechanized. According to the available information, no cracks were discovered in this dam. Afterwards, this construction method was widely applied in Russia.

Table 22.7 Relationship Between the Maximum Concrete Allowable Temperature and the Block Size

Size of Pouring Block (m)	Interval Between Layers (d)	Predicted Maximum Temperature in Summer of 1969 (°C)		Allowable Maximum Temperature of Concrete for Different Tensile Strength (°C)			
		(1) March–July	(2) July–August	1.3 MPa	1.6 MPa	1.8 MPa	2.0 MPa
16 × 15 × 1.5	15	20–30	32	24	27	29	31
16 × 30 × 1.5	15	29–30	32	21	24	26	28
16 × 15 × 0.5	15	22–23	26	24	27	29	31
16 × 15 × 0.5	7.5	22–23	26	21	23	25	26
16 × 30 × 0.5	5	22–23	26	21	24	26	28
16 × 30 × 0.5	7.5	22–23	26	19	21	23	24
16 × 30 × 0.5*	5–7	21–23	25	19	22	24	26

(1) Cooled mixing water, 12°C river water flows on placing blocks, temperature of concrete at exit is 16–17°C.
(2) Cooled mixing water, 16–18°C river water flow on placing blocks, temperature of concrete at exit is 20–22°C.
*Within the range of 3 m height of strong restraint.

22.7.3 Dworshak Gravity Dam

Dworshak is a gravity dam with the maximum height of 219 m and maximum bottom width of 152 m, the concrete volume is 525 × 10^4 m^3, and it is located at the northwest part of America in state of Iowa. It is the highest gravity dam in the United States, and also the highest gravity dam without longitudinal joint in the world. It was built during the years from 1968 to 1972. U.S. Army Corps of Engineers were in charge of the design and construction. As the tradition of U.S. Army Corps of Engineers, the dam was constructed without longitudinal joint. The thickness of pouring layer is 1.5 m. The main temperature control measures are as follows:

1. Rigorously control the temperature of concrete not higher than 6.7°C and not lower than 4.4°C when transported to the form (the local annual average temperature is 10°C, the air temperature in summer is about 26.7°C). So despite the water mixing with ice, precooling for aggregate is also needed. The concrete mixing plant locates at the left abutment and equipped with 10 3 m^3 concrete mixer. Aggregate processing plant and cooling plant are set around the mixing plant. Aggregate is sprayed with cold water on the conveying belt in precooling room. The belt is 1.52 m wide with a speed of 0.33 m/s. The flow of water sprayed is 95 L/s. The aggregate stays in the precooling room for 3 min, and then coarse aggregate is dehydrated and rescreened before entering the sealed tanks of mixing plant. After that, aggregate is cooled by cold wind and the water is mixed with ice. The total power of cooling plant is 4300 × 10^4 kJ/h, it can produce 360 t ice every day and 11 m^3 1.7°C cold water every minute.
2. Within the range of $0.4H$ (H is dam height) above bed rock and the range of 6.1 m above the old concrete with age longer than 28 d, cooling water pipes are embedded with a spacing of 1.5 m × 1.5 m. The water temperature at entrance of pipe is 5.0 ± 1.1°C, the average water temperature is 10°C, and the flow is 15 L/min. The water flow direction is changed every 12 h. The biggest length of pipe loop is 365 m and every pipe loop is equipped with indicator showing whether water is flowing inside the pipe. The concrete cooling speed is controlled as follows: for the first 14 days, the temperature drop should not exceed 6.6°C and the temperature drop for any 4 days should not exceed 4.4°C, so as to prevent cracks caused by fast cooling. The cooling process generally lasts for 21 days. If the cooling speed exceeds the index above, cooling process should be suspended for several days. The 21-day cooling process should be done in the first 30 days. When the concrete temperature is lower than 16°C, cooling process is stopped.
3. The maximum thickness of concrete lift is 1.5 m. The height difference between adjacent blocks should not exceed 4.5 m in winter (from December to next February) and 6.0 m for the rest month (from March to November). The height difference between the highest and lowest blocks of the dam should not exceed 12 m so that the dam body can rise evenly.
4. Surface protection: The required surface conductance of top and sides of concrete block in spring (1st March to 15th April) and autumn (15th September to 15th October) is $\beta \leq 10.1$ kJ/(m^2h °C), in winter (from 15th October to 1st March) is $\beta \leq 5.0$ kJ/(m^2h °C). Sometimes the surface thermal insulation board needs to be uncovered for the surface cleaning and other construction activity. In winter, the surface can be exposed for less than two 12 h, and the interval between them is at least 24 h, only when the temperature is higher than 7.2°C.

Various temperature control measures specified in design are implemented in the construction process. Measured concrete temperature: when air temperature is

23°C, the concrete forming temperature was 5.6°C, the maximum concrete temperature was 22°C after 21 days and 18°C after completion of concrete cooling process. The effect of superficial insulation: $\beta = 5.07$ kJ/(m^{2}h °C), initial temperature of concrete was 11.7°C, the initial air temperature was 2.2°C, after 21 days, the air temperature dropped 21°C, the surface temperature of concrete dropped 7.3°C.

There was not severe structural crack during the construction process of Dworshak gravity dam. It was considered to be a good example in temperature control of concrete dam, but after operation for several years, severe vertical cracks emerged at the upstream face of the dam body. For details, please refer to Section 13.5. It demonstrates that the protection at the upstream face of this dam was not effective enough.

22.8 Cooling Capacity

Cooling capacity is related to the local temperature, the construction process, the temperature control method, and the allowable temperature difference. Careful calculation–analysis, and comparison proposals need to be done before reaching a reasonable decision.

22.8.1 Calculation for the Total Cooling Capacity

The total cooling capacity is decided by the peak cooling load, including four parts which are the concrete precooling, and first stage, second stage, and late stage pipe cooling:

$$Q = Q_1 + Q_2 + Q_3 + Q_4 = k\rho C\left[S(T_a + T_s - T_0) + \frac{V_1 \Delta T_1}{\tau_1} + \frac{V_2 \Delta T_2}{\tau_2} + \frac{V_3 \Delta T_3}{\tau_3}\right] \tag{22.14}$$

where

Q—cooling capacity, kJ/h;
Q_1, Q_2, Q_3, Q_4—cooling capacity required during concrete precooling, first, second, and late stages of pipe cooling;
S—current concrete placing intensity, m^3/h;
T_a—current daily average temperature,°C;
T_s—influence of the sunlight to the concrete raw material, generally about 3–5°C, which is related to material storage condition, construction season, and some other factors;
T_0—required concrete temperature at the exit of mixer;
V_1—concrete volume of first stage pipe cooling, m^3;
ΔT_1—average temperature drop during first stage pipe cooling, which is calculated according to the spacing and length of the pipes and the water flow speed;
τ_1—average time spent on first stage pipe cooling, h;
V_2—concrete volume of second stage pipe cooling, m^3;

ΔT_2—average temperature drop during second stage pipe cooling, °C;
τ_2—average time spent on second pipe cooling, determined by calculation;
V_3—concrete volume of late stage pipe cooling, m^3;
ΔT_3—average temperature drop during late stage pipe cooling, °C;
τ_3—average time spent on late stage pipe cooling;
ρ—unit weight of concrete, kg/m^3;
c—concrete specific heat, kJ/(kg °C);
k—reserve coefficient, can be chosen from 1.1 to 1.3 considering the cooling lost during the construction.

The temperature drop of first sage pipe cooling can be calculated as follows:

$$\Delta T_1 = T_1 - T_2 \tag{22.15}$$

where T_1—temperature of the concrete lift after natural cooling at time τ_1 (no pipe cooling) and T_2—concrete temperature after combined action of natural cooling and pipe cooling at time τ_1.

T_1 and T_2 can all be calculated by one-dimensional finite difference method, the formula to calculate T_1 is

$$T_{i,\tau+\Delta\tau} = T_{i,\tau}(1-2r) + r(T_{i-1,\tau} + T_{i+1,\tau}) + \Delta\theta \tag{22.16}$$

and the formula to calculate T_2 is

$$T_{i,\tau+\Delta\tau} = T_{i,\tau}(1-2r) + r(T_{i-1,\tau} + T_{i+1,\tau}) + (T_0 - T_w)\Delta\phi + \theta_0\Delta\psi \tag{22.17}$$

where $r = a\Delta\tau/\Delta x^2$, $\Delta\theta = \theta(\tau+\Delta\tau) - \theta(\tau)$, $\Delta\varphi = \varphi(\tau+\Delta\tau) - \varphi(\tau)$, $\Delta\psi = \psi(\tau+\Delta\tau) - \psi(\tau)$, $\varphi(\tau)$ refers to formula (17.70), $\psi(\tau)$ refers to formula (17.74), T_0 is the initial concrete temperature, and T_w is the cooling water temperature.

Temperature drop caused by second pipe cooling is calculated by the formula:

$$\Delta T_2 = T_3 - T_4 \tag{22.18}$$

where T_3, T_4– concrete temperature before and after second stage pipe cooling

Temperature drop caused by third stage pipe cooling is given by

$$\Delta T_3 = T_5 - T_f \tag{22.19}$$

where T_5—concrete temperature before third stage cooling and T_f—grouting temperature of the dam body.

Concrete at the exit of mixer should be determined according to the allowable temperature, for example, for concrete above rock foundation

$$T_0 = \Delta T_a + T_f - T_r - T_v \tag{22.20}$$

where

T_0—temperature of concrete at exit of mixer;
ΔT_a—allowable temperature difference of concrete above foundation;
T_r—temperature rise caused by heat of hydration;
T_v—temperature rise during the transportation and pouring process.

Calculation process: The first step is to conduct the temperature control calculation to determine the concrete temperature at exit of mixer, the placing temperature, the time for first stage, second stage, and late stage pipe cooling; then, according to the process of concrete placing and pipe cooling, calculate the cooling load per month using the formula (22.14). Finally, determine the cooling capacity according to the peak load.

Here is an approximate calculation. In general, cooling capacity for concrete placing in summer is large, and precooling is implemented with the first stage cooling at the same time. So cooling capacity of aggregate precooling and first stage pipe cooling can be calculated together by the following formula:

$$Q = k\rho cS(T_a + T_s + T_r - T_f - \Delta T_{al}) \tag{22.21}$$

where

Q—cooling capacity, 10^4 kJ/h;
ρ—concrete density, kg/m^3;
c—concrete specific heat, kJ/(kg °C);
S—concrete placing intensity, m^3/h;
T_a—air temperature;
T_s—temperature rise caused by sunlight, including the impact on raw material, concrete transportation and pouring process;
T_r—temperature rise caused by heat of hydration without pipe cooling;
T_f—steady temperature of dam body;
ΔT_{al}—allowable temperature difference.

Calculation example Assume $k=1.3$, $c=1.0$ kJ/(kg °C), $\rho=2450$ kg/m^3, $T_a=30$°C, $T_s=8$°C, $T_r=18$°C, $T_f=14$°C, $\Delta T=16$°C, cooling load of precooling and first stage cooling is calculated using the formula (22.20):

$$Q = 8.28S(\times 10^4\ \text{kJ}/h) = 23.0S(\text{kW}) \tag{22.22}$$

If the concrete pouring intensity is $S = 250$ m^3/h,

$$Q = 8.28 \times 250 = 270 \times 10^4\ \text{kJ}/h = 5750\ \text{kW}$$

According to the cooling capacity used by practical concrete dam, the empirical relation between maximum concrete pouring intensity S and cooling capacity Q is

$$Q = \beta_1 S \tag{22.23}$$

where β_1 is an empirical coefficient which is the required cooling capacity of unity concrete pouring capacity. From practical value of 16 concrete dams $\beta_1 = 4-14 \times 10^4$ kJ/h, the average value is $\beta_1 = 7.92 \times 10^4$ kJ/h, whereas in formula (22.22), $\beta_1 = 8.28 \times 10^4$ kJ/h, these two values are close.

We can also establish the empirical relationship between volume V of concrete dam and cooling capacity Q.

$$Q = \beta_2 V \tag{22.24}$$

where $\beta_2 = 5-14$ kJ/h, average value $\beta_2 = 7.86$ kJ/h per unit volume of concrete.

22.8.2 Cooling Load for Different Cases

1. Cooling load for producing cooling water can be calculated by the following formula:

$$Q_1 = k_1 S_1 C_w (T_j - T_c)(\mathrm{kJ}/h) = k_1 S_1 C_w (T_j - T_c)/3600(\mathrm{kW}) \tag{22.25}$$

   where
   Q_1—load for producing cooling water (kJ/h) or (kW), 1 kW = 3600 kJ/h;
   k_1—reserve coefficient, $k_1 = 1.1-1.2$;
   S_1—production capacity of cooling water (kg/h);
   T_j—water temperature entering cooling machine (°C);
   T_c—water temperature exiting cooling machine (°C);
   C_w—water specific heat [kJ/(kg °C)].

2. Cooling load for producing cooling ice can be calculated using the following formula:

$$\begin{aligned} Q_2 &= k_2 S_2 [C_w T_w + C_b(0 - T_b) + 335](\mathrm{kJ}/h) \\ &= k_2 S_2 [C_w T_w + C_b(0 - T_b) + 335]/3600(\mathrm{kW}) \end{aligned} \tag{22.26}$$

   where
   Q_2—load of producing cooling ice (kJ/h or kW);
   k_2—reserve coefficient, $k_2 = 1.20-1.25$;
   S_2—production capacity of cooling ice (kg/h);
   T_w—water temperature for ice production (°C);
   T_b—ice temperature (°C);
   C_w, C_b—specific heat of ice and water, kg/(kg °C);
   335—latent heat for ice melting (kJ/kg).

Cooling load of air-cooled aggregate can be calculated by the following formula:

$$\begin{aligned} Q_3 &= \sum k_i S_i C_g \Delta T_i(\mathrm{kJ}/h) \\ &= \sum k_i S_i C_g \Delta T_i/3600(\mathrm{kW}) \end{aligned} \tag{22.27}$$

where

Q_3—cooling load of air-cooled aggregate (kJ/h or kW);
k_i—experience coefficient, about 1.50;
S_i—production capacity of i-th aggregate (kg/h);
C_g—aggregate specific heat (kJ/kg °C);
ΔT_i—temperature drop of i-th aggregate (°C).

22.9 Inspection and Classification of Concrete Cracks

22.9.1 Inspection of Concrete Cracks

With current level of temperature control, cracks in concrete dam can be prevented. Projects, such as the third stage of Three Gorges gravity dam and the Sanjianghe arch dam, have all been built without cracks, but it needs careful design and careful construction. Since practical conditions are complicated and changeable, actually some

Table 22.8 Classification of Crack in Mass Concrete Structure

Cracks Classification	Specification	Classification Criterion	
		Crack Width	Crack Depth
A-type cracks	Micro cracks	$\delta < 0.2$ mm	$h \leq 300$ mm
B-type cracks	Surface cracks or shallow cracks	0.2 mm $\leq \delta <$ 0.3 mm	300 mm $< h \leq$ 1000 mm
C-type cracks	Deep cracks	0.3 mm $\leq \delta <$ 0.5 mm	1000 mm $< h \leq$ 5000 mm
D-type cracks	Through cracks	$\delta \geq 0.5$ mm	$h > 5000$ mm

cracks may appear in some practical projects. Frequent inspections are needed in construction process in order to find and treat the cracks immediately, so that future big cracks or hidden cracks which may endanger the safety of structure can be avoided.

Cracks in mass concrete are mostly surface cracks. Deep cracks and through cracks are mostly developed from surface cracks. To inspect whether there are cracks or not, inspection for the concrete surface is the prime thing to be done. Cracks with a width bigger than 0.05 mm can be seen by naked eyes; to find out more thin cracks, concrete surface can be firstly wet with water and dried by wind. Since cracks wet with water are slower to be dried, it is easy to find thin cracks in this way. For cracks at high place, telescope can be used to do the inspection. The width of cracks can be measured by reading scale or thickness gauge. For important cracks, measuring apparatus should be setup to monitor the possible changes of these cracks with time.

For the depth inspection of cracks, following methods can be used:

1. Dig groove to check the depth of cracks, dig to the depth of no cracks are visible.
2. Ultrasonic detection (the maximum depth is about 0.8 m).
3. Surface wave detection.
4. Pouring water or air in drill hole to check the deep cracks.
5. Drill hole TV and photo. Put the probe into the drill holes with diameter larger than ϕ150 mm. The situation in drill hole can be displayed on TV screen or be shot.

22.9.2 Classification of Cracks in Mass Concrete

According to the width and depth, cracks in mass concrete can be classified as shown in Table 22.8:

22.10 Treatment of Concrete Cracks

22.10.1 Harm of Cracks

1. Overall failure of structure and deterioration of stress state

   Cracks may cause the overall failure of structure, deterioration of stress state, and decrease of the structure safety. The deterioration of stress state can be classified into two types. One is simple and visualized deterioration. Taking Norfork gravity dam for example, under the action of water load and dead weight, tensile stress should not exist at the dam heel, as shown in Figure 12.18(a). After through cracks occur inside dam body, tensile stress will be developed at the dam heel, as shown in Figure 12.3(a).

Another type of stress deterioration is more complicated. Take Kolnbrein arch dam for example, its dam body is thin and overhang degree is large. Under the action of dead weight and thermal stress, large horizontal crack occurs at the downstream side of cantilever during the construction process, and the section area was reduced for 40%. After reservoir impoundment, shearing stresses are concentrated at the upstream part of the uncracked section, which causes large principal tensile stress. As a result, two large inclined cracks occur at the dam heel.

2. Reduce the durability of structure

   If through cracks occur at the water retaining structure, leakage may happen. Normally the water leakage is limited, but the problem is water leakage may cause corrosion which directly reduces the durability of the structure. Deep cracks at the surface may also influence the structure durability.

22.10.2 Environmental Condition of Cracks

Environmental condition of mass concrete cracks can be classified into following three types:

Class I: Indoor or outdoor environment;

Class II: Upstream surface, water level variation zone and erosive groundwater environment;

Class III: Overflow surface, zone with sea water and salt mist action.

22.10.3 Principle of Crack Treatment

The purpose of crack treatment: (i) prevent leakage, (ii) restore integrity of structure, and (iii) prevent cracks from further development.

It is better to remove shallow surface cracks at the horizontal layer immediately during the construction. For deep cracks which are difficult to be removed, it is better to put crack resistant reinforcement above the fully cooled old concrete, before pouring of new concrete. Of course, it may largely influence the construction process.

In massive concrete structure, cracks are not static but dynamic. Small cracks today may develop into large cracks in the future. Whether cracks will develop depends on the stress condition. If the crack ends locate at the tensile region or the upstream side, after reservoir impoundment, cracks may develop because of the splitting action of the water pressure. So to cracks at the upstream side or strong restraint region should be paid more attention. Even though the cracks are A type or B type, they must be handled.

Principle of treatment of mass concrete cracks: (i) A-type and B-type cracks at class I condition can be not handled. A-type and B-type cracks at class II and class III conditions should be handled. (ii) C-type and D-type cracks at different environment conditions should all be handled.

22.10.4 Method of Crack Treatment

There are many methods for cracks treatment. The method should be chosen according to the depth, location, future stress condition, and the current construction condition.

1. Crack removal

   During the construction process, if shallow cracks are found on the horizontal layers, they can be removed by the air pick, air drill, or manual method. The gouging section should be a trapezoid with wide top side and narrow bottom side. The bottom should be wide enough to avoid new stress concentration. If the new concrete is poured after the cracks have been completely removed, further cracks may not occur at this place. Even though the crack removal is incomplete, such as the actual depth of crack is 30 cm, due to some reasons only 25 cm is removed and then poured with new concrete. The possibility that the 5 cm crack may develop in the future is relatively small comparing with the 30 cm deep crack. So when dealing with cracks at horizontal layers, removing is better than the crack resistant reinforcement.
2. Crack resistant reinforcement

   For deep cracks and through cracks occurred in construction process, it is hard to remove them by dig out method. Generally after the concrete is fully cooled down, one to two layers of crack resistant reinforcement should be set above the cracks before pouring of new concrete. Generally, the size of reinforcement is $\phi 25-32$ mm, spacing interval is 20 cm, length is 3–4 m. The reinforcements should not contain hooks and they should be staggered by length, to prevent cracks from emerging at the ends of the reinforcements. When cracks occur at the sides of the lower layer concrete, related part of the upper layer concrete should be set with crack resistant reinforcements. When the reinforcements are fully applied with stress, for example, the stress of reinforcement reaches 100 MPa, the tensile strain is 5×10^{-4}. At this point, since the extensibility of concrete is $\leq 1 \times 10^{-4}$ cracks may occur because concrete tries to maintain synchronous deformation with the steel reinforcement. Therefore, reinforcements cannot prevent concrete cracks, they can only limit the opening of cracks. In massive concrete structure, since the concrete section is large and lack of reinforcement, actually the crack resistant effect is limited. To prevent crack from developing upward, the prime treatment is to pour new concrete after old concrete is fully cooled and cracks are fully open, setting with reinforcement is just a supplementary measure. If lower layer concrete is not fully cooled and poured with new concrete, as a result, cracks may cross or bypass the reinforcement and develop upward. These cases are common in practical project.
3. Cement grouting

   For severe cracks, grouting should be done after the temperature of dam body has reduced to steady temperature. When the crack width is larger than 0.5 mm cement grouting can be done, otherwise chemical grouting should be done.
4. Chemical grouting

   When crack width is smaller than 0.5 mm, chemical grouting should be done. Generally epoxyresin is used for chemical grouting which can treat wet cracks.
5. Antiseepage measures for cracks at upstream surface

   Cracks at upstream side may easily develop to large cracks after reservoir impoundment, thereafter they are difficult to be treated. So, careful inspection is required before impoundment. All cracks including shallow surface cracks should be carefully treated to prevent the development of them after impoundment.

6. Drain hole

   For severe cracks at the upstream side, drainage holes must be drilled across the cracks to decrease the water pressure inside cracks. After severe upstream vertical cracks occurred at the Dworshak dam in the United States, epoxyresin is used to seal the crack at the upstream face, and then drill drainage holes to cross the cracks, the spacing between drain holes is 1.5 m. This treatment largely decreased the water leakage and the cracks become stable and never developed again.

7. Prestress anchoring

   For severe cracks at the upstream side, prestress anchoring can be used if the performance of all other methods is poor. Upstream vertical cracks at Zhaxi diamond-head buttress dam are reinforced by this method.

8. Gallery

   If a gallery is set at the top of a severe crack, it can help to reduce stress concentration and prevent the crack from developing upward.

23 Key Principles for Temperature Control of Mass Concrete

23.1 Selection of the Form of Structure

The mission of structural design is to obtain a rational structural form that can withstand the design load and satisfy the requirements of safety, durability, and economy. Three elements are included in the structural form: (i) reinforced or not, (ii) shape of the structure, and (iii) size of the structure.

The safety of structure depends on three factors: compressive strength, tensile strength, and shear strength, among which tensile strength is the main consideration for temperature control and crack prevention.

The tensile strength of concrete is very low, only about 8% of its compressive strength. Therefore, how to solve the problem of low tensile strength is the key point of concrete structural design.

Even though the thin-walled structure may cool in a short time and the temperature rise caused by the hydration heat of cement during the construction is not significant, the thin-walled structures are generally reinforced because the tensile stress caused by ambient temperature fluctuation and external load might be high. Tensile stress is borne by the steel and the concrete bears compressive stress only.

The tensile stress within the massive concrete structure caused by hydration heat of cement and the ambient temperature fluctuation might be significant, and it would use large amounts of steel if the tensile stress is borne by the steel because of the thick section. Therefore, in massive concrete structures, such as concrete dams, the tensile stress problem is usually solved by temperature control measures, which include the optimization of raw material, dividing the dam into blocks, precooling of concrete, pipe cooling, and surface thermal insulation instead of being reinforced.

Generally, massive concrete structures are known as the structures with the thickness above 0.8−1.0 m; in the author's opinion, this concept about massive concrete structures is too general and it should be further divided into massive hollow structures and massive solid structures, the former as the buttress dam, the latter as the gravity and arch dams.

It is possible to prevent cracks in massive concrete structures with thickness above 3 m, such as gravity dam and arch dam, by temperature control measures. It could also prevent cracks during the operating period if there are permanent insulating boards, otherwise there would be some surface cracks, among which the

Thermal Stresses and Temperature Control of Mass Concrete. DOI: http://dx.doi.org/10.1016/B978-0-12-407723-2.00023-3

upstream ones might develop into large cracks and do harm to the structure but the downstream ones are generally innocuous to the structure safety because of the large size of the structure itself.

There would be a large tensile stress caused by ambient temperature fluctuation within large hollow concrete structures, such as the buttress dam; many cracks may appear and some of which might be through cracks due to the thin section. Even though the dissipation of heat is quick in the hollow structure, it is hard to prevent cracks because the structure is sensitive to the ambient temperature fluctuation. Certainly, it is possible to prevent cracks if we make permanent insulation on the entire exposed surface, but it would increase the cost and might lose some original advantages of the hollow structure.

In conclusion, solid structure is more favorable in temperature control and crack prevention.

23.2 Optimization of Concrete Material

The purpose of choice of the concrete raw material and the optimization of concrete mix is to obtain larger crack resistance, which means the larger tensile strength and extensibility, smaller adiabatic temperature rise, modulus of elasticity and linear expansion coefficient, and larger semi-mature age of concrete.

The linear expansion coefficient of concrete mainly depends on aggregate varieties, as given in Tables 2.15 and 2.16, the linear expansion coefficients of different aggregate from small to large are limestone, basalt, granite, sandstone, and quartzite. Therefore, if there are different choices of aggregate in the construction site, more attention should be paid to this factor, and the aggregate should be selected after testing.

The tensile strength, extensibility, adiabatic temperature rise, and modulus of elasticity of concrete are all closely related to the water–cement ratio and compressive strength of concrete. There is a contradiction between high strength and low heat and small elastic modulus, and the contradiction would be relieved by adding appropriate amounts of fly ash, slag, and water-reducing agent; therefore, during the testing period, we should research all feasible measures to relieve this contradiction and select the primary material and mix design after a comprehensive consideration of strength, crack resistance, durability and workability of concrete.

23.3 Calculation of Crack Resistance of Concrete

Calculate the allowable temperature difference and surface insulation ability according to the following formulas:

$$\sigma \leq [\sigma] = \frac{E\varepsilon_{\mathrm{p}}}{K_1} \tag{23.1}$$

$$\sigma \leq [\sigma] = \frac{R_t}{K_2} \tag{23.2}$$

$$\sigma_y \leq [\sigma_y] = \frac{rR_t}{K_2} \tag{23.3}$$

where, σ is the maximum tensile stress, σ_y is the vertical tensile stress, r is the tensile strength reduction coefficient of the horizontal construction joint, ε_p is the tensile extensibility, R_t is the tensile strength.

Recommended safety factors are as follows:

$$\left.\begin{aligned} K_1 &= 1.6 - 2.2 \\ K_2 &= 1.4 - 1.9 \end{aligned}\right\} \tag{23.4}$$

23.4 Control of Temperature Difference of Mass Concrete

23.4.1 Temperature Difference Above Dam Foundation and Temperature Difference Between Upper and Lower Parts of Dam Block

1. The allowable temperature differences above rock foundation adopted by the gravity dam and arch dam design code of China are given in Table 22.3.

   The temperature difference between upper part and lower part of dam block is 15–20°C.
2. Suggestion of the author

   If the modulus of elasticity of concrete is equal to that of the bedrock, the restraint of old concrete would be smaller than that of bedrock; actually, the modulus of deformation of bedrock is smaller than concrete, so the restraint of old concrete to upper concrete might not be lower than that of the bedrock. The stress caused by temperature difference between upper part and lower part is also closely related to the length of concrete blocks. The author suggests that the allowable temperature difference above foundation takes the same value with the allowable temperature difference between upper part and lower part, as given in Table 23.1.

Table 23.1 The Allowable Temperature Difference ΔT Above Dam Foundation or Between Upper and Lower Parts of Dam

Height of New Concrete	**Long Edge of the Block L (m)**				
	<16 m	**17–20 m**	**21–30 m**	**31–40 m**	**>40 m**
0–0.1L	26–25	24–22	22–19	19–16	16–14
0.1–0.4L	33–31	31–28	28–26	24–20	20–18

23.4.2 Surface–Interior Temperature Difference

Most cracks in mass concrete are surface cracks but some of them may become large and deep cracks later on. Superficial thermal insulation is the most efficient measure for preventing surface cracks. Equations (11.11) and (11.23) may be used to determine the thickness of insulation layer. Simulation computation of thermal stress must be made for important massive concrete structures.

23.4.3 Maximum Temperature of Concrete

The maximum temperature of concrete determines the temperature difference above foundation, the temperature difference between upper and lower parts, and the surface–interior temperature difference.

In the past, there were several projects which set very low-allowable maximum temperature of concrete to solve the problem of surface–interior temperature difference, which caused some difficulties in construction and temperature control. The author thinks that this opinion is worth discussing. Three parts are included in the surface–interior temperature difference: temperature variations of air (annual variation and cold wave), temperature rise caused by hydration heat of cement, and the initial temperature difference. Since the annual change of air temperature is about 10–20°C and the temperature drop during the cold wave could be 10–20°C, even 20–40°C if in winter, it could reduce the maximum temperature by precooling the aggregate and pipe cooling, but it is difficult to gain significant reduction. Therefore, the reasonable method is setting the maximum temperature of concrete according to the allowable temperature difference above foundation and the temperature difference between upper and lower parts; the control of maximum temperature mainly depends on the heat dissipation of lift surface, pipe cooling and precooling of concrete; and control the surface–interior temperature difference mainly by the superficial thermal insulation, not by the reduction of maximum temperature.

23.5 Analysis of Thermal Stress of Mass Concrete

The analysis of thermal stress is divided into four kinds: (i) estimation, (ii) primary calculation, (iii) detailed calculation, and (iv) simulation calculation. We can choose one of them according to the importance of projects and the stage of design.

23.5.1 Estimation of Thermal Stress

The maximum temperature of concrete T_m could be calculated as Eq. (4.72).

The temperature difference above foundation is

$$\Delta T = T_m - T_f \tag{23.5}$$

where T_f is the steady temperature of dam; it could be estimated by the following formula:

$$T_f = \frac{T_W + T_d + \Delta T}{2} \tag{23.6}$$

where

T_W—the mean annual water temperature of upstream face
T_d—the mean annual water temperature or air temperature of downstream face
ΔT—the temperature rise caused by sunshine.

The first kind of temperature difference between the upper part and lower part is:

$$\Delta T_{u1} = T_{mu} - T_d \tag{23.7}$$

where

T_{mu}—the maximum mean temperature within the range of the upper L/4 of the block
T_d—the actual mean temperature of the lower concrete when placing new concrete.

The second kind of temperature difference between the upper part and lower part is:

$$\Delta T_{u2} = T_{mu} - T_{fd} \tag{23.8}$$

where T_{fd} is the steady temperature of the lower concrete before the grouting of joints.

The three kinds of above-mentioned temperature difference should not exceed the allowable temperature difference.

Calculate the necessary surface thermal insulation measures using the method mentioned in Chapter 10.

23.5.2 Primary Calculation of the Temperature Stress

Calculate the temperature difference above foundation and the temperature difference between upper part and lower part by the one-dimensional finite difference method, see Eq. (2.75); calculate the thermal stress of dam block by influence lines; calculate the cold wave and overwinter thermal stress by the method mentioned in Chapter 10.

23.5.3 Detailed Calculation of Thermal Stress

Calculate the thermal stress of the dam during the construction and operation period using the two-dimensional finite element method (FEM) and three-dimensional FEM (3D FEM).

23.5.4 Whole Process Simulation Calculation

Simulate the whole process of construction and whole dam by the 3D FEM; the influence of construction process, the influence of joint and the operating condition should be considered in the computation.

23.6 Dividing the Dam into Blocks

Concrete dams are very large structures; generally they are divided into blocks by joints to make the construction more convenient and reduce the thermal stress. The joints are grouted after the dam is cooled.

Transverse joints are perpendicular to dam axis. The spacing between them is about 15–20 m, which might be different in a same dam as hydraulic arrangement required. In the RCC dams, there is generally no longitudinal joint, but the transverse joints are still set by a vibrating grooving machine.

Longitudinal joints are parallel to dam axis, the spacing between them is about 15–40 m. With the development of dam construction technology, nowadays, there is generally no longitudinal joint in both gravity and arch dams.

The thickness of the concrete lift depends on three factors: (i) the effectiveness of superficial heat dissipation, (ii) the rising speed of the dam, and (iii) the convenience of construction.

For the multiple-arch dam and slab deck buttress dam, the thickness is less than 2 m, the dissipation of heat from the lateral surface is straightforward during construction, so the thickness of concrete lift usually takes about 5 m. For the diamond-head single buttress dam, which has a head of large cross section, the dissipation of heat from the lateral surface is difficult, so the concrete lift should not be too thick.

In the gravity dam and arch dam, which have a thick dam body, dissipation of heat is mainly from the horizontal construction joints, the effectiveness of heat dissipation is the determinant factor of the thickness of concrete lift.

During summer construction, if the concrete is precooled, the placing temperature T_p would be lower than the air temperature T_a, which causes an initial temperature difference $T_0 = T_a - T_p$; if $T_a > T_p$, there will be some heat absorbed in the first few days after placing—the thinner the lift is, the more the heat will be absorbed. The concrete temperature will rise due to the hydration heat of cement, and the heat in the interior will not dissipate until the internal temperature is higher than the outside. From this view, the thicker the better. According to the author's studies, when the $T_0/\theta_o = 0.80 - 0.85$, the total temperature rise is almost independent of the lift thickness; when the $T_0/\theta_o > 0.85$, the thicker the better, and when the $T_0/\theta_o < 0.80$, the thinner the better.

When the valley is wide, there are many dam blocks, the dam rising rate has little to do with the thickness of the lift, then the thin layer, short interval and even rising could be taken.

When the valley is narrow, there are fewer dam blocks, the thickness of the lift would be a constraint factor of dam rising rate.

During the dam construction, temperature control standards must be adhered to and could not be easily changed, but the thickness of the lift is changeable. Generally, 1.5 or 3.0 m is suitable; under special circumstances, thicker lift is adoptable, but the temperature control and management must be enhanced.

23.7 Temperature Control of Gravity Dam

Nowadays, the slotted gravity dam, buttress dam, and conventional gravity dam are basically not adopted, and the RCC gravity dam is usually built in wide river valleys.

In the early application period of the RCC gravity dam, some people thought that there was no need to consider the temperature control and crack prevention because in RCC less cement is used. Both theoretical analysis and practical experience have shown that it is not realistic; temperature control is also necessary for RCC gravity dams.

The temperature control of the RCC gravity dam has the following characteristics:

1. The adiabatic temperature rise of the roller compacted concrete is lower because it is mixed with more fly ash and less cement. Even so, the temperature rise caused by hydration heat of cement is not too low because the large amount of mixed fly ash would postpone the dissipation of hydration heat and the rising rate of the RCC dam is high, with less heat dissipation through horizontal lift surface.
2. Because the content of cement is less, the creep of the RCC is smaller and the extensibility is lower, which means lower crack resistance.
3. In addition to the hydration heat, the high placing temperature, cold wave, and low temperature in winter are also important factors that cause cracks. The influence to the RCC dam is as much as to the conventional concrete dam.
4. Generally, the RCC dams are poured without longitudinal joints and pipe cooling. The dam body would be already completed when the dam temperature drops to steady temperature, so the horizontal stress caused by self-weight and water pressure would offset some tensile stress caused by temperature reduction. But the internal temperature of dam body drops slowly; the low temperature in winter and cold wave may induce a large temperature difference between the surface and the interior of dam, which might cause horizontal and vertical cracks and transverse cracks on the upstream surface.
5. Generally, concrete is placed in horizontal lifts in the river channel; even though the concrete in the central part of riverbed has been out of the foundation restraint zone, the concrete near the river banks is still subjected to strong restraint of bedrock. There would be a "thin lift with long interval" problem when overflowing in the flood season or during winter layoff in a cold region.
6. When precooling the concrete, ice cannot be added into the mixture because of low water consumption; therefore, the fresh RCC temperature is usually higher than the conventional concrete. More heat absorbed during the paving and rolling. Thus, the placing temperature of the RCC is 15–17°C in summer construction, higher than the conventional concrete, which is about 12°C.

7. Generally, there should be a 2–4 m conventional concrete cushion plate on the bedrock and an interval of about 2 months for the bedrock consolidation grouting after placing the cushion. This is a typical "thin lift with long interval," which is most likely to produce through cracks. Because of locating in the strong constraint zone of bedrock, these kind of cracks are easy to extend upward, which is hard to prevent even though the steel bars are put across the joints when placing new concrete.
8. As the bottom outlet usually locates above the foundation, the extra cooling of the bottom outlet might cause cracks.

23.8 Temperature Control of Arch Dam

Generally, the arch dam is set with transverse joints, which means it needs artificial cooling before joint grouting to reduce the dam body temperature to the steady temperature. Compared to natural cooling, the artificial cooling time is relatively short, so there might be large tensile stress caused by the foundation restraint and upper–lower temperature differences. It would be more serious if the dam body were higher and thicker.

It is necessary to reduce the maximum temperature of concrete by early pipe cooling and precooling to satisfy the allowable temperature difference. When the dam body is thick, there should be one to two stages of mid-term pipe cooling between early and later pipe cooling to disperse the temperature difference. The height of the pipe cooling zone should not be less than 0.4 times the length of the dam block during the late and mid periods of pipe cooling, and it should form a certain temperature gradient in vertical direction to reduce the tensile stress.

It is better to set permanent insulation board on the upstream and downstream surfaces of a high arch dam. The horizontal construction joints and the lateral face of pouring blocks should be insulated by the foam insulation quilt.

When late pipe cooling is conducted in the hot season, the surface insulation must be enhanced; otherwise, the concrete temperature near the surface would be difficult to drop to the steady temperature.

The temperature load of the arch dam during the operating period may be calculated by the method mentioned in Chapter 13.

23.9 Control of Placing Temperature of Mass Concrete

When concrete is poured in hot weather, in order to satisfy the allowable temperature difference above foundation, the following measures could be adopted to reduce the placing temperature of concrete:

1. Reduce the aggregate temperature by building arbors or putting it under the ground,
2. Use cold water or ice when mixing the concrete,
3. Precool the aggregate,
4. Shorten the concrete transporting and concrete spreading time,
5. Timely coverage.

23.10 Pipe cooling of Mass Concrete

Pipe cooling could efficiently reduce the concrete maximum temperature and take the dam temperature down to the target temperature in a relatively short time. However, pipe cooling also has negative aspects: (i) there would be a large local tensile stress around the pipe and (ii) because pipe cooling makes the dam temperature decrease sharply, the creep of concrete could not fully develop, so larger tensile stress would appear in concrete compared with natural cooling.

Generally, pipe cooling should adhere to the following three principles:

1. As a rule, the initial cooling must be carried out immediately after placing the concrete to reduce the concrete maximum temperature; in the initial cooling of the strong restraint zone, dense cooling pipes could be taken to reduce the concrete maximum temperature to a lower value, which would reduce the foundation temperature difference and the temperature difference between the upper part and lower part. The cost will not be large as the range of dense pipes is small, but the effectiveness of crack resistance is significant. Part of the dense pipes could be shut down in 2 or 3 days after concrete reaches the maximum temperature to avoid faster cooling, which would cause a large tensile stress.
2. The height of pipe cooling zone should not be less than 0.4 times the length of dam block during the late and mid period, and it should form a certain temperature gradient in the vertical direction to reduce the tensile stress.
3. Disperse the temperature difference, which means that there must be several stages of cooling with small temperature differences. Between early cooling and late cooling, there should be at least one or two mid-cooling stages. So the difference between initial concrete temperature and water temperature $T_0 - T_w$ would ideally not exceed 10°C.

23.11 Surface Thermal Insulation

Practical experience shows that, initially, most of the cracks in massive concrete structures are surface cracks, some of which might develop into deep cracks, even through cracks. The factors that cause tensile stress are cold wave, annual air temperature variations, hydration heat of concrete, initial temperature difference, and desiccation; to prevent surface cracks, curing and surface insulation should be enhanced [31,32,42,64,66,71,72,89].

The upstream and downstream surfaces of a concrete dam may be insulated by expansive polystyrene foam board (EPS board) by the inside-paste or outside-paste method: the inside-paste method fixes the EPS boards to the concrete foam, leaving the EPS boards on the surface after form removing; the outside-paste method pastes the EPS boards to the concrete surface after form stripping. The horizontal lift and the lateral surface could be protected by EPS insulating quilt. The permanent insulation could also use polyurethane spray foam, the thickness of which should be determined by calculation, and there should be an antioxidant layer outside the insulation. The small- and medium-sized projects could also use the straw bag and sand layer as insulation.

For important concrete arch dams, it is recommended to set permanent insulating boards on both upstream and downstream surfaces; for normal concrete gravity dams and arch dams, it is recommended to set permanent insulating boards in both upstream and downstream surfaces in the strong constraint zone of rock foundation.

23.12 Winter Construction

Winter construction means the daily temperature is below 5°C or the minimum temperature is below −3°C.

Both freeze protection and crack prevention should be taken into account during massive concrete construction in winter, so the following three principles should be followed:

1. The optimum placing temperature of concrete is 5−12°C, therefore, the temperature of fresh concrete at batch plant may be decided according to the local climate and construction method, while also considering the heat loss during the transportation and pouring.
2. The fresh concrete could not suffer freeze injury until it reaches the 50% design strength, or it would lose strength because of the damage of internal structure.
3. To prevent cracks, the temperature difference above foundation, temperature difference between upper part and lower part, and temperature difference between surface and interior should not exceed the specified value.

When building a concrete dam in a cold region, there should be a specific study about whether or not to continue construction in winter in the schematic design stage. Generally speaking, for the conventional concrete dam, winter construction is feasible, but for the RCC dam, winter construction is not appropriate.

23.13 Conclusion

The theory in this book and the practical experience in China have shown that it is absolutely possible to prevent the cracks of the massive concrete structure by well-designed research and careful construction.

Appendix: Unit Conversion

1.00 kcal = 4.1868 kJ
1.00 kW = 3600 kJ/h
1.00 kg = 9.806 N
1.00 kg/cm^2 = 0.09806 MPa
1.00 Pa = 1.00 N/m^2
1.00 MPa = 1.00 N/mm^2 = 10^6 Pa
1.00 in. = 2.54 cm
1.00 lb = 0.454 kg = 4.45 N
1.00 $lb/in.^2$ = 0.07037 kg/cm^2 = 0.00690 MPa
1.00 ft = 12 in. = 0.3048 m
1 Btu = 0.252 kcal = 1.055 kJ

References

Part I. Monographs

[1] H.S. Carslaw, J.C. Jaeger, Conduction of Heat in Solids, second ed., Oxford University Press, Oxford, 1986.

[2] A.B. Laekove, Theory of Heat Conduction (in Russian), National Technical and Theoretical Press, Moscow, 1952.

[3] U.S. Bureau of Reclamation, Cooling of Concrete Dams, Denver, 1949.

[4] U.S. Bureau of Reclamation, Thermal Properties of Concrete, Denver, 1940.

[5] N.H. Aruchunion, Some Problems in the Theory of Creep (in Russian), National Technical and Theoretical Press, Moscow, 1952.

[6] Cao Zesheng, Xu Jinghua, Technology for MgO Concrete Dam (in Chinese), China Electric Power Press, Beijing, 2003.

[7] Zhu Bofang, Wang Tongshen, Ding Baoying, Guo Zhizhang, Thermal Stresses and Temperature Control of Hydraulic Concrete Structures (in Chinese), Beijing, China Water Power Press, 1976.

[8] Zhu Bofang, Thermal Stresses and Temperature Control of Mass Coucrete (in Chinese), China Electric Power Press, Beijing, 1999.

[9] Zhu Bofang, Finite Element Method, Theory and Applications, first ed. 1979, second ed. 1998, third ed, China Water Resources and Hydropower Press, Beijing, 2009.

[10] Zhu Bofang, Selected Papers of Academician Zhu Bofang, China Electric Power Press, Beijing, 1977.

[11] Zhu Bofang, New Developments in the Theory and Technology of Concrete Dams, China Water Resources and Hydropower Press, Beijing, 2009.

[12] DL 5108-1999, Design Specification for Concrete Gravity Dam, China Power Press, Beijing, 2000.

[13] DL/T 5346-2006, Design Specification for Concrete Arch Dam, China Power Press, Beijing, 2007.

[14] DL/T 5057-1996, Design Specification for Hydraulic Concrete Structures, China Power Press, Beijing, 1997.

[15] SL319-2005, Design Specification for Concrete Gravtity Dam, Beijing, China Water Resources and Hydropower Press.

[16] SL282-2003, Design Specification for Concrete Arch Dam, China Water Resources and Hydropower Press, Beijing, 2003.

Part II. Scientific Papers

[17] T. Alfrey, Non-homogeneous stresses in visco-elastic bodies, Q. Appl. Math. 2 (2) (1944) 113–119.

[18] Z.P. Bazant, S.T. Wu, Thermoviscoelasticity of aging concrete, J. Eng. Mech. Div., Proc. ASCE 100 (1974) 575–597.

[19] A.B. Belov, Mathematical theory of shrinkage of concrete, J. Nat. Res. Inst. Hydraulic Eng. Soviet Union, 35 (in Russian).
[20] H.L. Boggs, Cracking in concrete dams, in: USBR case histories, 15th ICOLD, II, pp. 173–190.
[21] Ding Baoying, Hu Ping, Huang Shuping, Approximate analysis of water temperatare in reservoir, J. Hydroelectric Eng. 4 (1984) 17–33 (in Chinese).
[22] C.F. Gytman, Analysis of thermal stresses due to harmonic variation of temperature, J. Nat. Res. Inst. Hydraulic Eng. Soviet Union 47 (1952) (in Russian).
[23] D.L. Houghton, Measures being taken for prevention of cracks in mass conerete at Deworshak and Libby dams, in: 10th ICOLD, IV, 1970, pp. 241–271.
[24] Hu Ping, Yang Ping, Zhang Guoxin, Researches on the measures of temperature control and crack prevention of Laxiwa arch dam, Water Power 11 (2007) 51–54 (in Chinese).
[25] Yang Bo, Xu Ping, Zhang Guoxin, D. Fupin, Research on temperature control measures of Jiangkou arch dam, J. China Inst. Water Resour. Hydropower Res. 2 (2003) (in Chinese).
[26] Yue Yaozhen, Thermal stress analysis taking into consideration of impact on concrete properties resulted from temperature, Water Resour. Hydropower Eng. 1 (1993) 15–21 (in Chinese).
[27] Zhang Guoxin, Lui Youzhi, Ma Xiaofang, Discussion on cracks caused by low-temperature placing of concrete dam, Water Resour. Hydropower Eng. 7 (2010) 45–48 (in Chinese).
[28] Zhang Guoxin, Zhu Bofang, Simulating analysis of the arch dam as a whole, Water Resour. Hydropower Eng. 12 (2002) 22–25 (in Chinese).
[29] Zhang Guoxin, Cheng Xianming, Du Lihui, Expansion model of dynamics describing MgO concrete, Water Resour. Hydropower Eng. 9 (2004) 88–91 (in Chinese).
[30] Zhang Guoxin, L. Yisheng, Theoretical analysis and measures to prevent cracking of the faceslab of rock fill dams, J. Hydroelectric Eng. 3 (2005) 30–33 (in Chinese).
[31] Zhang Guoxin, Yang Weizhong, Lo Heng, Yang Bo, Application of MgO micro-expanding concrete for construction of Sanjiang arch dam, Water Resour. Hydropower Eng. 8 (2006) 20–23.
[32] Zhu Bofang, The effect of pipe cooling in mass concrete with internal source of heat, J. Hydraulic Eng. 4 (1957) 87–106and Sci. Sin., X (4) (1961) 483–489 (in Chinese).
[33] Zhu Bofang, Stresses and deformations in the nonhomogeneous visco-elastic media under mixed boundary conditions, J. Mech. (2) (1964) 162–167 (in Chinese).
[34] Zhu Bofang, Thermal stresses in docks and sluices on soft foundations, J. Hydraulic Eng. (6) (1980) 23–33 (in Chinese).
[35] Zhu Bofang, An implicit method for the stress analysis of concrete structures considering the effect of creep, J. Hydraulic Eng. (5) (1983) 40–46 (in Chinese).
[36] Zhu Bofang, On the temperature loading of arch dams, Water Power (2) (1984) 23–29 (in Chinese).
[37] Zhu Bofang, Computation of thermal stresses in mass concrete with consideration of creep, in: Proceedings of the 15th International Congress on Large Dams, Lausanne, vol. II, 1985, pp. 529–546.
[38] Zhu Bofang, Computation of thermal stresses in mass concrete due to cold wave, Water Power 3 (1985) 13–17 (in Chinese).
[39] Zhu Bofang, Modulus of elasticity, unit creep and coefficient of stress relaxation of concrete, J. Hydraulic Eng. (9) (1985) 54–61 (in Chinese).

[40] Zhu Bofang, Method of equivalent modulus for analyzing stresses in matured concrete due to harmonic variation of temperatures, J. Hydraulic Eng. (8) (1986) 61266 (in Chinese).
[41] Zhu Bofang, Dissipation of heat in two directions and the protection of the corner of mass concrete in cold wave, Water Power 8 (1986) 21−24 (in Chinese).
[42] Zhu Bofang, Design of superficial thermal insulation of mass concrete, J. Hydraulic Eng. (2) (1987) 18−26 (in Chinese).
[43] Zhu Bofang, Temperature loads on arch dame, in: J.L. Serafim, R.W. Clough (Eds.), Proceedings International Workshop on Arch Dams, Coimbre, 1987 or "Arch Dams" Balkema, 1990, pp. 217−225.
[44] Zhu Bofang, C. Jiangbo, Finite element analysis of pipe cooling in mass concrete, a three dimensional problem, J. Constr. Eng., ASCE 115 (4) (1989) 487−498.
[45] Zhu Bofang, Thermal stresses in beams on elastic foundation, J. Mech. 3 (1977) 200−205 (in Chinese).
[46] Zhu Bofang, Equivalent equation of heat conduction considering the effect of pipe cooling, J. Hydraulic Eng. 3 (1991) 28−34 (in Chinese).
[47] Zhu Bofang, Temperature control and design of joint in RCC arch dams, Water Power 9 (1992) 11−17 (in Chinese).
[48] Zhu Bofang, Compound layer method for stress analysis simulating construction process, Dam Eng. 6 (2) (1995) 157−178.
[49] Zhu Bofang, Xu Ping, Thermal stresses in roller compacted concrete gravity dams, Dam Eng. 6 (3) (1995) 199−220.
[50] Zhu Bofang, Cracks on the upstream face of concrete gravity dams, J. Hydroelectric Eng. 4 (1997) 86−94 (in Chinese).
[51] Zhu Bofang, Prediction of water temperature in deep reservoir, Dam Eng. 18 (1) (1997) 13−26.
[52] Zhu Bofang, Xu Ping, New methods for thermal stress analysis simulating construction process of concrete dam, in: Proceedings of 10th International Conference for Numerical Methods in Thermal Problems, Swansea, UK, July 21−25, 1997, pp. 742−753.
[53] Zhu Bofang, Effect of cooling by water flowing in nommetal pipes embedded in mass concrete, J. Constr. Eng. ASCE 125 (1) (1999) 61−68.
[54] Zhu Bofang, Xu Ping, Wang Shuhe, Thermal stresses and temperature control of RCC gravity dams, in: Proceedings International Symposium on Roller Compacted Concrete Dam (Chinese), Chengdu, China, April 21−25, 1999, pp. 65−76.
[55] Zhu Bofang, Xu Ping, Thermal stresses and temperature control of concrete gravity dams without longitudinal joint including RCC gravity dams. Innovation in Concrete Structures: Design and Construction, Proceedings of International Conference on Creating in Concrete, Dundee, UK, September 8−10, 1999, pp. 127−133.
[56] Zhu Bofang, On the technology of contruction of dam by gentlely expansive concrete, J. Hydroelectric Eng. 3 (2000) 1−12 (in Chinese).
[57] Zhu Bofang, Finite element whole course simulation and sequential strength reduction method for safety appraisal of concrete dams, Water Resour. Hydropower Eng. 1 (2007) 1−6 (in Chinese).
[58] Zhu Bofang, A computing model for the volume expansion and some remarks on the method of experimentation of concrete with gentle volume expansion, J. Hydraulic Eng. 12 (2002) 18−21 (in Chinese).
[59] Zhu Bofang, Incremental type of computing model for the volume expansion of concrete with gentle volume expansion, Water Power 2 (2003) 20−23 (in Chinese).

[60] Zhu Bofang, On the time for joint grouting in arch dams, J. Hydroelectric Eng. (3) (2003) (in Chinese).
[61] Zhu Bofang, A new computing model for adiabatic temperature rise of concrete and back analysis, Water Power 4 (2003) 31–34 (in Chinese).
[62] Zhu Bofang, Computing model of MgO concrete considering effect of both the present temperature and the temperature history, Water Resour. Hydropower Eng. 4 (2003) 16–17 (in Chinese).
[63] Zhu Bofang, Temperature control and design of joints for RCC arch dams, Dam Eng. 14 (3) (2003) 205–226.
[64] Zhu Bofang, Temperature loads on arch dams with superficial thermal insulation layer in cold region, Water Resour. Hydropower Eng. 11 (2003) 46–49 (in Chinese).
[65] Zhu Bofang, Zhang Guoxin, Xu Lingxian, Yang Shuming, New concept and new techniques for solving the thermal stress problem in heightening of concrete gravity dam, Water Power 11 (2003) 26–30 (in Chinese).
[66] Zhu Bofang, Xu Ping, Strengthen superficial insulation of concrete dams to terminate the history of " no dam without crack", Water Power 3 (2004) 25–28 (in Chinese).
[67] Zhu Bofang, Equivalent finite element stresses and safety of arch dams under triaxial stresses, Water Resour. Hydropower Eng. 1 (2005) 43–47 (in Chinese).
[68] Zhu Bofang, Zhang Guoxin, Yang Weizhong, Yang Bo, Xu Ping, Two kinds of guiding thought and two results of practical engineering for application of MgO concrete to dams, Water Resour. Hydropower Eng. 6 (2005) 39–42 (in Chinese).
[69] Zhu Bofang, On the coefficients of safety for crack prevention of concrete dams, Water Resour. Hydropower Eng. 7 (2005) 36–40 (in Chinese).
[70] Zhu Bofang, Joint element for finite element analysis of temperature field, Water Resour. Hydropower Eng. 11 (2005) 45–47 (in Chinese).
[71] Zhu Bofang, M. Shufang, Permanent compound board of thermal and water insulation for concrete dams, Water Resour. Hydropower Eng. 4 (2006) 16–21 (in Chinese).
[72] Zhu Bofang, On necessity of long-time superficial thermal insulation of concrete arch dams, Water Power 8 (2006) 21–24 (in Chinese).
[73] Zhu Bofang, Thermal stresses in beams on elastic foundation, J. Civil Eng. 8 (2006) 99–104 (in Chinese).
[74] Zhu Bofang, On the feasibility of building high quality arch dams without cracking and the relevant techniques, J. Hydraulic Eng. 10 (2006) 1155–1162 (in Chinese).
[75] Zhu Bofang, Improvement of computing method for the temperature loads on arch dams, Water Resour, Hydropower Eng. 12 (2006) 22–25 (in Chinese).
[76] Zhu Bofang, Finite element method for analysing temperature field of non-homogeneous and anisotropic body and the influence of flow in cracks, Water Resour. Hydropower Eng. 3 (2007) 33–35.
[77] Zhu Bofang, Zhang Guoxin, Wu Longshen, Hu Ping, Researches on reducing the cracking of the surface between the new and the old concrete after the heightening of a gravity dam, J. Hydraulic Eng. 6 (2007) 639–645.
[78] Zhu Bofang, Zhang Guoxin, Xu Ping, Lu Zhenjiang, Decision making support system for temperature and stress control of high concrete dams in construction period, J. Hydraulic Eng. 1 (2008) 1–6 (in Chinese).
[79] Zhu Bofang, Yang Ping, Semi-mature age of concrete-a new method for improving the crack resistance of mass concrete, Water Resour. Hydropower Eng. 5 (2008) 30–35 (in Chinese).

[80] Zhu Bofang, Wu Longshen, Yang Ping, Zhang Guoxin, Strengthen the cooling of mass concrete by polyethylene pipes with small spacing, Water Resour. Hydropower Eng. 5 (2008) 36–39 (in Chinese).

[81] Zhu Bofang, Li Yue, Wu Longshen, Zhang Guoxin, Two principles for the allowable temperature differences of dam concrete above foundation, Water Resour. Hydropower Eng. 7 (2008) 21–26 (in Chinese).

[82] Zhu Bofang, Wu Longshen, Yang Ping, Zhang Guoxin, Planning of pipe cooling of concrete dams in later age, Water Resour. Hydropower Eng. 7 (2008) 27–31 (in Chinese).

[83] Zhu Bofang, New direction of pipe cooling of mass concrete—earlier and slow cooling with small temperature difference between water and concrete, Water Resour. Hydropower Eng. 1 (2009) 44–50 (in Chinese).

[84] Zhu Bofang, Wu Longshen, Li Yue, Zhang Guoxin, Thermal stresses in sluices, Water Resour. Hydropower Eng. 3 (2009) 30–3242, (in Chinese).

[85] Zhu Bofang, Wu Longshen, Zhang Guoxin, The virtues and defects of pipe cooling for concrete dams, Water Resour. Hydropower Eng. 12 (2009) 26–30 (in Chinese).

[86] Zhu Bofang, Theorectical solution of the temperature field in the construction lift of concrete dam under the simultaneous action of pipe cooling and natural cooling from the lift surface, Water Resour. Hydropower Eng. 7 (2010) 27–28 (in Chinese).

[87] Zhu Bofang, On the expected life span of concrete dams and the possibility of endlessly long life of solid concrete dams, J. Hydraulic Eng. 1 (2012) 1–9 (in Chinese).

[88] Zhu Bofang, On the finite element equivalent stress of concrete arch dams, Water Resour. Hydropower Eng. 4 (2012) 30–32 (in Chinese).

[89] O.C. Zienkiewicz, The computation of shrinkage and thermal stresses in massive structures, Proc. ICE (1955) 1–4.

[90] O.C. Zienkiewicz, D.V. Phillips, An automatic mesh generation scheme for plane and curved surface by isoparametric co-ordinates, Int. J. Num. Meth. Eng. 3 (1971) 519–528.

[91] O.C. Zienkiewicz, J.Z. Zhu, A simple error estimator and adaptive procedure for practical engineering analysis, Int. J. Num. Meth. Eng. 24 (1987) 337–357.

[92] W. Schleeh, Die Zwängspannungen in einseitig fetsgehaltenem Wandscheiben, Beton und Stahlbetonban 57, Jahrgang, Heft 3, März, 1962.

[93] J.M. Raphael, Prediction of temperature in rivers and reservoirs, J. Power Div., ASCE, PO2 (1962) 157–182.

[94] J.T. Orlob, L.G. Selna, Temperature variations in deep reservoirs, J. Hydraulic Div. ASCE, HY2 (1970).

[95] W.C. Huber, D.R.F. Hasleman, Temperature prediction in stratified reservoirs, J. Hydraulic Div. ASCE (4) (1972) 645–666.

[96] G.E. Troxell, J.M. Raphael, R.E. Davis, Long-time creep and shrinkage tests of plain and reinforced concrete, Proc. ASTM 58 (1958) 1101–1120.

[97] K.W. Nasser, A.M. Neville, Creep of concrete at elevated temperature, ACI J. 62 (12) (1965) 1567–1579.

[98] K.W. Masser, R.P. Lohtia, Creep of mass concrete at high temperature, ACI J. 68 (1971) 276–281.

[99] C.L. Townsend, U.S.B.R. practice for control of cracking in arch dams, Proc. ASCE, PO4 (Aug. 1959)PO4 Aug. (1960)

[100] C.M. Roberts, The heightening of a gravity dam, 5th International Congress On Large Dams, II, R 12, 1955.

[101] F.W. Sims, J.A. Rhodes, R.W. Clough, Cracking in Norfork dam, ACI J. 61 (3) (March 1964).

[102] W.J. Brunner, K.H. Wu, Cracking of Revelstoke concrete gravity dam, 15th International Congress on Large Dams, 2 Lausanne, 1985.

[103] G. Chavarri, A. De Fries, W.Y. Shiek, C.H. Yeh, Raising Guri gravity dam, stability and stress investigation, 13th International Congress on Large Dams, vol. 1, 1979 pp. 45–56.

[104] C.D. Norman, F.A. Anderson, Reanalysis of cracking in large concrete dams in the US Army Corps of Engineers, 15th International Congress on Large Dams 2 Lausanne, 1985.

[105] F. Hollingworth, D.J. Hooper, J.J. Beringer, Roller compacted concrete arch dams, Intern. Water Power Dam Construct. (Nov. 1989) 29–34.

[106] R.W. Carson, Drying shrinkage of large concrete member, ACI J. 33 (Jan.–Feb. 1937) 317–336.

[107] G. Pickett, Shrinkage stresses in concrete, ACI J. Part 1 (Jan. 1946) 165–195, Part 2 (Feb. 1946) 361–400

[108] T.C. Hsu, Microcracking of plain concrete and the shape of the stress-strain curve, ACI J. 60 (2) (1963) 209–224.

[109] M.F. Kaplan, Crack propagation and fracture of concrete, ACI J. 58 (5) (1961) 591–610.

[110] M.F. Kaplan, Strains and stresses of concrete at initiation of cracking and near failure, ACI J. 60 (2) (1963) 853.

[111] R.S. Barsoum, On the use of isoparametric finite elements in linear fracture mechanisms, Int. J. Num. Meth. Eng. 10 (1976) 25–37.

[112] R.D. Henshell, K.G. Shaw, Crack tip finite elements are unnecessary, Int. J. Num. Meth. Eng. 9 (1975) 495–507.

[113] R.S. Barsoum, Triangular quarter-point elements as elastic and perfectly-plastic crack tip elements, Int. J. Num. Meth. Eng. 11 (1977) 85–98.

[114] H.D. Hibbitt, Some properties of singular isoparametric elements, Int. J. Num. Meth. Eng. 11 (1977) 180–184.

[115] Z.P. Bazant, L. Cedolin, Blunt crack band propagation in finite element analysis, J. Eng. Mech. Div. ASCE 105 (2) (1979) 297–315.

Index

Note: Page numbers followed by "*f*" and "*t*" refer to figures and tables, respectively.

Steel bar, thermal stress of, 3